Student Solutions Manual

Applied Calculus for the Managerial, Life, and Social Sciences

A Brief Approach

TENTH EDITION

Soo T. Tan

Stonehill College

Prepared by

Andy Bulman-Fleming

Australia • Brazil • Mexico • Singapore • United Kingdom • United States

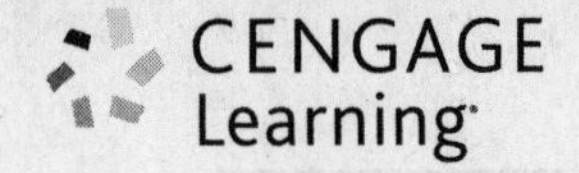

WCN: 01-100-101

For product information and technology assistance, contact us at **Cengage Learning Customer & Sales Support, 1-800-354-9706**.

For permission to use material from this text or product, submit all requests online at **www.cengage.com/permissions**
Further permissions questions can be emailed to **permissionrequest@cengage.com**.

ISBN-13: 978-1-285-85495-3
ISBN-10: 1-285-85495-0

Cengage Learning
200 First Stamford Place, 4th Floor
Stamford, CT 06902
USA

Cengage Learning is a leading provider of customized learning solutions with office locations around the globe, including Singapore, the United Kingdom, Australia, Mexico, Brazil, and Japan. Locate your local office at: **www.cengage.com/global**.

Cengage Learning products are represented in Canada by Nelson Education, Ltd.

To learn more about Cengage Learning Solutions, visit **www.cengage.com**.

Purchase any of our products at your local college store or at our preferred online store **www.cengagebrain.com**.

Printed in the United States of America
3 4 5 6 7 18 17 16 15

CONTENTS

CHAPTER 7 Additional Topics in Integration 207

CHAPTER 8 Calculus of Several Variables 229

1 PRELIMINARIES

1.1 Precalculus Review I

Exercises page 13

1. The interval (3, 6) is shown on the number line below. Note that this is an open interval indicated by "(" and ")".

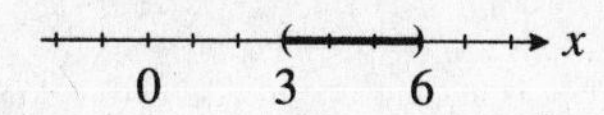

3. The interval $[-1, 4)$ is shown on the number line below. Note that this is a half-open interval indicated by "[" (closed) and ")" (open).

5. The infinite interval $(0, \infty)$ is shown on the number line below.

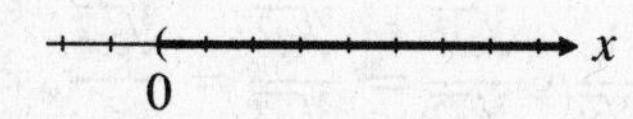

7. $27^{2/3} = \left(3^3\right)^{2/3} = 3^2 = 9.$

9. $\left(\frac{1}{\sqrt{3}}\right)^0 = 1.$ Recall that any number raised to the zeroth power is 1.

11. $\left[\left(\frac{1}{8}\right)^{1/3}\right]^{-2} = \left(\frac{1}{2}\right)^{-2} = \left(2^2\right) = 4.$

13. $\left(\dfrac{7^{-5} \cdot 7^2}{7^{-2}}\right)^{-1} = \left(7^{-5+2+2}\right)^{-1} = \left(7^{-1}\right)^{-1} = 7^1 = 7.$

15. $\left(125^{2/3}\right)^{-1/2} = 125^{(2/3)(-1/2)} = 125^{-1/3} = \dfrac{1}{125^{1/3}} = \frac{1}{5}.$

17. $\dfrac{\sqrt{32}}{\sqrt{8}} = \sqrt{\dfrac{32}{8}} = \sqrt{4} = 2.$

19. $\dfrac{16^{5/8}16^{1/2}}{16^{7/8}} = 16^{(5/8)+(1/2)-(7/8)} = 16^{1/4} = 2.$

21. $16^{1/4} \cdot 8^{-1/3} = 2 \cdot \left(\frac{1}{8}\right)^{1/3} = 2 \cdot \frac{1}{2} = 1.$

23. True.

25. False. $x^3 \times 2x^2 = 2x^{3+2} = 2x^5 \neq 2x^6.$

27. False. $\dfrac{2^{4x}}{1^{3x}} = \dfrac{2^{4x}}{1} = 2^{4x}.$

29. False. $\dfrac{1}{4^{-3}} = 4^3 = 64.$

31. False. $\left(1.2^{1/2}\right)^{-1/2} = (1.2)^{-1/4} \neq 1.$

33. $(xy)^{-2} = \dfrac{1}{(xy)^2}.$

35. $\dfrac{x^{-1/3}}{x^{1/2}} = x^{(-1/3)-(1/2)} = x^{-5/6} = \dfrac{1}{x^{5/6}}.$

37. $12^0 (s+t)^{-3} = 1 \cdot \dfrac{1}{(s+t)^3} = \dfrac{1}{(s+t)^3}.$

39. $\dfrac{x^{7/3}}{x^{-2}} = x^{(7/3)+2} = x^{(7/3)+(6/3)} = x^{13/3}.$

41. $\left(x^2y^{-3}\right)\left(x^{-5}y^3\right) = x^{2-5}y^{-3+3} = x^{-3}y^0 = x^{-3} = \dfrac{1}{x^3}$.

43. $\dfrac{x^{3/4}}{x^{-1/4}} = x^{(3/4)-(-1/4)} = x^{4/4} = x$.

45. $\left(\dfrac{x^3}{-27y^{-6}}\right)^{-2/3} = x^{3(-2/3)}\left(-\dfrac{1}{27}\right)^{-2/3} y^{6(-2/3)}$
$= x^{-2}\left(-\dfrac{1}{3}\right)^{-2} y^{-4} = \dfrac{9}{x^2y^4}$.

47. $\left(\dfrac{x^{-3}}{y^{-2}}\right)^2\left(\dfrac{y}{x}\right)^4 = \dfrac{x^{-3\cdot 2}y^4}{y^{-2\cdot 2}x^4} = \dfrac{y^{4+4}}{x^{4+6}} = \dfrac{y^8}{x^{10}}$.

49. $\sqrt[3]{x^{-2}}\cdot\sqrt{4x^5} = x^{-2/3}\cdot 4^{1/2}\cdot x^{5/2} = x^{(-2/3)+(5/2)}\cdot 2$
$= 2x^{11/6}$.

51. $-\sqrt[4]{16x^4y^8} = -\left(16^{1/4}\cdot x^{4/4}\cdot y^{8/4}\right) = -2xy^2$.

53. $\sqrt[6]{64x^8y^3} = 64^{1/6}\cdot x^{8/6}y^{3/6} = 2x^{4/3}y^{1/2}$.

55. $2^{3/2} = 2\left(2^{1/2}\right) \approx 2\,(1.414) = 2.828$.

57. $9^{3/4} = \left(3^2\right)^{3/4} = 3^{6/4} = 3^{3/2} = 3\cdot 3^{1/2}$
$\approx 3\,(1.732) = 5.196$.

59. $10^{3/2} = 10^{1/2}\cdot 10 \approx (3.162)\,(10) = 31.62$.

61. $10^{2.5} = 10^2\cdot 10^{1/2} \approx 100\,(3.162) = 316.2$.

63. $\dfrac{3}{2\sqrt{x}}\cdot\dfrac{\sqrt{x}}{\sqrt{x}} = \dfrac{3\sqrt{x}}{2x}$.

65. $\dfrac{2y}{\sqrt{3y}}\cdot\dfrac{\sqrt{3y}}{\sqrt{3y}} = \dfrac{2y\sqrt{3y}}{3y} = \dfrac{2\sqrt{3y}}{3}$.

67. $\dfrac{1}{\sqrt[3]{x}}\cdot\dfrac{\sqrt[3]{x^2}}{\sqrt[3]{x^2}} = \dfrac{\sqrt[3]{x^2}}{\sqrt[3]{x^3}} = \dfrac{\sqrt[3]{x^2}}{x}$.

69. $\dfrac{2\sqrt{x}}{3}\cdot\dfrac{\sqrt{x}}{\sqrt{x}} = \dfrac{2x}{3\sqrt{x}}$.

71. $\sqrt{\dfrac{2y}{x}} = \dfrac{\sqrt{2y}}{\sqrt{x}}\cdot\dfrac{\sqrt{2y}}{\sqrt{2y}} = \dfrac{2y}{\sqrt{2xy}}$

73. $\dfrac{\sqrt[3]{x^2z}}{y}\cdot\dfrac{\sqrt[3]{xz^2}}{\sqrt[3]{xz^2}} = \dfrac{\sqrt[3]{x^3z^3}}{y\sqrt[3]{xz^2}} = \dfrac{xz}{y\sqrt[3]{xz^2}}$.

75. $\left(7x^2 - 2x + 5\right) + \left(2x^2 + 5x - 4\right) = 7x^2 - 2x + 5 + 2x^2 + 5x - 4 = 9x^2 + 3x + 1$.

77. $\left(5y^2 - 2y + 1\right) - \left(y^2 - 3y - 7\right) = 5y^2 - 2y + 1 - y^2 + 3y + 7 = 4y^2 + y + 8$.

79. $x - \{2x - [-x - (1 - x)]\} = x - \{2x - [-x - 1 + x]\} = x - (2x + 1) = x - 2x - 1 = -x - 1$.

81. $\left(\dfrac{1}{3} - 1 + e\right) - \left(-\dfrac{1}{3} - 1 + e^{-1}\right) = \dfrac{1}{3} - 1 + e + \dfrac{1}{3} + 1 - \dfrac{1}{e} = \dfrac{2}{3} + e - \dfrac{1}{e} = \dfrac{3e^2 + 2e - 3}{3e}$.

83. $3\sqrt{8} + 8 - 2\sqrt{y} + \frac{1}{2}\sqrt{x} - \frac{3}{4}\sqrt{y} = 3\sqrt{8} + 8 + \frac{1}{2}\sqrt{x} - \frac{11}{4}\sqrt{y} = 6\sqrt{2} + 8 + \frac{1}{2}\sqrt{x} - \frac{11}{4}\sqrt{y}$.

85. $(x + 8)\,(x - 2) = x\,(x - 2) + 8\,(x - 2) = x^2 - 2x + 8x - 16 = x^2 + 6x - 16$.

87. $(a + 5)^2 = (a + 5)\,(a + 5) = a\,(a + 5) + 5\,(a + 5) = a^2 + 5a + 5a + 25 = a^2 + 10a + 25$.

89. $(x + 2y)^2 = (x + 2y)\,(x + 2y) = x\,(x + 2y) + 2y\,(x + 2y) = x^2 + 2xy + 2yx + 4y^2 = x^2 + 4xy + 4y^2$.

91. $(2x + y)\,(2x - y) = 2x\,(2x - y) + y\,(2x - y) = 4x^2 - 2xy + 2xy - y^2 = 4x^2 - y^2$.

93. $\left(2x^2 - 1\right)\left(3x^2\right) + \left(x^2 + 3\right)(4x) = 6x^4 - 3x^2 + 4x^3 + 12x = 6x^4 + 4x^3 - 3x^2 + 12x = x\left(6x^3 + 4x^2 - 3x + 12\right)$.

95. $6x\left(\frac{1}{2}\right)\left(2x^2+3\right)^{-1/2}(4x)+6\left(2x^2+3\right)^{1/2}=3\left(2x^2+3\right)^{-1/2}\left[x(4x)+2\left(2x^2+3\right)\right]=\dfrac{6\left(4x^2+3\right)}{\left(2x^2+3\right)^{1/2}}.$

97. $100\left(-10te^{-0.1t}-100e^{-0.1t}\right)=-1000(10+t)e^{-0.1t}.$

99. $4x^5-12x^4-6x^3=2x^3\left(2x^2-6x-3\right).$

101. $7a^4-42a^2b^2+49a^3b=7a^2\left(a^2+7ab-6b^2\right).$

103. $e^{-x}-xe^{-x}=e^{-x}(1-x).$

105. $2x^{-5/2}-\frac{3}{2}x^{-3/2}=\frac{1}{2}x^{-5/2}(4-3x).$

107. $6ac+3bc-4ad-2bd=3c(2a+b)-2d(2a+b)=(2a+b)(3c-2d).$

109. $4a^2-b^2=(2a+b)(2a-b)$, a difference of two squares.

111. $10-14x-12x^2=-2\left(6x^2+7x-5\right)=-2(3x+5)(2x-1).$

113. $3x^2-6x-24=3\left(x^2-2x-8\right)=3(x-4)(x+2).$

115. $12x^2-2x-30=2\left(6x^2-x-15\right)=2(3x-5)(2x+3).$

117. $9x^2-16y^2=(3x)^2-(4y)^2=(3x-4y)(3x+4y).$

119. $x^6+125=\left(x^2\right)^3+(5)^3=\left(x^2+5\right)\left(x^4-5x^2+25\right).$

121. $\left(x^2+y^2\right)x-xy(2y)=x^3+xy^2-2xy^2=x^3-xy^2.$

123.
$$\begin{aligned}2(x-1)(2x+2)^3[4(x-1)+(2x+2)]&=2(x-1)(2x+2)^3(4x-4+2x+2)\\&=2(x-1)(2x+2)^3(6x-2)=4(x-1)(3x-1)(2x+2)^3\\&=32(x-1)(3x-1)(x+1)^3.\end{aligned}$$

125.
$$\begin{aligned}4(x-1)^2(2x+2)^3(2)+(2x+2)^4(2)(x-1)&=2(x-1)(2x+2)^3[4(x-1)+(2x+2)]\\&=2(x-1)(2x+2)^3(6x-2)=4(x-1)(3x-1)(2x+2)^3\\&=32(x-1)(3x-1)(x+1)^3.\end{aligned}$$

127. $\left(x^2+2\right)^2\left[5\left(x^2+2\right)^2-3\right](2x)=\left(x^2+2\right)^2\left[5\left(x^4+4x^2+4\right)-3\right](2x)=(2x)\left(x^2+2\right)^2\left(5x^4+20x^2+17\right).$

129. We factor the left-hand side of $x^2+x-12=0$ to obtain $(x+4)(x-3)=0$, so $x=-4$ or $x=3$. We conclude that the roots are $x=-4$ and $x=3$.

131. $4t^2+2t-2=(2t-1)(2t+2)=0$. Thus, the roots are $t=\frac{1}{2}$ and $t=-1$.

133. $\frac{1}{4}x^2-x+1=\left(\frac{1}{2}x-1\right)\left(\frac{1}{2}x-1\right)=0$. Thus $\frac{1}{2}x=1$, and so $x=2$ is a double root of the equation.

135. We use the quadratic formula to solve the equation $4x^2 + 5x - 6 = 0$. In this case, $a = 4$, $b = 5$, and $c = -6$. Therefore, $x = \dfrac{-b \pm \sqrt{b^2 - 4ac}}{2a} = \dfrac{-5 \pm \sqrt{5^2 - 4(4)(-6)}}{2(4)} = \dfrac{-5 \pm \sqrt{121}}{8} = \dfrac{-5 \pm 11}{8}$. Thus, $x = -\frac{16}{8} = -2$ and $x = \frac{6}{8} = \frac{3}{4}$ are the roots of the equation.

137. We use the quadratic formula to solve the equation $8x^2 - 8x - 3 = 0$. Here $a = 8$, $b = -8$, and $c = -3$, so $x = \dfrac{-b \pm \sqrt{b^2 - 4ac}}{2a} = \dfrac{-(-8) \pm \sqrt{(-8)^2 - 4(8)(-3)}}{2(8)} = \dfrac{8 \pm \sqrt{160}}{16} = \dfrac{8 \pm 4\sqrt{10}}{16} = \dfrac{2 \pm \sqrt{10}}{4}$. Thus, $x = \frac{1}{2} + \frac{1}{4}\sqrt{10}$ and $x = \frac{1}{2} - \frac{1}{4}\sqrt{10}$ are the roots of the equation.

139. We use the quadratic formula to solve $2x^2 + 4x - 3 = 0$. Here $a = 2$, $b = 4$, and $c = -3$, so $x = \dfrac{-b \pm \sqrt{b^2 - 4ac}}{2a} = \dfrac{-4 \pm \sqrt{4^2 - 4(2)(-3)}}{2(2)} = \dfrac{-4 \pm \sqrt{40}}{4} = \dfrac{-4 \pm 2\sqrt{10}}{4} = \dfrac{-2 \pm \sqrt{10}}{2}$. Thus, $x = -1 + \frac{1}{2}\sqrt{10}$ and $x = -1 - \frac{1}{2}\sqrt{10}$ are the roots of the equation.

141. The total revenue is given by $\left(0.2t^2 + 150t\right) + \left(0.5t^2 + 200t\right) = 0.7t^2 + 350t$ thousand dollars t months from now, where $0 \le t \le 12$.

143. a. $f(30{,}000) = \left(5.6 \times 10^{11}\right)(30{,}000)^{-1.5} \approx 107{,}772$, or 107,772 families.

b. $f(60{,}000) = \left(5.6 \times 10^{11}\right)(60{,}000)^{-1.5} \approx 38{,}103$, or 38,103 families.

c. $f(150{,}000) = \left(5.6 \times 10^{11}\right)(150{,}000)^{-1.5} \approx 9639$, or 9639 families.

145. $8000x - 100x^2 = 100x(80 - x)$.

147. True. If $b^2 - 4ac < 0$, then $\sqrt{b^2 - 4ac}$ is not a real number.

1.2 Precalculus Review II

Exercises page 23

1. $\dfrac{x^2 + x - 2}{x^2 - 4} = \dfrac{(x+2)(x-1)}{(x+2)(x-2)} = \dfrac{x-1}{x-2}$.

3. $\dfrac{12t^2 + 12t + 3}{4t^2 - 1} = \dfrac{3\left(4t^2 + 4t + 1\right)}{4t^2 - 1} = \dfrac{3(2t+1)(2t+1)}{(2t+1)(2t-1)} = \dfrac{3(2t+1)}{2t-1}$.

5. $\dfrac{(4x-1)(3) - (3x+1)(4)}{(4x-1)^2} = \dfrac{12x - 3 - 12x - 4}{(4x-1)^2} = -\dfrac{7}{(4x-1)^2}$.

7. $\dfrac{2a^2 - 2b^2}{b - a} \cdot \dfrac{4a + 4b}{a^2 + 2ab + b^2} = \dfrac{2(a+b)(a-b)4(a+b)}{-(a-b)(a+b)(a+b)} = -8$.

9. $\dfrac{3x^2 + 2x - 1}{2x + 6} \div \dfrac{x^2 - 1}{x^2 + 2x - 3} = \dfrac{(3x-1)(x+1)}{2(x+3)} \cdot \dfrac{(x+3)(x-1)}{(x+1)(x-1)} = \dfrac{3x-1}{2}$.

11. $\dfrac{58}{3(3t+2)} + \dfrac{1}{3} = \dfrac{58 + 3t + 2}{3(3t+2)} = \dfrac{3t + 60}{3(3t+2)} = \dfrac{t + 20}{3t + 2}$.

13. $\dfrac{2x}{2x-1}-\dfrac{3x}{2x+5}=\dfrac{2x\,(2x+5)-3x\,(2x-1)}{(2x-1)\,(2x+5)}=\dfrac{4x^2+10x-6x^2+3x}{(2x-1)\,(2x+5)}=\dfrac{-2x^2+13x}{(2x-1)\,(2x+5)}$
$=-\dfrac{x\,(2x-13)}{(2x-1)\,(2x+5)}.$

15. $\dfrac{4}{x^2-9}-\dfrac{5}{x^2-6x+9}=\dfrac{4}{(x+3)\,(x-3)}-\dfrac{5}{(x-3)^2}=\dfrac{4\,(x-3)-5\,(x+3)}{(x-3)^2\,(x+3)}=-\dfrac{x+27}{(x-3)^2\,(x+3)}.$

17. $\dfrac{1+\frac{1}{x}}{1-\frac{1}{x}}=\dfrac{\frac{x+1}{x}}{\frac{x-1}{x}}=\dfrac{x+1}{x}\cdot\dfrac{x}{x-1}=\dfrac{x+1}{x-1}.$

19. $\dfrac{4x^2}{2\sqrt{2x^2+7}}+\sqrt{2x^2+7}=\dfrac{4x^2+2\sqrt{2x^2+7}\sqrt{2x^2+7}}{2\sqrt{2x^2+7}}=\dfrac{4x^2+4x^2+14}{2\sqrt{2x^2+7}}=\dfrac{4x^2+7}{\sqrt{2x^2+7}}.$

21. $5\left[\dfrac{(t^2+1)\,(1)-t\,(2t)}{(t^2+1)^2}\right]=\dfrac{5\,(t^2+1-2t^2)}{(t^2+1)^2}=\dfrac{5\,(1-t^2)}{(t^2+1)^2}=-\dfrac{5\,(t^2-1)}{(t^2+1)^2}.$

23. $\dfrac{(x^2+1)^2\,(-2)+(2x)\,2\,(x^2+1)\,(2x)}{(x^2+1)^4}=\dfrac{2\,(x^2+1)\left[-(x^2+1)+4x^2\right]}{(x^2+1)^4}=\dfrac{2\,(3x^2-1)}{(x^2+1)^3}.$

25. $3\left(\dfrac{2x+1}{3x+2}\right)^2\left[\dfrac{(3x+2)\,(2)-(2x+1)\,(3)}{(3x+2)^2}\right]=\dfrac{3\,(2x+1)^2\,(6x+4-6x-3)}{(3x+2)^4}=\dfrac{3\,(2x+1)^2}{(3x+2)^4}.$

27. $100\left[\dfrac{(t^2+20t+100)\,(2t+10)-(t^2+10t+100)\,(2t+20)}{(t^2+20t+100)^2}\right]$
$=100\left[\dfrac{2t^3+40t^2+200t+10t^2+200t+1000-2t^3-20t^2-200t-20t^2-200t-2000}{(t^2+20t+100)^2}\right]$
$=\dfrac{100\,(10t^2-1000)}{(t^2+20t+100)^2}=\dfrac{1000\,(t-10)}{(t+10)^3}.$

29. $\dfrac{1}{\sqrt{3}-1}\cdot\dfrac{\sqrt{3}+1}{\sqrt{3}+1}=\dfrac{\sqrt{3}+1}{3-1}=\dfrac{\sqrt{3}+1}{2}.$

31. $\dfrac{1}{\sqrt{x}-\sqrt{y}}\cdot\dfrac{\sqrt{x}+\sqrt{y}}{\sqrt{x}+\sqrt{y}}=\dfrac{\sqrt{x}+\sqrt{y}}{x-y}.$

33. $\dfrac{\sqrt{a}+\sqrt{b}}{\sqrt{a}-\sqrt{b}}\cdot\dfrac{\sqrt{a}+\sqrt{b}}{\sqrt{a}+\sqrt{b}}=\dfrac{\left(\sqrt{a}+\sqrt{b}\right)^2}{a-b}.$

35. $\dfrac{\sqrt{x}}{3}\cdot\dfrac{\sqrt{x}}{\sqrt{x}}=\dfrac{x}{3\sqrt{x}}.$

37. $\dfrac{1-\sqrt{3}}{3}\cdot\dfrac{1+\sqrt{3}}{1+\sqrt{3}}=\dfrac{1^2-\left(\sqrt{3}\right)^2}{3\left(1+\sqrt{3}\right)}=-\dfrac{2}{3\left(1+\sqrt{3}\right)}.$

39. $\dfrac{1+\sqrt{x+2}}{\sqrt{x+2}}\cdot\dfrac{1-\sqrt{x+2}}{1-\sqrt{x+2}}=\dfrac{1-(x+2)}{\sqrt{x+2}\,(1-\sqrt{x+2})}=-\dfrac{x+1}{\sqrt{x+2}\,(1-\sqrt{x+2})}.$

41. The statement is false because -3 is greater than -20. See the number line below.

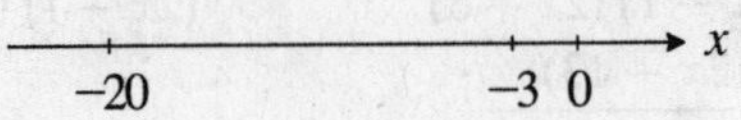

43. The statement is false because $\frac{2}{3} = \frac{4}{6}$ is less than $\frac{5}{6}$.

45. We are given $2x + 4 < 8$. Add -4 to each side of the inequality to obtain $2x < 4$, then multiply each side of the inequality by $\frac{1}{2}$ to obtain $x < 2$. We write this in interval notation as $(-\infty, 2)$.

47. We are given the inequality $-4x \geq 20$. Multiply both sides of the inequality by $-\frac{1}{4}$ and reverse the sign of the inequality to obtain $x \leq -5$. We write this in interval notation as $(-\infty, -5]$.

49. We are given the inequality $-6 < x - 2 < 4$. First add 2 to each member of the inequality to obtain $-6 + 2 < x < 4 + 2$ and $-4 < x < 6$, so the solution set is the open interval $(-4, 6)$.

51. We want to find the values of x that satisfy at least one of the inequalities $x + 1 > 4$ and $x + 2 < -1$. Adding -1 to both sides of the first inequality, we obtain $x + 1 - 1 > 4 - 1$, so $x > 3$. Similarly, adding -2 to both sides of the second inequality, we obtain $x + 2 - 2 < -1 - 2$, so $x < -3$. Therefore, the solution set is $(-\infty, -3) \cup (3, \infty)$.

53. We want to find the values of x that satisfy the inequalities $x + 3 > 1$ and $x - 2 < 1$. Adding -3 to both sides of the first inequality, we obtain $x + 3 - 3 > 1 - 3$, or $x > -2$. Similarly, adding 2 to each side of the second inequality, we obtain $x - 2 + 2 < 1 + 2$, so $x < 3$. Because both inequalities must be satisfied, the solution set is $(-2, 3)$.

55. We want to find the values of x that satisfy the inequality $(x + 3)(x - 5) \leq 0$. From the sign diagram, we see that the given inequality is satisfied when $-3 \leq x \leq 5$, that is, when the signs of the two factors are different or when one of the factors is equal to zero.

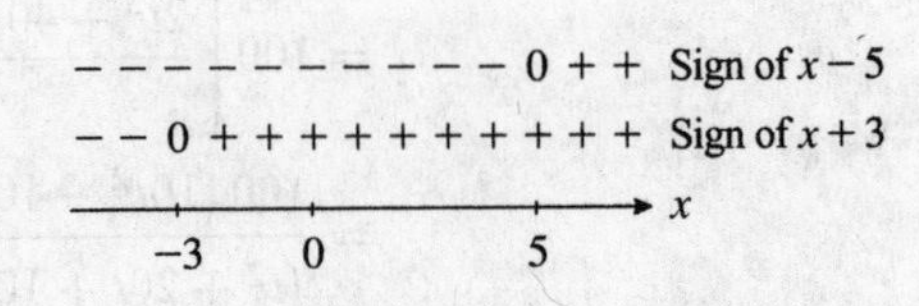

57. We want to find the values of x that satisfy the inequality $(2x - 3)(x - 1) \geq 0$. From the sign diagram, we see that the given inequality is satisfied when $x \leq 1$ or $x \geq \frac{3}{2}$; that is,when the signs of both factors are the same, or one of the factors is equal to zero.

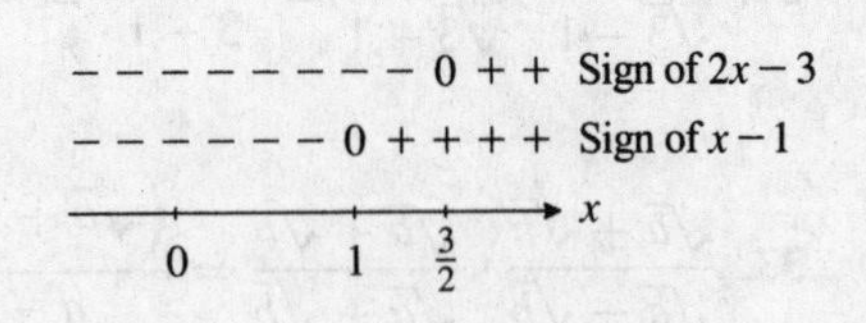

59. We want to find the values of x that satisfy the inequalities $\dfrac{x+3}{x-2} \geq 0$. From the sign diagram, we see that the given inequality is satisfied when $x \leq -3$ or $x > 2$, that is, when the signs of the two factors are the same. Notice that $x = 2$ is not included because the inequality is not defined at that value of x.

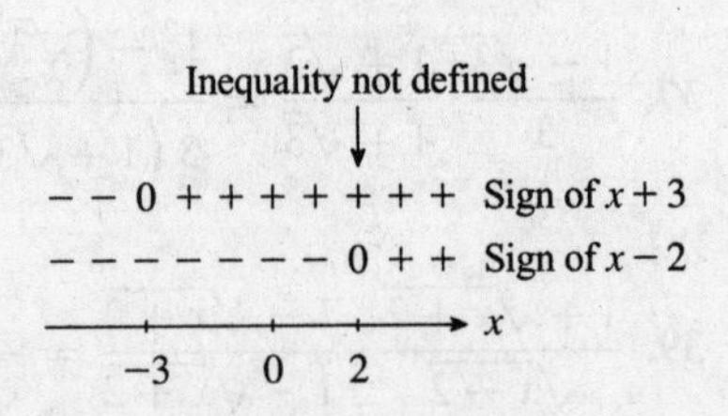

61. We want to find the values of x that satisfy the inequality $\frac{x-2}{x-1} \le 2$. Subtracting 2 from each side of the given inequality and simplifying gives $\frac{x-2}{x-1} - 2 \le 0$, $\frac{x-2-2(x-1)}{x-1} \le 0$, and $-\frac{x}{x-1} \le 0$. From the sign diagram, we see that the given inequality is satisfied when $x \le 0$ or $x > 1$; that is, when the signs of the two factors differ. Notice that $x = 1$ is not included because the inequality is undefined at that value of x.

Inequality not defined

+ + + 0 - - - - - - - - - Sign of $-x$

- - - - - - - - - 0 + + + Sign of $x-1$

0 1 x

63. $|-6+2| = 4$.

65. $\frac{|-12+4|}{|16-12|} = \frac{|-8|}{|4|} = 2$.

67. $\sqrt{3}|-2| + 3\left|-\sqrt{3}\right| = \sqrt{3}(2) + 3\sqrt{3} = 5\sqrt{3}$.

69. $|\pi - 1| + 2 = \pi - 1 + 2 = \pi + 1$.

71. $\left|\sqrt{2} - 1\right| + \left|3 - \sqrt{2}\right| = \sqrt{2} - 1 + 3 - \sqrt{2} = 2$.

73. False. If $a > b$, then $-a < -b$, $-a + b < -b + b$, and $b - a < 0$.

75. False. Let $a = -2$ and $b = -3$. Then $a^2 = 4$ and $b^2 = 9$, and $4 < 9$. Note that we need only to provide a counterexample to show that the statement is not always true.

77. True. There are three possible cases.
Case 1: If $a > 0$ and $b > 0$, then $a^3 > b^3$, since $a^3 - b^3 = (a-b)\left(a^2 + ab + b^2\right) > 0$.
Case 2: If $a > 0$ and $b < 0$, then $a^3 > 0$ and $b^3 < 0$, and it follows that $a^3 > b^3$.
Case 3: If $a < 0$ and $b < 0$, then $a^3 - b^3 = (a-b)\left(a^2 + ab + b^2\right) > 0$, and we see that $a^3 > b^3$. (Note that $a - b > 0$ and $ab > 0$.)

79. False. If we take $a = -2$, then $|-a| = |-(-2)| = |2| = 2 \ne a$.

81. True. If $a - 4 < 0$, then $|a - 4| = 4 - a = |4 - a|$. If $a - 4 > 0$, then $|4 - a| = a - 4 = |a - 4|$.

83. False. If we take $a = 3$ and $b = -1$, then $|a + b| = |3 - 1| = 2 \ne |a| + |b| = 3 + 1 = 4$.

85. If the car is driven in the city, then it can be expected to cover $(18.1)(20) = 362 \frac{\text{miles}}{\text{gal}} \cdot \text{gal}$, or 362 miles, on a full tank. If the car is driven on the highway, then it can be expected to cover $(18.1)(27) = 488.7 \frac{\text{miles}}{\text{gal}} \cdot \text{gal}$, or 488.7 miles, on a full tank. Thus, the driving range of the car may be described by the interval $[362, 488.7]$.

87. $6(P - 2500) \le 4(P + 2400)$ can be rewritten as $6P - 15{,}000 \le 4P + 9600$, $2P \le 24{,}600$, or $P \le 12{,}300$. Therefore, the maximum profit is \$12,300.

89. Let x represent the salesman's monthly sales in dollars. Then $0.15(x - 12{,}000) \ge 6000$, $15(x - 12{,}000) \ge 600{,}000$, $15x - 180{,}000 \ge 600{,}000$, $15x \ge 780{,}000$, and $x \ge 52{,}000$. We conclude that the salesman must have sales of at least \$52,000 to reach his goal.

91. The rod is acceptable if $0.49 \le x \le 0.51$ or $-0.01 \le x - 0.5 \le 0.01$. This gives the required inequality, $|x - 0.5| \le 0.01$.

93. We want to solve the inequality $-6x^2 + 30x - 10 \geq 14$. (Remember that x is expressed in thousands.) Adding -14 to both sides of this inequality, we have $-6x^2 + 30x - 10 - 14 \geq 14 - 14$, or $-6x^2 + 30x - 24 \geq 0$. Dividing both sides of the inequality by -6 (which reverses the sign of the inequality), we have $x^2 - 5x + 4 \leq 0$. Factoring this last expression, we have $(x - 4)(x - 1) \leq 0$.

From the sign diagram, we see that x must lie between 1 and 4. (The inequality is satisfied only when the two factors have opposite signs.) Because x is expressed in thousands of units, we see that the manufacturer must produce between 1000 and 4000 units of the commodity.

- - - - - - - - - 0 + + Sign of $x - 4$
- - 0 + + + + + + + + Sign of $x - 1$
0 1 4 x

95. We solve the inequalities $25 \leq \dfrac{0.5x}{100 - x} \leq 30$, obtaining $2500 - 25x \leq 0.5x \leq 3000 - 30x$, which is equivalent to $2500 - 25x \leq 0.5x$ and $0.5x \leq 3000 - 30x$. Simplifying further, $25.5x \geq 2500$ and $30.5x \leq 3000$, so $x \geq \dfrac{2500}{25.5} \approx 98.04$ and $x \leq \dfrac{3000}{30.5} \approx 98.36$. Thus, the city could expect to remove between 98.04% and 98.36% of the toxic pollutant.

97. We solve the inequality $\dfrac{136}{1 + 0.25(t - 4.5)^2} + 28 \geq 128$ or $\dfrac{136}{1 + 0.25(t - 4.5)^2} \geq 100$. Next, $136 \geq 100\left[1 + 0.25(t - 4.5)^2\right]$, so $136 \geq 100 + 25(t - 4.5)^2$, $36 \geq 25(t - 4.5)^2$, $(t - 4.5)^2 \leq \frac{36}{25}$, and $t - 4.5 \leq \pm\frac{6}{5}$. Solving this last inequality, we have $t \leq 5.7$ and $t \geq 3.3$. Thus, the amount of nitrogen dioxide is greater than or equal to 128 PSI between 10:18 a.m. and 12:42 p.m.

99. False. Take $a = 1$, $b = 2$, and $c = 3$. Then $a < b$, but $a - c = 1 - 3 = -2 \not> 2 - 3 = -1 = b - c$.

101. True. $|a - b| = |a + (-b)| \leq |a| + |-b| = |a| + |b|$.

1.3 The Cartesian Coordinate System

Problem-Solving Tips

Suppose you are asked to determine whether a given statement is true or false, and you are also asked to explain your answer. How would you answer the question?

If you think the statement is true, then prove it. On the other hand, if you think the statement is false, then give an example that disproves the statement. For example, the statement "If a and b are real numbers, then $a - b = b - a$" is false, and an example that disproves it may be constructed by taking $a = 3$ and $b = 5$. For these values of a and b, we find $a - b = 3 - 5 = -2$, but $b - a = 5 - 3 = 2$, and this shows that $a - b \neq b - a$. Such an example is called a **counterexample**.

Concept Questions page 29

1. a. $a < 0$ and $b > 0$ **b.** $a < 0$ and $b < 0$ **c.** $a > 0$ and $b < 0$

Exercises page 30

1. The coordinates of A are $(3, 3)$ and it is located in Quadrant I.

3. The coordinates of C are $(2, -2)$ and it is located in Quadrant IV.

5. The coordinates of E are $(-4, -6)$ and it is located in Quadrant III.

7. A

9. E, F, and G

11. F

For Exercises 13–19, refer to the following figure.

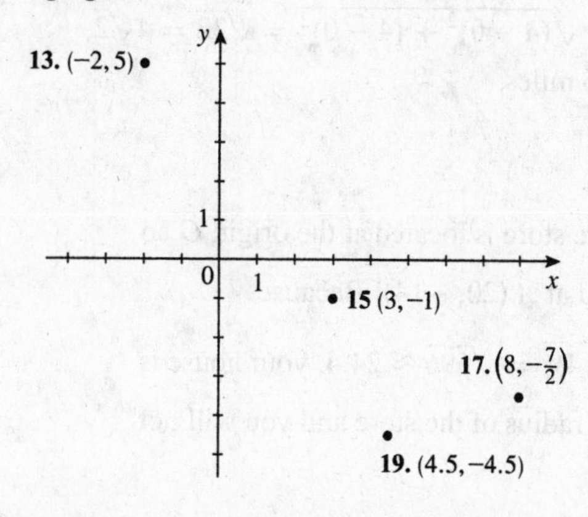

21. Using the distance formula, we find that $\sqrt{(4-1)^2+(7-3)^2}=\sqrt{3^2+4^2}=\sqrt{25}=5$.

23. Using the distance formula, we find that $\sqrt{[4-(-1)]^2+(9-3)^2}=\sqrt{5^2+6^2}=\sqrt{25+36}=\sqrt{61}$.

25. The coordinates of the points have the form $(x, -6)$. Because the points are 10 units away from the origin, we have $(x-0)^2+(-6-0)^2=10^2$, $x^2=64$, or $x=\pm 8$. Therefore, the required points are $(-8, -6)$ and $(8, -6)$.

27. The points are shown in the diagram. To show that the four sides are equal, we compute

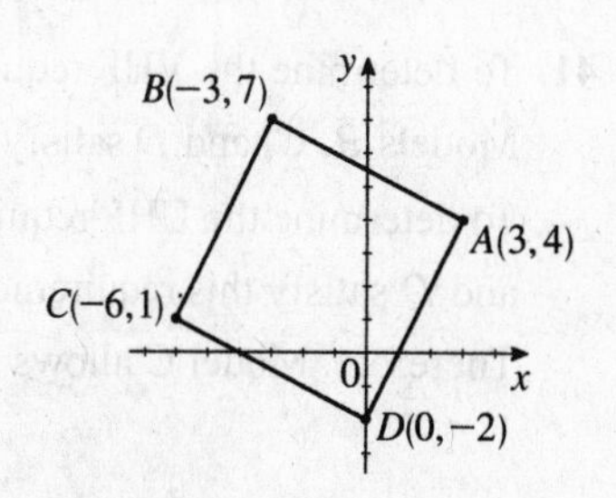

$d(A, B)=\sqrt{(-3-3)^2+(7-4)^2}=\sqrt{(-6)^2+3^2}=\sqrt{45}$,

$d(B, C)=\sqrt{[-6-(-3)]^2+(1-7)^2}=\sqrt{(-3)^2+(-6)^2}=\sqrt{45}$,

$d(C, D)=\sqrt{[0-(-6)]^2+[(-2)-1]^2}=\sqrt{(6)^2+(-3)^2}=\sqrt{45}$,

and $d(A, D)=\sqrt{(0-3)^2+(-2-4)^2}=\sqrt{(3)^2+(-6)^2}=\sqrt{45}$.

Next, to show that ΔABC is a right triangle, we show that it satisfies the Pythagorean Theorem. Thus, $d(A, C)=\sqrt{(-6-3)^2+(1-4)^2}=\sqrt{(-9)^2+(-3)^2}=\sqrt{90}=3\sqrt{10}$ and $[d(A, B)]^2+[d(B, C)]^2=90=[d(A, C)]^2$. Similarly, $d(B, D)=\sqrt{90}=3\sqrt{10}$, so ΔBAD is a right triangle as well. It follows that $\angle B$ and $\angle D$ are right angles, and we conclude that $ADCB$ is a square.

29. The equation of the circle with radius 5 and center $(2, -3)$ is given by $(x-2)^2+[y-(-3)]^2=5^2$, or $(x-2)^2+(y+3)^2=25$.

31. The equation of the circle with radius 5 and center $(0, 0)$ is given by $(x-0)^2+(y-0)^2=5^2$, or $x^2+y^2=25$.

33. The distance between the points $(5, 2)$ and $(2, -3)$ is given by $d = \sqrt{(5-2)^2 + [2-(-3)]^2} = \sqrt{3^2 + 5^2} = \sqrt{34}$. Therefore $r = \sqrt{34}$ and the equation of the circle passing through $(5, 2)$ and $(2, -3)$ is $(x-2)^2 + [y-(-3)]^2 = 34$, or $(x-2)^2 + (y+3)^2 = 34$.

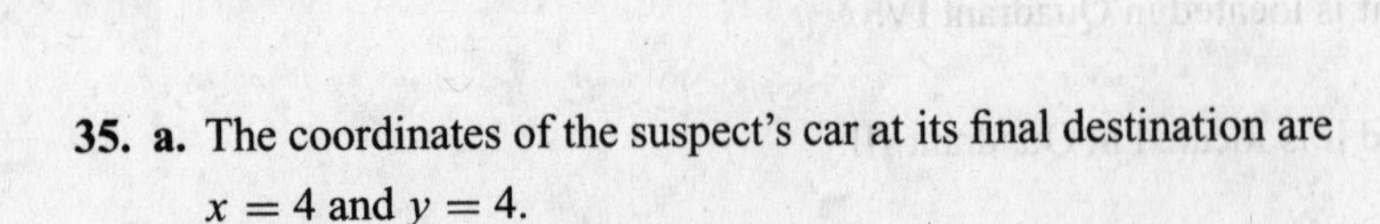

35. a. The coordinates of the suspect's car at its final destination are $x = 4$ and $y = 4$.

b. The distance traveled by the suspect was $5 + 4 + 1$, or 10 miles.

c. The distance between the original and final positions of the suspect's car was $d = \sqrt{(4-0)^2 + (4-0)^2} = \sqrt{32} = 4\sqrt{2}$, or approximately 5.66 miles.

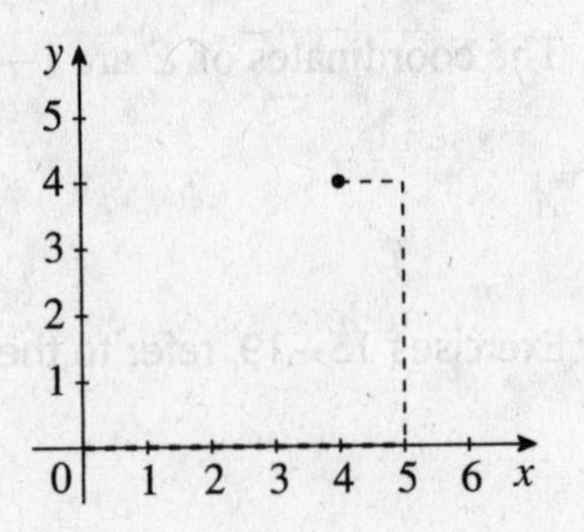

37. Suppose that the furniture store is located at the origin O so that your house is located at $A(20, -14)$. Because $d(O, A) = \sqrt{20^2 + (-14)^2} = \sqrt{596} \approx 24.4$, your house is located within a 25-mile radius of the store and you will not incur a delivery charge.

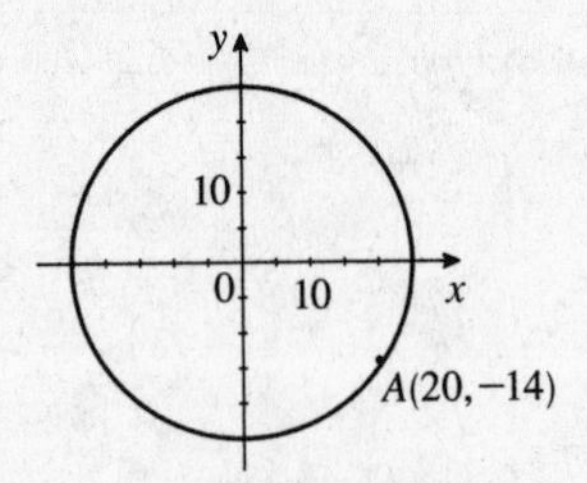

39. The cost of shipping by freight train is $(0.66)(2000)(100) = 132{,}000$, or \$132,000.
The cost of shipping by truck is $(0.62)(2200)(100) = 136{,}400$, or \$136,400.
Comparing these results, we see that the automobiles should be shipped by freight train. The net savings are $136{,}400 - 132{,}000 = 4400$, or \$4400.

41. To determine the VHF requirements, we calculate $d = \sqrt{25^2 + 35^2} = \sqrt{625 + 1225} = \sqrt{1850} \approx 43.01$. Models B, C, and D satisfy this requirement..
To determine the UHF requirements, we calculate $d = \sqrt{20^2 + 32^2} = \sqrt{400 + 1024} = \sqrt{1424} \approx 37.74$. Models C and D satisfy this requirement.
Therefore, Model C allows him to receive both channels at the least cost.

43. a. Let the positions of ships A and B be $(0, y)$ and $(x, 0)$, respectively. Then $y = 25\left(t + \frac{1}{2}\right)$ and $x = 20t$. The distance D in miles between the two ships is $D = \sqrt{(x-0)^2 + (0-y)^2} = \sqrt{x^2 + y^2} = \sqrt{400t^2 + 625\left(t + \frac{1}{2}\right)^2}$ (1).

b. The distance between the ships 2 hours after ship A has left port is obtained by letting $t = \frac{3}{2}$ in Equation (1), yielding $D = \sqrt{400\left(\frac{3}{2}\right)^2 + 625\left(\frac{3}{2} + \frac{1}{2}\right)^2} = \sqrt{3400}$, or approximately 58.31 miles.

45. a. Suppose that $P = (x_1, y_1)$ and $Q = (x_2, y_2)$ are endpoints of the line segment and that the point $M = \left(\frac{x_1 + x_2}{2}, \frac{y_1 + y_2}{2}\right)$ is the midpoint of the line segment PQ. The distance between P and Q is $\sqrt{(x_2 - x_1)^2 + (y_2 - y_1)^2}$. The distance between P and M is

$$\sqrt{\left(\frac{x_1 + x_2}{2} - x_1\right)^2 + \left(\frac{y_1 + y_2}{2} - y_1\right)^2} = \sqrt{\left(\frac{x_2 - x_1}{2}\right)^2 + \left(\frac{y_2 - y_1}{2}\right)^2} = \frac{1}{2}\sqrt{(x_2 - x_1)^2 + (y_2 - y_1)^2},$$

which is one-half the distance from P to Q. Similarly, we obtain the same expression for the distance from M to P.

b. The midpoint is given by $\left(\frac{4 - 3}{2}, \frac{-5 + 2}{2}\right)$, or $\left(\frac{1}{2}, -\frac{3}{2}\right)$.

47. True. Plot the points.

49. False. The distance between $P_1\,(a, b)$ and $P_3\,(kc, kd)$ is

$$d = \sqrt{(kc - a)^2 + (kd - b)^2}$$

$$\neq |k|\,D = |k|\,\sqrt{(c - a)^2 + (d - b)^2} = \sqrt{k^2\,(c - a)^2 + k^2\,(d - b)^2} = \sqrt{[k\,(c - a)]^2 + [k\,(d - b)]^2}.$$

51. Referring to the figure in the text, we see that the distance between the two points is given by the length of the hypotenuse of the right triangle. That is, $d = \sqrt{(x_2 - x_1)^2 + (y_2 - y_1)^2}$.

Problem-Solving Tips

When you solve a problem in the exercises that follow each section, first read the problem. Before you start computing or writing out a solution, try to formulate a strategy for solving the problem. Then proceed by using your strategy to solve the problem.

Here we summarize some general problem-solving techniques that are covered in this section.

1. **To show that two lines are parallel**, you need to show that the slopes of the two lines are equal or that their slopes are both undefined.

2. **To show that two lines L_1 and L_2 are perpendicular**, you need to show that the slope m_1 of L_1 is the negative reciprocal of the slope m_2 of L_2; that is, $m_1 = -1/m_2$.

3. **To find the equation of a line**, you need the slope of the line and a point lying on the line. You can then find the equation of the line using the point-slope form of the equation of a line: $(y - y_1) = m\,(x - x_1)$.

1.4 Straight Lines

Concept Questions

page 42

1. The slope is $m = \frac{y_2 - y_1}{x_2 - x_1}$, where $P\,(x_1, y_1)$ and $P\,(x_2, y_2)$ are any two distinct points on the nonvertical line. The slope of a vertical line is undefined.

3. a. $m_1 = m_2$ **b.** $m_2 = -\frac{1}{m_1}$

Exercises page 42

1. (e)

3. (a)

5. (f)

7. Referring to the figure shown in the text, we see that $m = \frac{2-0}{0-(-4)} = \frac{1}{2}$.

9. This is a vertical line, and hence its slope is undefined.

11. $m = \frac{y_2 - y_1}{x_2 - x_1} = \frac{8-3}{5-4} = 5$.

13. $m = \frac{y_2 - y_1}{x_2 - x_1} = \frac{8-3}{4-(-2)} = \frac{5}{6}$.

15. $m = \frac{y_2 - y_1}{x_2 - x_1} = \frac{d-b}{c-a}$, provided $a \neq c$.

17. Because the equation is already in slope-intercept form, we read off the slope $m = 4$.

a. If x increases by 1 unit, then y increases by 4 units.

b. If x decreases by 2 units, then y decreases by $4(-2) = -8$ units.

19. The slope of the line through A and B is $\frac{-10-(-2)}{-3-1} = \frac{-8}{-4} = 2$. The slope of the line through C and D is $\frac{1-5}{-1-1} = \frac{-4}{-2} = 2$. Because the slopes of these two lines are equal, the lines are parallel.

21. The slope of the line through A and B is $\frac{2-5}{4-(-2)} = -\frac{3}{6} = -\frac{1}{2}$. The slope of the line through C and D is $\frac{6-(-2)}{3-(-1)} = \frac{8}{4} = 2$. Because the slopes of these two lines are the negative reciprocals of each other, the lines are perpendicular.

23. The slope of the line through the point $(1, a)$ and $(4, -2)$ is $m_1 = \frac{-2-a}{4-1}$ and the slope of the line through $(2, 8)$ and $(-7, a+4)$ is $m_2 = \frac{a+4-8}{-7-2}$. Because these two lines are parallel, m_1 is equal to m_2. Therefore, $\frac{-2-a}{3} = \frac{a-4}{-9}$, $-9(-2-a) = 3(a-4)$, $18 + 9a = 3a - 12$, and $6a = -30$, so $a = -5$.

25. An equation of a horizontal line is of the form $y = b$. In this case $b = -3$, so $y = -3$ is an equation of the line.

27. We use the point-slope form of an equation of a line with the point $(3, -4)$ and slope $m = 2$. Thus $y - y_1 = m(x - x_1)$ becomes $y - (-4) = 2(x - 3)$. Simplifying, we have $y + 4 = 2x - 6$, or $y = 2x - 10$.

29. Because the slope $m = 0$, we know that the line is a horizontal line of the form $y = b$. Because the line passes through $(-3, 2)$, we see that $b = 2$, and an equation of the line is $y = 2$.

31. We first compute the slope of the line joining the points $(2, 4)$ and $(3, 7)$ to be $m = \frac{7-4}{3-2} = 3$. Using the point-slope form of an equation of a line with the point $(2, 4)$ and slope $m = 3$, we find $y - 4 = 3(x - 2)$, or $y = 3x - 2$.

33. We first compute the slope of the line joining the points $(1, 2)$ and $(-3, -2)$ to be $m = \dfrac{-2-2}{-3-1} = \dfrac{-4}{-4} = 1$. Using the point-slope form of an equation of a line with the point $(1, 2)$ and slope $m = 1$, we find $y - 2 = x - 1$, or $y = x + 1$.

35. We use the slope-intercept form of an equation of a line: $y = mx + b$. Because $m = 3$ and $b = 4$, the equation is $y = 3x + 4$.

37. We use the slope-intercept form of an equation of a line: $y = mx + b$. Because $m = 0$ and $b = 5$, the equation is $y = 5$.

39. We first write the given equation in the slope-intercept form: $x - 2y = 0$, so $-2y = -x$, or $y = \frac{1}{2}x$. From this equation, we see that $m = \frac{1}{2}$ and $b = 0$.

41. We write the equation in slope-intercept form: $2x - 3y - 9 = 0$, $-3y = -2x + 9$, and $y = \frac{2}{3}x - 3$. From this equation, we see that $m = \frac{2}{3}$ and $b = -3$.

43. We write the equation in slope-intercept form: $2x + 4y = 14$, $4y = -2x + 14$, and $y = -\frac{2}{4}x + \frac{14}{4} = -\frac{1}{2}x + \frac{7}{2}$. From this equation, we see that $m = -\frac{1}{2}$ and $b = \frac{7}{2}$.

45. We first write the equation $2x - 4y - 8 = 0$ in slope-intercept form: $2x - 4y - 8 = 0$, $4y = 2x - 8$, $y = \frac{1}{2}x - 2$. Now the required line is parallel to this line, and hence has the same slope. Using the point-slope form of an equation of a line with $m = \frac{1}{2}$ and the point $(-2, 2)$, we have $y - 2 = \frac{1}{2}[x - (-2)]$ or $y = \frac{1}{2}x + 3$.

47. The midpoint of the line segment joining $P_1(-2, -4)$ and $P_2(3, 6)$ is $M\left(\dfrac{-2+3}{2}, \dfrac{-4+6}{2}\right)$ or $M\left(\frac{1}{2}, 1\right)$. Using the point-slope form of the equation of a line with $m = -2$, we have $y - 1 = -2\left(x - \frac{1}{2}\right)$ or $y = -2x + 2$.

49. A line parallel to the x-axis has slope 0 and is of the form $y = b$. Because the line is 6 units below the axis, it passes through $(0, -6)$ and its equation is $y = -6$.

51. We use the point-slope form of an equation of a line to obtain $y - b = 0\,(x - a)$, or $y = b$.

53. Because the required line is parallel to the line joining $(-3, 2)$ and $(6, 8)$, it has slope $m = \dfrac{8-2}{6-(-3)} = \dfrac{6}{9} = \dfrac{2}{3}$. We also know that the required line passes through $(-5, -4)$. Using the point-slope form of an equation of a line, we find $y - (-4) = \frac{2}{3}[x - (-5)]$, $y = \frac{2}{3}x + \frac{10}{3} - 4$, and finally $y = \frac{2}{3}x - \frac{2}{3}$.

55. Because the point $(-3, 5)$ lies on the line $kx + 3y + 9 = 0$, it satisfies the equation. Substituting $x = -3$ and $y = 5$ into the equation gives $-3k + 15 + 9 = 0$, or $k = 8$.

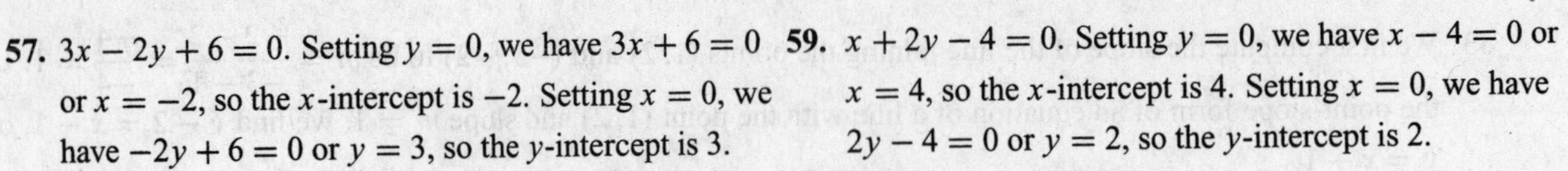

57. $3x - 2y + 6 = 0$. Setting $y = 0$, we have $3x + 6 = 0$ or $x = -2$, so the x-intercept is -2. Setting $x = 0$, we have $-2y + 6 = 0$ or $y = 3$, so the y-intercept is 3.

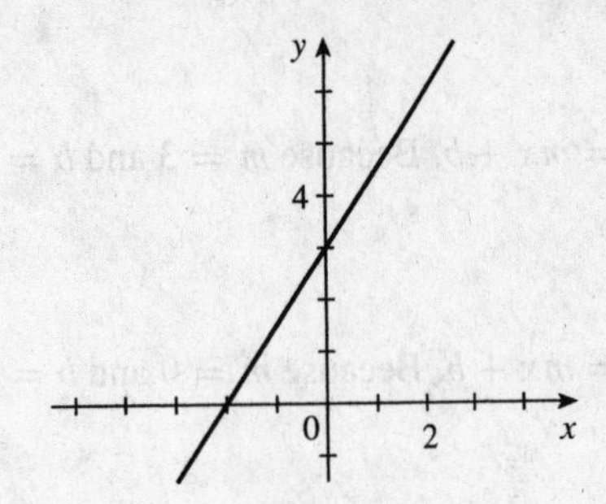

59. $x + 2y - 4 = 0$. Setting $y = 0$, we have $x - 4 = 0$ or $x = 4$, so the x-intercept is 4. Setting $x = 0$, we have $2y - 4 = 0$ or $y = 2$, so the y-intercept is 2.

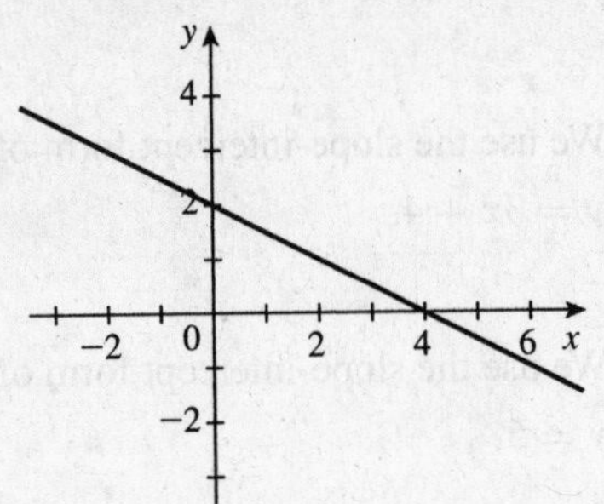

61. $y + 5 = 0$. Setting $y = 0$, we have $0 + 5 = 0$, which has no solution, so there is no x-intercept. Setting $x = 0$, we have $y + 5 = 0$ or $y = -5$, so the y-intercept is -5.

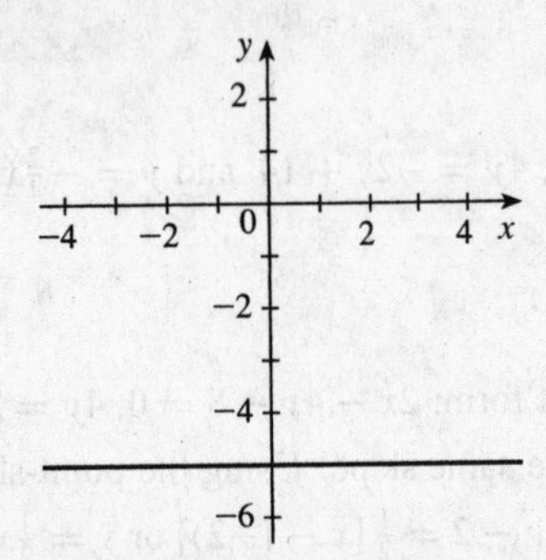

63. Because the line passes through the points $(a, 0)$ and $(0, b)$, its slope is $m = \frac{b - 0}{0 - a} = -\frac{b}{a}$. Then, using the point-slope form of an equation of a line with the point $(a, 0)$, we have $y - 0 = -\frac{b}{a}(x - a)$ or $y = -\frac{b}{a}x + b$, which may be written in the form $\frac{b}{a}x + y = b$. Multiplying this last equation by $\frac{1}{b}$, we have $\frac{x}{a} + \frac{y}{b} = 1$.

65. Using the equation $\frac{x}{a} + \frac{y}{b} = 1$ with $a = -2$ and $b = -4$, we have $-\frac{x}{2} - \frac{y}{4} = 1$. Then $-4x - 2y = 8$, $2y = -8 - 4x$, and finally $y = -2x - 4$.

67. Using the equation $\frac{x}{a} + \frac{y}{b} = 1$ with $a = 4$ and $b = -\frac{1}{2}$, we have $\frac{x}{4} + \frac{y}{-1/2} = 1$, $-\frac{1}{4}x + 2y = -1$, $2y = \frac{1}{4}x - 1$, and so $y = \frac{1}{8}x - \frac{1}{2}$.

69. The slope of the line passing through A and B is $m = \frac{7 - 1}{1 - (-2)} = \frac{6}{3} = 2$, and the slope of the line passing through B and C is $m = \frac{13 - 7}{4 - 1} = \frac{6}{3} = 2$. Because the slopes are equal, the points lie on the same line.

71. The slope of the line L passing through P_1 (1.8, −6.44) and P_2 (2.4, −5.72) is $m = \frac{-5.72 - (-6.44)}{2.4 - 1.8} = 1.2$, so an equation of L is $y - (-6.44) = 1.2(x - 1.8)$ or $y = 1.2x - 8.6$.
Substituting $x = 5.0$ into this equation gives $y = 1.2(5) - 8.6 = -2.6$. This shows that the point P_3 (5.0, −2.72) does not lie on L, and we conclude that Alison's claim is not valid.

73. a.

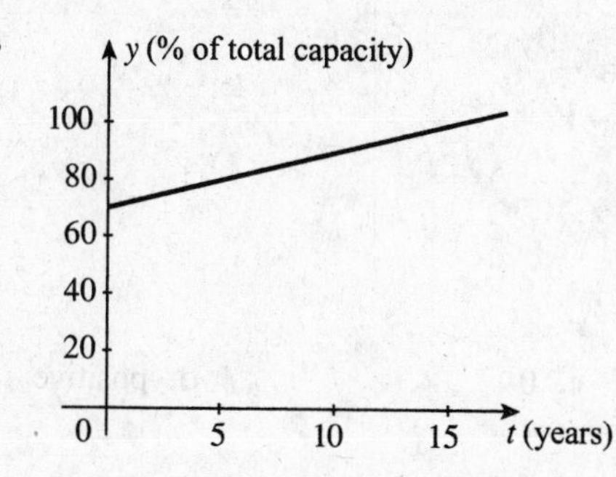

b. The slope is 1.9467 and the y-intercept is 70.082.

c. The output is increasing at the rate of 1.9467% per year. The output at the beginning of 1990 was 70.082%.

d. We solve the equation $1.9467t + 70.082 = 100$, obtaining $t \approx 15.37$. We conclude that the plants were generating at maximum capacity during April 2005.

75. a. $y = 0.55x$ **b.** Solving the equation $1100 = 0.55x$ for x, we have $x = \dfrac{1100}{0.55} = 2000$.

77. Using the points $(0, 0.68)$ and $(10, 0.80)$, we see that the slope of the required line is $m = \dfrac{0.80 - 0.68}{10 - 0} = \dfrac{0.12}{10} = 0.012$. Next, using the point-slope form of the equation of a line, we have $y - 0.68 = 0.012\,(t - 0)$ or $y = 0.012t + 0.68$. Therefore, when $t = 18$, we have $y = 0.012\,(18) + 0.68 = 0.896$, or 89.6%. That is, in 2008 women's wages were expected to be 89.6% of men's wages.

79. a, b.

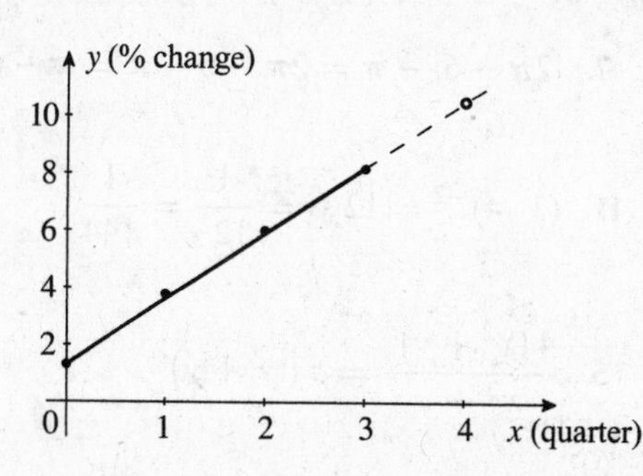

c. The slope of L is $m = \dfrac{8.2 - 1.3}{3 - 0} = 2.3$, so an equation of L is $y - 1.3 = 2.3\,(x - 0)$ or $y = 2.3x + 1.3$.

d. The change in spending in the first quarter of 2014 is estimated to be $2.3\,(4) + 1.3$, or 10.5%.

81. a. The slope of the line L passing through $A\,(0, 545)$ and $B\,(4, 726)$ is $m = \dfrac{726 - 545}{4 - 0} = \dfrac{181}{4}$, so an equation of L is $y - 545 = \frac{181}{4}\,(t - 0)$ or $y = \frac{181}{4}t + 545$.

b.

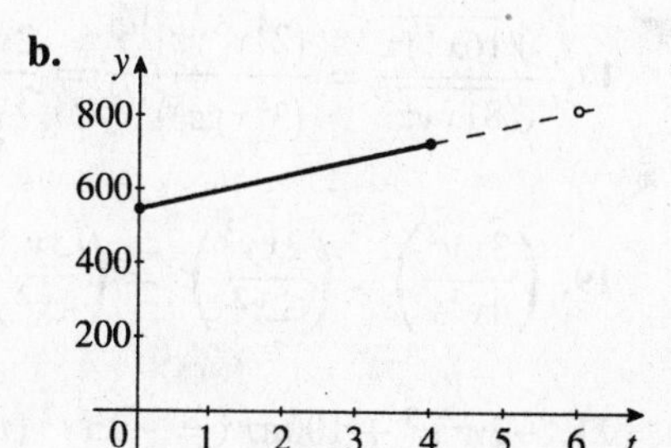

c. The number of corporate fraud cases pending at the beginning of 2014 is estimated to be $\frac{181}{4}\,(6) + 545$, or approximately 817.

83. True. The slope of the line is given by $-\frac{2}{4} = -\frac{1}{2}$.

85. False. Let the slope of L_1 be $m_1 > 0$. Then the slope of L_2 is $m_2 = -\dfrac{1}{m_1} < 0$.

87. True. Set $y = 0$ and we have $Ax + C = 0$ or $x = -C/A$, and this is where the line intersects the x-axis.

89. Writing each equation in the slope-intercept form, we have $y = -\dfrac{a_1}{b_1}x - \dfrac{c_1}{b_1}$ $(b_1 \neq 0)$ and $y = -\dfrac{a_2}{b_2}x - \dfrac{c_2}{b_2}$ $(b_2 \neq 0)$. Because two lines are parallel if and only if their slopes are equal, we see that the lines are parallel if and only if $-\dfrac{a_1}{b_1} = -\dfrac{a_2}{b_2}$, or $a_1b_2 - b_1a_2 = 0$.

CHAPTER 1 Concept Review Questions page 48

1. ordered, abscissa or x-coordinate, ordinate or y-coordinate

3. $\sqrt{(c-a)^2+(d-b)^2}$

5. a. $\dfrac{y_2-y_1}{x_2-x_1}$ **b.** undefined **c.** 0 **d.** positive

7. a. $y-y_1=m(x-x_1)$, point-slope form **b.** $y=mx+b$, slope-intercept

CHAPTER 1 Review Exercises page 48

1. Adding x to both sides yields $3 \le 3x+9$, $3x \ge -6$, or $x \ge -2$. We conclude that the solution set is $[-2,\infty)$.

3. The inequalities imply $x>5$ or $x<-4$, so the solution set is $(-\infty,-4)\cup(5,\infty)$.

5. $|-5+7|+|-2|=|2|+|-2|=2+2=4$.

7. $|2\pi-6|-\pi=2\pi-6-\pi=\pi-6$.

9. $\left(\dfrac{9}{4}\right)^{3/2}=\dfrac{9^{3/2}}{4^{3/2}}=\dfrac{27}{8}$.

11. $(3\cdot4)^{-2}=12^{-2}=\dfrac{1}{12^2}=\dfrac{1}{144}$.

13. $\dfrac{(3\cdot2^{-3})(4\cdot3^5)}{2\cdot9^3}=\dfrac{3\cdot2^{-3}\cdot2^2\cdot3^5}{2\cdot(3^2)^3}=\dfrac{2^{-1}\cdot3^6}{2\cdot3^6}=\dfrac{1}{4}$.

15. $\dfrac{4(x^2+y)^3}{x^2+y}=4(x^2+y)^2$.

17. $\dfrac{\sqrt[4]{16x^5yz}}{\sqrt[4]{81xyz^5}}=\dfrac{(2^4x^5yz)^{1/4}}{(3^4xyz^5)^{1/4}}=\dfrac{2x^{5/4}y^{1/4}z^{1/4}}{3x^{1/4}y^{1/4}z^{5/4}}=\dfrac{2x}{3z}$.

19. $\left(\dfrac{3xy^2}{4x^3y}\right)^{-2}\left(\dfrac{3xy^3}{2x^2}\right)^3=\left(\dfrac{3y}{4x^2}\right)^{-2}\left(\dfrac{3y^3}{2x}\right)^3=\left(\dfrac{4x^2}{3y}\right)^2\left(\dfrac{3y^3}{2x}\right)^3=\dfrac{(16x^4)(27y^9)}{(9y^2)(8x^3)}=6xy^7$.

21. $-2\pi^2r^3+100\pi r^2=-2\pi r^2(\pi r-50)$.

23. $16-x^2=4^2-x^2=(4-x)(4+x)$.

25. $8x^2+2x-3=(4x+3)(2x-1)=0$, so $x=-\frac{3}{4}$ and $x=\frac{1}{2}$ are the roots of the equation.

27. $-x^3-2x^2+3x=-x(x^2+2x-3)=-x(x+3)(x-1)=0$, and so the roots of the equation are $x=0$, $x=-3$, and $x=1$.

29. Factoring the given expression, we have $(2x-1)(x+2)\le0$. From the sign diagram, we conclude that the given inequality is satisfied when $-2\le x\le\frac{1}{2}$.

- - - - - - - 0 + + Sign of $2x-1$
- - 0 + + + + + + + Sign of $x+2$
x-axis: -2, 0, $\frac{1}{2}$

31. The given inequality is equivalent to $|2x-3|<5$ or $-5<2x-3<5$. Thus, $-2<2x<8$, or $-1<x<4$.

33. We use the quadratic formula to solve the equation $x^2 - 2x - 5 = 0$. Here $a = 1$, $b = -2$, and $c = -5$, so

$$x = \frac{-b \pm \sqrt{b^2 - 4ac}}{2a} = \frac{-(-2) \pm \sqrt{(-2)^2 - 4(1)(-5)}}{2(1)} = \frac{2 \pm \sqrt{24}}{2} = 1 \pm \sqrt{6}.$$

35. $\dfrac{(t+6)(60) - (60t+180)}{(t+6)^2} = \dfrac{60t + 360 - 60t - 180}{(t+6)^2} = \dfrac{180}{(t+6)^2}.$

37. $\dfrac{2}{3}\left(\dfrac{4x}{2x^2-1}\right) + 3\left(\dfrac{3}{3x-1}\right) = \dfrac{8x}{3(2x^2-1)} + \dfrac{9}{3x-1} = \dfrac{8x(3x-1) + 27(2x^2-1)}{3(2x^2-1)(3x-1)} = \dfrac{78x^2 - 8x - 27}{3(2x^2-1)(3x-1)}.$

39. $\dfrac{\sqrt{x}-1}{x-1} = \dfrac{\sqrt{x}-1}{x-1} \cdot \dfrac{\sqrt{x}+1}{\sqrt{x}+1} = \dfrac{\left(\sqrt{x}\right)^2 - 1}{(x-1)\left(\sqrt{x}+1\right)} = \dfrac{x-1}{(x-1)\left(\sqrt{x}+1\right)} = \dfrac{1}{\sqrt{x}+1}.$

41. The distance is $d = \sqrt{[1-(-2)]^2 + [-7-(-3)]^2} = \sqrt{3^2 + (-4)^2} = \sqrt{9+16} = \sqrt{25} = 5.$

43. The distance is $d = \sqrt{\left(-\frac{1}{2} - \frac{1}{2}\right)^2 + \left(2\sqrt{3} - \sqrt{3}\right)^2} = \sqrt{1+3} = \sqrt{4} = 2.$

45. An equation is $y = 4$.

47. The line passes through the points $(-2, 4)$ and $(3, 0)$, so its slope is $m = \dfrac{4-0}{-2-3} = -\dfrac{4}{5}$. An equation is $y - 0 = -\frac{4}{5}(x-3)$, or $y = -\frac{4}{5}x + \frac{12}{5}$.

49. Writing the given equation in the form $y = -\frac{4}{3}x + 2$, we see that the slope of the given line is $-\frac{4}{3}$. Therefore, the slope of the required line is $\frac{3}{4}$ and an equation of the line is $y - 4 = \frac{3}{4}(x+2)$ or $y = \frac{3}{4}x + \frac{11}{2}$.

51. The slope of the line joining the points $(-3, 4)$ and $(2, 1)$ is $m = \dfrac{1-4}{2-(-3)} = -\dfrac{3}{5}$. Using the point-slope form of the equation of a line with the point $(-1, 3)$ and slope $-\frac{3}{5}$, we have $y - 3 = -\frac{3}{5}[x-(-1)]$. Therefore, $y = -\frac{3}{5}(x+1) + 3 = -\frac{3}{5}x + \frac{12}{5}$.

53. Rewriting the given equation in the slope-intercept form $y = \frac{2}{3}x - 8$, we see that the slope of the line with this equation is $\frac{2}{3}$. The slope of a line perpendicular to this line is thus $-\frac{3}{2}$. Using the point-slope form of the equation of a line with the point $(-2, -4)$ and slope $-\frac{3}{2}$, we have $y - (-4) = -\frac{3}{2}[x-(-2)]$ or $y = -\frac{3}{2}x - 7$. The general form of this equation is $3x + 2y + 14 = 0$.

55. Substituting $x = 2$ and $y = -4$ into the equation, we obtain $2(2) + k(-4) = -8$, so $-4k = -12$ and $k = 3$.

57. Using the point-slope form of an equation of a line, we have $y - 2 = -\frac{2}{3}(x-3)$ or $y = -\frac{2}{3}x + 4$. If $y = 0$, then $x = 6$, and if $x = 0$, then $y = 4$. A sketch of the line is shown.

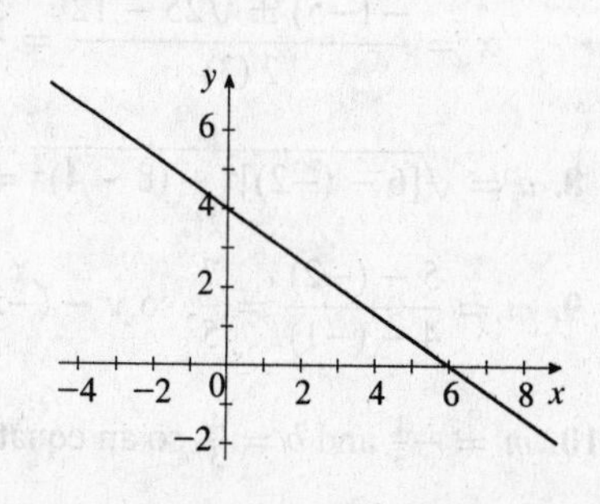

59. $3(2R-320) \le 3R+240$ gives $6R-960 \le 3R+240$, $3R \le 1200$ and finally $R \le 400$. We conclude that the maximum revenue is $400.

61. a, b.

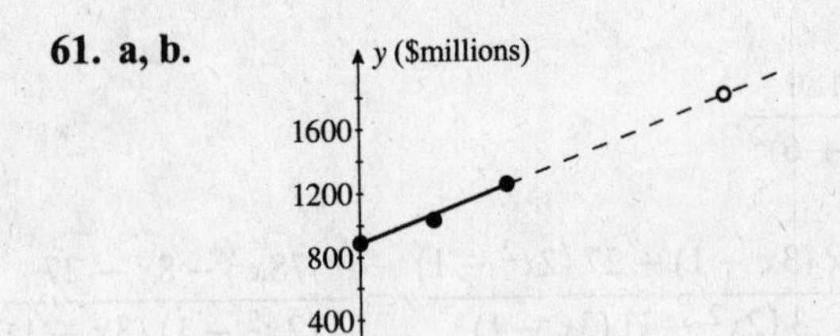

c. The slope of L is $\dfrac{1251-887}{2-0} = 182$, so an equation of L is $y-887 = 182(t-0)$ or $y = 182t+887$.

d. The amount consumers are projected to spend on Cyber Monday, 2014 ($t=5$) is $182(5)+887$, or $1.797 billion.

CHAPTER 1 Before Moving On... page 50

1. a. $\left|\pi-2\sqrt{3}\right| - \left|\sqrt{3}-\sqrt{2}\right| = -\left(\pi-2\sqrt{3}\right) - \left(\sqrt{3}-\sqrt{2}\right) = \sqrt{3}+\sqrt{2}-\pi$.

b. $\left[\left(-\frac{1}{3}\right)^{-3}\right]^{1/3} = \left(-\frac{1}{3}\right)^{(-3)\left(\frac{1}{3}\right)} = \left(-\frac{1}{3}\right)^{-1} = -3.$

2. a. $\sqrt[3]{64x^6}\cdot\sqrt{9y^2x^6} = (4x^2)(3yx^3) = 12x^5y.$

b. $\left(\dfrac{a^{-3}}{b^{-4}}\right)^2\left(\dfrac{b}{a}\right)^{-3} = \dfrac{a^{-6}}{b^{-8}}\cdot\dfrac{b^{-3}}{a^{-3}} = \dfrac{b^8}{a^6}\cdot\dfrac{a^3}{b^3} = \dfrac{b^5}{a^3}.$

3. a. $\dfrac{2x}{3\sqrt{y}}\cdot\dfrac{\sqrt{y}}{\sqrt{y}} = \dfrac{2x\sqrt{y}}{3y}.$

b. $\dfrac{x}{\sqrt{x}-4}\cdot\dfrac{\sqrt{x}+4}{\sqrt{x}+4} = \dfrac{x\left(\sqrt{x}+4\right)}{x-16}.$

4. a. $\dfrac{(x^2+1)\left(\frac{1}{2}x^{-1/2}\right) - x^{1/2}(2x)}{(x^2+1)^2} = \dfrac{\frac{1}{2}x^{-1/2}\left[(x^2+1)-4x^2\right]}{(x^2+1)^2} = \dfrac{1-3x^2}{2x^{1/2}(x^2+1)^2}.$

b. $-\dfrac{3x}{\sqrt{x+2}} + 3\sqrt{x+2} = \dfrac{-3x+3(x+2)}{\sqrt{x+2}} = \dfrac{6}{\sqrt{x+2}} = \dfrac{6\sqrt{x+2}}{x+2}.$

5. $\dfrac{\sqrt{x}+\sqrt{y}}{\sqrt{x}-\sqrt{y}} = \dfrac{\sqrt{x}+\sqrt{y}}{\sqrt{x}-\sqrt{y}}\cdot\dfrac{\sqrt{x}-\sqrt{y}}{\sqrt{x}-\sqrt{y}} = \dfrac{x-y}{\left(\sqrt{x}-\sqrt{y}\right)^2}.$

6. a. $12x^3-10x^2-12x = 2x(6x^2-5x-6) = 2x(2x-3)(3x+2).$

b. $2bx-2by+3cx-3cy = 2b(x-y)+3c(x-y) = (2b+3c)(x-y)$

7. a. $12x^2-9x-3=0$, so $3(4x^2-3x-1)=0$ and $3(4x+1)(x-1)=0$. Thus, $x=-\frac{1}{4}$ or $x=1$.

b. $3x^2-5x+1=0$. Using the quadratic formula with $a=3$, $b=-5$, and $c=1$, we have $x = \dfrac{-(-5)\pm\sqrt{25-12}}{2(3)} = \dfrac{5\pm\sqrt{13}}{6}.$

8. $d = \sqrt{[6-(-2)]^2+(8-4)^2} = \sqrt{64+16} = \sqrt{80} = 4\sqrt{5}.$

9. $m = \dfrac{5-(-2)}{4-(-1)} = \dfrac{7}{5}$, so $y-(-2) = \frac{7}{5}[x-(-1)]$, $y+2 = \frac{7}{5}x+\frac{7}{5}$, or $y = \frac{7}{5}x-\frac{3}{5}$.

10. $m=-\frac{1}{3}$ and $b=\frac{4}{3}$, so an equation is $y=-\frac{1}{3}x+\frac{4}{3}$.

2 FUNCTIONS, LIMITS, AND THE DERIVATIVE

2.1 Functions and Their Graphs

Problem-Solving Tips

New mathematical terms in each section appear in **boldface** type along with their definition, or they are defined in green boxes. Each time you encounter a new term, read through the definition and then try to express the definition in your own words without looking at the book. Once you understand these definitions, it will be easier for you to work the exercise sets that follow each section.

Here are some tips for solving the problems in the exercises that follow:

1. To find the domain of a function $f(x)$, find all values of x for which $f(x)$ is a real number.

a. If the function involves a quotient, check to see if there are any values of x at which the denominator is equal to zero. (Remember, division by zero is not allowed.) Then exclude those points from the domain.

b. If the function involves the root of a real number, check to see if the root is an even or an odd root. If n is even, the nth root of a negative number is not defined, so values of x yielding the nth root of a negative number must be excluded from the domain of f. For example, $\sqrt{x-1}$ is defined only for $x \geq 1$, so the domain of $f(x) = \sqrt{x-1}$ is $[1, \infty)$.

2. To evaluate a piecewise-defined function $f(x)$ at a specific value of x, check to see which subdomain x lies in. Then evaluate the function using the rule for that subdomain.

3. To determine whether a curve is the graph of a function, use the vertical line test. If you can draw a vertical line through the curve that intersects the curve at more than one point, then the curve is not the graph of a function.

Concept Questions page 59

1. a. A function is a rule that associates with each element in a set A exactly one element in a set B.

b. The domain of a function f is the set of all elements x in the set such that $f(x)$ is an element in B. The range of f is the set of all elements $f(x)$ whenever x is an element in its domain.

c. An independent variable is a variable in the domain of a function f. The dependent variable is $y = f(x)$.

3. a. Yes, every vertical line intersects the curve in at most one point.

b. No, a vertical line intersects the curve at more than one point.

c. No, a vertical line intersects the curve at more than one point.

d. Yes, every vertical line intersects the curve in at most one point.

Exercises page 59

1. $f(x) = 5x + 6$. Therefore $f(3) = 5(3) + 6 = 21$, $f(-3) = 5(-3) + 6 = -9$, $f(a) = 5(a) + 6 = 5a + 6$, $f(-a) = 5(-a) + 6 = -5a + 6$, and $f(a + 3) = 5(a + 3) + 6 = 5a + 15 + 6 = 5a + 21$.

3. $g(x) = 3x^2 - 6x - 3$, so $g(0) = 3(0) - 6(0) - 3 = -3$, $g(-1) = 3(-1)^2 - 6(-1) - 3 = 3 + 6 - 3 = 6$, $g(a) = 3(a)^2 - 6(a) - 3 = 3a^2 - 6a - 3$, $g(-a) = 3(-a)^2 - 6(-a) - 3 = 3a^2 + 6a - 3$, and $g(x + 1) = 3(x + 1)^2 - 6(x + 1) - 3 = 3(x^2 + 2x + 1) - 6x - 6 - 3 = 3x^2 + 6x + 3 - 6x - 9 = 3x^2 - 6$.

5. $f(x) = 2x + 5$, so $f(a + h) = 2(a + h) + 5 = 2a + 2h + 5$, $f(-a) = 2(-a) + 5 = -2a + 5$, $f(a^2) = 2(a^2) + 5 = 2a^2 + 5$, $f(a - 2h) = 2(a - 2h) + 5 = 2a - 4h + 5$, and $f(2a - h) = 2(2a - h) + 5 = 4a - 2h + 5$

7. $s(t) = \dfrac{2t}{t^2 - 1}$. Therefore, $s(4) = \dfrac{2(4)}{(4)^2 - 1} = \dfrac{8}{15}$, $s(0) = \dfrac{2(0)}{0^2 - 1} = 0$,

$s(a) = \dfrac{2(a)}{a^2 - 1} = \dfrac{2a}{a^2 - 1}$; $s(2 + a) = \dfrac{2(2 + a)}{(2 + a)^2 - 1} = \dfrac{2(2 + a)}{a^2 + 4a + 4 - 1} = \dfrac{2(2 + a)}{a^2 + 4a + 3}$, and

$s(t + 1) = \dfrac{2(t + 1)}{(t + 1)^2 - 1} = \dfrac{2(t + 1)}{t^2 + 2t + 1 - 1} = \dfrac{2(t + 1)}{t(t + 2)}$.

9. $f(t) = \dfrac{2t^2}{\sqrt{t - 1}}$. Therefore, $f(2) = \dfrac{2(2^2)}{\sqrt{2 - 1}} = 8$, $f(a) = \dfrac{2a^2}{\sqrt{a - 1}}$, $f(x + 1) = \dfrac{2(x + 1)^2}{\sqrt{(x + 1) - 1}} = \dfrac{2(x + 1)^2}{\sqrt{x}}$,

and $f(x - 1) = \dfrac{2(x - 1)^2}{\sqrt{(x - 1) - 1}} = \dfrac{2(x - 1)^2}{\sqrt{x - 2}}$.

11. Because $x = -2 \le 0$, we calculate $f(-2) = (-2)^2 + 1 = 4 + 1 = 5$. Because $x = 0 \le 0$, we calculate $f(0) = (0)^2 + 1 = 1$. Because $x = 1 > 0$, we calculate $f(1) = \sqrt{1} = 1$.

13. Because $x = -1 < 1$, $f(-1) = -\frac{1}{2}(-1)^2 + 3 = \frac{5}{2}$. Because $x = 0 < 1$, $f(0) = -\frac{1}{2}(0)^2 + 3 = 3$. Because $x = 1 \ge 1$, $f(1) = 2(1^2) + 1 = 3$. Because $x = 2 \ge 1$, $f(2) = 2(2^2) + 1 = 9$.

15. **a.** $f(0) = -2$.

b. **(i)** $f(x) = 3$ when $x \approx 2$. **(ii)** $f(x) = 0$ when $x = 1$.

c. $[0, 6]$

d. $[-2, 6]$

17. $g(2) = \sqrt{2^2 - 1} = \sqrt{3}$, so the point $\left(2, \sqrt{3}\right)$ lies on the graph of g.

19. $f(-2) = \dfrac{|-2 - 1|}{-2 + 1} = \dfrac{|-3|}{-1} = -3$, so the point $(-2, -3)$ does lie on the graph of f.

21. Because the point $(1, 5)$ lies on the graph of f it satisfies the equation defining f. Thus, $f(1) = 2(1)^2 - 4(1) + c = 5$, or $c = 7$.

23. Because $f(x)$ is a real number for any value of x, the domain of f is $(-\infty, \infty)$.

25. $f(x)$ is not defined at $x = 0$ and so the domain of f is $(-\infty, 0) \cup (0, \infty)$.

27. $f(x)$ is a real number for all values of x. Note that $x^2 + 1 \ge 1$ for all x. Therefore, the domain of f is $(-\infty, \infty)$.

29. Because the square root of a number is defined for all real numbers greater than or equal to zero, we have $5 - x \geq 0$, or $-x \geq -5$ and so $x \leq 5$. (Recall that multiplying by -1 reverses the sign of an inequality.) Therefore, the domain of f is $(-\infty, 5]$.

31. The denominator of f is zero when $x^2 - 1 = 0$, or $x = \pm 1$. Therefore, the domain of f is $(-\infty, -1) \cup (-1, 1) \cup (1, \infty)$.

33. f is defined when $x + 3 \geq 0$, that is, when $x \geq -3$. Therefore, the domain of f is $[-3, \infty)$.

35. The numerator is defined when $1 - x \geq 0$, $-x \geq -1$ or $x \leq 1$. Furthermore, the denominator is zero when $x = \pm 2$. Therefore, the domain is the set of all real numbers in $(-\infty, -2) \cup (-2, 1]$.

37. a. The domain of f is the set of all real numbers.

b. $f(x) = x^2 - x - 6$, so

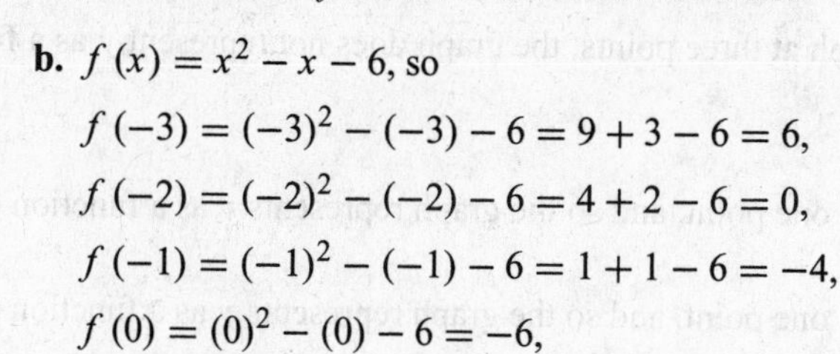

$f(-3) = (-3)^2 - (-3) - 6 = 9 + 3 - 6 = 6$,

$f(-2) = (-2)^2 - (-2) - 6 = 4 + 2 - 6 = 0$,

$f(-1) = (-1)^2 - (-1) - 6 = 1 + 1 - 6 = -4$,

$f(0) = (0)^2 - (0) - 6 = -6$,

$f\left(\frac{1}{2}\right) = \left(\frac{1}{2}\right)^2 - \left(\frac{1}{2}\right) - 6 = \frac{1}{4} - \frac{2}{4} - \frac{24}{4} = -\frac{25}{4}$, $f(1) = (1)^2 - 1 - 6 = -6$,

$f(2) = (2)^2 - 2 - 6 = 4 - 2 - 6 = -4$, and $f(3) = (3)^2 - 3 - 6 = 9 - 3 - 6 = 0$.

c.

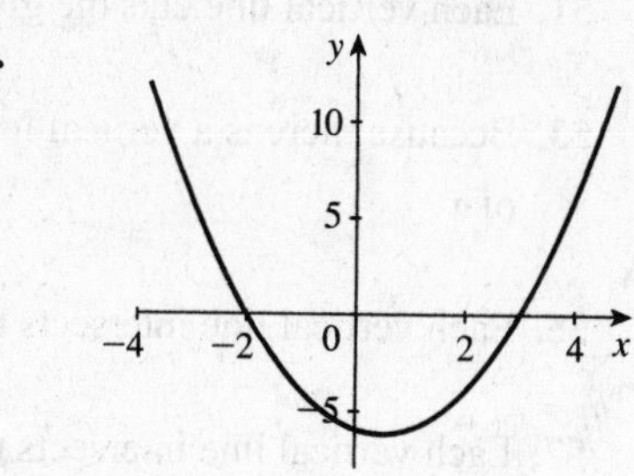

39. $f(x) = 2x^2 + 1$ has domain $(-\infty, \infty)$ and range $[1, \infty)$.

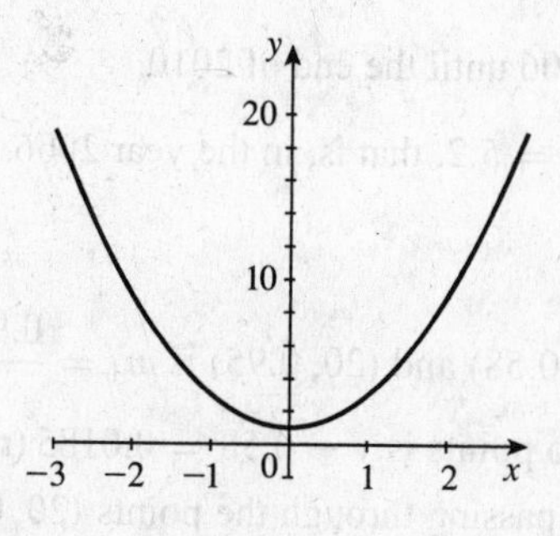

41. $f(x) = 2 + \sqrt{x}$ has domain $[0, \infty)$ and range $[2, \infty)$.

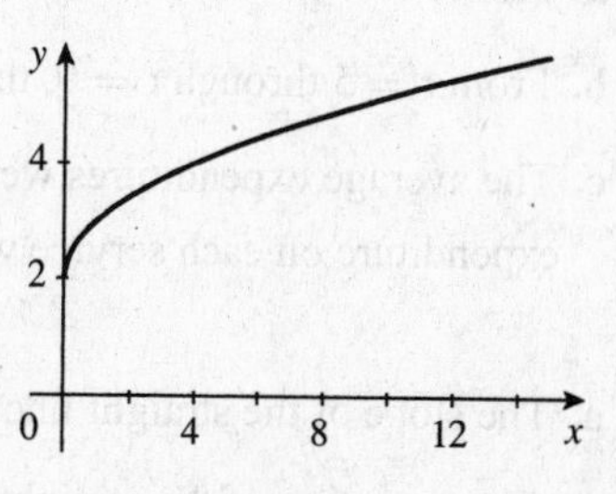

43. $f(x) = \sqrt{1 - x}$ has domain $(-\infty, 1]$ and range $[0, \infty)$

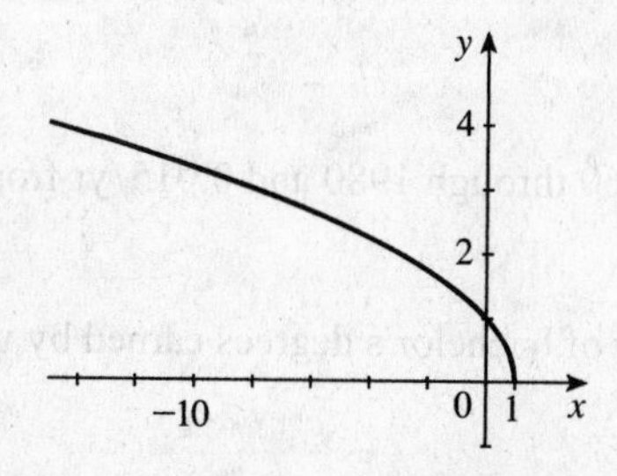

45. $f(x) = |x| - 1$ has domain $(-\infty, \infty)$ and range $[-1, \infty)$.

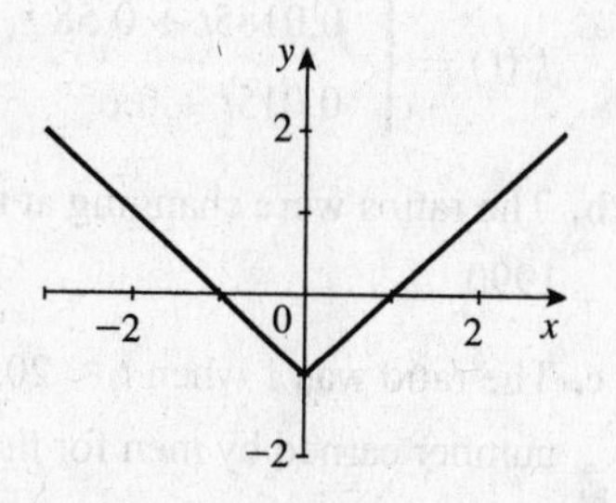

47. $f(x) = \begin{cases} x & \text{if } x < 0 \\ 2x + 1 & \text{if } x \geq 0 \end{cases}$ has domain $(-\infty, \infty)$ and range $(-\infty, 0) \cup [1, \infty)$.

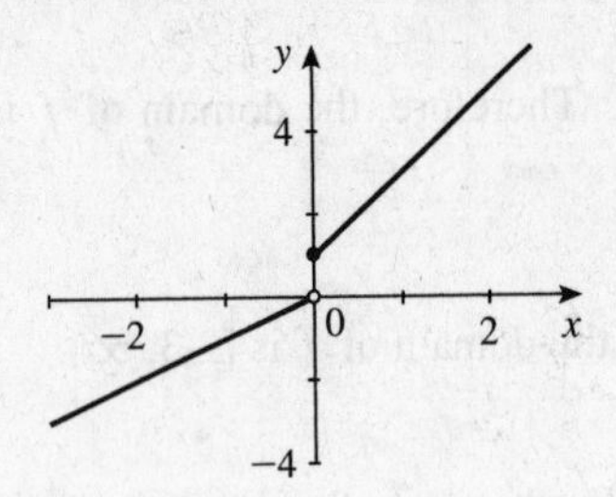

49. If $x \leq 1$, the graph of f is the half-line $y = -x + 1$. For $x > 1$, we calculate a few points: $f(2) = 3$, $f(3) = 8$, and $f(4) = 15$. f has domain $(-\infty, \infty)$ and range $[0, \infty)$.

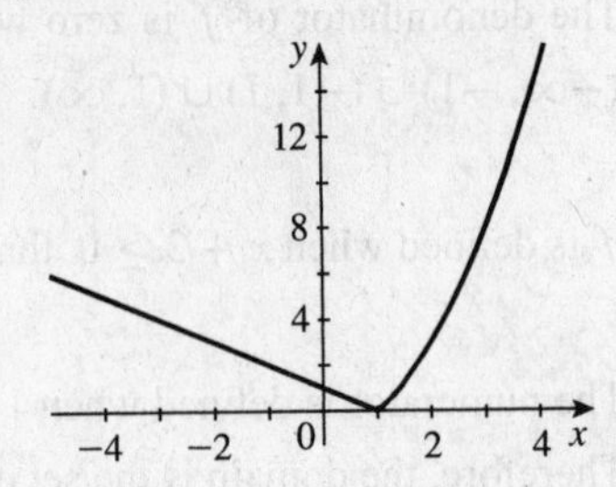

51. Each vertical line cuts the given graph at exactly one point, and so the graph represents y as a function of x.

53. Because there is a vertical line that intersects the graph at three points, the graph does not represent y as a function of x.

55. Each vertical line intersects the graph of f at exactly one point, and so the graph represents y as a function of x.

57. Each vertical line intersects the graph of f at exactly one point, and so the graph represents y as a function of x.

59. The circumference of a circle with a 5-inch radius is given by $C(5) = 2\pi(5) = 10\pi$, or 10π inches.

61. $C(0) = 6$, or 6 billion dollars; $C(50) = 0.75(50) + 6 = 43.5$, or 43.5 billion dollars; and $C(100) = 0.75(100) + 6 = 81$, or 81 billion dollars.

63. **a.** From $t = 0$ through $t = 5$, that is, from the beginning of 2001 until the end of 2005.

b. From $t = 5$ through $t = 9$, that is, from the beginning of 2006 until the end of 2010.

c. The average expenditures were the same at approximately $t = 5.2$, that is, in the year 2006. The level of expenditure on each service was approximately \$900.

65. **a.** The slope of the straight line passing through the points (0, 0.58) and (20, 0.95) is $m_1 = \dfrac{0.95 - 0.58}{20 - 0} = 0.0185$, so an equation of the straight line passing through these two points is $y - 0.58 = 0.0185(t - 0)$ or $y = 0.0185t + 0.58$. Next, the slope of the straight line passing through the points (20, 0.95) and (30, 1.1) is $m_2 = \dfrac{1.1 - 0.95}{30 - 20} = 0.015$, so an equation of the straight line passing through the two points is $y - 0.95 = 0.015(t - 20)$ or $y = 0.015t + 0.65$. Therefore, a rule for f is

$$f(t) = \begin{cases} 0.0185t + 0.58 & \text{if } 0 \leq t \leq 20 \\ 0.015t + 0.65 & \text{if } 20 < t \leq 30 \end{cases}$$

b. The ratios were changing at the rates of 0.0185/yr from 1960 through 1980 and 0.015/yr from 1980 through 1990.

c. The ratio was 1 when $t \approx 20.3$. This shows that the number of bachelor's degrees earned by women equaled the number earned by men for the first time around 1983.

67. a. $I(x) = 1.053x$

b. $I(1520) = 1.053(1520) = 1600.56$, or \$1600.56.

69. $S(r) = 4\pi r^2$.

71. a. The median age was changing at the rate of 0.3 years/year.

b. The median age in 2011 was $M(11) = 0.3(11) + 37.9 = 41.2$ (years).

c. The median age in 2015 is projected to be $M(5) = 0.3(15) + 37.9 = 42.4$ (years).

73. a. The graph of the function is a straight line passing through (0, 120000) and (10, 0). Its slope is $m = \dfrac{0 - 120{,}000}{10 - 0} = -12{,}000$. The required equation is $V = -12{,}000n + 120{,}000$.

b.

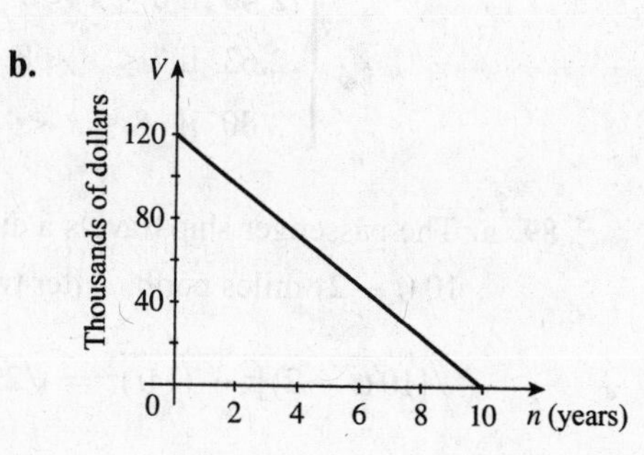

c. $V = -12{,}000(6) + 120{,}000 = 48{,}000$, or \$48,000.

d. This is given by the slope, that is, \$12,000 per year.

75. a. The number of incidents in 2009 was $f(0) = 0.46$ (million).

b. The number of incidents in 2013 was $f(4) = 0.2(4^2) - 0.14(4) + 0.46 = 3.1$ (million).

77. a. The life expectancy of a male whose current age is 65 is $f(65) = 0.0069502(65)^2 - 1.6357(65) + 93.76 \approx 16.80$, or approximately 16.8 years.

b. The life expectancy of a male whose current age is 75 is $f(75) = 0.0069502(75)^2 - 1.6357(75) + 93.76 \approx 10.18$, or approximately 10.18 years.

79. The projected number in 2030 is $P(20) = -0.0002083(20)^3 + 0.0157(20)^2 - 0.093(20) + 5.2 = 7.9536$, or approximately 8 million.
The projected number in 2050 is $P(40) = -0.0002083(40)^3 + 0.0157(40)^2 - 0.093(40) + 5.2 = 13.2688$, or approximately 13.3 million.

81. When the proportion of popular votes won by the Democratic presidential candidate is 0.60, the proportion of seats in the House of Representatives won by Democratic candidates is given by
$$s(0.6) = \frac{(0.6)^3}{(0.6)^3 + (1 - 0.6)^3} = \frac{0.216}{0.216 + 0.064} = \frac{0.216}{0.280} \approx 0.77.$$

83. The domain of the function f is the set of all real positive numbers where $V \neq 0$; that is, $(0, \infty)$.

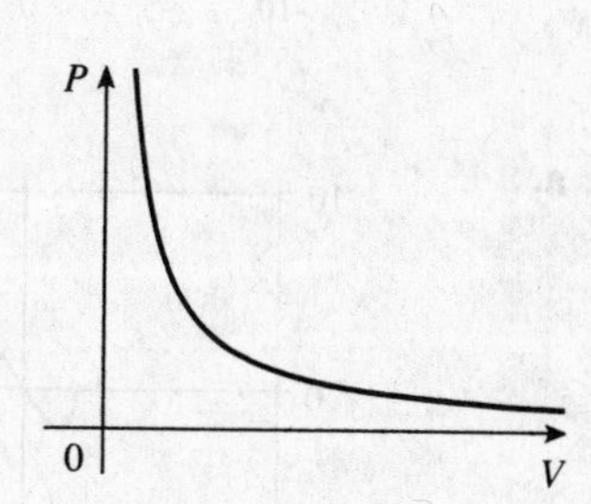

85. a. The assets at the beginning of 2002 were \$0.6 trillion. At the beginning of 2003, they were $f(1) = 0.6$, or \$0.6 trillion.

b. The assets at the beginning of 2005 were $f(3) = 0.6(3)^{0.43} \approx 0.96$, or \$0.96 trillion. At the beginning of 2007, they were $f(5) = 0.6(5)^{0.43} \approx 1.20$, or \$1.2 trillion.

87. a. The domain of f is $(0, 13]$.

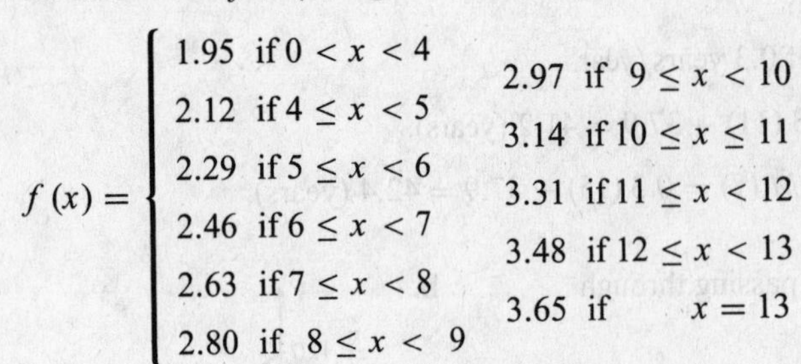

$$f(x) = \begin{cases} 1.95 & \text{if } 0 < x < 4 \\ 2.12 & \text{if } 4 \le x < 5 \\ 2.29 & \text{if } 5 \le x < 6 \\ 2.46 & \text{if } 6 \le x < 7 \\ 2.63 & \text{if } 7 \le x < 8 \\ 2.80 & \text{if } 8 \le x < 9 \\ 2.97 & \text{if } 9 \le x < 10 \\ 3.14 & \text{if } 10 \le x \le 11 \\ 3.31 & \text{if } 11 \le x < 12 \\ 3.48 & \text{if } 12 \le x < 13 \\ 3.65 & \text{if } x = 13 \end{cases}$$

b.

y (¢)
300
200
100
0 4 8 12 x (oz)

89. a. The passenger ship travels a distance given by $14t$ miles east and the cargo ship travels a distance of $10(t-2)$ miles north. After two hours have passed, the distance between the two ships is given by $\sqrt{[10(t-2)]^2 + (14t)^2} = \sqrt{296t^2 - 400t + 400}$ miles, so $D(t) = \begin{cases} 14t & \text{if } 0 \le t \le 2 \\ 2\sqrt{74t^2 - 100t + 100} & \text{if } t > 2 \end{cases}$

b. Three hours after the cargo ship leaves port the value of t is 5. Therefore, $D = 2\sqrt{74(5)^2 - 100(5) + 100} \approx 76.16$, or 76.16 miles.

91. False. Take $f(x) = x^2$, $a = 1$, and $b = -1$. Then $f(1) = 1 = f(-1)$, but $a \ne b$.

93. False. It intersects the graph of a function in at most one point.

95. False. Take $f(x) = x^2$ and $k = 2$. Then $f(x) = (2x)^2 = 4x^2 \ne 2x^2 = 2f(x)$.

97. False. They are equal everywhere except at $x = 0$, where g is not defined.

Using Technology page 68

1. a.

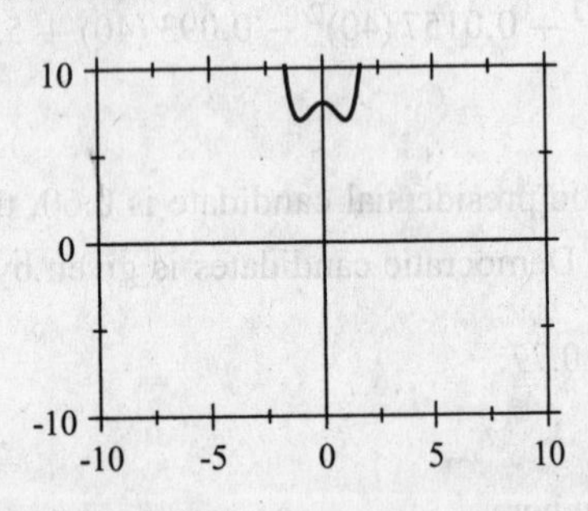

b.

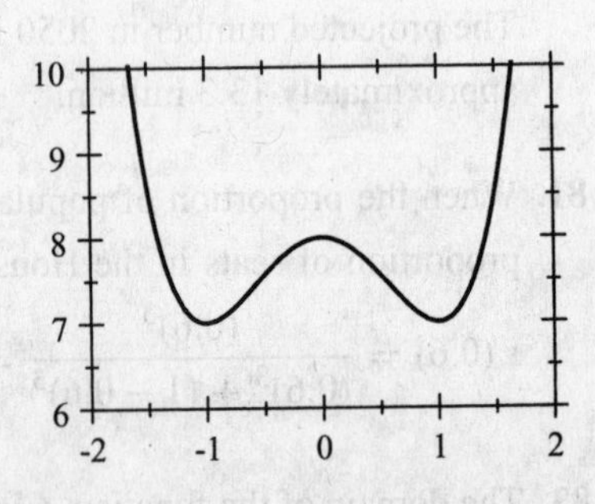

3. a.

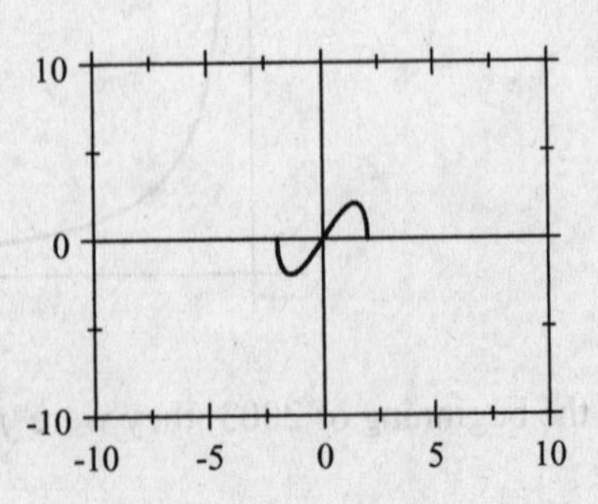

b.

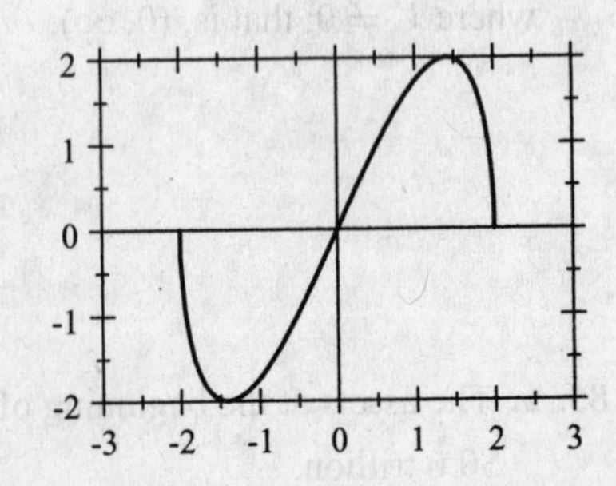

5.

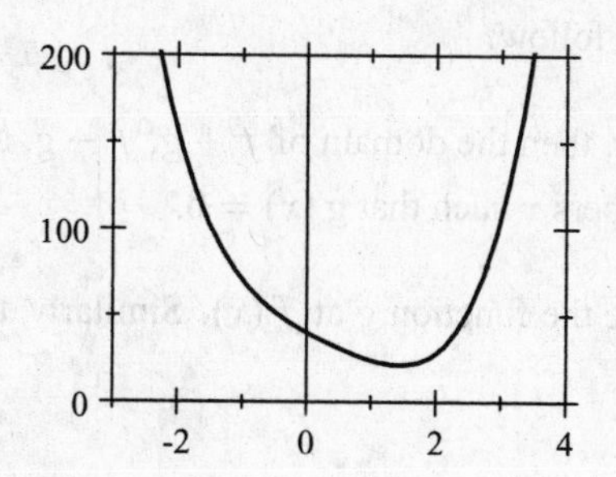

7.

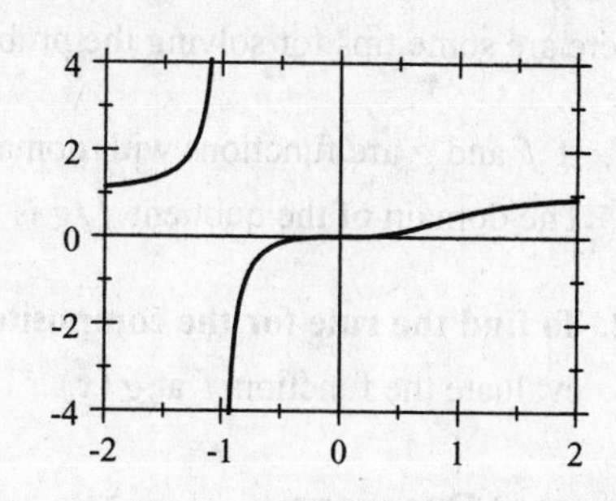

9. $f(2.145) \approx 18.5505$.

11. $f(2.41) \approx 4.1616$.

13. a.

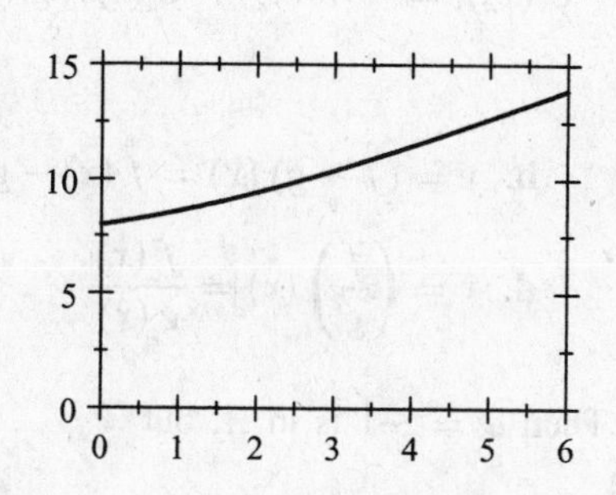

b. The amount spent in the year 2005 was $f(2) \approx 9.42$, or approximately \$9.4 billion. In 2009, it was $f(6) \approx 13.88$, or approximately \$13.9 billion.

15. a.

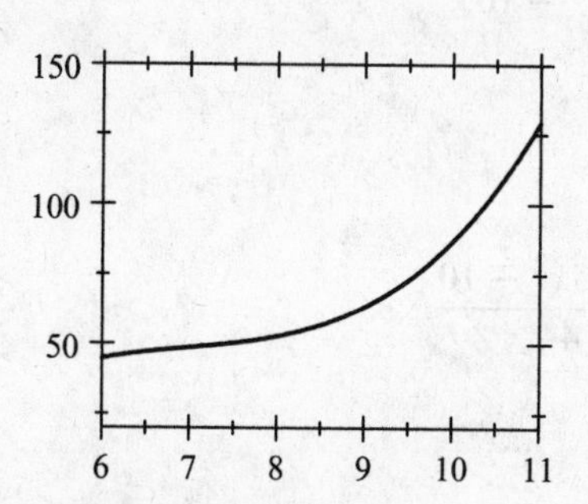

b. $f(6) = 44.7$, $f(8) = 52.7$, and $f(11) = 129.2$.

2.2 The Algebra of Functions

Problem-Solving Tips

When you come across new notation, make sure that you understand that notation. If you can't express the notation verbally, you haven't yet grasped its meaning. For example, in this section we introduced the notation $g \circ f$, read "g circle f."We use this notation to describe the composition of the functions g and f. Note that $g \circ f$ is not the same as $f \circ g$.

Here are some tips for solving the problems in the exercises that follow:

1. If f and g are functions with domains A and B, respectively, then the domain of $f+g$, $f-g$, and fg is $A \cap B$. The domain of the quotient f/g is $A \cap B$ excluding all numbers x such that $g(x) = 0$.

2. **To find the rule for the composite function** $g \circ f$, evaluate the function g at $f(x)$. Similarly, to find $f \circ g$, evaluate the function f at $g(x)$.

Concept Questions page 73

1. a. $P(x_1) = R(x_1) - C(x_1)$ gives the profit if x_1 units are sold.

b. $P(x_2) = R(x_2) - C(x_2)$. Because $P(x_2) < 0$, $|R(x_2) - C(x_2)| = -[R(x_2) - C(x_2)]$ gives the loss sustained if x_2 units are sold.

3. a. $y = (f+g)(x) = f(x) + g(x)$

b. $y = (f-g)(x) = f(x) - g(x)$

c. $y = (fg)(x) = f(x)g(x)$

d. $y = \left(\frac{f}{g}\right)(x) = \frac{f(x)}{g(x)}$

5. No. Let $A = (-\infty, \infty)$, $f(x) = x$, and $g(x) = \sqrt{x}$. Then $a = -1$ is in A, but $(g \circ f)(-1) = g(f(-1)) = g(-1) = \sqrt{-1}$ is not defined.

Exercises page 74

1. $(f+g)(x) = f(x) + g(x) = (x^3+5) + (x^2-2) = x^3 + x^2 + 3$.

3. $fg(x) = f(x)g(x) = (x^3+5)(x^2-2) = x^5 - 2x^3 + 5x^2 - 10$.

5. $\frac{f}{g}(x) = \frac{f(x)}{g(x)} = \frac{x^3+5}{x^2-2}$.

7. $\frac{fg}{h}(x) = \frac{f(x)g(x)}{h(x)} = \frac{(x^3+5)(x^2-2)}{2x+4} = \frac{x^5-2x^3+5x^2-10}{2x+4}$.

9. $(f+g)(x) = f(x) + g(x) = x - 1 + \sqrt{x+1}$.

11. $(fg)(x) = f(x)g(x) = (x-1)\sqrt{x+1}$.

13. $\frac{g}{h}(x) = \frac{g(x)}{h(x)} = \frac{\sqrt{x+1}}{2x^3-1}$.

15. $\frac{fg}{h}(x) = \frac{(x-1)(\sqrt{x+1})}{2x^3-1}$.

17. $\frac{f-h}{g}(x) = \frac{x-1-(2x^3-1)}{\sqrt{x+1}} = \frac{x-2x^3}{\sqrt{x+1}}$.

19. $(f+g)(x) = x^2 + 5 + \sqrt{x} - 2 = x^2 + \sqrt{x} + 3$, $(f-g)(x) = x^2 + 5 - (\sqrt{x} - 2) = x^2 - \sqrt{x} + 7$, $(fg)(x) = (x^2+5)(\sqrt{x}-2)$, and $\left(\frac{f}{g}\right)(x) = \frac{x^2+5}{\sqrt{x}-2}$.

21. $(f+g)(x) = \sqrt{x+3} + \frac{1}{x-1} = \frac{(x-1)\sqrt{x+3}+1}{x-1}$, $(f-g)(x) = \sqrt{x+3} - \frac{1}{x-1} = \frac{(x-1)\sqrt{x+3}-1}{x-1}$, $(fg)(x) = \sqrt{x+3}\left(\frac{1}{x-1}\right) = \frac{\sqrt{x+3}}{x-1}$, and $\left(\frac{f}{g}\right) = \sqrt{x+3}\,(x-1)$.

23. $(f+g)(x) = \frac{x+1}{x-1} + \frac{x+2}{x-2} = \frac{(x+1)(x-2)+(x+2)(x-1)}{(x-1)(x-2)} = \frac{x^2-x-2+x^2+x-2}{(x-1)(x-2)}$

$= \frac{2x^2-4}{(x-1)(x-2)} = \frac{2(x^2-2)}{(x-1)(x-2)}$,

$(f-g)(x) = \frac{x+1}{x-1} - \frac{x+2}{x-2} = \frac{(x+1)(x-2)-(x+2)(x-1)}{(x-1)(x-2)} = \frac{x^2-x-2-x^2-x+2}{(x-1)(x-2)}$

$= \frac{-2x}{(x-1)(x-2)}$,

$(fg)(x) = \frac{(x+1)(x+2)}{(x-1)(x-2)}$, and $\left(\frac{f}{g}\right)(x) = \frac{(x+1)(x-2)}{(x-1)(x+2)}$.

25. $(f\circ g)(x) = f(g(x)) = f(x^2) = (x^2)^2 + x^2 + 1 = x^4 + x^2 + 1$ and

$(g\circ f)(x) = g(f(x)) = g(x^2+x+1) = (x^2+x+1)^2$.

27. $(f\circ g)(x) = f(g(x)) = f(x^2-1) = \sqrt{x^2-1}+1$ and

$(g\circ f)(x) = g(f(x)) = g(\sqrt{x}+1) = (\sqrt{x}+1)^2 - 1 = x+2\sqrt{x}+1-1 = x+2\sqrt{x}$.

29. $(f\circ g)(x) = f(g(x)) = f\left(\frac{1}{x}\right) = \frac{1}{x} \div \left(\frac{1}{x^2}+1\right) = \frac{1}{x}\cdot\frac{x^2}{x^2+1} = \frac{x}{x^2+1}$ and

$(g\circ f)(x) = g(f(x)) = g\left(\frac{x}{x^2+1}\right) = \frac{x^2+1}{x}$.

31. $h(2) = g(f(2))$. But $f(2) = 2^2+2+1 = 7$, so $h(2) = g(7) = 49$.

33. $h(2) = g(f(2))$. But $f(2) = \frac{1}{2(2)+1} = \frac{1}{5}$, so $h(2) = g\left(\frac{1}{5}\right) = \frac{1}{\sqrt{5}} = \frac{\sqrt{5}}{5}$.

35. $f(x) = 2x^3 + x^2 + 1$, $g(x) = x^5$.

37. $f(x) = x^2 - 1$, $g(x) = \sqrt{x}$.

39. $f(x) = x^2 - 1$, $g(x) = \frac{1}{x}$.

41. $f(x) = 3x^2 + 2$, $g(x) = \frac{1}{x^{3/2}}$.

43. $f(a+h) - f(a) = [3(a+h)+4] - (3a+4) = 3a+3h+4-3a-4 = 3h$.

45. $f(a+h) - f(a) = 4-(a+h)^2 - (4-a^2) = 4-a^2-2ah-h^2-4+a^2 = -2ah-h^2 = -h(2a+h)$.

47. $\frac{f(a+h)-f(a)}{h} = \frac{\left[(a+h)^2+1\right]-(a^2+1)}{h} = \frac{a^2+2ah+h^2+1-a^2-1}{h} = \frac{2ah+h^2}{h}$

$= \frac{h(2a+h)}{h} = 2a+h$.

49. $\frac{f(a+h)-f(a)}{h} = \frac{\left[(a+h)^3-(a+h)\right]-(a^3-a)}{h} = \frac{a^3+3a^2h+3ah^2+h^3-a-h-a^3+a}{h}$

$= \frac{3a^2h+3ah^2+h^3-h}{h} = 3a^2+3ah+h^2-1$.

51. $\frac{f(a+h)-f(a)}{h} = \frac{\frac{1}{a+h}-\frac{1}{a}}{h} = \frac{\frac{a-(a+h)}{a(a+h)}}{h} = -\frac{1}{a(a+h)}$.

53. $F(t)$ represents the total revenue for the two restaurants at time t.

55. $f(t)g(t)$ represents the dollar value of Nancy's holdings at time t.

57. $g \circ f$ is the function giving the amount of carbon monoxide pollution from cars in parts per million at time t.

59. $C(x) = 0.6x + 12{,}100$.

61. $D(t) = (D_2 - D_1)(t) = D_2(t) - D_1(t) = (0.035t^2 + 0.21t + 0.24) - (0.0275t^2 + 0.081t + 0.07)$

$\approx 0.0075t^2 + 0.129t + 0.17$.

The function D gives the difference in year t between the deficit without the \$160 million rescue package and the deficit with the rescue package.

63. a. $(g \circ f)(1) = g(f(1)) = g(406) = 23$. So in 2002, the percentage of reported serious crimes that end in arrests or in the identification of suspects was 23.

b. $(g \circ f)(6) = g(f(6)) = g(326) = 18$. In 2007, 18% of reported serious crimes ended in arrests or in the identification of suspects.

c. Between 2002 and 2007, the total number of detectives had dropped from 406 to 326 and as a result, the percentage of reported serious crimes that ended in arrests or in the identification of suspects dropped from 23 to 18.

65. a. $C(x) = V(x) + 20000 = 0.000001x^3 - 0.01x^2 + 50x + 20000 = 0.000001x^3 - 0.01x^2 + 50x + 20{,}000$.

b. $P(x) = R(x) - C(x) = -0.02x^2 + 150x - 0.000001x^3 + 0.01x^2 - 50x - 20{,}000$

$= -0.000001x^3 - 0.01x^2 + 100x - 20{,}000$.

c. $P(2000) = -0.000001(2000)^3 - 0.01(2000)^2 + 100(2000) - 20{,}000 = 132{,}000$, or \$132,000.

67. a. $h(t) = f(t) + g(t) = (4.389t^3 - 47.833t^2 + 374.49t + 2390) + (13.222t^3 - 132.524t^2 + 757.9t + 7481)$

$= 17.611t^3 - 180.357t^2 + 1132.39t + 9871,\ 1 \le t \le 7$.

b. $f(6) = 3862.976$ and $g(6) = 10{,}113.488$, so $f(6) + g(6) = 13{,}976.464$. The worker's contribution was approximately \$3862.98, the employer's contribution was approximately \$10,113.49, and the total contributions were approximately \$13,976.46.

c. $h(6) = 13{,}976 = f(6) + g(6)$, as expected.

69. a. The occupancy rate at the beginning of January is $r(0) = \frac{10}{81}(0)^3 - \frac{10}{3}(0)^2 + \frac{200}{9}(0) + 55 = 55$, or 55%.

$r(5) = \frac{10}{81}(5)^3 - \frac{10}{3}(5)^2 + \frac{200}{9}(5) + 55 \approx 98.2$, or approximately 98.2%.

b. The monthly revenue at the beginning of January is $R(55) = -\frac{3}{5000}(55)^3 + \frac{9}{50}(55)^2 \approx 444.68$, or approximately \$444,700.

The monthly revenue at the beginning of June is $R(98.2) = -\frac{3}{5000}(98.2)^3 + \frac{9}{50}(98.2)^2 \approx 1167.6$, or approximately \$1,167,600.

71. a. $s = f + g + h = (f + g) + h = f + (g + h)$. This suggests we define the sum s by $s(x) = (f + g + h)(x) = f(x) + g(x) + h(x)$.

b. Let f, g, and h define the revenue (in dollars) in week t of three branches of a store. Then its total revenue (in dollars) in week t is $s(t) = (f + g + h)(t) = f(t) + g(t) + h(t)$.

73. True. $(f+g)(x) = f(x) + g(x) = g(x) + f(x) = (g+f)(x)$.

75. False. Take $f(x) = \sqrt{x}$ and $g(x) = x+1$. Then $(g \circ f)(x) = \sqrt{x}+1$, but $(f \circ g)(x) = \sqrt{x+1}$.

77. True. $(h \circ (g \circ f))(x) = h((g \circ f)(x)) = h(g(f(x)))$ and $((h \circ g) \circ f)(x) = (h \circ g)(f(x)) = h(g(f(x)))$.

2.3 Functions and Mathematical Models

Problem-Solving Tips

When you solve a problem involving a function, it is helpful to identify the type of function you are working with. For example, there are no restrictions on the domain of a polynomial function. If you want to find the domain of a rational function, you have to check to see if there are any values for which the denominator is equal to 0.
Here are some tips for solving the problems in the exercises that follow:

1. **To find the market equilibrium of a commodity**, find the point of intersection of the supply and demand equations for the commodity. (Market equilibrium prevails when the quantity produced is equal to the quantity demanded.).
2. **To construct a mathematical model**, follow the guidelines given in the text on page 86. First try solving Examples 5 and 6 in the text without looking at the solutions. Then go on to try a few similar problems (Exercises 76–84 on pages 94–95 of the text).

Concept Questions page 88

1. See page 78 of the text. Answers will vary.

3. a. A demand function $p = D(x)$ gives the relationship between the unit price of a commodity p and the quantity x demanded. A supply function $p = S(x)$ gives the relationship between the unit price of a commodity p and the quantity x the supplier will make available in the marketplace.

b. Market equilibrium occurs when the quantity produced is equal to the quantity demanded. To find the market equilibrium, we solve the equations $p = D(x)$ and $p = S(x)$ simultaneously.

Exercises page 88

1. Yes. $2x + 3y = 6$ and so $y = -\frac{2}{3}x + 2$.

3. Yes. $2y = x + 4$ and so $y = \frac{1}{2}x + 2$.

5. Yes. $4y = 2x + 9$ and so $y = \frac{1}{2}x + \frac{9}{4}$.

7. No, because of the term x^2.

9. f is a polynomial function in x of degree 6.

11. Expanding $G(x) = 2\left(x^2 - 3\right)^3$, we have $G(x) = 2x^6 - 18x^4 + 54x^2 - 54$, and we conclude that G is a polynomial function of degree 6 in x.

13. f is neither a polynomial nor a rational function.

15. $f(0) = 2$ gives $f(0) = m(0) + b = b = 2$. Next, $f(3) = -1$ gives $f(3) = m(3) + b = -1$. Substituting $b = 2$ in this last equation, we have $3m + 2 = -1$, or $3m = -3$, and therefore, $m = -1$ and $b = 2$.

17. a. $C(x) = 8x + 40{,}000$.

b. $R(x) = 12x$.

c. $P(x) = R(x) - C(x) = 12x - (8x + 40{,}000) = 4x - 40{,}000$.

d. $P(8000) = 4(8000) - 40{,}000 = -8000$, or a loss of \$8000. $P(12{,}000) = 4(12{,}000) - 40{,}000 = 8000$, or a profit of \$8000.

19. The individual's disposable income is $D = (1 - 0.28) \cdot 60{,}000 = 43{,}200$, or \$43,200.

21. The child should receive $D(4) = \left(\frac{4+1}{24}\right)(500) \approx 104.17$, or approximately 104 mg.

23. a. The slope of the graph of f is a line with slope -13.2 passing through the point $(0, 400)$, so an equation of the line is $y - 400 = -13.2(t - 0)$ or $y = -13.2t + 400$, and the required function is $f(t) = -13.2t + 400$.

b. The emissions cap is projected to be $f(2) = -13.2(2) + 400 = 373.6$, or 373.6 million metric tons of carbon dioxide equivalent.

25. a.

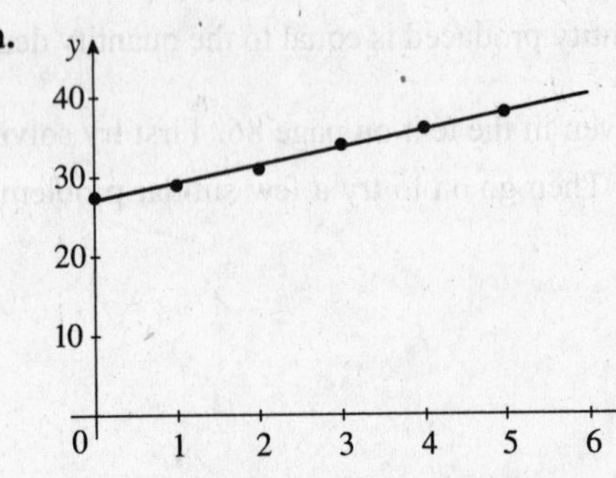

b. The projected revenue in 2010 is projected to be $f(6) = 2.19(6) + 27.12 = 40.26$, or \$40.26 billion.

c. The rate of increase is the slope of the graph of f, that is, 2.19 (billion dollars per year).

27. $P(28) = -\frac{1}{8}(28)^2 + 7(28) + 30 = 128$, or \$128,000.

29. a. The average time spent per day in 2009 was $f(0) = 21.76$ (minutes).

b. The average time spent per day in 2013 is projected to be $f(4) = 2.25(4)^2 + 13.41(4) + 21.76 = 111.4$ (minutes).

31. a. The GDP per capita in 2000 was $f(10) = 1.86251(10)^2 - 28.08043(10) + 884 = 789.4467$, or \$789.45.

b. The GDP per capita in 2030 is projected to be $f(40) = 1.86251(40)^2 - 28.08043(40) + 884 = 2740.7988$, or \$2740.80.

33. $S(6) = 0.73(6)^2 + 15.8(6) + 2.7 = 123.78$ million kilowatt-hr.
$S(8) = 0.73(8)^2 + 15.8(8) + 2.7 = 175.82$ million kilowatt-hr.

35. The percentage who expected to work past age 65 in 1991 was $f(0) = 11$, or 11%. The percentage in 2013 was $f(22) = 0.004545(22)^3 - 0.1113(22)^2 + 1.385(22) + 11 = 35.99596$, or approximately 36%.

37. a. Total global mobile data traffic in 2009 was $f(0) = 0.06$, or 60,000 terabytes.

b. The total in 2014 will be $f(5) = 0.021(5)^3 + 0.015(5)^2 + 0.12(5) + 0.06 = 3.66$, or 3.66 million terabytes.

39. a. We first construct a table.

t	$N(t)$
1	52
2	75
3	93
4	109
5	122

t	$N(t)$
6	135
7	146
8	157
9	167
10	177

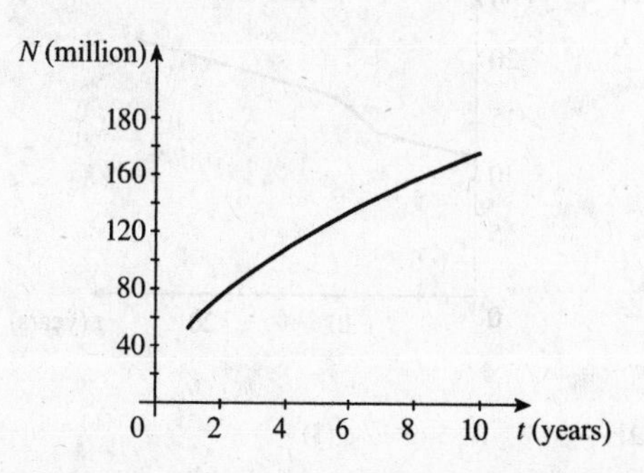

b. The number of viewers in 2012 is given by $N(10) = 52(10)^{0.531} \approx 176.61$, or approximately 177 million viewers.

41. $N(5) = 0.0018425(10)^{2.5} \approx 0.58265$, or approximately 0.583 million. $N(13) = 0.0018425(18)^{2.5} \approx 2.5327$, or approximately 2.5327 million.

43. a. We are given that $f(1) = 5240$ and $f(4) = 8680$. This leads to the system of equations $a + b = 5240$, $11a + b = 8680$. Solving, we find $a = 344$ and $b = 4896$.

b. From part (a), we have $f(t) = 344t + 4896$, so the approximate per capita costs in 2005 were $f(5) = 344(5) + 4896 = 6616$, or \$6616.

45. a. $f(0) = 6.85$, $g(0) = 16.58$. Because $g(0) > f(0)$, we see that more film cameras were sold in 2001 (when $t = 0$).

b. We solve the equation $f(t) = g(t)$, that is, $3.05t + 6.85 = -1.85t + 16.58$, so $4.9t = 9.73$ and $t = 1.99 \approx 2$. So sales of digital cameras first exceed those of film cameras in approximately 2003.

47. a. We are given that $T = aN + b$ where a and b are constants to be determined. The given conditions imply that $70 = 120a + b$ and $80 = 160a + b$. Subtracting the first equation from the second gives $10 = 40a$, or $a = \frac{1}{4}$. Substituting this value of a into the first equation gives $70 = 120\left(\frac{1}{4}\right) + b$, or $b = 40$. Therefore, $T = \frac{1}{4}N + 40$.

b. Solving the equation in part (a) for N, we find $\frac{1}{4}N = T - 40$, or $N = f(t) = 4T - 160$. When $T = 102$, we find $N = 4(102) - 160 = 248$, or 248 times per minute.

49. a. We have $f(0) = c = 1547$, $f(2) = 4a + 2b + c = 1802$, and $f(4) = 16a + 4b + c = 2403$. Solving this system of equations gives $a = 43.25$, $b = 41$, and $c = 1547$.

b. From part (a), we have $f(t) = 43.25t^2 + 41t + 1547$, so the number of craft-beer breweries in 2014 is projected to be $f(6) = 43.25(6)^2 + 41(6) + 1547 = 3350$.

51. Using the formula given in Exercise 50, we have $V(2) = 100{,}000 - \dfrac{100{,}000 - 30{,}000}{5}(2) = 100{,}000 - \dfrac{70{,}000}{5}(2) = 72{,}000$, or \$72,000.

53. The total cost by 2011 is given by $f(1) = 5$, or \$5 billion. The total cost by 2015 is given by $f(5) = -0.5278(5^3) + 3.012(5^2) + 49.23(5) - 103.29 = 152.185$, or approximately \$152 billion.

55. a.

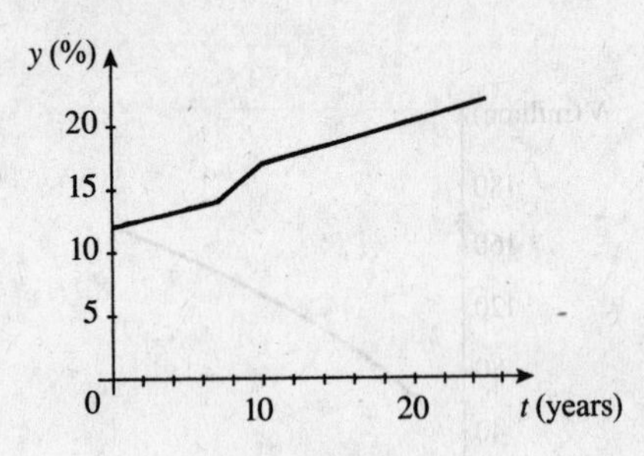

b. $f(5) = \frac{2}{7}(5) + 12 = \frac{10}{7} + 12 \approx 13.43$, or approximately 13.43%. $f(25) = \frac{1}{3}(25) + \frac{41}{3} = 22$, or 22%.

57. a.

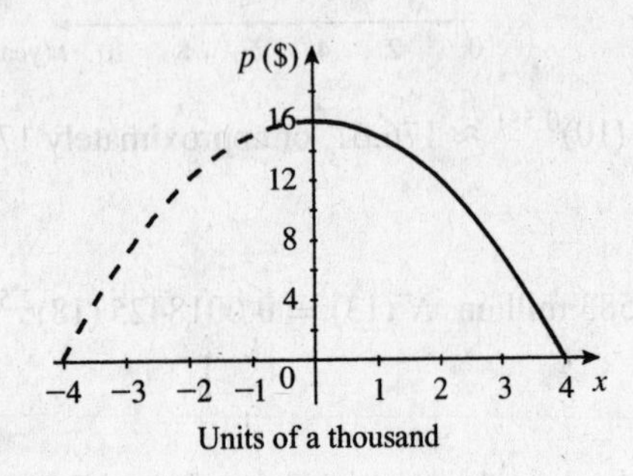

b. If $p = 7$, we have $7 = -x^2 + 16$, or $x^2 = 9$, so that $x = \pm 3$. Therefore, the quantity demanded when the unit price is \$7 is 3000 units.

59. a.

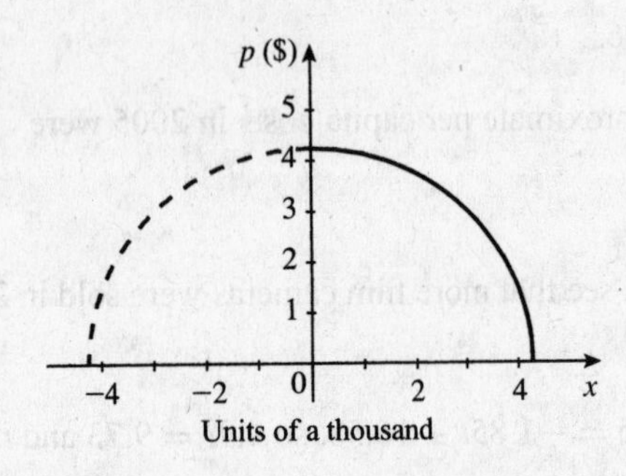

b. If $p = 3$, then $3 = \sqrt{18 - x^2}$, and $9 = 18 - x^2$, so that $x^2 = 9$. and $x = \pm 3$. Therefore, the quantity demanded when the unit price is \$3 is 3000 units.

61. a.

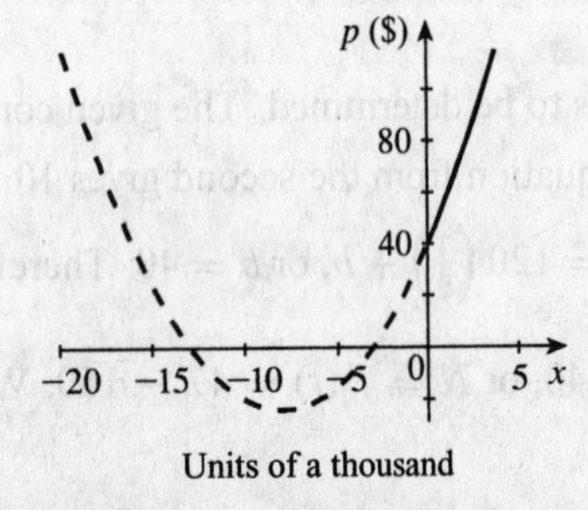

b. If $x = 2$, then $p = 2^2 + 16(2) + 40 = 76$, or \$76.

63. a.

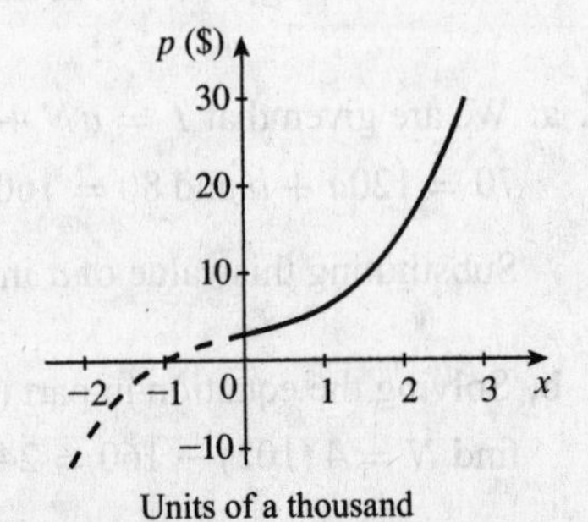

b. $p = 2^3 + 2(2) + 3 = 15$, or \$15.

65. a.

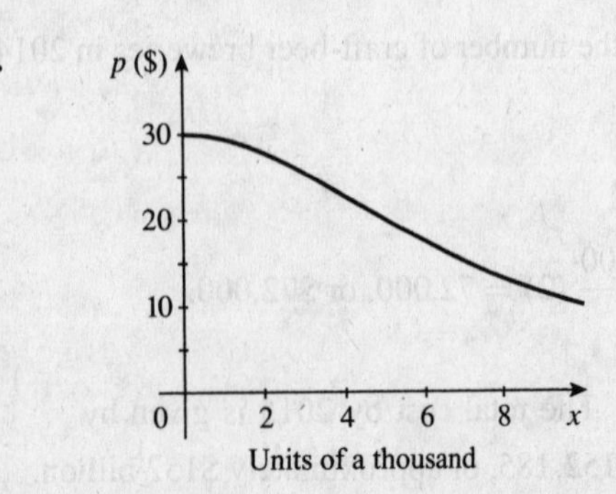

b. Substituting $x = 10$ into the demand function, we have

$$p = \frac{30}{0.02(10)^2 + 1} = \frac{30}{3} = 10, \text{ or } \$10.$$

67. a.

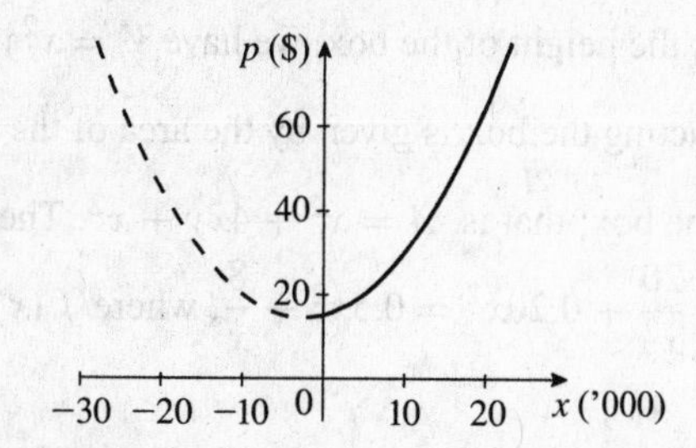

b. If $x = 5$, then

$p = 0.1(5)^2 + 0.5(5) + 15 = 20$, or \$20.

69. a. We solve the system of equations $p = cx + d$ and $p = ax + b$. Substituting the first equation into the second gives $cx + d = ax + d$, so $(c - a)x = b - d$ and $x = \dfrac{b-d}{c-a}$. Because $a < 0$ and $c > 0$, $c - a \neq 0$ and x is well-defined. Substituting this value of x into the second equation, we obtain $p = a\left(\dfrac{b-d}{c-a}\right) + b = \dfrac{ab - ad + bc - ab}{c-a} = \dfrac{bc - ad}{c-a}$. Therefore, the equilibrium quantity is $\dfrac{b-d}{c-a}$ and the equilibrium price is $\dfrac{bc-ad}{c-a}$.

b. If c is increased, the denominator in the expression for x increases and so x gets smaller. At the same time, the first term in the first equation for p decreases and so p gets larger. This analysis shows that if the unit price for producing the product is increased then the equilibrium quantity decreases while the equilibrium price increases.

c. If b is decreased, the numerator of the expression for x decreases while the denominator stays the same. Therefore, x decreases. The expression for p also shows that p decreases. This analysis shows that if the (theoretical) upper bound for the unit price of a commodity is lowered, then both the equilibrium quantity and the equilibrium price drop.

71. We solve the equation $-2x^2 + 80 = 15x + 30$, or $2x^2 + 15x - 50 = 0$ for x. Thus, $(2x - 5)(x + 10) = 0$, and so $x = \frac{5}{2}$ or $x = -10$. Rejecting the negative root, we have $x = \frac{5}{2}$. The corresponding value of p is $p = -2\left(\frac{5}{2}\right)^2 + 80 = 67.5$. We conclude that the equilibrium quantity is 2500 and the equilibrium price is \$67.50.

73. Solving both equations for x, we have $x = -\frac{11}{3}p + 22$ and $x = 2p^2 + p - 10$. Equating the right-hand sides, we have $-\frac{11}{3}p + 22 = 2p^2 + p - 10$, or $-11p + 66 = 6p^2 + 3p - 30$, and so $6p^2 + 14p - 96 = 0$. Dividing this last equation by 2 and then factoring, we have $(3p + 16)(p - 3) = 0$, so $p = 3$ is the only valid solution. The corresponding value of x is $2(3)^2 + 3 - 10 = 11$. We conclude that the equilibrium quantity is 11,000 and the equilibrium price is \$3.

75. Equating the right-hand sides of the two equations, we have $144 - x^2 = 48 + \frac{1}{2}x^2$, so $288 - 2x^2 = 96 + x^2$, $3x^2 = 192$, and $x^2 = 64$. Therefore, $x = \pm 8$. We take $x = 8$, and the corresponding value of p is $144 - 8^2 = 80$. We conclude that the equilibrium quantity is 8000 tires and the equilibrium price is \$80.

77. The area of Juanita's garden is 250 ft^2. Therefore $xy = 250$ and $y = \dfrac{250}{x}$. The amount of fencing needed is given by $2x + 2y$. Therefore, $f = 2x + 2\left(\dfrac{250}{x}\right) = 2x + \dfrac{500}{x}$. The domain of f is $x > 0$.

79. Because the volume of the box is the area of the base times the height of the box, we have $V = x^2 y = 20$. Thus, we have $y = \frac{20}{x^2}$. Next, the amount of material used in constructing the box is given by the area of the base of the box, plus the area of the four sides, plus the area of the top of the box; that is, $A = x^2 + 4xy + x^2$. Then, the cost of constructing the box is given by $f(x) = 0.30x^2 + 0.40x \cdot \frac{20}{x^2} + 0.20x^2 = 0.5x^2 + \frac{8}{x}$, where $f(x)$ is measured in dollars and $f(x) > 0$.

81. The average yield of the apple orchard is 36 bushels/tree when the density is 22 trees/acre. Let x be the unit increase in tree density beyond 22. Then the yield of the apple orchard in bushels/acre is given by $(22 + x)(36 - 2x)$.

83. a. Let x denote the number of bottles sold beyond 10,000 bottles. Then $P(x) = (10{,}000 + x)(5 - 0.0002x) = -0.0002x^2 + 3x + 50{,}000$.

b. He can expect a profit of $P(6000) = -0.0002(6000^2) + 3(6000) + 50{,}000 = 60{,}800$, or \$60,800.

85. False. $f(x) = 3x^{3/4} + x^{1/2} + 1$ is not a polynomial function. The powers of x must be nonnegative integers.

87. False. $f(x) = x^{1/2}$ is not defined for negative values of x.

Using Technology page 98

1. $(-3.0414, 0.1503)$, $(3.0414, 7.4497)$.

3. $(-2.3371, 2.4117)$, $(6.0514, -2.5015)$.

5. $(-1.0219, -6.3461)$, $(1.2414, -1.5931)$, and $(5.7805, 7.9391)$.

7. a.

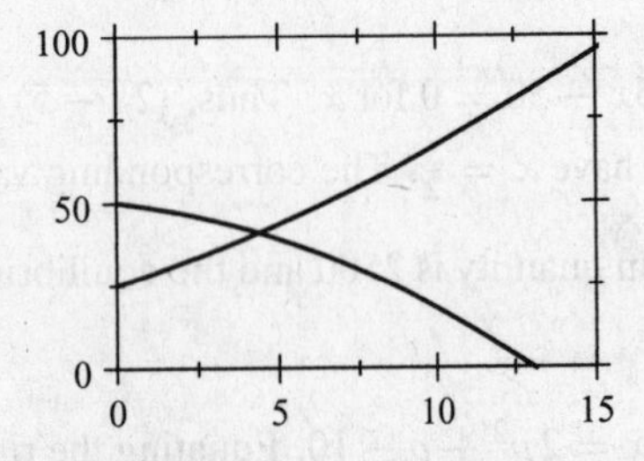

b. 438 wall clocks; \$40.92.

9. a. $f(t) = 1.85t + 16.9$.

b.

c.

t	y
1	18.8
2	20.6
3	22.5
4	24.3
5	26.2
6	28.0

These values are close to the given data.

d. $f(8) = 1.85(8) + 16.9 = 31.7$ gallons.

11. a. $f(t) = -0.221t^2 + 4.14t + 64.8.$

b.

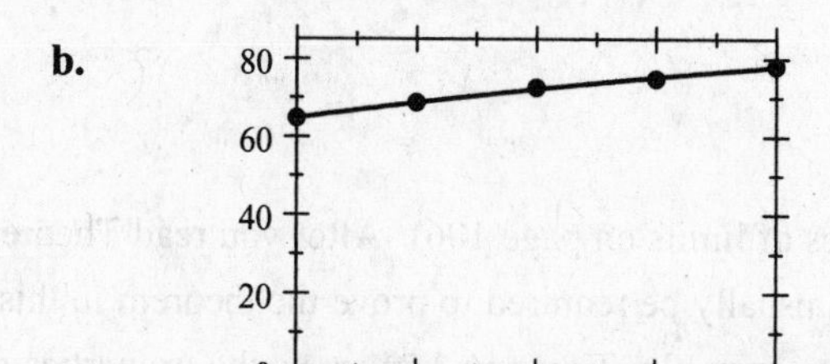

c. 77.8 million

13. a. $f(t) = 2.4t^2 + 15t + 31.4.$

b.

15. a. $f(t) = -0.00081t^3 + 0.0206t^2 + 0.125t + 1.69.$

b.

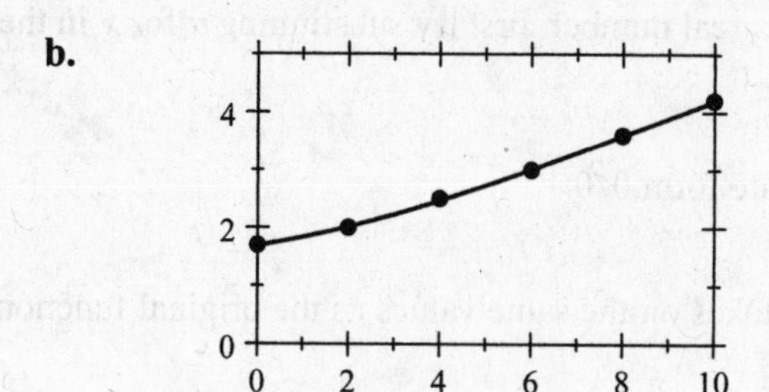

c.

t	y
1	1.8
5	2.7
10	4.2

The revenues were $1.8 trillion in 2001, $2.7 trillion in 2005, and $4.2 trillion in 2010.

17. a. $f(t) = -0.0056t^3 + 0.112t^2 + 0.51t + 8.$

b.

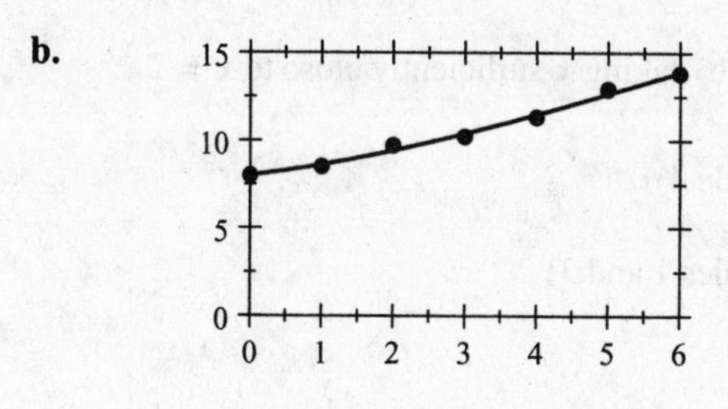

c.

t	0	3	6
$f(t)$	8	10.4	13.9

19. a. $f(t) = 0.00125t^4 + 0.0051t^3 - 0.0243t^2 + 0.129t + 1.71.$

b.

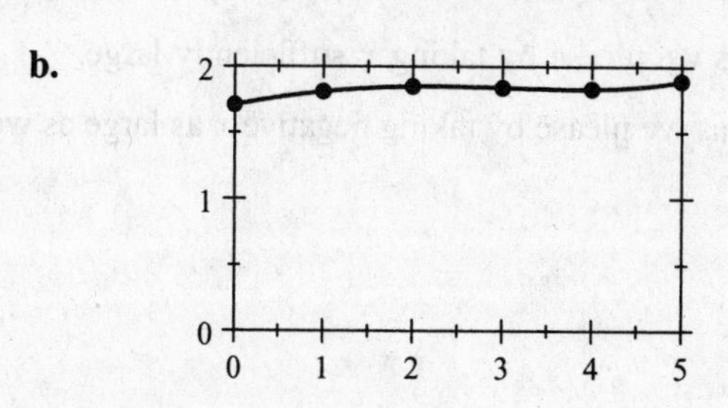

c.

t	0	1	2	3	4	5
$f(t)$	1.71	1.81	1.85	1.84	1.83	1.89

d. The average amount of nicotine in 2005 is $f(6) = 2.128$, or approximately 2.13 mg/cigarette.

2.4 Limits

Problem-Solving Tips

In this section, an important theorem was introduced (properties of limits on page 106). After you read Theorem 1, try to express the theorem in your own words. While you will not usually be required to prove the theorem in this course, you will be asked to understand the results of the theorem. For example, Theorem 1 gives us the properties of limits that allow us to evaluate sums, differences, products, quotients, powers, and constant multiples of functions at specified values, with certain restrictions. You should be able to use limit notation to write out each of these properties. You should also be able to use these properties to evaluate the limits of functions.

Here are some tips for solving the problems in the exercises that follow:

1. **To find the limit of a function** $f(x)$ **as** $x \to a$, where a is a real number, first try substituting a for x in the rule for f and simplify the result.

2. **To evaluate the limit of a quotient** that has the indeterminate form $0/0$:

   a. Replace the given function with an appropriate one that takes on the same values as the original function everywhere except at $x = a$.

   b. Evaluate the limit of this function as x approaches a.

Concept Questions page 115

1. The values of $f(x)$ can be made as close to 3 as we please by taking x sufficiently close to $x = 2$.

3. **a.** $\lim_{x\to 4} \sqrt{x}\,(2x^2+1) = \lim_{x\to 4} (\sqrt{x}) \lim_{x\to 4} (2x^2+1)$ (Rule 4)

$$= \sqrt{4}\left[2(4)^2+1\right] \quad \text{(Rules 1 and 3)}$$

$$= 66$$

b. $\lim_{x\to 1}\left(\frac{2x^2+x+5}{x^4+1}\right)^{3/2} = \left(\lim_{x\to 1}\frac{2x^2+x+5}{x^4+1}\right)^{3/2}$ (Rule 1)

$$= \left(\frac{2+1+5}{1+1}\right)^{3/2} \quad \text{(Rules 2, 3, and 5)}$$

$$= 4^{3/2} = 8$$

5. $\lim_{x\to\infty} f(x) = L$ means $f(x)$ can be made as close to L as we please by taking x sufficiently large. $\lim_{x\to-\infty} f(x) = M$ means $f(x)$ can be made as close to M as we please by taking negative x as large as we please in absolute value.

Exercises page 115

1. $\lim_{x\to -2} f(x) = 3$.

3. $\lim_{x\to 3} f(x) = 3$.

5. $\lim_{x\to -2} f(x) = 3$.

7. The limit does not exist. If we consider any value of x to the right of $x = -2$, $f(x) \le 2$. If we consider values of x to the left of $x = -2$, $f(x) \ge -2$. Because $f(x)$ does not approach any one number as x approaches $x = -2$, we conclude that the limit does not exist.

9.

x	1.9	1.99	1.999	2.001	2.01	2.1
$f(x)$	4.61	4.9601	4.9960	5.004	5.0401	5.41

$\lim_{x \to 2} (x^2 + 1) = 5.$

11.

x	-0.1	-0.01	-0.001	0.001	0.01	0.1
$f(x)$	-1	-1	-1	1	1	1

The limit does not exist.

13.

x	0.9	0.99	0.999	1.001	1.01	1.1
$f(x)$	100	10,000	1,000,000	1,000,000	10,000	100

The limit does not exist.

15.

x	0.9	0.99	0.999	1.001	1.01	1.1
$f(x)$	2.9	2.99	2.999	3.001	3.01	3.1

$$\lim_{x \to 1} \frac{x^2 + x - 2}{x - 1} = 3.$$

17.

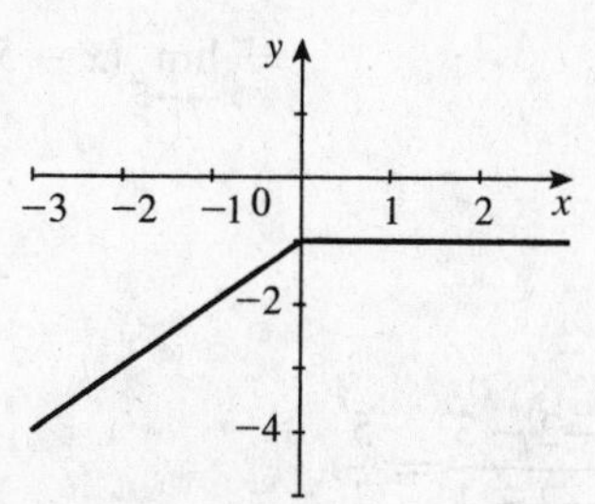

$\lim_{x \to 0} f(x) = -1.$

19.

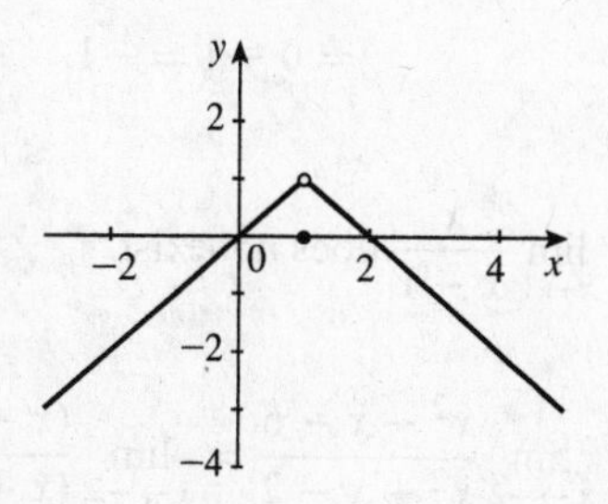

$\lim_{x \to 1} f(x) = 1.$

21.

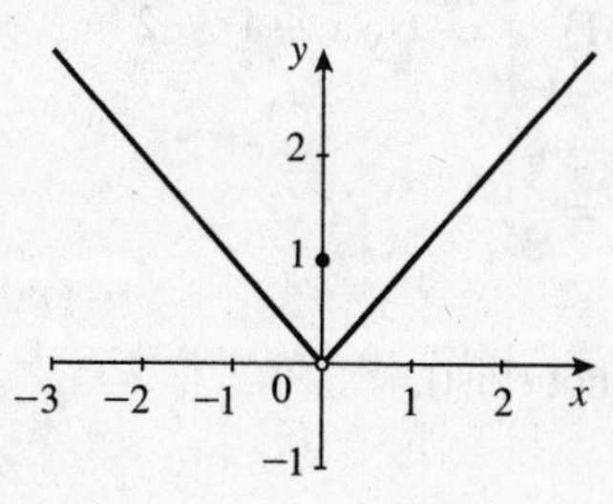

$\lim_{x \to 0} f(x) = 0.$

23. $\lim_{x \to 2} 3 = 3.$

25. $\lim_{x \to 3} x = 3.$

27. $\lim_{x\to 1}\left(1-2x^2\right)=1-2(1)^2=-1.$

29. $\lim_{x\to 1}\left(2x^3-3x^2+x+2\right)=2(1)^3-3(1)^2+1+2=2.$

31. $\lim_{s\to 0}\left(2s^2-1\right)(2s+4)=(-1)(4)=-4.$

33. $\lim_{x\to 2}\frac{2x+1}{x+2}=\frac{2(2)+1}{2+2}=\frac{5}{4}.$

35. $\lim_{x\to 2}\sqrt{x+2}=\sqrt{2+2}=2.$

37. $\lim_{x\to -3}\sqrt{2x^4+x^2}=\sqrt{2(-3)^4+(-3)^2}=\sqrt{162+9}=\sqrt{171}=3\sqrt{19}.$

39. $\lim_{x\to -1}\frac{\sqrt{x^2+8}}{2x+4}=\frac{\sqrt{(-1)^2+8}}{2(-1)+4}=\frac{\sqrt{9}}{2}=\frac{3}{2}.$

41. $\lim_{x\to a}\left[f(x)-g(x)\right]=\lim_{x\to a}f(x)-\lim_{x\to a}g(x)=3-4=-1.$

43. $\lim_{x\to a}\left[2f(x)-3g(x)\right]=\lim_{x\to a}2f(x)-\lim_{x\to a}3g(x)=2(3)-3(4)=-6.$

45. $\lim_{x\to a}\sqrt{g(x)}=\lim_{x\to a}\sqrt{4}=2.$

47. $\lim_{x\to a}\frac{2f(x)-g(x)}{f(x)g(x)}=\frac{2(3)-(4)}{(3)(4)}=\frac{2}{12}=\frac{1}{6}.$

49. $\lim_{x\to 1}\frac{x^2-1}{x-1}=\lim_{x\to 1}\frac{(x-1)(x+1)}{x-1}=\lim_{x\to 1}(x+1)=1+1=2.$

51. $\lim_{x\to 0}\frac{x^2-x}{x}=\lim_{x\to 0}\frac{x(x-1)}{x}=\lim_{x\to 0}(x-1)=0-1=-1.$

53. $\lim_{x\to -5}\frac{x^2-25}{x+5}=\lim_{x\to -5}\frac{(x+5)(x-5)}{x+5}=\lim_{x\to -5}(x-5)=-10.$

55. $\lim_{x\to 1}\frac{x}{x-1}$ does not exist.

57. $\lim_{x\to -2}\frac{x^2-x-6}{x^2+x-2}=\lim_{x\to -2}\frac{(x-3)(x+2)}{(x+2)(x-1)}=\lim_{x\to -2}\frac{x-3}{x-1}=\frac{-2-3}{-2-1}=\frac{5}{3}.$

59. $\lim_{x\to 1}\frac{\sqrt{x}-1}{x-1}=\lim_{x\to 1}\frac{\sqrt{x}-1}{x-1}\cdot\frac{\sqrt{x}+1}{\sqrt{x}+1}=\lim_{x\to 1}\frac{x-1}{(x-1)\left(\sqrt{x}+1\right)}=\lim_{x\to 1}\frac{1}{\sqrt{x}+1}=\frac{1}{2}.$

61. $\lim_{x\to 1}\frac{x-1}{x^3+x^2-2x}=\lim_{x\to 1}\frac{x-1}{x(x-1)(x+2)}=\lim_{x\to 1}\frac{1}{x(x+2)}=\frac{1}{3}.$

63. $\lim_{x\to\infty}f(x)=\infty$ (does not exist) and $\lim_{x\to-\infty}f(x)=\infty$ (does not exist).

65. $\lim_{x\to\infty}f(x)=0$ and $\lim_{x\to-\infty}f(x)=0.$

67. $\lim_{x\to\infty}f(x)=-\infty$ (does not exist) and $\lim_{x\to-\infty}f(x)=-\infty$ (does not exist).

69. $f(x) = \dfrac{1}{x^2+1}$.

x	1	10	100	1000
$f(x)$	0.5	0.009901	0.0001	0.000001

x	−1	−10	−100	−1000
$f(x)$	0.5	0.009901	0.0001	0.000001

$\lim_{x\to\infty} f(x) = \lim_{x\to-\infty} f(x) = 0$.

71. $f(x) = 3x^3 - x^2 + 10$.

x	1	5	10	100	1000
$f(x)$	12	360	2910	2.99×10^6	2.999×10^9

x	−1	−5	−10	−100	−1000
$f(x)$	6	−390	−3090	-3.01×10^6	-3.0×10^9

$\lim_{x\to\infty} f(x) = \infty$ (does not exist) and $\lim_{x\to-\infty} f(x) = -\infty$ (does not exist).

73. $\lim_{x\to\infty} \dfrac{3x+2}{x-5} = \lim_{x\to\infty} \dfrac{3+\frac{2}{x}}{1-\frac{5}{x}} = \dfrac{3}{1} = 3$.

75. $\lim_{x\to-\infty} \dfrac{3x^3+x^2+1}{x^3+1} = \lim_{x\to-\infty} \dfrac{3+\frac{1}{x}+\frac{1}{x^3}}{1+\frac{1}{x^3}} = 3$.

77. $\lim_{x\to-\infty} \dfrac{x^4+1}{x^3-1} = \lim_{x\to-\infty} \dfrac{x+\frac{1}{x^3}}{1-\frac{1}{x^3}} = -\infty$; that is, the limit does not exist.

79. $\lim_{x\to\infty} \dfrac{x^5-x^3+x-1}{x^6+2x^2+1} = \lim_{x\to\infty} \dfrac{\frac{1}{x}-\frac{1}{x^3}+\frac{1}{x^5}-\frac{1}{x^6}}{1+\frac{2}{x^4}+\frac{1}{x^6}} = 0$.

81. a. The cost of removing 50% of the pollutant is $C(50) = \dfrac{0.5(50)}{100-50} = 0.5$, or \$500,000. Similarly, we find that the cost of removing 60%, 70%, 80%, 90%, and 95% of the pollutant is \$750,000, \$1,166,667, \$2,000,000, \$4,500,000, and \$9,500,000, respectively.

b. $\lim_{x\to100} \dfrac{0.5x}{100-x} = \infty$, which means that the cost of removing the pollutant increases without bound if we wish to remove almost all of the pollutant.

83. $\lim_{x\to\infty} \overline{C}(x) = \lim_{x\to\infty} 2.2 + \dfrac{2500}{x} = 2.2$, or \$2.20 per DVD. In the long run, the average cost of producing x DVDs approaches \$2.20/disc.

85. a. $T(1) = \dfrac{120}{1+4} = 24$, or \$24 million. $T(2) = \dfrac{120(2)^2}{8} = 60$, or \$60 million. $T(3) = \dfrac{120(3)^2}{13} = 83.1$, or \$83.1 million.

b. In the long run, the movie will gross $\lim_{x\to\infty} \frac{120x^2}{x^2+4} = \lim_{x\to\infty} \frac{120}{1+\frac{4}{x^2}} = 120$, or \$120 million.

87. a. The average cost of driving 5000 miles per year is $C(5) = \frac{2410}{5^{1.95}} + 32.8 \approx 137.28$, or 137.3 cents per mile. Similarly, we see that the average costs of driving 10, 15, 20, and 25 thousand miles per year are 59.8, 45.1, 39.8, and 37.3 cents per mile, respectively.

b.

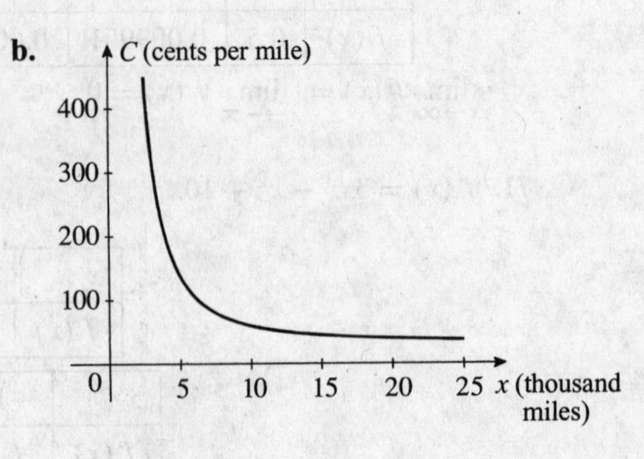

c. It approaches 32.8 cents per mile.

89. False. Let $f(x) = \begin{cases} -1 & \text{if } x < 0 \\ 1 & \text{if } x \geq 0 \end{cases}$ Then $\lim_{x\to 0} f(x) = 1$, but $f(1)$ is not defined.

91. True. Division by zero is not permitted.

93. True. Each limit in the sum exists. Therefore, $\lim_{x\to 2}\left(\frac{x}{x+1} + \frac{3}{x-1}\right) = \lim_{x\to 2} \frac{x}{x+1} + \lim_{x\to 2} \frac{3}{x-1} = \frac{2}{3} + \frac{3}{1} = \frac{11}{3}$.

95. $\lim_{x\to\infty} \frac{ax}{x+b} = \lim_{x\to\infty} \frac{a}{1+\frac{b}{x}} = a$. As the amount of substrate becomes very large, the initial speed approaches the constant a moles per liter per second.

97. Consider the functions $f(x) = \begin{cases} -1 & \text{if } x < 0 \\ 1 & \text{if } x \geq 0 \end{cases}$ and $g(x) = \begin{cases} 1 & \text{if } x < 0 \\ -1 & \text{if } x \geq 0 \end{cases}$ Then $\lim_{x\to 0} f(x)$ and $\lim_{x\to 0} g(x)$ do not exist, but $\lim_{x\to 0} [f(x) g(x)] = \lim_{x\to 0} (-1) = -1$. This example does not contradict Theorem 1 because the hypothesis of Theorem 1 is that $\lim_{x\to 0} f(x)$ and $\lim_{x\to 0} g(x)$ both exist. It does not say anything about the situation where one or both of these limits fails to exist.

Using Technology

page 121

1. 5

3. 3

5. $\frac{2}{3}$

7. $e^2 \approx 7.38906$

9.

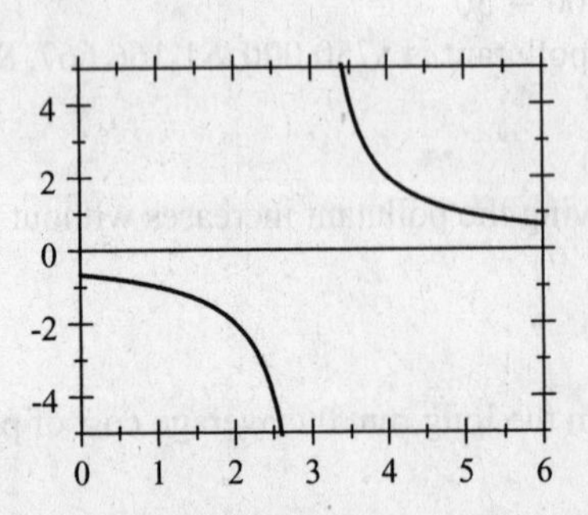

From the graph we see that $f(x)$ does not approach any finite number as x approaches 3.

11. a.

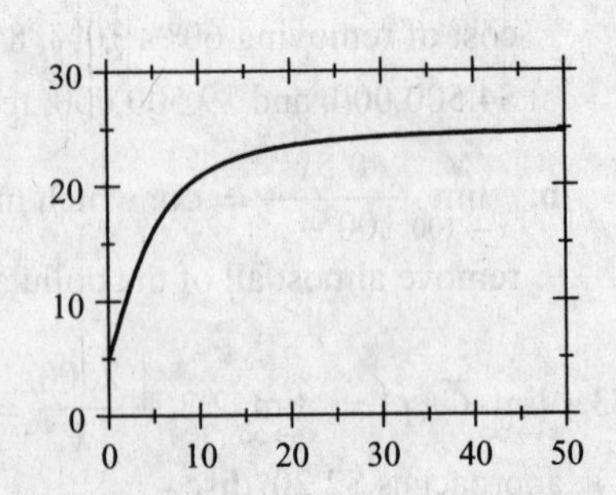

b. $\lim_{t\to\infty} \frac{25t^2 + 125t + 200}{t^2 + 5t + 40} = 25$, so in the long run the population will approach 25,000.

2.5 One-Sided Limits and Continuity

Problem-Solving Tips

The problem-solving skills that you learned in earlier sections are building blocks for the rest of the course. You can't skip a section or a concept and hope to understand the material in a new section. It just won't work. If you don't build a strong foundation, you won't be able to understand the later concepts. For example, in this section we discussed one-sided limits. You need to understand the definition of a limit before you can understand what is meant by a one-sided limit. That means you should be able to express the definition of a limit in your own words. If you can't grasp a new concept, it may well be that you still don't understand a previous concept. If so, you need to go back and review the earlier section before you go on.

As another example, the continuity of polynomial and rational functions is discussed on page 125. If you don't remember how to identify polynomial and rational functions, go back to Section 2.3 and review this material.

Here are some tips for solving the problems in the exercises that follow:

1. **To evaluate the limit of a piecewise-defined function** at a real number a, follow the same procedure that you used to evaluate a piecewise-defined function. First find the subdomain that a lies in, then use the rule for that subdomain to find the limit of f at a.

2. **To determine the values of** x **at which a function is continuous**, check to see if the function is a polynomial or rational function. A polynomial function $y = P(x)$ is continuous at every value of x and a rational function is continuous at every value of x where the denominator is nonzero.

Concept Questions page 129

1. $\lim_{x\to 3^-} f(x) = 2$ means $f(x)$ can be made as close to 2 as we please by taking x sufficiently close to but to the left of $x = 3$. $\lim_{x\to 3^+} f(x) = 4$ means $f(x)$ can be made as close to 4 as we please by taking x sufficiently close to but to the right of $x = 3$.

3. a. f is continuous at a if $\lim_{x\to a} f(x) = f(a)$.

b. f is continuous on an interval I if f is continuous at each point in I.

5. a. f is continuous because the plane does not suddenly jump from one point to another.

b. f is continuous.

c. f is discontinuous because the fare "jumps"after the cab has covered a certain distance or after a certain amount of time has elapsed.

d. f is discontinuous because the rates "jump"by a certain amount (up or down) when it is adjusted at certain times.

Exercises page 130

1. $\lim_{x\to 2^-} f(x) = 3$ and $\lim_{x\to 2^+} f(x) = 2$, so $\lim_{x\to 2} f(x)$ does not exist.

3. $\lim_{x\to -1^-} f(x) = \infty$ and $\lim_{x\to -1^+} f(x) = 2$, so $\lim_{x\to -1} f(x)$ does not exist.

5. $\lim_{x\to 1^-} f(x) = 0$ and $\lim_{x\to 1^+} f(x) = 2$, so $\lim_{x\to 1} f(x)$ does not exist.

7. $\lim_{x\to 0^-} f(x) = -2$ and $\lim_{x\to 0^+} f(x) = 2$, so $\lim_{x\to 0} f(x)$ does not exist.

9. True. **11.** True. **13.** False. **15.** True. **17.** False. **19.** True.

21. $\lim_{x\to 1^+} (2x+4) = 6$.

23. $\lim_{x\to 2^-} \frac{x-3}{x+2} = \frac{2-3}{2+2} = -\frac{1}{4}$.

25. $\lim_{x\to 0^+} \frac{1}{x}$ does not exist because $\frac{1}{x} \to \infty$ as $x \to 0$ from the right.

27. $\lim_{x\to 0^+} \frac{x-1}{x^2+1} = \frac{-1}{1} = -1$.

29. $\lim_{x\to 0^+} \sqrt{x} = \sqrt{\lim_{x\to 0^+} x} = 0$.

31. $\lim_{x\to -2^+} \left(2x+\sqrt{2+x}\right) = \lim_{x\to -2^+} 2x + \lim_{x\to -2^+} \sqrt{2+x} = -4+0 = -4$.

33. $\lim_{x\to 1^-} \frac{1+x}{1-x} = \infty$, that is, the limit does not exist.

35. $\lim_{x\to 2^-} \frac{x^2-4}{x-2} = \lim_{x\to 2^-} \frac{(x+2)(x-2)}{x-2} = \lim_{x\to 2^-} (x+2) = 4$.

37. $\lim_{x\to 0^+} f(x) = \lim_{x\to 0^+} x^2 = 0$ and $\lim_{x\to 0^-} f(x) = \lim_{x\to 0^-} 2x = 0$.

39. The function is discontinuous at $x = 0$. Conditions 2 and 3 are violated.

41. The function is continuous everywhere.

43. The function is discontinuous at $x = 0$. Condition 3 is violated.

45. f is continuous for all values of x.

47. f is continuous for all values of x. Note that $x^2+1 \geq 1 > 0$.

49. f is discontinuous at $x = \frac{1}{2}$, where the denominator is 0. Thus, f is continuous on $\left(-\infty, \frac{1}{2}\right)$ and $\left(\frac{1}{2}, \infty\right)$.

51. Observe that $x^2+x-2 = (x+2)(x-1) = 0$ if $x = -2$ or $x = 1$, so f is discontinuous at these values of x. Thus, f is continuous on $(-\infty, -2)$, $(-2, 1)$, and $(1, \infty)$.

53. f is continuous everywhere since all three conditions are satisfied.

55. f is continuous everywhere since all three conditions are satisfied.

57. Because the denominator $x^2-1 = (x-1)(x+1) = 0$ if $x = -1$ or 1, we see that f is discontinuous at -1 and 1.

59. Because $x^2-3x+2 = (x-2)(x-1) = 0$ if $x = 1$ or 2, we see that the denominator is zero at these points and so f is discontinuous at these numbers.

61. The function f is discontinuous at $x = 4, 5, 6, \ldots, 13$ because the limit of f does not exist at these points.

63. Having made steady progress up to $x = x_1$, Michael's progress comes to a standstill at that point. Then at $x = x_2$ a sudden breakthrough occurs and he then continues to solve the problem.

65. Conditions 2 and 3 are not satisfied at any of these points.

67.

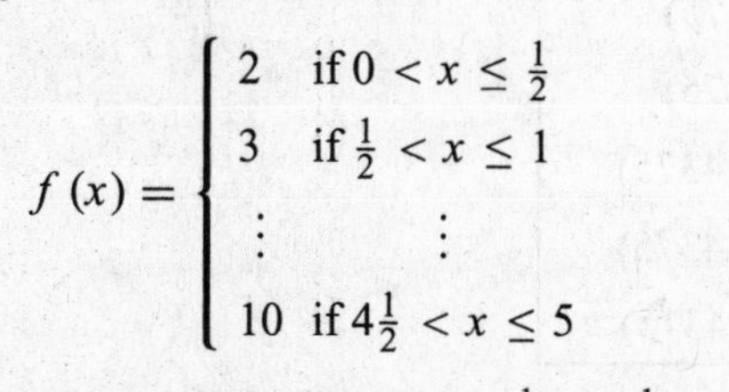

$$f(x) = \begin{cases} 2 & \text{if } 0 < x \le \frac{1}{2} \\ 3 & \text{if } \frac{1}{2} < x \le 1 \\ \vdots & \vdots \\ 10 & \text{if } 4\frac{1}{2} < x \le 5 \end{cases}$$

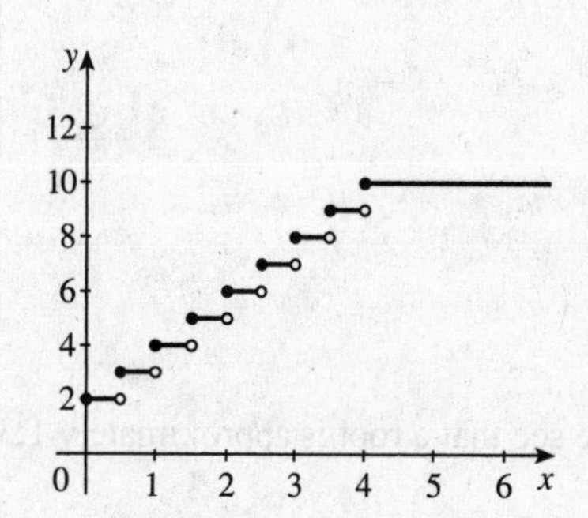

f is discontinuous at $x = \frac{1}{2}, 1, 1\frac{1}{2}, \ldots, 4$.

69.

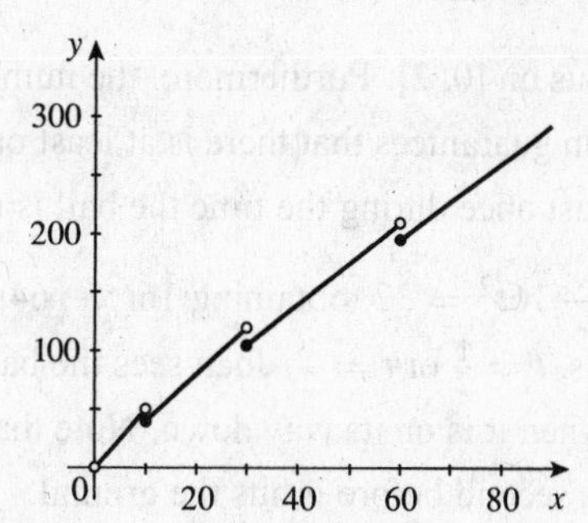

C is discontinuous at $x = 0$, 10, 30, and 60.

71. a. $\lim_{t\to 0^+} S(t) = \lim_{t\to 0^+} \frac{a}{t} + b = \infty$. As the time taken to excite the tissue is made shorter and shorter, the electric current gets stronger and stronger.

b. $\lim_{t\to\infty} \frac{a}{t} + b = b$. As the time taken to excite the tissue is made longer and longer, the electric current gets weaker and weaker and approaches b.

73. We require that $f(1) = 1 + 2 = 3 = \lim_{x\to 1^+} kx^2 = k$, so $k = 3$.

75. a. f is a polynomial of degree 2 and is therefore continuous everywhere, including the interval $[1, 3]$.

b. $f(1) = 3$ and $f(3) = -1$ and so f must have at least one zero in $(1, 3)$.

77. a. f is a polynomial of degree 3 and is therefore continuous on $[-1, 1]$.

b. $f(-1) = (-1)^3 - 2(-1)^2 + 3(-1) + 2 = -1 - 2 - 3 + 2 = -4$ and $f(1) = 1 - 2 + 3 + 2 = 4$. Because $f(-1)$ and $f(1)$ have opposite signs, we see that f has at least one zero in $(-1, 1)$.

79. $f(0) = 6$, $f(3) = 3$, and f is continuous on $[0, 3]$. Thus, the Intermediate Value Theorem guarantees that there is at least one value of x for which $f(x) = 4$. Solving $f(x) = x^2 - 4x + 6 = 4$, we find $x^2 - 4x + 2 = 0$. Using the quadratic formula, we find that $x = 2 \pm \sqrt{2}$. Because $2 + \sqrt{2}$ does not lie in $[0, 3]$, we see that $x = 2 - \sqrt{2} \approx 0.59$.

81. $x^5 + 2x - 7 = 0$

Step	Interval in which a root lies
1	(1, 2)
2	(1, 1.5)
3	(1.25, 1.5)
4	(1.25, 1.375)
5	(1.3125, 1.375)
6	(1.3125, 1.34375)
7	(1.328125, 1.34375)
8	(1.3359375, 1.34375)

We see that a root is approximately 1.34.

83. a. $h(0) = 4 + 64(0) - 16(0) = 4$ and $h(2) = 4 + 64(2) - 16(4) = 68$.

b. The function h is continuous on $[0, 2]$. Furthermore, the number 32 lies between 4 and 68. Therefore, the Intermediate Value Theorem guarantees that there is at least one value of t in $(0, 2]$ such that $h(t) = 32$, that is, Joan must see the ball at least once during the time the ball is in the air.

c. We solve $h(t) = 4 + 64t - 16t^2 = 32$, obtaining $16t^2 - 64t + 28 = 0$, $4t^2 - 16t + 7 = 0$, and $(2t - 1)(2t - 7) = 0$. Thus, $t = \frac{1}{2}$ or $t = \frac{7}{2}$. Joan sees the ball on its way up half a second after it was thrown and again 3 seconds later when it is on its way down. Note that the ball hits the ground when $t \approx 4.06$, but Joan sees it approximately half a second before it hits the ground.

85. False. Take $f(x) = \begin{cases} -1 & \text{if } x < 2 \\ 4 & \text{if } x = 2 \\ 1 & \text{if } x > 2 \end{cases}$ Then $f(2) = 4$, but $\lim_{x \to 2} f(x)$ does not exist.

87. False. Consider $f(x) = \begin{cases} 0 & \text{if } x < 2 \\ 3 & \text{if } x \geq 2 \end{cases}$ Then $\lim_{x \to 2^+} f(x) = f(2) = 3$, but $\lim_{x \to 2^-} f(x) = 0$.

89. False. Consider $f(x) = \begin{cases} 2 & \text{if } x < 5 \\ 3 & \text{if } x > 5 \end{cases}$ Then $f(5)$ is not defined, but $\lim_{x \to 5^-} f(x) = 2$.

91. False. Let $f(x) = \begin{cases} x & \text{if } x \neq 0 \\ 1 & \text{if } x = 0 \end{cases}$ Then $\lim_{x \to 0^+} f(x) = \lim_{x \to 0^-} f(x)$, but $f(0) = 1$.

93. False. Let $f(x) = \begin{cases} 1/x & \text{if } x \neq 0 \\ 0 & \text{if } x = 0 \end{cases}$ Then f is continuous for all $x \neq 0$ and $f(0) = 0$, but $\lim_{x \to 0} f(x)$ does not exist.

95. False. Consider $f(x) = \begin{cases} -1 & \text{if } -1 \leq x \leq 0 \\ 1 & \text{if } 0 < x \leq 1 \end{cases}$ and $g(x) = \begin{cases} 1 & \text{if } -1 \leq x \leq 0 \\ -1 & \text{if } 0 < x \leq 1 \end{cases}$

97. False. Consider $f(x) = \begin{cases} -1 & \text{if } x < 0 \\ 1 & \text{if } x \geq 0 \end{cases}$ and $g(x) = \begin{cases} x+1 & \text{if } x < 0 \\ x-1 & \text{if } x \geq 0 \end{cases}$

99. a. f is a rational function whose denominator is never zero, and so it is continuous for all values of x.

b. Because the numerator x^2 is nonnegative and the denominator is $x^2 + 1 \geq 1$ for all values of x, we see that $f(x)$ is nonnegative for all values of x.

c. $f(0) = \frac{0}{0+1} = \frac{0}{1} = 0$, and so f has a zero at $x = 0$. This does not contradict Theorem 5.

101. a. (i) Repeated use of Property 3 shows that $g(x) = x^n = \underbrace{x \cdot x \cdot \cdots \cdot x}_{n \text{ times}}$ is a continuous function, since $f(x) = x$ is continuous by Property 1.

(ii) Properties 1 and 5 combine to show that $c \cdot x^n$ is continuous using the results of part (a)(i).

(iii) Each of the terms of $p(x) = a_n x^n + a_{n-1} x^{n-1} + \cdots + a_0$ is continuous and so Property 4 implies that p is continuous.

b. Property 6 now shows that $R(x) = \frac{p(x)}{q(x)}$ is continuous if $q(a) \neq 0$, since p and q are continuous at $x = a$.

Using Technology page 136

1.

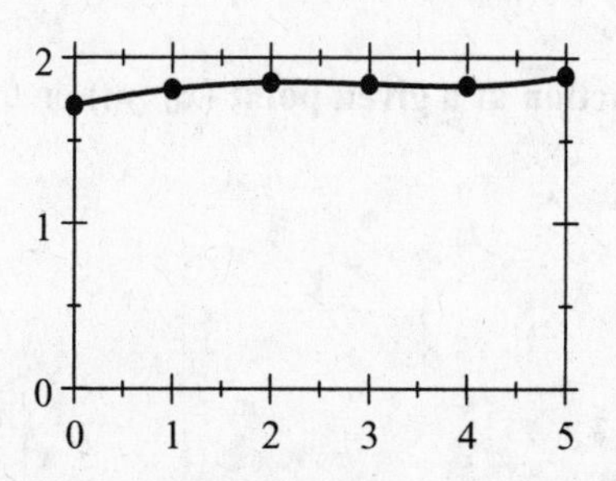

The function is discontinuous at $x = 0$ and $x = 1$.

3.

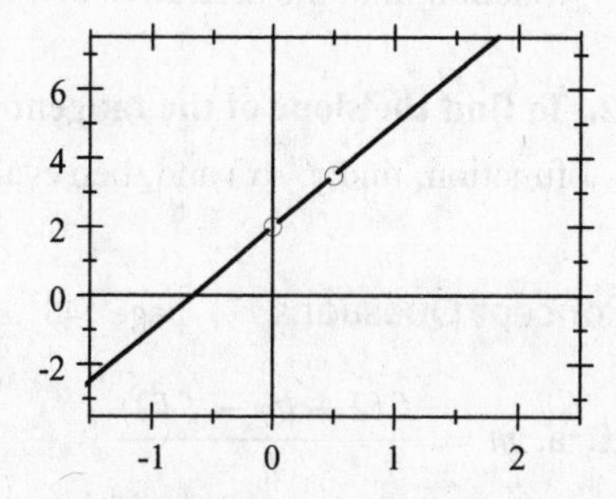

The function is discontinuous at $x = 0$ and $\frac{1}{2}$.

5.

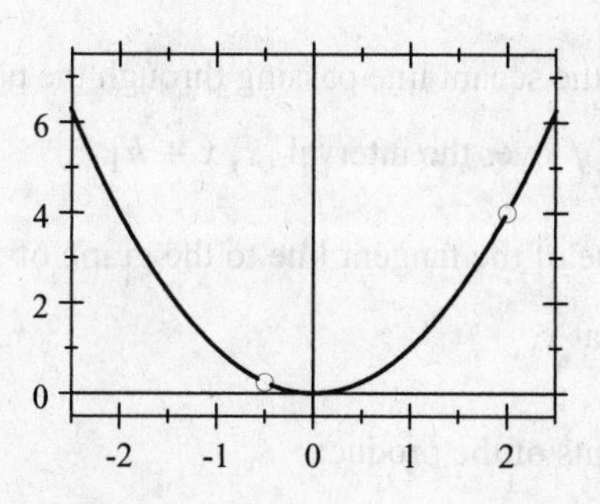

The function is discontinuous at $x = -\frac{1}{2}$ and 2.

7.

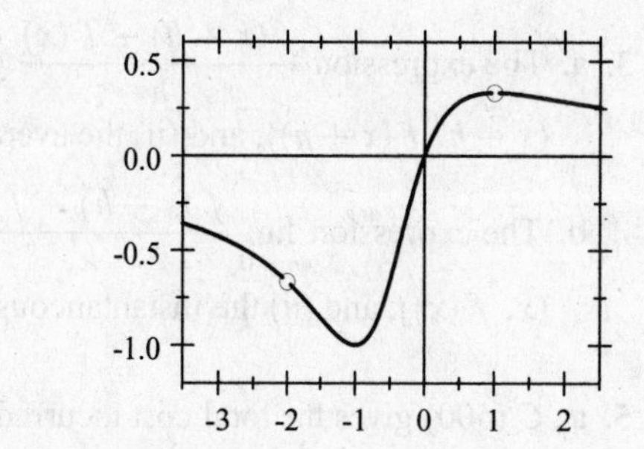

The function is discontinuous at $x = -2$ and 1.

9.

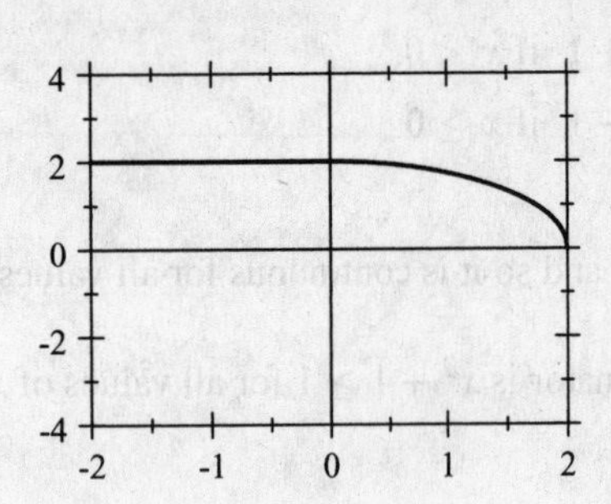

11.

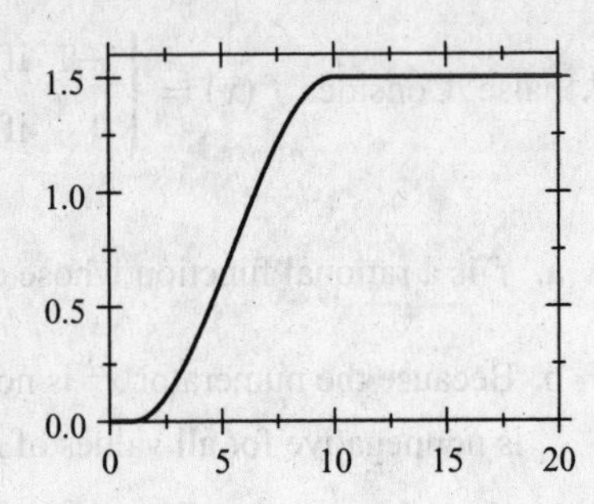

2.6 The Derivative

Problem-Solving Tips

When you solve an applied problem, it is important to understand the question in mathematical terms. For example, if you are given a function $f(t)$ describing the size of a country's population at any time t and asked to find the rate of change of that country's population at any time t, this means that you need to find the derivative of the given function; that is, find $f'(t)$. If you are then asked to find the population of the country at a specified time, say $t = 2$, you need to evaluate the function at $t = 2$; that is, find $f(2)$. On the other hand, if you are asked to find the rate of change of the population at time $t = 2$, then you need to evaluate the derivative of the function at the value $t = 2$; that is find $f'(2)$. Here again, the key is to be familiar with the terminology and notation introduced in the chapter.

Here are some tips for solving the problems in the exercises that follow:

1. **To find the slope of the tangent line to the graph of a function at an arbitrary point** on the graph of that function, find the derivative of f; that is, find $f'(x)$.

2. **To find the slope of the tangent line to the graph of a function at a given point** (x_0, y_0) on the graph of that function, find $f'(x)$ and then evaluate $f'(x_0)$.

Concept Questions

page 148

1. a. $m = \dfrac{f(2+h) - f(2)}{h}$

b. The slope of the tangent line is $\lim\limits_{h\to 0} \dfrac{f(2+h) - f(2)}{h}$.

3. a. The expression $\dfrac{f(x+h) - f(x)}{h}$ gives (i) the slope of the secant line passing through the points $(x, f(x))$ and $(x+h, f(x+h))$, and (ii) the average rate of change of f over the interval $[x, x+h]$.

b. The expression $\lim\limits_{h\to 0} \dfrac{f(x+h) - f(x)}{h}$ gives (i) the slope of the tangent line to the graph of f at the point $(x, f(x))$, and (ii) the instantaneous rate of change of f at x.

5. a. $C(500)$ gives the total cost incurred in producing 500 units of the product.

b. $C'(500)$ gives the rate of change of the total cost function when the production level is 500 units.

Exercises page 149

1. The rate of change of the average infant's weight when $t = 3$ is $\frac{7.5}{5}$, or 1.5 lb/month. The rate of change of the average infant's weight when $t = 18$ is $\frac{3.5}{6}$, or approximately 0.58 lb/month. The average rate of change over the infant's first year of life is $\frac{22.5-7.5}{12}$, or 1.25 lb/month.

3. The rate of change of the percentage of households watching television at 4 p.m. is $\frac{12.3}{4}$, or approximately 3.1 percent per hour. The rate at 11 p.m. is $\frac{-42.3}{2} = -21.15$, that is, it is dropping off at the rate of 21.15 percent per hour.

5. a. Car A is travelling faster than Car B at t_1 because the slope of the tangent line to the graph of f is greater than the slope of the tangent line to the graph of g at t_1.

b. Their speed is the same because the slope of the tangent lines are the same at t_2.

c. Car B is travelling faster than Car A.

d. They have both covered the same distance and are once again side by side at t_3.

7. a. P_2 is decreasing faster at t_1 because the slope of the tangent line to the graph of g at t_1 is greater than the slope of the tangent line to the graph of f at t_1.

b. P_1 is decreasing faster than P_2 at t_2.

c. Bactericide B is more effective in the short run, but bactericide A is more effective in the long run.

9. $f(x) = 13$.

Step 1 $f(x+h) = 13$.

Step 2 $f(x+h) - f(x) = 13 - 13 = 0$.

Step 3 $\dfrac{f(x+h) - f(x)}{h} = \dfrac{0}{h} = 0$.

Step 4 $f'(x) = \lim\limits_{h \to 0} \dfrac{f(x+h) - f(x)}{h} = \lim\limits_{h \to 0} 0 = 0$.

11. $f(x) = 2x + 7$.

Step 1 $f(x+h) = 2(x+h) + 7$.

Step 2 $f(x+h) - f(x) = 2(x+h) + 7 - (2x+7) = 2h$.

Step 3 $\dfrac{f(x+h) - f(x)}{h} = \dfrac{2h}{h} = 2$.

Step 4 $f'(x) = \lim\limits_{h \to 0} \dfrac{f(x+h) - f(x)}{h} = \lim\limits_{h \to 0} 2 = 2$.

13. $f(x) = 3x^2$.

Step 1 $f(x+h) = 3(x+h)^2 = 3x^2 + 6xh + 3h^2$.

Step 2 $f(x+h) - f(x) = (3x^2 + 6xh + 3h^2) - 3x^2 = 6xh + 3h^2 = h(6x + 3h)$.

Step 3 $\dfrac{f(x+h) - f(x)}{h} = \dfrac{h(6x+3h)}{h} = 6x + 3h$.

Step 4 $f'(x) = \lim\limits_{h \to 0} \dfrac{f(x+h) - f(x)}{h} = \lim\limits_{h \to 0} (6x + 3h) = 6x$.

15. $f(x) = -x^2 + 3x$.

Step 1 $f(x+h) = -(x+h)^2 + 3(x+h) = -x^2 - 2xh - h^2 + 3x + 3h$.

Step 2 $f(x+h) - f(x) = (-x^2 - 2xh - h^2 + 3x + 3h) - (-x^2 + 3x) = -2xh - h^2 + 3h = h(-2x - h + 3)$.

Step 3 $\dfrac{f(x+h) - f(x)}{h} = \dfrac{h(-2x - h + 3)}{h} = -2x - h + 3$.

Step 4 $f'(x) = \lim_{h\to 0} \dfrac{f(x+h) - f(x)}{h} = \lim_{h\to 0} (-2x - h + 3) = -2x + 3$.

17. $f(x) = 2x + 7$.

Step 1 $f(x+h) = 2(x+h) + 7 = 2x + 2h + 7$.

Step 2 $f(x+h) - f(x) = 2x + 2h + 7 - 2x - 7 = 2h$.

Step 3 $\dfrac{f(x+h) - f(x)}{h} = \dfrac{2h}{h} = 2$.

Step 4 $f'(x) = \lim_{h\to 0} \dfrac{f(x+h) - f(x)}{h} = \lim_{h\to 0} 2 = 2$.

Therefore, $f'(x) = 2$. In particular, the slope at $x = 2$ is 2. Therefore, an equation of the tangent line is $y - 11 = 2(x - 2)$ or $y = 2x + 7$.

19. $f(x) = 3x^2$. We first compute $f'(x) = 6x$ (see Exercise 13). Because the slope of the tangent line is $f'(1) = 6$, we use the point-slope form of the equation of a line and find that an equation is $y - 3 = 6(x - 1)$, or $y = 6x - 3$.

21. $f(x) = -1/x$. We first compute $f'(x)$ using the four-step process:

Step 1 $f(x+h) = -\dfrac{1}{x+h}$.

Step 2 $f(x+h) - f(x) = -\dfrac{1}{x+h} + \dfrac{1}{x} = \dfrac{-x + (x+h)}{x(x+h)} = \dfrac{h}{x(x+h)}$.

Step 3 $\dfrac{f(x+h) - f(x)}{h} = \dfrac{\frac{h}{x(x+h)}}{h} = \dfrac{1}{x(x+h)}$.

Step 4 $f'(x) = \lim_{h\to 0} \dfrac{f(x+h) - f(x)}{h} = \lim_{h\to 0} \dfrac{1}{x(x+h)} = \dfrac{1}{x^2}$.

The slope of the tangent line is $f'(3) = \frac{1}{9}$. Therefore, an equation is $y - \left(-\frac{1}{3}\right) = \frac{1}{9}(x - 3)$, or $y = \frac{1}{9}x - \frac{2}{3}$.

23. a. $f(x) = 2x^2 + 1$.

Step 1 $f(x+h) = 2(x+h)^2 + 1 = 2x^2 + 4xh + 2h^2 + 1$.

Step 2 $f(x+h) - f(x) = (2x^2 + 4xh + 2h^2 + 1) - (2x^2 + 1) = 4xh + 2h^2 = h(4x + 2h)$.

Step 3 $\dfrac{f(x+h) - f(x)}{h} = \dfrac{h(4x + 2h)}{h} = 4x + 2h$.

Step 4 $f'(x) = \lim_{h\to 0} \dfrac{f(x+h) - f(x)}{h} = \lim_{h\to 0} (4x + 2h) = 4x$.

b. The slope of the tangent line is $f'(1) = 4(1) = 4$. Therefore, an equation is $y - 3 = 4(x - 1)$ or $y = 4x - 1$.

c.

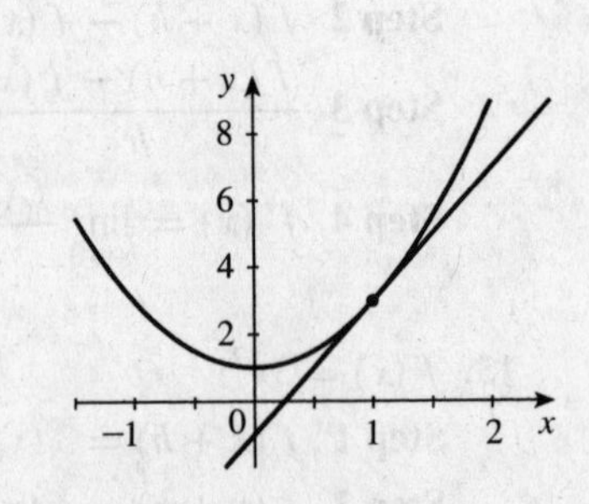

25. a. $f(x) = x^2 - 2x + 1$. We use the four-step process:

Step 1 $f(x+h) = (x+h)^2 - 2(x+h) + 1 = x^2 + 2xh + h^2 - 2x - 2h + 1$.

Step 2 $f(x+h) - f(x) = (x^2 + 2xh + h^2 - 2x - 2h + 1) - (x^2 - 2x + 1) = 2xh + h^2 - 2h$
$= h(2x + h - 2)$.

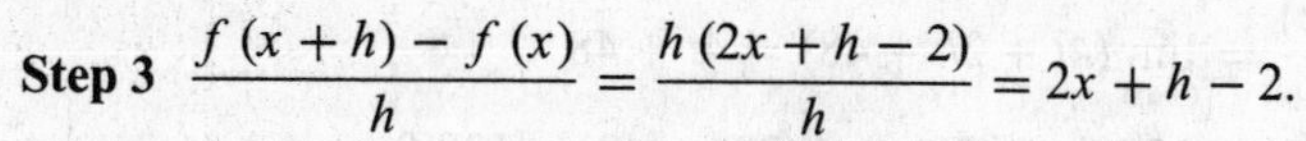

Step 3 $\dfrac{f(x+h) - f(x)}{h} = \dfrac{h(2x+h-2)}{h} = 2x + h - 2$.

Step 4 $f'(x) = \lim_{h\to 0} \dfrac{f(x+h) - f(x)}{h} = \lim_{h\to 0} (2x + h - 2)$
$= 2x - 2$.

c.

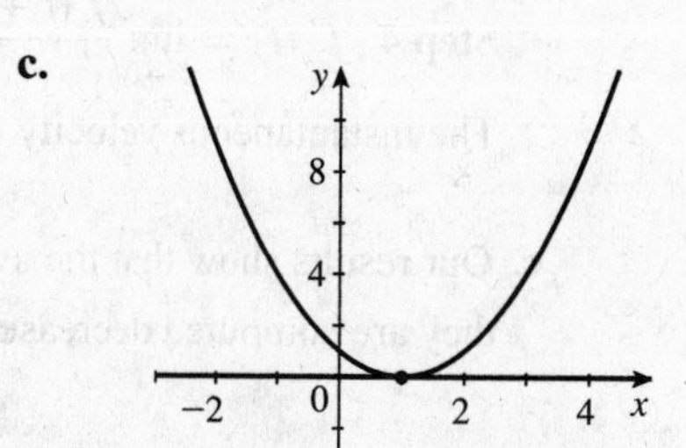

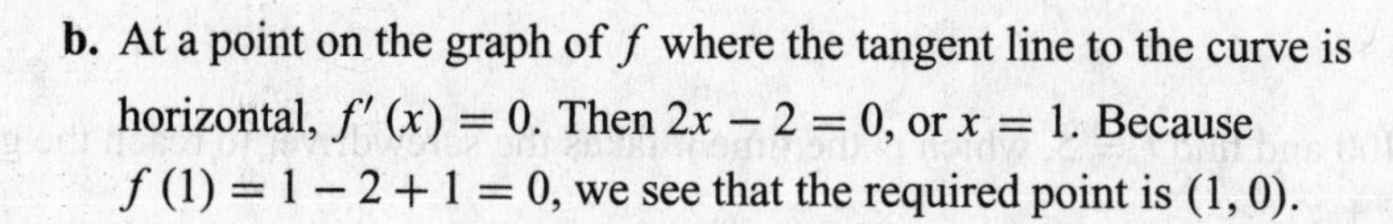

b. At a point on the graph of f where the tangent line to the curve is horizontal, $f'(x) = 0$. Then $2x - 2 = 0$, or $x = 1$. Because $f(1) = 1 - 2 + 1 = 0$, we see that the required point is $(1, 0)$.

d. It is changing at the rate of 0 units per unit change in x.

27. a. $f(x) = x^2 + x$, so $\dfrac{f(3) - f(2)}{3 - 2} = \dfrac{(3^2 + 3) - (2^2 + 2)}{1} = 6$,

$\dfrac{f(2.5) - f(2)}{2.5 - 2} = \dfrac{(2.5^2 + 2.5) - (2^2 + 2)}{0.5} = 5.5$, and $\dfrac{f(2.1) - f(2)}{2.1 - 2} = \dfrac{(2.1^2 + 2.1) - (2^2 + 2)}{0.1} = 5.1$.

b. We first compute $f'(x)$ using the four-step process.

Step 1 $f(x+h) = (x+h)^2 + (x+h) = x^2 + 2xh + h^2 + x + h$.

Step 2 $f(x+h) - f(x) = (x^2 + 2xh + h^2 + x + h) - (x^2 + x) = 2xh + h^2 + h = h(2x + h + 1)$.

Step 3 $\dfrac{f(x+h) - f(x)}{h} = \dfrac{h(2x+h+1)}{h} = 2x + h + 1$.

Step 4 $f'(x) = \lim_{h\to 0} \dfrac{f(x+h) - f(x)}{h} = \lim_{h\to 0} (2x + h + 1) = 2x + 1$.

The instantaneous rate of change of y at $x = 2$ is $f'(2) = 2(2) + 1$, or 5 units per unit change in x.

c. The results of part (a) suggest that the average rates of change of f at $x = 2$ approach 5 as the interval $[2, 2+h]$ gets smaller and smaller ($h = 1$, 0.5, and 0.1). This number is the instantaneous rate of change of f at $x = 2$ as computed in part (b).

29. a. $f(t) = 2t^2 + 48t$. The average velocity of the car over the time interval $[20, 21]$ is

$\dfrac{f(21) - f(20)}{21 - 20} = \dfrac{[2(21)^2 + 48(21)] - [2(20)^2 + 48(20)]}{1} = 130 \dfrac{\text{ft}}{\text{s}}$. Its average velocity over $[20, 20.1]$ is

$\dfrac{f(20.1) - f(20)}{20.1 - 20} = \dfrac{[2(20.1)^2 + 48(20.1)] - [2(20)^2 + 48(20)]}{0.1} = 128.2 \dfrac{\text{ft}}{\text{s}}$. Its average velocity over

$[20, 20.01]$ is $\dfrac{f(20.01) - f(20)}{20.01 - 20} = \dfrac{[2(20.01)^2 + 48(20.01)] - [2(20)^2 + 48(20)]}{0.01} = 128.02 \dfrac{\text{ft}}{\text{s}}$.

b. We first compute $f'(t)$ using the four-step process.

Step 1 $f(t+h) = 2(t+h)^2 + 48(t+h) = 2t^2 + 4th + 2h^2 + 48t + 48h.$

Step 2 $f(t+h) - f(t) = (2t^2 + 4th + 2h^2 + 48t + 48h) - (2t^2 + 48t) = 4th + 2h^2 + 48h$
$= h(4t + 2h + 48).$

Step 3 $\dfrac{f(t+h) - f(t)}{h} = \dfrac{h(4t + 2h + 48)}{h} = 4t + 2h + 48.$

Step 4 $f'(t) = \lim_{t \to 0} \dfrac{f(t+h) - f(t)}{h} = \lim_{t \to 0} (4t + 2h + 48) = 4t + 48.$

The instantaneous velocity of the car at $t = 20$ is $f'(20) = 4(20) + 48$, or 128 ft/s.

c. Our results show that the average velocities do approach the instantaneous velocity as the intervals over which they are computed decreases.

31. a. We solve the equation $16t^2 = 400$ and find $t = 5$, which is the time it takes the screwdriver to reach the ground.

b. The average velocity over the time interval $[0, 5]$ is $\dfrac{f(5) - f(0)}{5 - 0} = \dfrac{16(25) - 0}{5} = 80$, or 80 ft/s.

c. The velocity of the screwdriver at time t is

$$v(t) = \lim_{h \to 0} \frac{f(t+h) - f(t)}{h} = \lim_{h \to 0} \frac{16(t+h)^2 - 16t^2}{h} = \lim_{h \to 0} \frac{16t^2 + 32th + 16h^2 - 16t^2}{h}$$
$$= \lim_{h \to 0} \frac{(32t + 16h)h}{h} = 32t.$$

In particular, the velocity of the screwdriver when it hits the ground (at $t = 5$) is $v(5) = 32(5) = 160$, or 160 ft/s.

33. a. We write $V = f(p) = \dfrac{1}{p}$. The average rate of change of V is $\dfrac{f(3) - f(2)}{3 - 2} = \dfrac{\frac{1}{3} - \frac{1}{2}}{1} = -\dfrac{1}{6}$, a decrease of $\frac{1}{6}$ liter/atmosphere.

b. $V'(t) = \lim_{h \to 0} \dfrac{\frac{f(p+h) - f(p)}{h}}{h} = \lim_{h \to 0} \dfrac{\frac{1}{p+h} - \frac{1}{p}}{h} = \lim_{h \to 0} \dfrac{p - (p+h)}{hp(p+h)} = \lim_{h \to 0} -\dfrac{1}{p(p+h)} = -\dfrac{1}{p^2}$. In particular, the rate of change of V when $p = 2$ is $V'(2) = -\dfrac{1}{2^2}$, a decrease of $\frac{1}{4}$ liter/atmosphere.

35. a. $P(x) = -\frac{1}{3}x^2 + 7x + 30$. Using the four-step process, we find that

$$P'(x) = \lim_{h \to 0} \frac{P(x+h) - P(x)}{h} = \lim_{h \to 0} \frac{-\frac{1}{3}(x^2 + 2xh + h^2) + 7x + 7h + 30 - \left(-\frac{1}{3}x^2 + 7x + 30\right)}{h}$$
$$= \lim_{h \to 0} \frac{-\frac{2}{3}xh - \frac{1}{3}h^2 + 7h}{h} = \lim_{h \to 0} \left(-\tfrac{2}{3}x - \tfrac{1}{3}h + 7\right) = -\tfrac{2}{3}x + 7.$$

b. $P'(10) = -\frac{2}{3}(10) + 7 \approx 0.333$, or approximately \$333 per \$1000 spent on advertising.
$P'(30) = -\frac{2}{3}(30) + 7 = -13$, a decrease of \$13,000 per \$1000 spent on advertising.

37. $N(t) = t^2 + 2t + 50$. We first compute $N'(t)$ using the four-step process.

Step 1 $N(t+h) = (t+h)^2 + 2(t+h) + 50 = t^2 + 2th + h^2 + 2t + 2h + 50$.

Step 2 $N(t+h) - N(t) = (t^2 + 2th + h^2 + 2t + 2h + 50) - (t^2 + 2t + 50) = 2th + h^2 + 2h = h(2t + h + 2)$.

Step 3 $\dfrac{N(t+h) - N(t)}{h} = 2t + h + 2$.

Step 4 $N'(t) = \lim_{h \to 0} (2t + h + 2) = 2t + 2$.

The rate of change of the country's GNP two years from now is $N'(2) = 2(2) + 2 = 6$, or \$6 billion/yr. The rate of change four years from now is $N'(4) = 2(4) + 2 = 10$, or \$10 billion/yr.

39. a. $f'(h)$ gives the instantaneous rate of change of the temperature with respect to height at a given height h, in °F per foot.

b. Because the temperature decreases as the altitude increases, the sign of $f'(h)$ is negative.

c. Because $f'(1000) = -0.05$, the change in the air temperature as the altitude changes from 1000 ft to 1001 ft is approximately $-0.05°$ F.

41. $\dfrac{f(a+h) - f(a)}{h}$ gives the average rate of change of the seal population over the time interval $[a, a+h]$. $\lim_{h \to 0} \dfrac{f(a+h) - f(a)}{h}$ gives the instantaneous rate of change of the seal population at $x = a$.

43. $\dfrac{f(a+h) - f(a)}{h}$ gives the average rate of change of the country's industrial production over the time interval $[a, a+h]$. $\lim_{h \to 0} \dfrac{f(a+h) - f(a)}{h}$ gives the instantaneous rate of change of the country's industrial production at $x = a$.

45. $\dfrac{f(a+h) - f(a)}{h}$ gives the average rate of change of the atmospheric pressure over the altitudes $[a, a+h]$. $\lim_{h \to 0} \dfrac{f(a+h) - f(a)}{h}$ gives the instantaneous rate of change of the atmospheric pressure with respect to altitude at $x = a$.

47. a. f has a limit at $x = a$.

b. f is not continuous at $x = a$ because $f(a)$ is not defined.

c. f is not differentiable at $x = a$ because it is not continuous there.

49. a. f has a limit at $x = a$.

b. f is continuous at $x = a$.

c. f is not differentiable at $x = a$ because f has a kink at the point $x = a$.

51. a. f does not have a limit at $x = a$ because it is unbounded in the neighborhood of a.

b. f is not continuous at $x = a$.

c. f is not differentiable at $x = a$ because it is not continuous there.

53. $s(t) = -0.1t^3 + 2t^2 + 24t$. Our computations yield the following results: 32.1, 30.939, 30.814, 30.8014, 30.8001, and 30.8000. The motorcycle's instantaneous velocity at $t = 2$ is approximately 30.8 ft/s.

55. False. Let $f(x) = |x|$. Then f is continuous at $x = 0$, but is not differentiable there.

57. Observe that the graph of f has a kink at $x = -1$. We have $\dfrac{f(-1+h) - f(-1)}{h} = 1$ if $h > 0$, and -1 if $h < 0$, so that $\lim_{h \to 0} \dfrac{f(-1+h) - f(-1)}{h}$ does not exist.

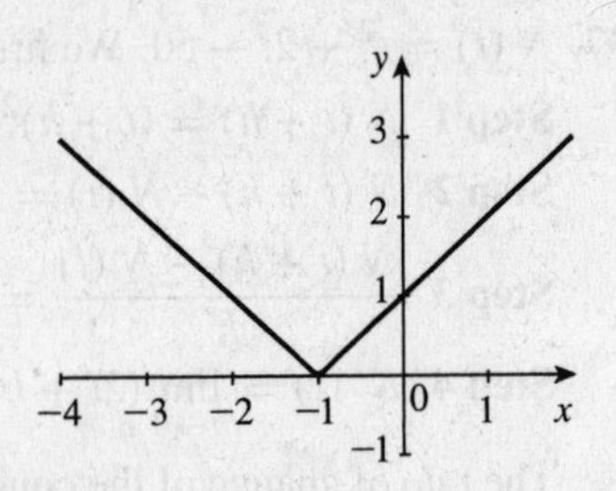

59. For continuity, we require that $f(1) = 1 = \lim_{x \to 1^+} (ax + b) = a + b$, or $a + b = 1$. Next, using the four-step process, we have $f'(x) = \begin{cases} 2x & \text{if } x < 1 \\ a & \text{if } x > 1 \end{cases}$ In order that the derivative exist at $x = 1$, we require that $\lim_{x \to 1^-} 2x = \lim_{x \to 1^+} a$, or $2 = a$. Therefore, $b = -1$ and so $f(x) = \begin{cases} x^2 & \text{if } x \leq 1 \\ 2x - 1 & \text{if } x > 1 \end{cases}$

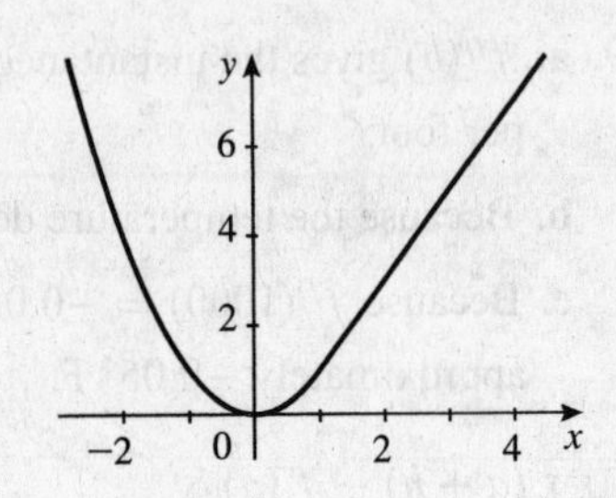

61. We have $f(x) = x$ if $x > 0$ and $f(x) = -x$ if $x < 0$. Therefore, when $x > 0$, $f'(x) = \lim_{h \to 0} \dfrac{f(x+h) - f(x)}{h} = \lim_{h \to 0} \dfrac{x + h - x}{h} = \lim_{h \to 0} \dfrac{h}{h} = 1$, and when $x < 0$, $f'(x) = \lim_{h \to 0} \dfrac{f(x+h) - f(x)}{h} = \lim_{h \to 0} \dfrac{-x - h - (-x)}{h} = \lim_{h \to 0} \dfrac{-h}{h} = -1$. Because the right-hand limit does not equal the left-hand limit, we conclude that $\lim_{h \to 0} f(x)$ does not exist.

Using Technology page 155

1. a. 9

b.

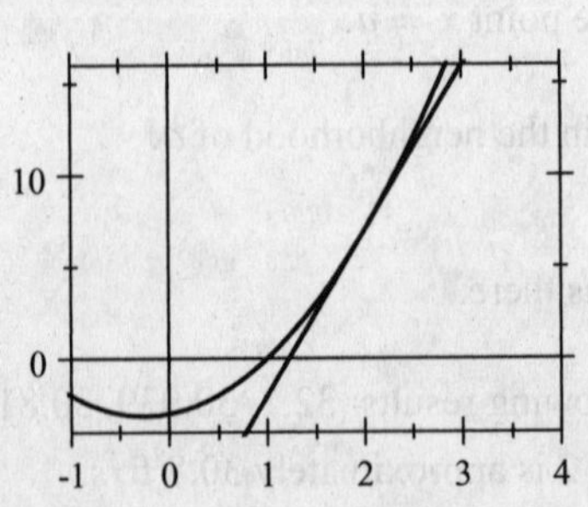

c. $y = 9x - 11$

3. a. 0.083

b.

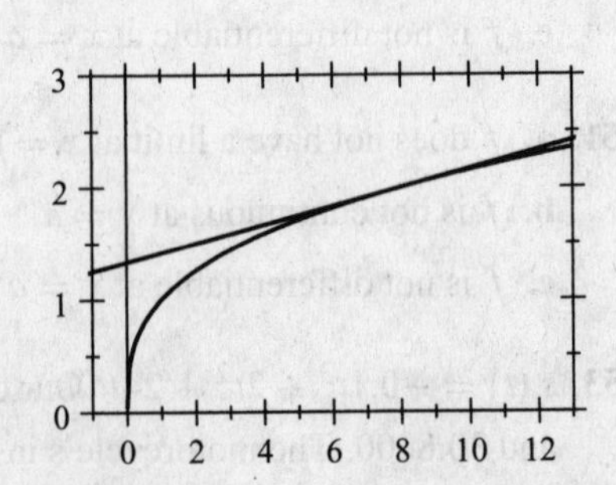

c. $y = \frac{1}{12}x + \frac{4}{3}$

5. a. 4

b.

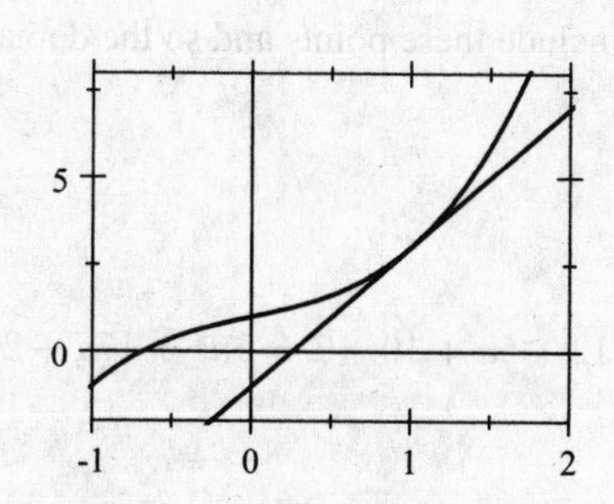

c. $y = 4x - 1$

7. a. 4.02

b.

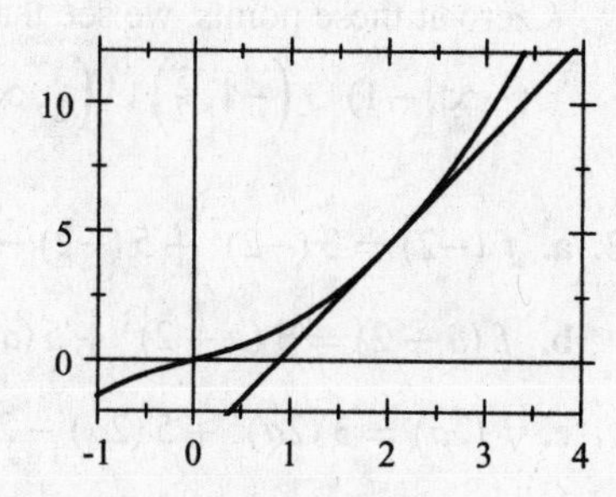

c. $y = 4.02x - 3.57$

9. a.

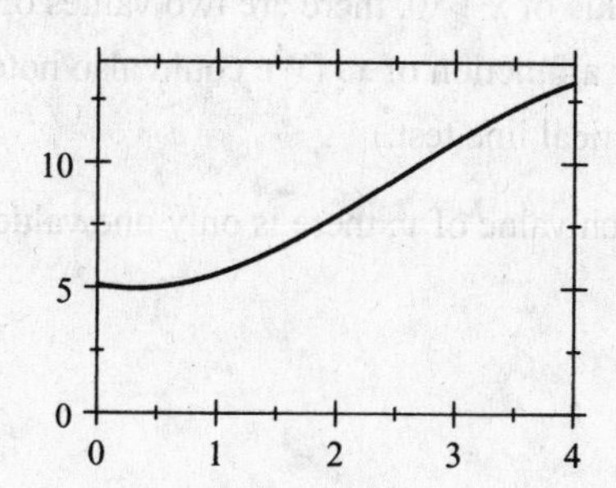

b. $f'(3) = 2.8826$ (million per decade)

CHAPTER 2 Concept Review Questions page 156

1. domain, range, B

3. $f(x) \pm g(x)$, $f(x)g(x)$, $\dfrac{f(x)}{g(x)}$, $A \cap B$, $A \cap B$, 0

5. a. $P(x) = a_n x^n + a_{n-1}x^{n-1} + \cdots + a_1 x + a_0$, where $a_n \neq 0$ and n is a positive integer

b. linear, quadratic, cubic

c. quotient, polynomials

d. x^r, where r is a real number

7. a. L^r

b. $L \pm M$

c. LM

d. $\dfrac{L}{M}$, $M \neq 0$

9. a. right

b. left

c. L, L

11. a. a, a, $g(a)$

b. everywhere

c. Q

13. a. $f'(a)$

b. $y - f(a) = m(x - a)$

CHAPTER 2 Review Exercises page 157

1. a. $9 - x \geq 0$ gives $x \leq 9$, and the domain is $(-\infty, 9]$.

b. $2x^2 - x - 3 = (2x-3)(x+1)$, and $x = \frac{3}{2}$ or -1. Because the denominator of the given expression is zero at these points, we see that the domain of f cannot include these points and so the domain of f is $(-\infty, -1) \cup \left(-1, \frac{3}{2}\right) \cup \left(\frac{3}{2}, \infty\right)$.

3. a. $f(-2) = 3(-2)^2 + 5(-2) - 2 = 0$.

b. $f(a+2) = 3(a+2)^2 + 5(a+2) - 2 = 3a^2 + 12a + 12 + 5a + 10 - 2 = 3a^2 + 17a + 20$.

c. $f(2a) = 3(2a)^2 + 5(2a) - 2 = 12a^2 + 10a - 2$.

d. $f(a+h) = 3(a+h)^2 + 5(a+h) - 2 = 3a^2 + 6ah + 3h^2 + 5a + 5h - 2$.

5. a.

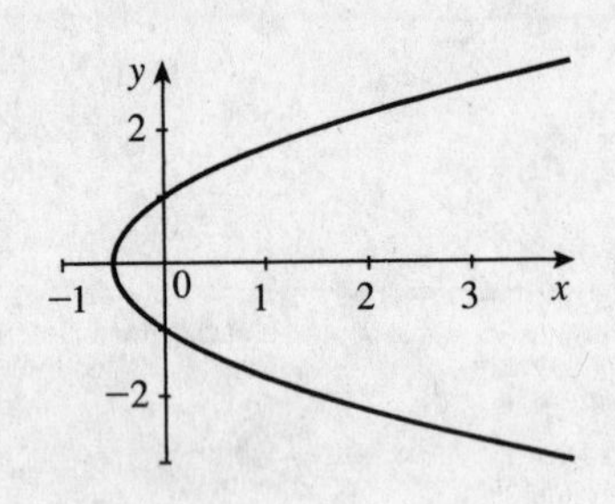

b. For each value of $x > 0$, there are two values of y. We conclude that y is not a function of x. (We could also note that the function fails the vertical line test.)

c. Yes. For each value of y, there is only one value of x.

7. a. $f(x)g(x) = \frac{2x+3}{x}$.

b. $\frac{f(x)}{g(x)} = \frac{1}{x(2x+3)}$.

c. $f(g(x)) = \frac{1}{2x+3}$.

d. $g(f(x)) = 2\left(\frac{1}{x}\right) + 3 = \frac{2}{x} + 3$.

9. a. Take $f(x) = 2x^2 + x + 1$ and $g(x) = \frac{1}{x^3}$.

b. Take $f(x) = x^2 + x + 4$ and $g(x) = \sqrt{x}$.

11. $\lim_{x\to 0}(5x-3) = 5(0) - 3 = -3$.

13. $\lim_{x\to -1}(3x^2+4)(2x-1) = \left[3(-1)^2+4\right][2(-1)-1] = -21$.

15. $\lim_{x\to 2}\frac{x+3}{x^2-9} = \frac{2+3}{4-9} = -1$.

17. $\lim_{x\to 3}\sqrt{2x^3-5} = \sqrt{2(27)-5} = 7$.

19. $\lim_{x\to 1^+}\frac{x-1}{x(x-1)} = \lim_{x\to 1^+}\frac{1}{x} = 1$.

21. $\lim_{x\to\infty}\frac{x^2}{x^2-1} = \lim_{x\to\infty}\frac{1}{1-\frac{1}{x^2}} = 1$.

23. $\lim_{x\to\infty}\frac{3x^2+2x+4}{2x^2-3x+1} = \lim_{x\to\infty}\frac{3+\frac{2}{x}+\frac{4}{x^2}}{2-\frac{3}{x}+\frac{1}{x^2}} = \frac{3}{2}$.

25. $\lim_{x\to 2^+} f(x) = \lim_{x\to 2^+} (-x+3) = -2+3 = 1$ and
$\lim_{x\to 2^-} f(x) = \lim_{x\to 2^-} (2x-3) = 2(2) - 3 = 4 - 3 = 1.$
Therefore, $\lim_{x\to 2} f(x) = 1.$

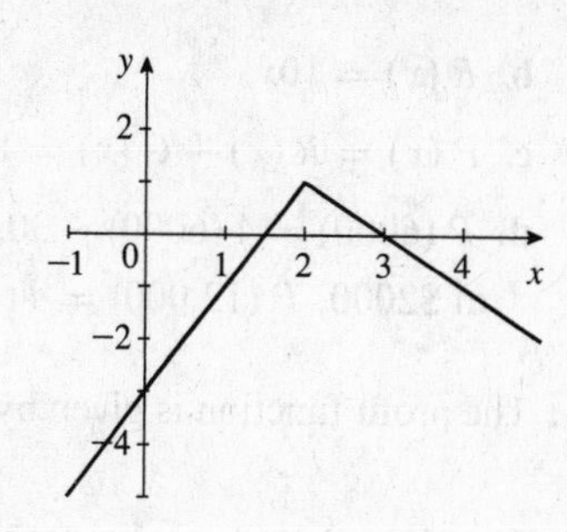

27. The function is discontinuous at $x = 2$.

29. Because $\lim_{x\to -1} f(x) = \lim_{x\to -1} \frac{1}{(x+1)^2} = \infty$ (does not exist), we see that f is discontinuous at $x = -1$.

31. a. Let $f(x) = x^2 + 2$. Then the average rate of change of y over $[1, 2]$ is $\frac{f(2) - f(1)}{2-1} = \frac{(4+2) - (1+2)}{1} = 3.$
Over $[1, 1.5]$, it is $\frac{f(1.5) - f(1)}{1.5 - 1} = \frac{(2.25+2) - (1+2)}{0.5} = 2.5$. Over $[1, 1.1]$, it is
$\frac{f(1.1) - f(1)}{1.1 - 1} = \frac{(1.21+2) - (1+2)}{0.1} = 2.1.$

b. Computing $f'(x)$ using the four-step process., we obtain
$f'(x) = \lim_{h\to 0} \frac{f(x+h) - f(x)}{h} = \lim_{h\to 0} \frac{h(2x+h)}{h} = \lim_{h\to 0} (2x+h) = 2$. Therefore, the instantaneous rate of change of f at $x = 1$ is $f'(1) = 2$, or 2 units per unit change in x.

33. $f(x) = \frac{3}{2}x + 5$. We use the four-step process:

Step 1 $f(x+h) = \frac{3}{2}(x+h) + 5 = \frac{3}{2}x + \frac{3}{2}h + 5.$

Step 2 $f(x+h) - f(x) = \frac{3}{2}x + \frac{3}{2}h + 5 - \frac{3}{2}x - 5 = \frac{3}{2}h.$

Step 3 $\frac{f(x+h) - f(x)}{h} = \frac{3}{2}.$

Step 4 $f'(x) = \lim_{h\to 0} \frac{f(x+h) - f(x)}{h} = \lim_{h\to 0} \frac{3}{2} = \frac{3}{2}.$

Therefore, the slope of the tangent line to the graph of the function f at the point $(-2, 2)$ is $\frac{3}{2}$. To find the equation of the tangent line to the curve at the point $(-2, 2)$, we use the point-slope form of the equation of a line, obtaining $y - 2 = \frac{3}{2}[x - (-2)]$ or $y = \frac{3}{2}x + 5$.

35. $f(x) = -\frac{1}{x}$. We use the four-step process:

Step 1 $f(x+h) = -\frac{1}{x+h}.$

Step 2 $f(x+h) - f(x) = -\frac{1}{x+h} - \left(-\frac{1}{x}\right) = -\frac{1}{x+h} + \frac{1}{x} = \frac{h}{x(x+h)}.$

Step 3 $\frac{f(x+h) - f(x)}{h} = \frac{1}{x(x+h)}.$

Step 4 $f'(x) = \lim_{h\to 0} \frac{f(x+h) - f(x)}{h} = \lim_{h\to 0} \frac{1}{x(x+h)} = \frac{1}{x^2}.$

37. $S(4) = 6000(4) + 30{,}000 = 54{,}000.$

39. a. $C(x) = 6x + 30{,}000.$

b. $R(x) = 10x$.

c. $P(x) = R(x) - C(x) = 10x - (6x + 30{,}000) = 4x - 30{,}000$.

d. $P(6000) = 4(6000) - 30{,}000 = -6000$, or a loss of \$6000. $P(8000) = 4(8000) - 30{,}000 = 2000$, or a profit of \$2000. $P(12{,}000) = 4(12{,}000) - 30{,}000 = 18{,}000$, or a profit of \$18,000.

41. The profit function is given by $P(x) = R(x) - C(x) = 20x - (12x + 20{,}000) = 8x - 20{,}000$.

43. The child should receive $D(35) = \dfrac{500(35)}{150} \approx 117$, or approximately 117 mg.

45. $R(30) = -\frac{1}{2}(30)^2 + 30(30) = 450$, or \$45,000.

47. The population will increase by $P(9) - P(0) = \left[50{,}000 + 30(9)^{3/2} + 20(9)\right] - 50{,}000$, or 990, during the next 9 months. The population will increase by $P(16) - P(0) = \left[50{,}000 + 30(16)^{3/2} + 20(16)\right] - 50{,}000$, or 2240 during the next 16 months.

49. We need to find the point of intersection of the two straight lines representing the given linear functions. We solve the equation $2.3 + 0.4t = 1.2 + 0.6t$, obtaining $1.1 = 0.2t$ and thus $t = 5.5$. This tells us that the annual sales of the Cambridge Drug Store first surpasses that of the Crimson Drug store $5\frac{1}{2}$ years from now.

51. The life expectancy of a female whose current age is 65 is $C(65) \approx 16.80$ (years).
The life expectancy of a female whose current age is 75 is $C(75) \approx 10.18$ (years).

53. $N(0) = 648$, or 648,000, $N(1) = -35.8 + 202 + 87.7 + 648 \approx 902$ or 902,000,
$N(2) = -35.8(2)^3 + 202(2)^2 + 87.8(2) + 648 = 1345.2$ or 1,345,200, and
$N(3) = -35.8(3)^3 + 202(3)^2 + 87.8(3) + 648 = 1762.8$ or 1,762,800.

55. a. $f(t) = 267$; $g(t) = 2t^2 + 46t + 733$.

b. $h(t) = (f + g)(t) = f(t) + g(t) = 267 + \left(2t^2 + 46t + 733\right) = 2t^2 + 46t + 1000$.

c. $h(13) = 2(13)^2 + 46(13) + 1000 = 1936$, or 1936 tons.

57. a. $f(r) = \pi r^2$.

b. $g(t) = 2t$.

c. $h(t) = (f \circ g)(t) = f(g(t)) = \pi\left[g(t)\right]^2 = 4\pi t^2$.

d. $h(30) = 4\pi\left(30^2\right) = 3600\pi$, or 3600π ft^2.

59. Let h denote the height of the box. Then its volume is $V = (x)(2x)h = 30$, so that $h = \dfrac{15}{x^2}$. Thus, the cost is

$$\begin{aligned} C(x) &= 30(x)(2x) + 15[2xh + 2(2x)h] + 20(x)(2x) \\ &= 60x^2 + 15(6xh) + 40x^2 = 100x^2 + (15)(6)x\left(\frac{15}{x^2}\right) \\ &= 100x^2 + \frac{1350}{x}. \end{aligned}$$

61. $\lim\limits_{x\to\infty} \overline{C}(x) = \lim\limits_{x\to\infty}\left(20 + \dfrac{400}{x}\right) = 20$. As the level of production increases without bound, the average cost of producing the commodity steadily decreases and approaches \$20 per unit.

63. True. If $x < 0$, then $\sqrt{x}$ is not defined, and if $x > 0$, then $\sqrt{-x}$ is not defined. Therefore $f(x)$ is defined nowhere, and is not a function.

CHAPTER 2 Before Moving On... page 160

1. a. $f(-1) = -2(-1) + 1 = 3.$

b. $f(0) = 2.$

c. $f\left(\frac{3}{2}\right) = \left(\frac{3}{2}\right)^2 + 2 = \frac{17}{4}.$

2. a. $(f+g)(x) = f(x) + g(x) = \dfrac{1}{x+1} + x^2 + 1.$

b. $(fg)(x) = f(x)g(x) = \dfrac{x^2+1}{x+1}.$

c. $(f \circ g)(x) = f(g(x)) = \dfrac{1}{g(x)+1} = \dfrac{1}{x^2+2}.$

d. $(g \circ f)(x) = g(f(x)) = [f(x)]^2 + 1 = \dfrac{1}{(x+1)^2} + 1.$

3. $4x + h = 108$, so $h = 108 - 4x$. The volume is $V = x^2h = x^2(108 - 4x) = 108x^2 - 4x^3$.

4. $\displaystyle\lim_{x\to-1} \frac{x^2+4x+3}{x^2+3x+2} = \lim_{x\to-1} \frac{(x+3)(x+1)}{(x+2)(x+1)} = 2.$

5. a. $\displaystyle\lim_{x\to1^-} f(x) = \lim_{x\to1^-} \left(x^2 - 1\right) = 0.$

b. $\displaystyle\lim_{x\to1^+} f(x) = \lim_{x\to1^+} x^3 = 1.$

Because $\displaystyle\lim_{x\to1^-} f(x) \neq \lim_{x\to1^+} f(x)$, f is not continuous at 1.

6. The slope of the tangent line at any point is

$$\begin{aligned}
\lim_{h\to0} \frac{f(x+h) - f(x)}{h} &= \lim_{h\to0} \frac{(x+h)^2 - 3(x+h) + 1 - \left(x^2 - 3x + 1\right)}{h} \\
&= \lim_{h\to0} \frac{x^2 + 2xh + h^2 - 3x - 3h - x^2 + 3x - 1}{h} \\
&= \lim_{h\to0} \frac{h(2x + h - 3)}{h} = \lim_{h\to0} (2x + h - 3) = 2x - 3.
\end{aligned}$$

Therefore, the slope at 1 is $2(1) - 3 = -1$. An equation of the tangent line is $y - (-1) = -1(x - 1)$, or $y + 1 = -x + 1$, or $y = -x$.

3 DIFFERENTIATION

3.1 Basic Rules of Differentiation

Problem-Solving Tips

In this section, you are given four basic rules for finding the derivative of a function. As you work through the exercises that follow, first decide which rule(s) you need to find the derivative of the given function. Then write out your solution. After doing this a few times, you should have the formulas memorized. The key here is to try not to look at the formula in the text, and to work the problem just as if you were taking a test. If you train yourself to work in this manner, writing tests will become a lot easier. Also, make sure to distinguish between the notation dy/dx and d/dx. The first notation is used for the derivative of a function y, where as the second notation tells us to find the derivative of the function that follows with respect to x.

Here are some tips for solving the problems in the exercises that follow:

1. **To find the derivative of a function involving radicals**, first rewrite the expression in exponential form. For example, if $f(x) = 2x - 5\sqrt{x}$, rewrite the function in the form $f(x) = 2x - 5x^{1/2}$.

2. **To find the point on the graph of f where the tangent line is horizontal**, set $f'(x) = 0$ and solve for x. (Here we are making use of the fact that the slope of a horizontal line is zero.) This yields the x-value of the point on the graph where the tangent line is horizontal. To find the corresponding y-value, evaluate the function f at this value of x.

Concept Questions page 169

1. a. The derivative of a constant is zero.

b. The derivative of $f(x) = x^n$ is n times x raised to the $(n-1)$th power.

c. The derivative of a constant times a function is the constant times the derivative of the function.

d. The derivative of the sum is the sum of the derivatives.

3. a. $F'(x) = \frac{d}{dx}[af(x) + bg(x)] = \frac{d}{dx}[af(x)] + \frac{d}{dx}[bg(x)] = af'(x) + bg'(x)$.

b. $F'(x) = \frac{d}{dx}\left[\frac{f(x)}{a}\right] = \frac{1}{a}\frac{d}{dx}[f(x)] = \frac{f'(x)}{a}$.

Exercises page 169

1. $f'(x) = \frac{d}{dx}(-3) = 0$.

3. $f'(x) = \frac{d}{dx}(x^5) = 5x^4$.

5. $f'(x) = \frac{d}{dx}(x^{3.1}) = 3.1x^{2.1}$.

7. $f'(x) = \frac{d}{dx}(3x^2) = 6x$.

9. $f'(r) = \frac{d}{dr}(\pi r^2) = 2\pi r$.

11. $f'(x) = \frac{d}{dx}(9x^{1/3}) = \frac{1}{3}(9)x^{(1/3-1)} = 3x^{-2/3}$.

13. $f'(x) = \frac{d}{dx}(3\sqrt{x}) = \frac{d}{dx}(3x^{1/2}) = \frac{1}{2}(3)x^{-1/2} = \frac{3}{2}x^{-1/2} = \frac{3}{2\sqrt{x}}$.

15. $f'(x) = \frac{d}{dx}\left(7x^{-12}\right) = (-12)(7)x^{-12-1} = -84x^{-13}$.

17. $f'(x) = \frac{d}{dx}\left(5x^2 - 3x + 7\right) = 10x - 3$.

19. $f'(x) = \frac{d}{dx}\left(-x^3 + 2x^2 - 6\right) = -3x^2 + 4x$.

21. $f'(x) = \frac{d}{dx}\left(0.03x^2 - 0.4x + 10\right) = 0.06x - 0.4$.

23. $f(x) = \dfrac{2x^3 - 4x^2 + 3}{x} = 2x^2 - 4x + \dfrac{3}{x}$, so $f'(x) = \dfrac{d}{dx}\left(2x^2 - 4x + 3x^{-1}\right) = 4x - 4 - \dfrac{3}{x^2}$.

25. $f'(x) = \frac{d}{dx}\left(4x^4 - 3x^{5/2} + 2\right) = 16x^3 - \frac{15}{2}x^{3/2}$.

27. $f'(x) = \frac{d}{dx}\left(5x^{-1} + 4x^{-2}\right) = -5x^{-2} - 8x^{-3} = \dfrac{-5}{x^2} - \dfrac{8}{x^3}$.

29. $f'(t) = \frac{d}{dt}\left(4t^{-4} - 3t^{-3} + 2t^{-1}\right) = -16t^{-5} + 9t^{-4} - 2t^{-2} = -\dfrac{16}{t^5} + \dfrac{9}{t^4} - \dfrac{2}{t^2}$.

31. $f'(x) = \frac{d}{dx}\left(3x - 5x^{1/2}\right) = 3 - \frac{5}{2}x^{-1/2} = 3 - \dfrac{5}{2\sqrt{x}}$.

33. $f'(x) = \dfrac{d}{dx}\left(2x^{-2} - 3x^{-1/3}\right) = -4x^{-3} + x^{-4/3} = -\dfrac{4}{x^3} + \dfrac{1}{x^{4/3}}$.

35. $f'(x) = \frac{d}{dx}\left(2x^3 - 4x\right) = 6x^2 - 4$.

a. $f'(-2) = 6(-2)^2 - 4 = 20$. **b.** $f'(0) = 6(0) - 4 = -4$. **c.** $f'(2) = 6(2)^2 - 4 = 20$.

37. The given limit is $f'(1)$, where $f(x) = x^3$. Because $f'(x) = 3x^2$, we have $\lim\limits_{h\to 0} \dfrac{(1+h)^3 - 1}{h} = f'(1) = 3$.

39. Let $f(x) = 3x^2 - x$. Then $\lim\limits_{h\to 0} \dfrac{3(2+h)^2 - (2+h) - 10}{h} = \lim\limits_{h\to 0} \dfrac{f(2+h) - f(2)}{h}$ because $f(2+h) - f(2) = 3(2+h)^2 - (2+h) - [3(4) - 2] = 3(2+h)^2 - (2+h) - 10$. But the last limit is simply $f'(2)$. Because $f'(x) = 6x - 1$, we have $f'(2) = 11$. Therefore, $\lim\limits_{h\to 0} \dfrac{3(2+h)^2 - (2+h) - 10}{h} = 11$.

41. $f(x) = 2x^2 - 3x + 4$. The slope of the tangent line at any point $(x, f(x))$ on the graph of f is $f'(x) = 4x - 3$. In particular, the slope of the tangent line at the point $(2, 6)$ is $f'(2) = 4(2) - 3 = 5$. An equation of the required tangent line is $y - 6 = 5(x - 2)$ or $y = 5x - 4$.

43. $f(x) = x^4 - 3x^3 + 2x^2 - x + 1$, so $f'(x) = 4x^3 - 9x^2 + 4x - 1$. The slope is $f'(2) = 4(2)^3 - 9(2)^2 + 4(2) - 1 = 3$. An equation of the tangent line is $y - (-1) = 3(x - 2)$ or $y = 3x - 7$.

45. a. $f'(x) = 3x^2$. At a point where the tangent line is horizontal, $f'(x) = 0$, or $3x^2 = 0$, and so $x = 0$. Therefore, the point is $(0, 0)$.

b.

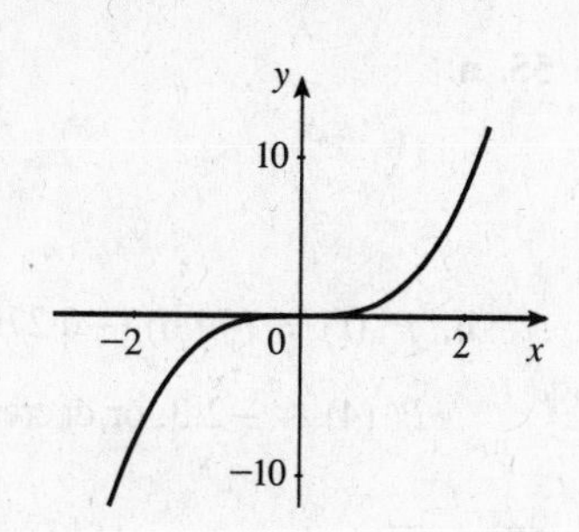

47. a. $f(x) = x^3 + 1$. The slope of the tangent line at any point $(x, f(x))$ on the graph of f is $f'(x) = 3x^2$. At the point(s) where the slope is 12, we have $3x^2 = 12$, so $x = \pm 2$. The required points are $(-2, -7)$ and $(2, 9)$.

b. The tangent line at $(-2, -7)$ has equation $y - (-7) = 12[x - (-2)]$, or $y = 12x + 17$, and the tangent line at $(2, 9)$ has equation $y - 9 = 12(x - 2)$, or $y = 12x - 15$.

c.

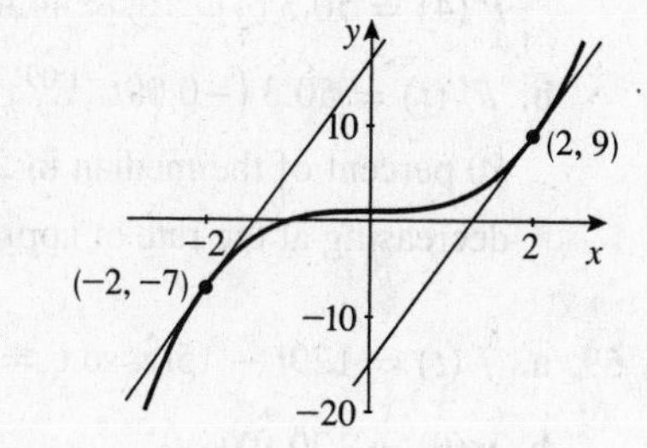

49. $f(x) = \frac{1}{4}x^4 - \frac{1}{3}x^3 - x^2$, so $f'(x) = x^3 - x^2 - 2x$.

a. $f'(x) = x^3 - x^2 - 2x = -2x$ implies $x^3 - x^2 = 0$, so $x^2(x - 1) = 0$. Thus, $x = 0$ or $x = 1$. $f(1) = \frac{1}{4}(1)^4 - \frac{1}{3}(1)^3 - (1)^2 = -\frac{13}{12}$ and $f(0) = \frac{1}{4}(0)^4 - \frac{1}{3}(0)^3 - (0)^2 = 0$. We conclude that the corresponding points on the graph are $\left(1, -\frac{13}{12}\right)$ and $(0, 0)$.

b. $f'(x) = x^3 - x^2 - 2x = 0$ implies $x(x^2 - x - 2) = 0$, $x(x - 2)(x + 1) = 0$, and so $x = 0, 2$, or -1. $f(0) = 0$, $f(2) = \frac{1}{4}(2)^4 - \frac{1}{3}(2)^3 - (2)^2 = 4 - \frac{8}{3} - 4 = -\frac{8}{3}$, and $f(-1) = \frac{1}{4}(-1)^4 - \frac{1}{3}(-1)^3 - (-1)^2 = \frac{1}{4} + \frac{1}{3} - 1 = -\frac{5}{12}$. We conclude that the corresponding points are $(0, 0)$, $\left(2, -\frac{8}{3}\right)$, and $\left(-1, -\frac{5}{12}\right)$.

c. $f'(x) = x^3 - x^2 - 2x = 10x$ implies $x^3 - x^2 - 12x = 0$, $x(x^2 - x - 12) = 0$, $x(x - 4)(x + 3) = 0$, so $x = 0, 4$, or -3. $f(0) = 0$, $f(4) = \frac{1}{4}(4)^4 - \frac{1}{3}(4)^3 - (4)^2 = 48 - \frac{64}{3} = \frac{80}{3}$, and $f(-3) = \frac{1}{4}(-3)^4 - \frac{1}{3}(-3)^3 - (-3)^2 = \frac{81}{4} + 9 - 9 = \frac{81}{4}$. We conclude that the corresponding points are $(0, 0)$, $\left(4, \frac{80}{3}\right)$, and $\left(-3, \frac{81}{4}\right)$.

51. $V(r) = \frac{4}{3}\pi r^3$, so $V'(r) = 4\pi r^2$.

a. $V'\left(\frac{2}{3}\right) = 4\pi\left(\frac{4}{9}\right) = \frac{16}{9}\pi$ cm³/cm.

b. $V'\left(\frac{5}{4}\right) = 4\pi\left(\frac{25}{16}\right) = \frac{25}{4}\pi$ cm³/cm.

53. a. The number of tablets and smartphones in use in 2011 was $f(2) = 128.1(2)^{1.94} \approx 491.5269$, or approximately 491.5 million.

b. $f'(t) = 128.1 \cdot 1.94t^{0.94} = 248.514t^{0.94}$, so the number of tablets and smartphones in 2011 was changing at the rate of $f'(2) = 248.514(2)^{0.94} \approx 476.7811$, or approximately 476.8 million/year.

55. a.

1970 ($t = 1$)	1990 ($t = 3$)	2010 ($t = 5$)
49.6%	36.9%	32.1%

b. $P'(t) = (49.6)\left(-0.27t^{-1.27}\right) = -\frac{13.392}{t^{1.27}}$. In 1990, $P'(3) \approx -3.3$, or decreasing at 3.3%/decade. In 2000, $P'(4) \approx -2.3$, or decreasing at 2.3%/decade.

57. a. The percentage of households with annual incomes within 50 percent of the median in 2010 was $P(4) = 50.3(4)^{-0.09} \approx 44.40$ (percent).

b. $P'(t) = 50.3\left(-0.09t^{-1.09}\right) \approx -4.527t^{-1.09}$, so the percentage of households with annual incomes within 50 percent of the median in 2010 was changing at the rate of $P'(4) = -4.527(4)^{-1.09} \approx -1.00$; that is, decreasing at the rate of approximately 1%/decade.

59. a. $f(t) = 120t - 15t^2$, so $v = f'(t) = 120 - 30t$.

b. $v(0) = 120$ ft/sec

c. Setting $v = 0$ gives $120 - 30t = 0$, or $t = 4$. Therefore, the stopping distance is $f(4) = 120(4) - 15(16)$ or 240 ft.

61. a. The approximate average medical cost for a family of four in 2010 was $C(10) = 22.9883(10)^2 + 830.358(10) + 7513 = 18{,}115.41$, or \$18,115.41.

b. $C'(t) = 45.9766t + 830.358$, so the rate at which the average medical cost for a family of four was increasing in 2010 was approximately $C'(10) = 45.9766(10) + 830.358 = 1290.124$, or \$1290.12/year.

63. $I(t) = -0.2t^3 + 3t^2 + 100$, so $I'(t) = -0.6t^2 + 6t$.

a. In 2008, it was changing at a rate of $I'(5) = -0.6(25) + 6(5)$, or 15 points/yr. In 2010, it is $I'(7) = -0.6(49) + 6(7)$, or 12.6 points/yr. In 2013, it is $I'(10) = -0.6(100) + 6(10)$, or 0 points/yr.

b. The average rate of increase of the CPI over the period from 2008 to 2013 is
$\frac{I(10) - I(5)}{5} = \frac{[-0.2(1000) + 3(100) + 100] - [-0.2(125) + 3(25) + 100]}{5} = \frac{200 - 150}{5} = 10$, or 10 points/yr.

65. $P(t) = -\frac{1}{3}t^3 + 64t + 3000$, so $P'(t) = -t^2 + 64$. The rates of change at the end of years one, two, three and four are $P'(1) = -1 + 64 = 63$, or 63,000 people/yr; $P'(2) = -4 + 64 = 60$, or 60,000 people/yr; $P'(3) = -9 + 64 = 55$, or 55,000 people/yr; and $P'(4) = -16 + 64 = 48$, or 48,000 people/yr. It appears that the plan is working.

67. a. $f(t) = -2t^3 + 12t^2 + 5$, so $v = f'(t) = -6t^2 + 24t$.

b. $f'(0) = 0$, or 0 ft/sec; $f'(2) = -6(4) + 24(2) = 24$, or 24 ft/sec; $f'(4) = -6(16) + 24(4) = 0$, or 0 ft/sec; and $f'(6) = -6(36) + 24(6) = -72$, or -72 ft/sec. The rocket starts out at an initial velocity of 0 ft/sec. It climbs upward until a maximum altitude is attained 4 seconds into flight. It then descends until it hits the ground.

c. At the highest point, $v = 0$. But this occurs when $t = 4$ (see part (b)). The maximum altitude is $f(4) = -2(4)^3 + 12(4)^2 + 5 = 69$, or 69 feet.

69. $P(t) = 50{,}000 + 30t^{3/2} + 20t$. The rate at which the population is increasing at any time t is $P'(t) = 45t^{1/2} + 20$. Nine months from now the population will be increasing at the rate of $P'(9) = 45(9)^{1/2} + 20$, or 155 people/month. Sixteen months from now the population will be increasing at the rate of $P'(16) = 45(16)^{1/2} + 20$, or 200 people/month.

71. $S(x) = -0.002x^3 + 0.6x^2 + x + 500$, so $S'(x) = -0.006x^2 + 1.2x + 1$.

a. When $x = 100$, $S'(100) = -0.006(100)^2 + 1.2(100) + 1 = 61$, or \$61,000 per thousand dollars.

b. When $x = 150$, $S'(150) = -0.006(150)^2 + 1.2(150) + 1 = 46$, or \$46,000 per thousand dollars. We conclude that the company's total sales increase at a faster rate with option (a); that is, when \$100,000 is spent on advertising.

73. a. $P(t) = 0.0004t^3 + 0.0036t^2 + 0.8t + 12$. At the beginning of 1991, $P(0) = 12\%$. At the beginning of 2010, $P(19) = 0.0004(19)^3 + 0.0036(19)^2 + 0.8(19) + 12 \approx 31.2$, or approximately 31.2%.

b. $P'(t) = 0.0012t^2 + 0.0072t + 0.8$. At the beginning of 1991, $P'(0) = 0.8$, or 0.8%/yr. At the beginning of 2010, $P'(19) = 0.0012(19)^2 + 0.0072(19) + 0.8 \approx 1.4$, or approximately 1.4%/yr.

75. a. $G(t) = J(t) - N(t) = \begin{cases} -0.0002t^2 + 0.032t + 0.1 & \text{if } 0 \le t < 5 \\ 0.0002t^2 - 0.006t + 0.28 & \text{if } 5 \le t < 10 \\ -0.0012t^2 + 0.082t - 0.46 & \text{if } 10 \le t < 15 \end{cases}$

b. In 2008, where $t = 8$, the gap is changing at a rate of $G'(8) = \left[\frac{d}{dt}(0.0002t^2 - 0.006t + 0.28)\right]_{t=8} = (0.0004t - 0.006)|_{t=8} = -0.0028$; that is, the gap is narrowing at a rate of 2800 jobs/yr. In 2012, where $t = 12$, the gap is changing at a rate of $G'(12) = \left[\frac{d}{dt}(-0.0012t^2 + 0.082t - 0.46)\right]_{t=12} = (-0.0024t + 0.082)|_{t=12} = 0.0532$; that is, the gap is increasing at a rate of 53,200 jobs/yr.

77. False. f is not a power function.

Using Technology page 175

1. 1 **3.** 0.4226 **5.** 0.1613

7. a.

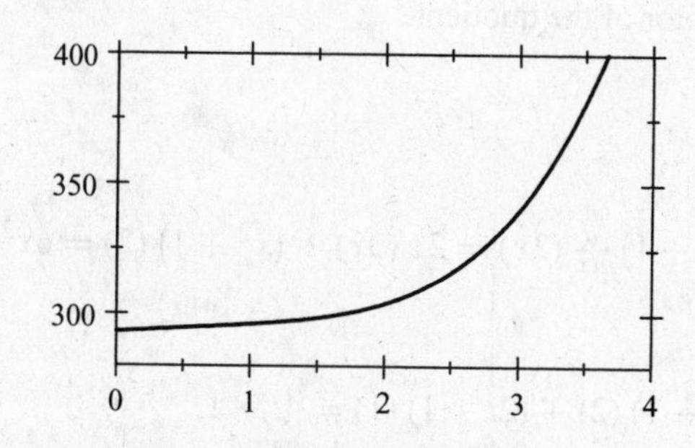

b. 3.4295 parts/million per 40 years;
164.239 parts/million per 40 years

3.2 The Product and Quotient Rules

Problem-Solving Tips

The answers at the back of the book for the exercises in this section are given in both simplified and unsimplified terms. Here, as with all of your homework, you should make it a practice to analyze your errors. If you do not get the right answer for the unsimplified form, it means that you are not applying the rules for differentiating correctly. In this case you need to review the rules, making sure that you can write out each rule. If you have the correct answer for the unsimplified form but the incorrect answer for the simplified form, it probably means that you have made an algebraic error. You may need to review the rules for simplifying algebraic expressions given on page 16 of the text and then work some of the exercises given in Section 1.2 to get back into practice. In any case, you will need to simplify your answers when you work the problems on the applications of the derivative in the next chapter, so you should get in the habit of doing so now.

Here are some tips for solving the problems in the exercises that follow:

1. **To find the derivative of a function involving radicals**, first rewrite the expression in exponential form. For example, if $f(x) = 2x - 5\sqrt{x}$, rewrite the function in the form $f'^{1/2}$.

2. **To find the point on the graph of f where the tangent line is horizontal**, set $f'(x) = 0$ and solve for x. (Here we are making use of the fact that the slope of a horizontal line is zero.) This yields the x-value of the point on the graph where the tangent line is horizontal. To find the corresponding y-value, evaluate the function f at this value of x.

Concept Questions page 181

1. **a.** The derivative of the product of two functions is equal to the first function times the derivative of the second function plus the second function times the derivative of the first function.

   b. The derivative of the quotient of two functions is equal to the quotient whose numerator is given by the denominator times the derivative of the numerator minus the numerator times the derivative of the denominator and whose denominator is the square of the denominator of the quotient.

Exercises page 181

1. $f(x) = 2x(x^2+1)$, so $f'(x) = 2x\frac{d}{dx}(x^2+1) + (x^2+1)\frac{d}{dx}(2x) = 2x(2x) + (x^2+1)(2) = 6x^2+2$.

3. $f(t) = (t-1)(2t+1)$, so
$f'(t) = (t-1)\frac{d}{dt}(2t+1) + (2t+1)\frac{d}{dt}(t-1) = (t-1)(2) + (2t+1)(1) = 4t-1$.

5. $f(x) = (3x+1)(x^2-2)$, so
$f'(x) = (3x+1)\frac{d}{dx}(x^2-2) + (x^2-2)\frac{d}{dx}(3x+1) = (3x+1)(2x) + (x^2-2)(3) = 9x^2+2x-6$.

7. $f(x) = (x^3-1)(x+1)$, so
$f'(x) = (x^3-1)\frac{d}{dx}(x+1) + (x+1)\frac{d}{dx}(x^3-1) = (x^3-1)(1) + (x+1)(3x^2) = 4x^3+3x^2-1$.

9. $f(w) = (w^3 - w^2 + w - 1)(w^2 + 2)$, so

$$\begin{aligned} f'(w) &= (w^3 - w^2 + w - 1)\frac{d}{dw}(w^2 + 2) + (w^2 + 2)\frac{d}{dw}(w^3 - w^2 + w - 1) \\ &= (w^3 - w^2 + w - 1)(2w) + (w^2 + 2)(3w^2 - 2w + 1) \\ &= 2w^4 - 2w^3 + 2w^2 - 2w + 3w^4 - 2w^3 + w^2 + 6w^2 - 4w + 2 = 5w^4 - 4w^3 + 9w^2 - 6w + 2. \end{aligned}$$

11. $f(x) = (5x^2 + 1)(2\sqrt{x} - 1)$, so

$$\begin{aligned} f'(x) &= (5x^2 + 1)\tfrac{d}{dx}(2x^{1/2} - 1) + (2x^{1/2} - 1)\tfrac{d}{dx}(5x^2 + 1) = (5x^2 + 1)(x^{-1/2}) + (2x^{1/2} - 1)(10x) \\ &= 5x^{3/2} + x^{-1/2} + 20x^{3/2} - 10x = \frac{25x^2 - 10x\sqrt{x} + 1}{\sqrt{x}}. \end{aligned}$$

13. $f(x) = (x^2 - 5x + 2)\left(x - \dfrac{2}{x}\right)$, so

$$\begin{aligned} f'(x) &= (x^2 - 5x + 2)\frac{d}{dx}\left(x - \frac{2}{x}\right) + \left(x - \frac{2}{x}\right)\frac{d}{dx}(x^2 - 5x + 2) \\ &= \frac{(x^2 - 5x + 2)(x^2 + 2)}{x^2} + \frac{(x^2 - 2)(2x - 5)}{x} = \frac{(x^2 - 5x + 2)(x^2 + 2) + x(x^2 - 2)(2x - 5)}{x^2} \\ &= \frac{x^4 + 2x^2 - 5x^3 - 10x + 2x^2 + 4 + 2x^4 - 5x^3 - 4x^2 + 10x}{x^2} = \frac{3x^4 - 10x^3 + 4}{x^2}. \end{aligned}$$

15. $f(x) = \dfrac{1}{x - 2}$, so $f'(x) = \dfrac{(x - 2)\frac{d}{dx}(1) - (1)\frac{d}{dx}(x - 2)}{(x - 2)^2} = \dfrac{0 - 1(1)}{(x - 2)^2} = -\dfrac{1}{(x - 2)^2}$.

17. $f(x) = \dfrac{2x - 1}{2x + 1}$, so

$$f'(x) = \frac{(2x + 1)\frac{d}{dx}(2x - 1) - (2x - 1)\frac{d}{dx}(2x + 1)}{(2x + 1)^2} = \frac{2(2x + 1) - (2x - 1)(2)}{(2x + 1)^2} = \frac{4}{(2x + 1)^2}.$$

19. $f(x) = \dfrac{1}{x^2 + x + 2}$, so $f'(x) = \dfrac{(x^2 + x + 2)(0) - (1)(2x + 1)}{(x^2 + x + 2)^2} = -\dfrac{2x + 1}{(x^2 + x + 2)^2}$.

21. $f(s) = \dfrac{s^2 - 4}{s + 1}$, so

$$f'(s) = \frac{(s + 1)\dfrac{d}{ds}(s^2 - 4) - (s^2 - 4)\dfrac{d}{ds}(s + 1)}{(s + 1)^2} = \frac{(s + 1)(2s) - (s^2 - 4)(1)}{(s + 1)^2} = \frac{s^2 + 2s + 4}{(s + 1)^2}.$$

23. $f(x) = \dfrac{\sqrt{x} + 1}{x^2 + 1}$, so

$$\begin{aligned} f'(x) &= \frac{(x^2 + 1)\frac{d}{dx}(x^{1/2}) - (x^{1/2} + 1)\frac{d}{dx}(x^2 + 1)}{(x^2 + 1)^2} = \frac{(x^2 + 1)\left(\frac{1}{2}x^{-1/2}\right) - (x^{1/2} + 1)(2x)}{(x^2 + 1)^2} \\ &= \frac{\left(\frac{1}{2}x^{-1/2}\right)\left[(x^2 + 1) - (x^{1/2} + 1)4x^{3/2}\right]}{(x^2 + 1)^2} = \frac{1 - 3x^2 - 4x^{3/2}}{2\sqrt{x}(x^2 + 1)^2}. \end{aligned}$$

25. $f(x) = \dfrac{x^2+2}{x^2+x+1}$, so

$$f'(x) = \frac{(x^2+x+1)\frac{d}{dx}(x^2+2) - (x^2+2)\frac{d}{dx}(x^2+x+1)}{(x^2+x+1)^2}$$

$$= \frac{(x^2+x+1)(2x) - (x^2+2)(2x+1)}{(x^2+x+1)^2} = \frac{2x^3+2x^2+2x-2x^3-x^2-4x-2}{(x^2+x+1)^2} = \frac{x^2-2x-2}{(x^2+x+1)^2}.$$

27. $f(x) = \dfrac{(x+1)(x^2+1)}{x-2} = \dfrac{(x^3+x^2+x+1)}{x-2}$, so

$$f'(x) = \frac{(x-2)\frac{d}{dx}(x^3+x^2+x+1) - (x^3+x^2+x+1)\frac{d}{dx}(x-2)}{(x-2)^2}$$

$$= \frac{(x-2)(3x^2+2x+1) - (x^3+x^2+x+1)}{(x-2)^2}$$

$$= \frac{3x^3+2x^2+x-6x^2-4x-2-x^3-x^2-x-1}{(x-2)^2} = \frac{2x^3-5x^2-4x-3}{(x-2)^2}.$$

29. $f(x) = \dfrac{x}{x^2-4} - \dfrac{x-1}{x^2+4} = \dfrac{x(x^2+4)-(x-1)(x^2-4)}{(x^2-4)(x^2+4)} = \dfrac{x^2+8x-4}{(x^2-4)(x^2+4)}$, so

$$f'(x) = \frac{(x^2-4)(x^2+4)\frac{d}{dx}(x^2+8x-4) - (x^2+8x-4)\frac{d}{dx}(x^4-16)}{(x^2-4)^2(x^2+4)^2}$$

$$= \frac{(x^2-4)(x^2+4)(2x+8) - (x^2+8x-4)(4x^3)}{(x^2-4)^2(x^2+4)^2}$$

$$= \frac{2x^5+8x^4-32x-128-4x^5-32x^4+16x^3}{(x^2-4)^2(x^2+4)^2} = \frac{-2x^5-24x^4+16x^3-32x-128}{(x^2-4)^2(x^2+4)^2}.$$

31. $h(x) = f(x)g(x)$, so $h'(x) = f(x)g'(x) + f'(x)g(x)$ by the Product Rule. Therefore, $h'(1) = f(1)g'(1) + f'(1)g(1) = (2)(3) + (-1)(-2) = 8$.

33. $h(x) = \dfrac{xf(x)}{x+g(x)}$. Using the Quotient Rule followed by the Product Rule, we obtain

$$h'(x) = \frac{[x+g(x)]\frac{d}{dx}[xf(x)] - xf(x)\frac{d}{dx}[x+g(x)]}{[x+g(x)]^2} = \frac{[x+g(x)][xf'(x)+f(x)] - xf(x)[1+g'(x)]}{[x+g(x)]^2}.$$

Therefore,

$$h'(1) = \frac{[1+g(1)][f'(1)+f(1)] - f(1)[1+g'(1)]}{[1+g(1)]^2} = \frac{(1-2)(-1+2)-2(1+3)}{(1-2)^2} = \frac{-1-8}{1} = -9.$$

35. $f(x) = (2x-1)(x^2+3)$, so

$$f'(x) = (2x-1)\frac{d}{dx}(x^2+3) + (x^2+3)\frac{d}{dx}(2x-1) = (2x-1)(2x) + (x^2+3)(2)$$

$$= 6x^2-2x+6 = 2(3x^2-x+3).$$

At $x=1$, $f'(1) = 2\left[3(1)^2-(1)+3\right] = 2(5) = 10$.

37. $f(x) = \dfrac{x}{x^4 - 2x^2 - 1}$, so

$$f'(x) = \frac{\left(x^4 - 2x^2 - 1\right)\frac{d}{dx}(x) - x\frac{d}{dx}\left(x^4 - 2x^2 - 1\right)}{\left(x^4 - 2x^2 - 1\right)^2} = \frac{\left(x^4 - 2x^2 - 1\right)(1) - x\left(4x^3 - 4x\right)}{\left(x^4 - 2x^2 - 1\right)^2}$$

$$= \frac{-3x^4 + 2x^2 - 1}{\left(x^4 - 2x^2 - 1\right)^2}.$$

Therefore, $f'(-1) = \dfrac{-3 + 2 - 1}{(1 - 2 - 1)^2} = -\dfrac{2}{4} = -\dfrac{1}{2}$.

39. $f(x) = \left(x^3 + 1\right)\left(x^2 - 2\right)$, so

$f'(x) = \left(x^3 + 1\right)\frac{d}{dx}\left(x^2 - 2\right) + \left(x^2 - 2\right)\frac{d}{dx}\left(x^3 + 1\right) = \left(x^3 + 1\right)(2x) + \left(x^2 - 2\right)\left(3x^2\right)$. The slope of the tangent line at $(2, 18)$ is $f'(2) = (8 + 1)(4) + (4 - 2)(12) = 60$. An equation of the tangent line is $y - 18 = 60(x - 2)$, or $y = 60x - 102$.

41. $f(x) = \dfrac{x + 1}{x^2 + 1}$, so

$f'(x) = \dfrac{\left(x^2 + 1\right)\frac{d}{dx}(x + 1) - (x + 1)\frac{d}{dx}\left(x^2 + 1\right)}{\left(x^2 + 1\right)^2} = \dfrac{\left(x^2 + 1\right)(1) - (x + 1)(2x)}{\left(x^2 + 1\right)^2} = \dfrac{-x^2 - 2x + 1}{\left(x^2 + 1\right)^2}$. At $x = 1$, $f'(1) = \dfrac{-1 - 2 + 1}{4} = -\dfrac{1}{2}$. Therefore, the slope of the tangent line at $x = 1$ is $-\frac{1}{2}$ and an equation is $y - 1 = -\frac{1}{2}(x - 1)$ or $y = -\frac{1}{2}x + \frac{3}{2}$.

43. Using the Product Rule, we find

$g'(x) = \frac{d}{dx}\left[x^2 f(x)\right] = x^2\frac{d}{dx}\left[f(x)\right] + f(x)\frac{d}{dx}\left(x^2\right) = x^2 f'(x) + 2xf(x)$. Therefore, $g'(2) = 2^2 f'(2) + 2(2)f(2) = (4)(-1) + 4(3) = 8$.

45. $f(x) = \left(x^3 + 1\right)\left(3x^2 - 4x + 2\right)$, so

$$f'(x) = \left(x^3 + 1\right)\frac{d}{dx}\left(3x^2 - 4x + 2\right) + \left(3x^2 - 4x + 2\right)\frac{d}{dx}\left(x^3 + 1\right)$$

$$= \left(x^3 + 1\right)(6x - 4) + \left(3x^2 - 4x + 2\right)\left(3x^2\right)$$

$$= 6x^4 + 6x - 4x^3 - 4 + 9x^4 - 12x^3 + 6x^2 = 15x^4 - 16x^3 + 6x^2 + 6x - 4.$$

At $x = 1$, $f'(1) = 15(1)^4 - 16(1)^3 + 6(1) + 6(1) - 4 = 7$. Thus, the slope of the tangent line at the point $x = 1$ is 7 and an equation is $y - 2 = 7(x - 1)$, or $y = 7x - 5$.

47. $f(x) = \left(x^2 + 1\right)(2 - x)$, so

$f'(x) = \left(x^2 + 1\right)\frac{d}{dx}(2 - x) + (2 - x)\frac{d}{dx}\left(x^2 + 1\right) = \left(x^2 + 1\right)(-1) + (2 - x)(2x) = -3x^2 + 4x - 1$. At a point where the tangent line is horizontal, we have $f'(x) = -3x^2 + 4x - 1 = 0$ or $3x^2 - 4x + 1 = (3x - 1)(x - 1) = 0$, giving $x = \frac{1}{3}$ or $x = 1$. Because $f\left(\frac{1}{3}\right) = \left(\frac{1}{9} + 1\right)\left(2 - \frac{1}{3}\right) = \frac{50}{27}$ and $f(1) = 2(2 - 1) = 2$, we see that the required points are $\left(\frac{1}{3}, \frac{50}{27}\right)$ and $(1, 2)$.

49. $f(x) = (x^2+6)(x-5)$, so

$$f'(x) = (x^2+6)\frac{d}{dx}(x-5) + (x-5)\frac{d}{dx}(x^2+6) = (x^2+6)(1) + (x-5)(2x)$$
$$= x^2+6+2x^2-10x = 3x^2-10x+6.$$

At a point where the slope of the tangent line is -2, we have $f'(x) = 3x^2-10x+6 = -2$. This gives $3x^2-10x+8 = (3x-4)(x-2) = 0$, so $x = \frac{4}{3}$ or $x = 2$. Because $f\left(\frac{4}{3}\right) = \left(\frac{16}{9}+6\right)\left(\frac{4}{3}-5\right) = -\frac{770}{27}$ and $f(2) = (4+6)(2-5) = -30$, the required points are $\left(\frac{4}{3}, -\frac{770}{27}\right)$ and $(2, -30)$.

51. $y = \dfrac{1}{1+x^2}$, so $y' = \dfrac{(1+x^2)\frac{d}{dx}(1) - (1)\frac{d}{dx}(1+x^2)}{(1+x^2)^2} = \dfrac{-2x}{(1+x^2)^2}$. Thus, the slope of the tangent line at $\left(1, \frac{1}{2}\right)$ is $y'|_{x=1} = \left.\dfrac{-2x}{(1+x^2)^2}\right|_{x=1} = \dfrac{-2}{4} = -\dfrac{1}{2}$ and an equation of the tangent line is $y - \frac{1}{2} = -\frac{1}{2}(x-1)$, or $y = -\frac{1}{2}x + 1$. Next, the slope of the required normal line is 2 and its equation is $y - \frac{1}{2} = 2(x-1)$, or $y = 2x - \frac{3}{2}$.

53. $C(x) = \dfrac{0.5x}{100-x}$, so $C'(x) = \dfrac{(100-x)(0.5) - 0.5x(-1)}{(100-x)^2} = \dfrac{50}{(100-x)^2}$. $C'(80) = \dfrac{50}{20^2} = 0.125$, $C'(90) = \dfrac{50}{10^2} = 0.5$, $C'(95) = \dfrac{50}{5^2} = 2$, and $C'(99) = \dfrac{50}{1} = 50$. The rates of change of the cost of removing 80%, 90%, 95%, and 99% of the toxic waste are 0.125, 0.5, 2, and 50 million dollars per 1% increase in waste removed. It is too costly to remove all of the pollutant.

55. $N(t) = \dfrac{10{,}000}{1+t^2} + 2000$, so $N'(t) = \dfrac{d}{dt}\left[10{,}000\left(1+t^2\right)^{-1} + 2000\right] = -\dfrac{10{,}000}{\left(1+t^2\right)^2}(2t) = -\dfrac{20{,}000t}{\left(1+t^2\right)^2}$. The rates of change after 1 minute and 2 minutes are $N'(1) = -\dfrac{20{,}000}{\left(1+1^2\right)^2} = -5000$ and $N'(2) = -\dfrac{20{,}000(2)}{\left(1+2^2\right)^2} = -1600$.

The population of bacteria after one minute is $N(1) = \dfrac{10{,}000}{1+1} + 2000 = 7000$, and the population after two minutes is $N(2) = \dfrac{10{,}000}{1+4} + 2000 = 4000$.

57. a. $R(x) = xd(x) = \dfrac{50x}{0.01x^2+1}$.

b. $R'(x) = \dfrac{d}{dx}\left(\dfrac{50x}{0.01x^2+1}\right) = 50\dfrac{d}{dx}\left(\dfrac{x}{0.01x^2+1}\right) = 50 \cdot \dfrac{(0.01x^2+1)(1) - x(0.02x)}{\left(0.01x^2+1\right)^2} = \dfrac{50\left(1-0.01x^2\right)}{\left(0.01x^2+1\right)^2}$.

c. $R'(8) \approx 6.69$, $R'(10) = 0$, and $R'(12) \approx -3.70$, so the revenue is increasing at the rate of approximately \$6700 per thousand watches at a sales level of 8000 watches per week, the revenue is stable at a sales level of 10,000 watches per week, and the revenue is decreasing by approximately \$3700 per thousand watches at a sales level of 12,000 watches per week.

59. a. The average 30-year fixed mortgage rate in the first week of May in 2010 was

$$M(1) = \frac{55.9}{1-0.31+11.2} \approx 4.701, \text{ or approximately } 4.7\% \text{ per year.}$$

b. $M'(t) = \dfrac{(t^2 - 0.31t + 11.2)(0) - 55.9(2t - 0.31)}{(t^2 - 0.31t + 11.2)^2} = \dfrac{-55.9(2t - 0.31)}{(t^2 - 0.31t + 11.2)^2}$. Thus, the 30-year fixed mortgage rate was changing at the rate of $M'(1) = \dfrac{-55.9(2 - 0.31)}{(1 - 0.31 + 11.2)^2} \approx -0.668$ in the first week of May in 2010. That is, it was decreasing at approximately 0.67% per year.

61. a. $N(t) = \dfrac{60t + 180}{t + 6}$, so

$$N'(t) = \frac{(t+6)\frac{d}{dt}(60t+180) - (60t+180)\frac{d}{dt}(t+6)}{(t+6)^2} = \frac{(t+6)(60) - (60t+180)(1)}{(t+6)^2} = \frac{180}{(t+6)^2}.$$

b. $N'(1) = \dfrac{180}{(1+6)^2} \approx 3.7$, $N'(3) = \dfrac{180}{(3+6)^2} \approx 2.2$, $N'(4) = \dfrac{180}{(4+6)^2} = 1.8$, and $N'(7) = \dfrac{180}{(7+6)^2} \approx 1.1$. We conclude that the rates at which the average student is increasing his or her speed one week, three weeks, four weeks, and seven weeks into the course are approximately 3.7, 2.2, 1.8, and 1.1 words per minute, respectively.

c. Yes.

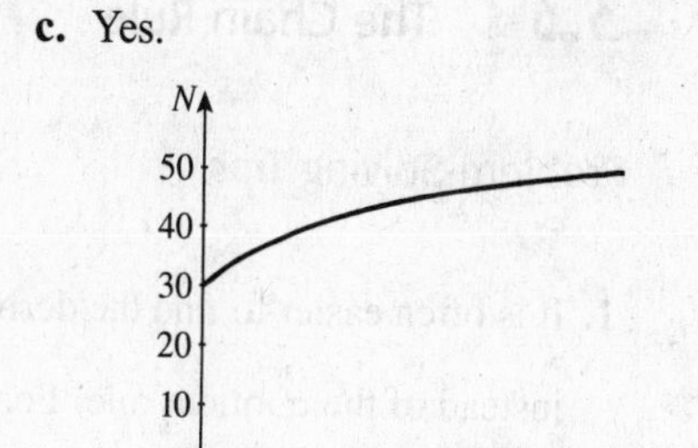

d. $N(12) = \dfrac{60(12) + 180}{12 + 6} = 50$, or 50 words/minute.

63. $f(t) = \dfrac{0.055t + 0.26}{t + 2}$, so $f'(t) = \dfrac{(t+2)(0.055) - (0.055t + 0.26)(1)}{(t+2)^2} = -\dfrac{0.15}{(t+2)^2}$. At the beginning, the formaldehyde level is changing at the rate of $f'(0) = -\dfrac{0.15}{4} = -0.0375$; that is, it is decreasing at the rate of 0.0375 parts per million per year. Next, $f'(3) = -\dfrac{0.15}{5^2} = -0.006$, and so the level is decreasing at the rate of 0.006 parts per million per year at the beginning of the fourth year (when $t = 3$).

65. a. $R'(x) = \frac{d}{dx}[xD(x)] = xD'(x) + (1)D(x) = xD'(x) + D(x)$.

b. Here $p = D(x) = a - bx$, so $D'(x) = -b$. Therefore, $R'(x) = x(-b) + (a - bx) = a - 2bx$.

c. $R(x) = xD(x) = x(a - bx) = ax - bx^2$, so $R'(x) = a - 2bx$, as obtained in part (a).

67. a. If there is a substrate present, then the relative growth rate is $R(0) = 0$.

b. $\displaystyle\lim_{s\to\infty} R(s) = \lim_{s\to\infty} \frac{cs}{k+s} = \lim_{s\to\infty} \frac{c}{\frac{k}{s}+1} = c$. Thus, the relative growth rate approaches c when the substrate is present in great excess.

c. $R'(s) = \dfrac{(k+s)(c) - cs(1)}{(k+s)^2} = \dfrac{kc}{(k+s)^2}$.

69. False. Take $f(x) = x$ and $g(x) = x$. Then $f(x)g(x) = x^2$, so $\frac{d}{dx}[f(x)g(x)] = \frac{d}{dx}(x^2) = 2x \neq f'(x)g'(x) = 1$.

71. False. Let $f(x) = x^3$. Then $\dfrac{d}{dx}\left[\dfrac{f(x)}{x^2}\right] = \dfrac{d}{dx}\left(\dfrac{x^3}{x^2}\right) = \dfrac{d}{dx}(x) = 1 \neq \dfrac{f'(x)}{2x} = \dfrac{3x^2}{2x} = \frac{3}{2}x$.

73. Let $f(x) = u(x)v(x)$ and $g(x) = w(x)$. Then $h(x) = f(x)g(x)$. Therefore, $h'(x) = f'(x)g(x) + f(x)g'(x)$. But $f'(x) = u(x)v'(x) + u'(x)v(x)$, so

$$h'(x) = [u(x)v'(x) + u'(x)v(x)]g(x) + u(x)v(x)w'(x)$$
$$= u(x)v(x)w'(x) + u(x)v'(x)w(x) + u'(x)v(x)w(x).$$

Using Technology page 186

1. 0.8750 **3.** 0.0774 **5.** −0.5000 **7.** 31,312 per year

3.3 The Chain Rule

Problem-Solving Tips

1. It is often easier to find the derivative of a quotient when the numerator is a constant by using the general power rule instead of the quotient rule. For example, to find the derivative of $f(x) = -\frac{1}{\sqrt{2x^2-1}}$ in Self-Check Exercise 1 of this section, we first rewrite the function in the form $f(x) = -(2x^2-1)^{-1/2}$ and then use the general power rule to find the derivative.

2. To simplify a function involving the powers of an expression, factor out the lowest power of the expression. For example, to simplify $5(x+1)^{1/2} - 3(x+1)^{-1/2}$, factor out $(x+1)^{-1/2}$, which is the lowest power of $x+1$ in the expression.

Concept Questions page 194

1. The derivative of $h(x) = g(f(x))$ is equal to the derivative of g evaluated at $f(x)$ times the derivative of f.

3. $(g \circ f)'(t) = [(g \circ f)(t)]' = g'(f(t))f'(t)$ describes the rate of change of the revenue as a function of time.

Exercises page 194

1. $f(x) = (2x-1)^3$, so $f'(x) = 3(2x-1)^2 \frac{d}{dx}(2x-1) = 3(2x-1)^2(2) = 6(2x-1)^2$.

3. $f(x) = (x^2+2)^5$, so $f'(x) = 5(x^2+2)^4(2x) = 10x(x^2+2)^4$.

5. $f(x) = (2x-x^2)^3$, so $f'(x) = 3(2x-x^2)^2 \frac{d}{dx}(2x-x^2) = 3(2x-x^2)^2(2-2x) = 6x^2(1-x)(2-x)^2$.

7. $f(x) = (2x+1)^{-2}$, so $f'(x) = -2(2x+1)^{-3} \frac{d}{dx}(2x+1) = -2(2x+1)^{-3}(2) = -4(2x+1)^{-3}$.

9. $f(x) = (x^2-4)^{5/2}$, so $f'(x) = \frac{5}{2}(x^2-4)^{3/2} \frac{d}{dx}(x^2-4) = \frac{5}{2}(x^2-4)^{3/2}(2x) = 5x(x^2-4)^{3/2}$.

11. $f(x) = \sqrt{3x-2} = (3x-2)^{1/2}$, so $f'(x) = \frac{1}{2}(3x-2)^{-1/2}(3) = \frac{3}{2}(3x-2)^{-1/2} = \frac{3}{2\sqrt{3x-2}}$.

13. $f(x) = \sqrt[3]{1-x^2}$, so

$$f'(x) = \tfrac{d}{dx}\left(1-x^2\right)^{1/3} = \tfrac{1}{3}\left(1-x^2\right)^{-2/3}\tfrac{d}{dx}\left(1-x^2\right) = \tfrac{1}{3}\left(1-x^2\right)^{-2/3}(-2x) = -\tfrac{2}{3}x\left(1-x^2\right)^{-2/3}$$
$$= \frac{-2x}{3\left(1-x^2\right)^{2/3}}.$$

15. $f(x) = \dfrac{1}{(2x+3)^3} = (2x+3)^{-3}$, so $f'(x) = -3(2x+3)^{-4}(2) = -6(2x+3)^{-4} = -\dfrac{6}{(2x+3)^4}$.

17. $f(t) = \dfrac{1}{\sqrt{2t-4}}$, so $f'(t) = \tfrac{d}{dt}(2t-4)^{-1/2} = -\tfrac{1}{2}(2t-4)^{-3/2}(2) = -(2t-4)^{-3/2} = -\dfrac{1}{(2t-4)^{3/2}}$.

19. $y = \dfrac{1}{\left(4x^4+x\right)^{3/2}}$, so $\dfrac{dy}{dx} = \tfrac{d}{dx}\left(4x^4+x\right)^{-3/2} = -\tfrac{3}{2}\left(4x^4+x\right)^{-5/2}\left(16x^3+1\right) = -\tfrac{3}{2}\left(16x^3+1\right)\left(4x^4+x\right)^{-5/2}$.

21. $f(x) = \left(3x^2+2x+1\right)^{-2}$, so

$$f'(x) = -2\left(3x^2+2x+1\right)^{-3}\tfrac{d}{dx}\left(3x^2+2x+1\right) = -2\left(3x^2+2x+1\right)^{-3}(6x+2)$$
$$= -4(3x+1)\left(3x^2+2x+1\right)^{-3}.$$

23. $f(x) = \left(x^2+1\right)^3 - \left(x^3+1\right)^2$, so

$$f'(x) = 3\left(x^2+1\right)^2\tfrac{d}{dx}\left(x^2+1\right) - 2\left(x^3+1\right)\tfrac{d}{dx}\left(x^3+1\right) = 3\left(x^2+1\right)^2(2x) - 2\left(x^3+1\right)\left(3x^2\right)$$
$$= 6x\left[\left(x^2+1\right)^2 - x\left(x^3+1\right)\right] = 6x\left(2x^2-x+1\right).$$

25. $f(t) = \left(t^{-1}-t^{-2}\right)^3$, so $f'(t) = 3\left(t^{-1}-t^{-2}\right)^2\tfrac{d}{dt}\left(t^{-1}-t^{-2}\right) = 3\left(t^{-1}-t^{-2}\right)^2\left(-t^{-2}+2t^{-3}\right)$.

27. $f(x) = \sqrt{x+1} + \sqrt{x-1} = (x+1)^{1/2} + (x-1)^{1/2}$, so

$$f'(x) = \tfrac{1}{2}(x+1)^{-1/2}(1) + \tfrac{1}{2}(x-1)^{-1/2}(1) = \tfrac{1}{2}\left[(x+1)^{-1/2} + (x-1)^{-1/2}\right].$$

29. $f(x) = 2x^2(3-4x)^4$, so

$$f'(x) = 2x^2(4)(3-4x)^3(-4) + (3-4x)^4(4x) = 4x(3-4x)^3(-8x+3-4x)$$
$$= 4x(3-4x)^3(-12x+3) = (-12x)(4x-1)(3-4x)^3.$$

31. $f(x) = (x-1)^2(2x+1)^4$, so

$$f'(x) = (x-1)^2\tfrac{d}{dx}(2x+1)^4 + (2x+1)^4\tfrac{d}{dx}(x-1)^2 \quad \text{(by the Product Rule)}$$
$$= (x-1)^2(4)(2x+1)^3\tfrac{d}{dx}(2x+1) + (2x+1)^4(2)(x-1)\tfrac{d}{dx}(x-1)$$
$$= 8(x-1)^2(2x+1)^3 + 2(x-1)(2x+1)^4 = 2(x-1)(2x+1)^3(4x-4+2x+1)$$
$$= 6(x-1)(2x-1)(2x+1)^3.$$

33. $f(x) = \left(\frac{x+3}{x-2}\right)^3$, so

$$f'(x) = 3\left(\frac{x+3}{x-2}\right)^2 \frac{d}{dx}\left(\frac{x-3}{x-2}\right) = 3\left(\frac{x+3}{x-2}\right)^2 \left[\frac{(x-2)(1)-(x+3)(1)}{(x-2)^2}\right]$$

$$= 3\left(\frac{x+3}{x-2}\right)^2 \left[-\frac{5}{(x-2)^2}\right] = -\frac{15(x+3)^2}{(x-2)^4}.$$

35. $s(t) = \left(\frac{t}{2t+1}\right)^{3/2}$, so

$$s'(t) = \frac{3}{2}\left(\frac{t}{2t+1}\right)^{1/2} \frac{d}{dt}\left(\frac{t}{2t+1}\right) = \frac{3}{2}\left(\frac{t}{2t+1}\right)^{1/2} \left[\frac{(2t+1)(1)-t(2)}{(2t+1)^2}\right]$$

$$= \frac{3}{2}\left(\frac{t}{2t+1}\right)^{1/2} \left[\frac{1}{(2t+1)^2}\right] = \frac{3t^{1/2}}{2(2t+1)^{5/2}}.$$

37. $g(u) = \left(\frac{u+1}{3u+2}\right)^{1/2}$, so

$$g'(u) = \frac{1}{2}\left(\frac{u+1}{3u+2}\right)^{-1/2} \frac{d}{du}\left(\frac{u+1}{3u+2}\right) = \frac{1}{2}\left(\frac{u+1}{3u+2}\right)^{-1/2} \left[\frac{(3u+2)(1)-(u+1)(3)}{(3u+2)^2}\right]$$

$$= -\frac{1}{2\sqrt{u+1)}(3u+2)^{3/2}}.$$

39. $f(x) = \frac{x^2}{(x^2-1)^4}$, so

$$f'(x) = \frac{(x^2-1)^4 \frac{d}{dx}(x^2) - (x^2)\frac{d}{dx}(x^2-1)^4}{\left[(x^2-1)^4\right]^2} = \frac{(x^2-1)^4(2x) - x^2(4)(x^2-1)^3(2x)}{(x^2-1)^8}$$

$$= \frac{(x^2-1)^3(2x)(x^2-1-4x^2)}{(x^2-1)^8} = \frac{(-2x)(3x^2+1)}{(x^2-1)^5}.$$

41. $h(x) = \frac{(3x^2+1)^3}{(x^2-1)^4}$, so

$$h'(x) = \frac{(x^2-1)^4(3)(3x^2+1)^2(6x) - (3x^2+1)^3(4)(x^2-1)^3(2x)}{(x^2-1)^8}$$

$$= \frac{2x(x^2-1)^3(3x^2+1)^2[9(x^2-1)-4(3x^2+1)]}{(x^2-1)^8} = -\frac{2x(3x^2+13)(3x^2+1)^2}{(x^2-1)^5}.$$

43. $f(x) = \frac{\sqrt{2x+1}}{x^2-1}$, so

$$f'(x) = \frac{(x^2-1)\left(\frac{1}{2}\right)(2x+1)^{-1/2}(2) - (2x+1)^{1/2}(2x)}{(x^2-1)^2} = \frac{(2x+1)^{-1/2}[(x^2-1)-(2x+1)(2x)]}{(x^2-1)^2}$$

$$= -\frac{3x^2+2x+1}{\sqrt{2x+1}(x^2-1)^2}.$$

45. $g(t) = \dfrac{(t+1)^{1/2}}{(t^2+1)^{1/2}}$, so

$$g'(t) = \frac{(t^2+1)^{1/2}\dfrac{d}{dt}(t+1)^{1/2} - (t+1)^{1/2}\dfrac{d}{dt}(t^2+1)^{1/2}}{t^2+1}$$
$$= \frac{(t^2+1)^{1/2}\left(\frac{1}{2}\right)(t+1)^{-1/2}(1) - (t+1)^{1/2}\left(\frac{1}{2}\right)(t^2+1)^{-1/2}(2t)}{t^2+1}$$
$$= \frac{\frac{1}{2}(t+1)^{-1/2}(t^2+1)^{-1/2}\left[(t^2+1) - 2t(t+1)\right]}{t^2+1} = -\frac{t^2+2t-1}{2\sqrt{t+1}\,(t^2+1)^{3/2}}.$$

47. $f(x) = (3x+1)^4(x^2-x+1)^3$, so

$$f'(x) = (3x+1)^4\frac{d}{dx}(x^2-x+1)^3 + (x^2-x+1)^3\frac{d}{dx}(3x+1)^4$$
$$= (3x+1)^4\cdot 3(x^2-x+1)^2(2x-1) + (x^2-x+1)^3\cdot 4(3x+1)^3\cdot 3$$
$$= 3(3x+1)^3(x^2-x+1)^2\left[(3x+1)(2x-1) + 4(x^2-x+1)\right]$$
$$= 3(3x+1)^3(x^2-x+1)^2(6x^2-3x+2x-1+4x^2-4x+4)$$
$$= 3(3x+1)^3(x^2-x+1)^2(10x^2-5x+3).$$

49. $y = g(u) = u^{4/3}$, so $\frac{dy}{du} = \frac{4}{3}u^{1/3}$, and $u = f(x) = 3x^2-1$, so $\dfrac{du}{dx} = 6x$. Thus,

$$\frac{dy}{dx} = \frac{dy}{du}\cdot\frac{du}{dx} = \tfrac{4}{3}u^{1/3}(6x) = \tfrac{4}{3}(3x^2-1)^{1/3}6x = 8x(3x^2-1)^{1/3}.$$

51. $y = u^{-2/3}$ and $u = 2x^3-x+1$, so $\dfrac{dy}{du} = -\frac{2}{3}u^{-5/3} = -\dfrac{2}{3u^{5/3}}$ and $\dfrac{du}{dx} = 6x^2-1$. Thus,

$$\frac{dy}{dx} = \frac{dy}{du}\,\frac{du}{dx} = -\frac{2(6x^2-1)}{3u^{5/3}} = -\frac{2(6x^2-1)}{3(2x^3-x+1)^{5/3}}.$$

53. $y = \sqrt{u} + \dfrac{1}{\sqrt{u}}$ and $u = x^3-x$, so $\dfrac{dy}{du} = \frac{1}{2}u^{-1/2} - \frac{1}{2}u^{-3/2}$ and $\dfrac{du}{dx} = 3x^2-1$. Thus,

$$\frac{dy}{dx} = \frac{dy}{du}\cdot\frac{du}{dx} = \left[\frac{1}{2\sqrt{x^3-x}} - \frac{1}{2(x^3-x)^{3/2}}\right](3x^2-1) = \frac{(3x^2-1)(x^3-x-1)}{2(x^3-x)^{3/2}}.$$

55. $g(x) = f(2x+1)$. Let $u = 2x+1$, so $\dfrac{du}{dx} = 2$. Using the Chain Rule, we have

$$g'(x) = f'(u)\frac{du}{dx} = f'(2x+1)\cdot 2 = 2f'(2x+1).$$

57. $F(x) = g(f(x))$, so $F'(x) = g'(f(x))f'(x)$. Thus, $F'(2) = g'(3)(-3) = (4)(-3) = -12$.

59. Let $g(x) = x^2+1$. Then $F(x) = f(g(x))$. Next, $F'(x) = f'(g(x))g'(x)$ and $F'(1) = f'(2)(2x) = (3)(2) = 6$.

61. No. Suppose $h = g(f(x))$. Let $f(x) = x$ and $g(x) = x^2$. Then $h = g(f(x)) = g(x) = x^2$ and $h'(x) = 2x \neq g'(f'(x)) = g'(1) = 2(1) = 2$.

63. $f(x) = (1-x)(x^2-1)^2$, so
$f'(x) = (1-x)2(x^2-1)(2x) + (-1)(x^2-1)^2 = (x^2-1)(4x-4x^2-x^2+1) = (x^2-1)(-5x^2+4x+1)$.
Therefore, the slope of the tangent line at $(2,-9)$ is $f'(2) = \left[(2)^2-1\right]\left[-5(2)^2+4(2)+1\right] = -33$. Then an equation of the line is $y+9=-33(x-2)$, or $y=-33x+57$.

65. $f(x) = x\sqrt{2x^2+7}$, so $f'(x) = \sqrt{2x^2+7} + x\left(\frac{1}{2}\right)(2x^2+7)^{-1/2}(4x)$. The slope of the tangent line at $x=3$ is $f'(3) = \sqrt{25} + \left(\frac{3}{2}\right)(25)^{-1/2}(12) = \frac{43}{5}$, so an equation is $y-15 = \frac{43}{5}(x-3)$, or $y = \frac{43}{5}x - \frac{54}{5}$.

67. $N(t) = (60+2t)^{2/3}$, so $N'(t) = \frac{2}{3}(60+2t)^{-1/3}\frac{d}{dt}(60+2t) = \frac{4}{3}(60+2t)^{-1/3}$. The rate of increase at the end of the second week is $N'(2) = \frac{4}{3}(64)^{-1/3} = \frac{1}{3}$, or $\frac{1}{3}$ million/week. At the end of the 12th week, $N'(12) = \frac{4}{3}(84)^{-1/3} \approx 0.3$, or 0.3 million/week. The number of viewers in the 2nd week is $N(2) = (60+4)^{2/3} = 16$, or 16 million, and the number of viewers in the 24th week is $N(24) = (60+48)^{2/3} \approx 22.7$, or approximately 22.7 million.

69. $N(t) = -0.05(t+1.1)^{2.2} + 0.7t + 0.9$, so in 2008, the cumulative number of jobs that were outsourced was changing at the rate of $N'(3) = \left[-0.05(2.2)(t+1.1)^{1.2} + 0.7\right]_{t=3} = -0.05(2.2)(3+1.1)^{1.2} + 0.7 \approx 0.101956$, or approximately 102,000 per year.

71. a. $P(1) \approx 30.0$ and $P(2) \approx 10.3$. The probability of survival at the moment of diagnosis is 100%, the probability of survival 1 year after diagnosis is approximately 30%, and the probability after 2 years is approximately 10.3%.

b. $P'(t) = \frac{d}{dt}\left[100(1+0.14t)^{-9.2}\right] = 100(-9.2)(1+0.14t)^{-10.2}(0.14) = -\frac{920 \cdot 0.14}{(1+0.14t)^{10.2}} = -\frac{128.8}{(1+0.14t)^{10.2}}$.
Thus, $P'(1) \approx -33.84$ and $P'(2) \approx 10.38$. After 1 year, the probability of survival is dropping at the rate of approximately 34% per year, and after 2 years, it is dropping at approximately 10.4% per year.

73. $C(t) = 0.01\left(0.2t^2+4t+64\right)^{2/3}$.

a. $C'(t) = 0.01\left(\frac{2}{3}\right)\left(0.2t^2+4t+64\right)^{-1/3}\frac{d}{dt}\left(0.2t^2+4t+64\right)$
$= (0.01)(0.667)(0.4t+4)\left(0.2t^2+4t+64\right)^{-1/3} \approx 0.027(0.1t+1)\left(0.2t^2+4t+64\right)^{-1/3}$.

b. $C'(5) = 0.027(0.5+1)\left[0.2(25)+4(5)+64\right]^{-1/3} \approx 0.009$, or 0.009 parts per million per year.

75. a. $A(t) = 0.03t^3(t-7)^4 + 60.2$, so
$A'(t) = 0.03\left[3t^2(t-7)^4 + t^3(4)(t-7)^3\right] = 0.03t^2(t-7)^3\left[3(t-7)+4t\right] = 0.21t^2(t-3)(t-7)^3$.

b. $A'(1) = 0.21(-2)(-6)^3 = 90.72$, $A'(3) = 0$, and $A'(4) = 0.21(16)(1)(-3)^3 = -90.72$. The amount of pollutant is increasing at the rate of 90.72 units/hr at 8 a.m. The rate of change is 0 units/hr at 10 a.m. and -90.72 units/hr at 11 a.m.

77. $P(t) = \dfrac{300\sqrt{\frac{1}{2}t^2 + 2t + 25}}{t + 25} = \dfrac{300\left(\frac{1}{2}t^2 + 2t + 25\right)^{1/2}}{t + 25}$, so

$$P'(t) = 300\left[\frac{(t+25)\frac{1}{2}\left(\frac{1}{2}t^2 + 2t + 25\right)^{-1/2}(t+2) - \left(\frac{1}{2}t^2 + 2t + 25\right)^{1/2}(1)}{(t+25)^2}\right]$$

$$= 300\left[\frac{\left(\frac{1}{2}t^2 + 2t + 25\right)^{-1/2}\left[(t+25)(t+2) - 2\left(\frac{1}{2}t^2 + 2t + 25\right)\right]}{2(t+25)^2}\right] = \frac{3450t}{(t+25)^2\sqrt{\frac{1}{2}t^2 + 2t + 25}}.$$

Ten seconds into the run, the athlete's pulse rate is increasing at $P'(10) = \dfrac{3450(10)}{(35)^2\sqrt{50 + 20 + 25}} \approx 2.9$, or approximately 2.9 beats per minute per second. Sixty seconds into the run, it is increasing at $P'(60) = \dfrac{3450(60)}{(85)^2\sqrt{1800 + 120 + 25}} \approx 0.65$, or approximately 0.7 beats per minute per second. Two minutes into the run, it is increasing at $P'(120) = \dfrac{3450(120)}{(145)^2\sqrt{7200 + 240 + 25}} \approx 0.23$, or approximately 0.2 beats per minute per second. The pulse rate two minutes into the run is given by $P(120) = \dfrac{300\sqrt{7200 + 240 + 25}}{120 + 25} \approx 178.8$, or approximately 179 beats per minute.

79. The area is given by $A = \pi r^2$. The rate at which the area is increasing is given by dA/dt, that is, $\dfrac{dA}{dt} = \dfrac{d}{dt}(\pi r^2) = \dfrac{d}{dt}(\pi r^2)\dfrac{dr}{dt} = 2\pi r\dfrac{dr}{dt}$. If $r = 40$ and $dr/dt = 2$, then $\dfrac{dA}{dt} = 2\pi(40)(2) = 160\pi$, that is, it is increasing at the rate of 160π, or approximately 503 ft^2/sec.

81. $N(x) = 1.42x$ and $x(t) = \dfrac{7t^2 + 140t + 700}{3t^2 + 80t + 550}$. The number of construction jobs as a function of time is $n(t) = N(x(t))$. Using the Chain Rule,

$$n'(t) = \frac{dN}{dx}\cdot\frac{dx}{dt} = 1.42\frac{dx}{dt} = (1.42)\left[\frac{(3t^2 + 80t + 550)(14t + 140) - (7t^2 + 140t + 700)(6t + 80)}{(3t^2 + 80t + 550)^2}\right]$$

$$= \frac{1.42(140t^2 + 3500t + 21000)}{(3t^2 + 80t + 550)^2}.$$

$n'(12) = \dfrac{1.42\left[140(12)^2 + 3500(12) + 21000\right]}{\left[3(12)^2 + 80(12) + 550\right]^2} \approx 0.0313115$, or approximately 31,312 jobs/year/month.

83. $x = f(p) = \frac{100}{9}\sqrt{810{,}000 - p^2}$ and $p(t) = \dfrac{400}{1 + \frac{1}{8}\sqrt{t}} + 200$. We want to find

$\dfrac{dx}{dt} = \dfrac{dx}{dp} \cdot \dfrac{dp}{dt}$. But $\dfrac{dx}{dp} = \dfrac{100}{9}\left(\dfrac{1}{2}\right)(810{,}000 - p^2)^{-1/2}(-2p) = -\dfrac{100p}{9\sqrt{810{,}000 - p^2}}$ and

$\dfrac{dp}{dt} = 400\dfrac{d}{dt}\left(1 + \frac{1}{8}t^{1/2}\right)^{-1} + \dfrac{d}{dt}(200) = -400\left(1 + \frac{1}{8}t^{1/2}\right)^{-2}\left(\frac{1}{8}\right)\left(\frac{1}{2}t^{-1/2}\right) = -\dfrac{25}{\sqrt{t}\left(1 + \frac{1}{8}\sqrt{t}\right)^2}$,

so $\dfrac{dx}{dt} = \dfrac{2500p}{9\sqrt{t}\sqrt{810{,}000 - p^2}\left(1 + \frac{1}{8}\sqrt{t}\right)^2}$ and when $t = 16$, $p = \dfrac{400}{1 + \frac{1}{8}\sqrt{16}} + 200 = \frac{1400}{3}$. Therefore,

$\dfrac{dx}{dt} = \dfrac{2500\left(\frac{1400}{3}\right)}{9\sqrt{16}\sqrt{810{,}000 - \left(\frac{1400}{3}\right)^2}\left(1 + \frac{1}{8}\sqrt{16}\right)^2} \approx 18.7$. The quantity demanded will be changing at the rate of approximately 19 computers/month.

85. True. This is just the statement of the Chain Rule.

87. True. $\frac{d}{dx}\sqrt{f(x)} = \frac{d}{dx}[f(x)]^{1/2} = \frac{1}{2}[f(x)]^{-1/2} f'(x) = \dfrac{f'(x)}{2\sqrt{f(x)}}$.

89. Let $f(x) = x^{1/n}$ so that $[f(x)]^n = x$. Differentiating both sides with respect to x, we get $n[f(x)]^{n-1} f'(x) = 1$, so $f'(x) = \dfrac{1}{n[f(x)]^{n-1}} = \dfrac{1}{n\left[x^{1/n}\right]^{n-1}} = \dfrac{1}{nx^{1-(1/n)}} = \dfrac{1}{n}x^{(1/n)-1}$, as was to be shown.

Using Technology

page 198

1. 0.5774

3. 0.9390

5. −4.9498

7. a. Using the numerical derivative operation, we find that $N'(0) = 5.41450$, so the rate of change of the number of people watching TV on mobile phones at the beginning of 2007 is approximately 5.415 million/year.

b. $N'(4) \approx 2.5136$, so the corresponding rate of change at the beginning of 2011 is expected to be approximately 2.5136 million/year.

3.4 Marginal Functions in Economics

Problem-Solving Tips

1. **The marginal cost function is the derivative of the cost function**. Similarly, the marginal profit function and the marginal revenue function are the derivatives of the profit function and the revenue function, respectively. The key word here is "marginal": it indicates that we are dealing with the derivative of a function.

2. **The average cost function** is given by $\overline{C}(x) = \frac{C(x)}{x}$ and the **marginal average cost function** is given by $\overline{C}'(x)$.

3. Remember that the revenue is increasing on an interval where the demand is inelastic, decreasing on an interval where the demand is elastic, and stationary at the point where the demand is unitary.

Concept Questions

page 209

1. **a.** The marginal cost function is the derivative of the cost function.

   b. The average cost function is equal to the total cost function divided by the total number of the commodity produced.

   c. The marginal average cost function is the derivative of the average cost function.

   d. The marginal revenue function is the derivative of the revenue function.

   e. The marginal profit function is the derivative of the profit function.

3. $P(x) = R(x) - C(x)$, so $P'(x) = R'(x) - C'(x)$. Using the given information, we find $P'(500) = R'(500) - C'(500) = 3 - 2.8 = 0.2$. Thus, if the level of production is 500, then the marginal profit is \$0.20 per unit. This tells us that the proprietor should increase production in order to increase the company's profit.

Exercises

page 209

1. **a.** $C(x)$ is always increasing because as x, the number of units produced, increases, the amount of money that must be spent on production also increases.

   b. This occurs at $x = 4$, a production level of 4000. You can see this by looking at the slopes of the tangent lines for x less than, equal to, and a little larger then $x = 4$.

3. **a.** The actual cost incurred in the production of the 1001st disc is given by

$$C(1001) - C(1000) = \left[2000 + 2(1001) - 0.0001(1001)^2\right] - \left[2000 + 2(1000) - 0.0001(1000)^2\right]$$
$$= 3901.7999 - 3900 = 1.7999, \text{ or approximately } \$1.80.$$

The actual cost incurred in the production of the 2001st disc is given by

$$C(2001) - C(2000) = \left[2000 + 2(2001) - 0.0001(2001)^2\right] - \left[2000 + 2(2000) - 0.0001(2000)^2\right]$$
$$= 5601.5999 - 5600 = 1.5999, \text{ or approximately } \$1.60.$$

   b. The marginal cost is $C'(x) = 2 - 0.0002x$. In particular, $C'(1000) = 2 - 0.0002(1000) = 1.80$ and $C'(2000) = 2 - 0.0002(2000) = 1.60$.

5. **a.** $\overline{C}(x) = \dfrac{C(x)}{x} = \dfrac{100x + 200{,}000}{x} = 100 + \dfrac{200{,}000}{x}$.

   b. $\overline{C}'(x) = \dfrac{d}{dx}(100) + \dfrac{d}{dx}\left(200{,}000x^{-1}\right) = -200{,}000x^{-2} = -\dfrac{200{,}000}{x^2}$.

   c. $\lim_{x\to\infty} \overline{C}(x) = \lim_{x\to\infty}\left(100 + \dfrac{200{,}000}{x}\right) = 100$. This says that the average cost approaches \$100 per unit if the production level is very high.

7. $\overline{C}(x) = \dfrac{C(x)}{x} = \dfrac{2000 + 2x - 0.0001x^2}{x} = \dfrac{2000}{x} + 2 - 0.0001x$, so

$$\overline{C}'(x) = -\frac{2000}{x^2} + 0 - 0.0001 = -\frac{2000}{x^2} - 0.0001.$$

9. **a.** $R'(x) = \dfrac{d}{dx}\left(8000x - 100x^2\right) = 8000 - 200x$.

   b. $R'(39) = 8000 - 200(39) = 200$, $R'(40) = 8000 - 200(40) = 0$, and $R'(41) = 8000 - 200(41) = -200$.

   c. This suggests the total revenue is maximized if the price charged per passenger is \$40.

11. a. $P(x) = R(x) - C(x) = (-0.04x^2 + 800x) - (200x + 300{,}000) = -0.04x^2 + 600x - 300{,}000.$

b. $P'(x) = -0.08x + 600.$

c. $P'(5000) = -0.08(5000) + 600 = 200$ and $P'(8000) = -0.08(8000) + 600 = -40.$

d.

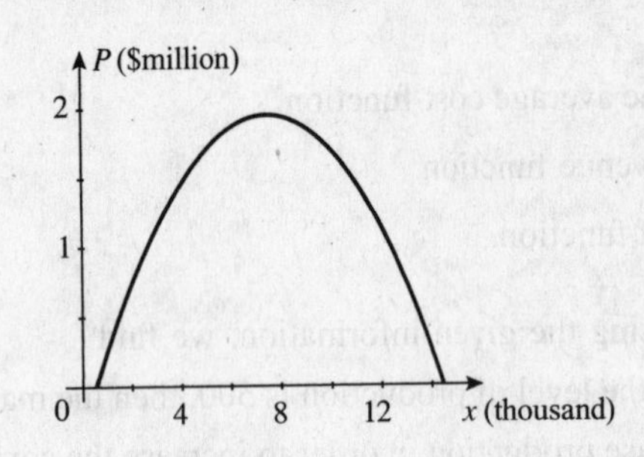

The profit realized by the company increases as production increases, peaking at a production level of 7500 units. Beyond this level of production, the profit begins to fall.

13. a. The revenue function is $R(x) = px = (600 - 0.05x)x = 600x - 0.05x^2$ and the profit function is

$$P(x) = R(x) - C(x) = (600x - 0.05x^2) - (0.000002x^3 - 0.03x^2 + 400x + 80{,}000)$$
$$= -0.000002x^3 - 0.02x^2 + 200x - 80{,}000.$$

b. $C'(x) = \dfrac{d}{dx}(0.000002x^3 - 0.03x^2 + 400x + 80{,}000) = 0.000006x^2 - 0.06x + 400,$

$R'(x) = \dfrac{d}{dx}(600x - 0.05x^2) = 600 - 0.1x,$ and

$P'(x) = \dfrac{d}{dx}(-0.000002x^3 - 0.02x^2 + 200x - 80{,}000) = -0.000006x^2 - 0.04x + 200.$

c. $C'(2000) = 0.000006(2000)^2 - 0.06(2000) + 400 = 304$, and this says that at a production level of 2000 units, the cost for producing the 2001st unit is \$304. $R'(2000) = 600 - 0.1(2000) = 400$, and this says that the revenue realized in selling the 2001st unit is \$400. $P'(2000) = R'(2000) - C'(2000) = 400 - 304 = 96$, and this says that the revenue realized in selling the 2001st unit is \$96.

d.

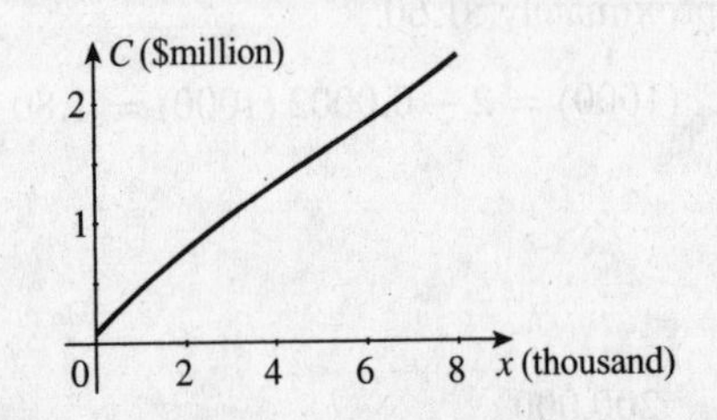

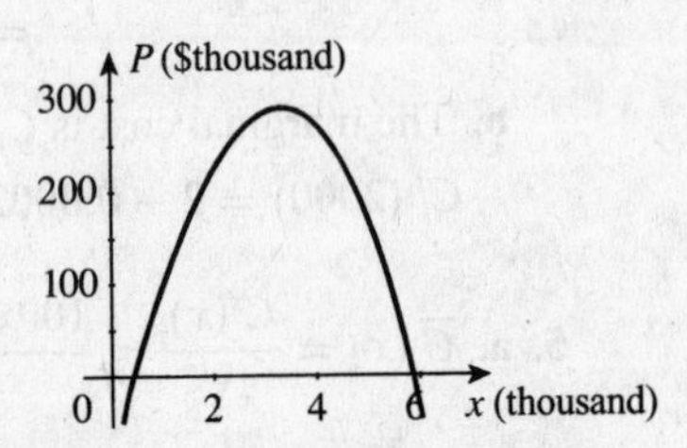

15. a. $\overline{C}(x) = \dfrac{C(x)}{x} = \dfrac{0.000002x^3 - 0.03x^2 + 400x + 80{,}000}{x} = 0.000002x^2 - 0.03x + 400 + \dfrac{80{,}000}{x}.$

b. $\overline{C}'(x) = 0.000004x - 0.03 - \dfrac{80{,}000}{x^2}.$

c. $\overline{C}'(5000) = 0.000004(5000) - 0.03 - \dfrac{80{,}000}{5000^2} \approx -0.0132$, and this says that at a production level of 5000 units, the average cost of production is dropping at the rate of approximately a penny per unit.

$\overline{C}'(10{,}000) = 0.000004(10000) - 0.03 - \dfrac{80{,}000}{10{,}000^2} \approx 0.0092,$

and this says that, at a production level of 10,000 units, the average cost of production is increasing at the rate of approximately a penny per unit.

d.

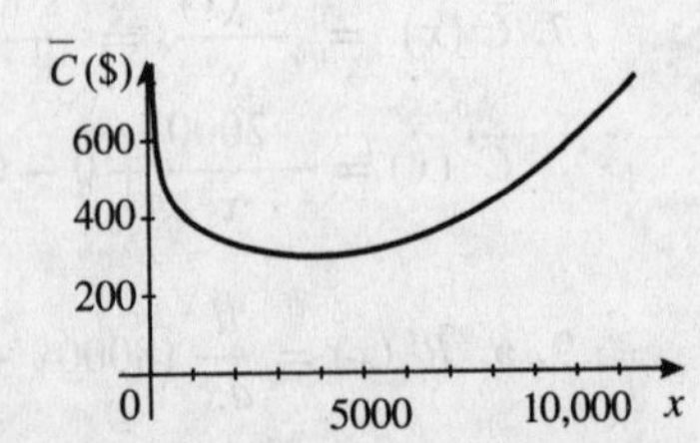

17. a. $R(x) = px = \dfrac{50x}{0.01x^2+1}$.

b. $R'(x) = \dfrac{(0.01x^2+1)\,50 - 50x\,(0.02x)}{(0.01x^2+1)^2} = \dfrac{50-0.5x^2}{(0.01x^2+1)^2}$.

c. $R'(2) = \dfrac{50-0.5\,(4)}{[0.01\,(4)+1]^2} \approx 44.379$. This result says that at a sales level of 2000 units, the revenue increases at the rate of approximately \$44,379 per 1000 units.

19. $C(x) = 0.873x^{1.1} + 20.34$, so $C'(x) = 0.873\,(1.1)\,x^{0.1}$. $C'(10) = 0.873\,(1.1)\,(10)^{0.1} = 1.21$.

21. The consumption function is given by $C(x) = 0.712x + 95.05$. The marginal propensity to consume is given by $\frac{dC}{dx} = 0.712$. The marginal propensity to save is given by $\frac{dS}{dx} = 1 - \frac{dC}{dx} = 1 - 0.712 = 0.288$.

23. $f(x) = 2x^2 + x + 1$, so $f'(x) = 4x + 1$. The percentage rate of change of f at $x = 2$ is $100\dfrac{f'(2)}{f(2)} = 100\left[\dfrac{4x+1}{2x^2+x+1}\right]_{x=2} = 100\left(\dfrac{8+1}{8+2+1}\right) \approx 81.82$ (percent per unit change in x).

25. $f(x) = \dfrac{x+1}{x^3+x+1}$, so $f'(x) = \dfrac{(x^3+x+1)(1) - (x+1)(3x^2+1)}{(x^3+x+1)^2}$. Then $f(2) = \frac{3}{11}$ and $f'(2) = -\frac{28}{121}$, so the percentage rate of change of f at $x = 2$ is $100 \cdot \dfrac{-\frac{28}{121}}{\frac{3}{11}} \approx -84.85$ (percent per unit change in x).

27. Here $x = f(p) = -\frac{5}{4}p + 20$ and so $f'(p) = -\frac{5}{4}$. Therefore, $E(p) = -\dfrac{pf'(p)}{f(p)} = -\dfrac{p\left(-\frac{5}{4}\right)}{-\frac{5}{4}p+20} = \dfrac{5p}{80-5p}$.

$E(10) = \dfrac{5\,(10)}{80-5\,(10)} = \dfrac{50}{30} = \dfrac{5}{3} > 1$, and so the demand is elastic.

29. $f(p) = -\frac{1}{3}p + 20$, so $f'(p) = -\frac{1}{3}$. Then the elasticity of demand is given by $E(p) = -\dfrac{p\left(-\frac{1}{3}\right)}{-\frac{1}{3}p+20}$, and $E(30) = -\dfrac{30\left(-\frac{1}{3}\right)}{-\frac{1}{3}\,(30)+20} = 1$, and we conclude that the demand is unitary at this price.

31. $x^2 = 169 - p$ and $f(p) = (169-p)^{1/2}$. Next, $f'(p) = \frac{1}{2}(169-p)^{-1/2}(-1) = -\frac{1}{2}(169-p)^{-1/2}$. Then the elasticity of demand is given by $E(p) = -\dfrac{pf'(p)}{f(p)} = -\dfrac{p\left(-\frac{1}{2}\right)(169-p)^{-1/2}}{(169-p)^{1/2}} = \dfrac{\frac{1}{2}p}{169-p}$. Therefore, when $p = 29$, $E(p) = \dfrac{\frac{1}{2}(29)}{169-29} = \dfrac{14.5}{140} \approx 0.104$. Because $E(p) < 1$, we conclude that demand is inelastic at this price.

33. a. The percentage rate of change in per capita income in year t is

$$100\frac{C'(t)}{C(t)} = 100 \cdot \frac{\frac{d}{dt}\left[\frac{I(t)}{P(t)}\right]}{C(t)} = 100 \cdot \frac{P(t)\,I'(t) - I(t)\,P'(t)}{[P(t)]^2} \cdot \frac{P(t)}{I(t)} = 100 \cdot \frac{P(t)\,I'(t) - I(t)\,P'(t)}{P(t)\,I(t)}.$$

b. Here $I(t) = 10^9(300 + 12t)$ and $P(t) = 2 \times 10^7 e^{0.02t}$, so $I'(t) = 12 \times 10^9$ and $P'(t) = 4 \cdot 10^5 e^{0.02t}$. Therefore, the percentage rate of change in per capita income in year t is

$$100 \cdot \frac{3 \times 10^7 e^{0.02t}\left(12 \times 10^9\right) - 10^9(300 + 12t)\left(4 \cdot 10^5 e^{0.02t}\right)}{\left(2 \cdot 10^7 e^{0.02t}\right)10^9(300 + 12t)} = \frac{2400 \times 10^{16} e^{0.02t} - 48 \times 10^{16} e^{0.02t}(25 + t)}{24 \cdot 10^{16} e^{0.02t}(25 + t)} = \frac{50 - 2t}{25 + t}.$$

c. The percentage rate of change in per capita income 2 years from now is projected to be $\frac{50 - 2(2)}{25 + 2} = \frac{46}{27}$, or approximately 1.7%/yr.

35. $f(p) = \frac{1}{5}(225 - p^2)$, so $f'(p) = \frac{1}{5}(-2p) = -\frac{2}{5}p$. Then the elasticity of demand is given by

$$E(p) = -\frac{pf'(p)}{f(p)} = -\frac{p\left(-\frac{2}{5}p\right)}{\frac{1}{5}(225 - p^2)} = \frac{2p^2}{225 - p^2}.$$

a. When $p = 8$, $E(8) = \frac{2(64)}{225 - 64} \approx 0.8 < 1$ and the demand is inelastic. When $p = 10$, $E(10) = \frac{2(100)}{225 - 100} = 1.6 > 1$ and the demand is elastic.

b. The demand is unitary when $E = 1$. Solving $\frac{2p^2}{225 - p^2} = 1$, we find $2p^2 = 225 - p^2$, $3p^2 = 225$, and $p \approx 8.66$. So the demand is unitary when $p \approx 8.66$.

c. Because demand is elastic when $p = 10$, lowering the unit price will cause the revenue to increase.

d. Because the demand is inelastic at $p = 8$, a slight increase in the unit price will cause the revenue to increase.

37. $f(p) = \frac{2}{3}(36 - p^2)^{1/2}$. $f'(p) = \frac{2}{3}\left(\frac{1}{2}\right)(36 - p^2)^{-1/2}(-2p) = -\frac{2}{3}p(36 - p^2)^{-1/2}$. Then the elasticity of demand is given by $E(p) = -\frac{pf'(p)}{f(p)} = -\frac{-\frac{2}{3}p(36 - p^2)^{-1/2}p}{\frac{2}{3}(36 - p^2)^{1/2}} = \frac{p^2}{36 - p^2}$.

a. When $p = 2$, $E(2) = \frac{4}{36 - 4} = \frac{1}{8} < 1$, and we conclude that the demand is inelastic.

b. Because the demand is inelastic, the revenue will increase when the rental price is increased.

39. We first solve the demand equation for x in terms of p. Thus, $p = \sqrt{9 - 0.02x}$, and $p^2 = 9 - 0.02x$, or $x = -50p^2 + 450$. With $f(p) = -50p^2 + 450$, we find $E(p) = -\frac{pf'(p)}{f(p)} = -\frac{p(-100p)}{-50p^2 + 450} = \frac{2p^2}{9 - p^2}$.

Setting $E(p) = 1$ gives $2p^2 = 9 - p^2$, so $p = \sqrt{3}$. So the demand is inelastic in $\left(0, \sqrt{3}\right)$, unitary when $p = \sqrt{3}$, and elastic in $\left(\sqrt{3}, 3\right)$.

41. True. $\overline{C}'(x) = \frac{d}{dx}\left[\frac{C(x)}{x}\right] = \frac{xC'(x) - C(x)\frac{d}{dx}(x)}{x^2} = \frac{xC'(x) - C(x)}{x^2}$.

3.5 Higher-Order Derivatives

Problem-Solving Tips

When you work applied problems, keep track of the units of measure used. For example, if velocity is measured in ft/sec, then the units of acceleration will be ft/sec^2. If you are working an applied problem and the units in your answer are not appropriate, it may indicate that you have made an error in your calculations or in the formulation of the problem.

Here are some tips for solving the problems in the exercises that follow:

1. Make sure to simplify expressions before finding higher order derivatives.
2. The velocity of an object moving in a straight path is given by the derivative of the position function for that object. The acceleration of the object is given by the derivative of the velocity function.

Concept Questions page 217

1. a. The second derivative of f is the derivative of f'.

b. To find the second derivative of f, we differentiate f'.

3. a. $f'(t) > 0$ and $f''(t) > 0$ in (a, b).

b. $f'(t) > 0$ and $f''(t) < 0$ in (a, b).

c. $f'(t) < 0$ and $f''(t) < 0$ in (a, b).

d. $f'(t) < 0$ and $f''(t) > 0$ in (a, b).

Exercises page 218

1. $f(x) = 4x^2 - 2x + 1$, so $f'(x) = 8x - 2$ and $f''(x) = 8$.

3. $f(x) = 2x^3 - 3x^2 + 1$, so $f'(x) = 6x^2 - 6x$ and $f''(x) = 12x - 6 = 6(2x - 1)$.

5. $h(t) = t^4 - 2t^3 + 6t^2 - 3t + 10$, so $h'(t) = 4t^3 - 6t^2 + 12t - 3$ and $h''(t) = 12t^2 - 12t + 12 = 12(t^2 - t + 1)$.

7. $f(x) = (x^2+2)^5$, so $f'(x) = 5(x^2+2)^4(2x) = 10x(x^2+2)^4$ and

$f''(x) = 10(x^2+2)^4 + 10x(4)(x^2+2)^3(2x) = 10(x^2+2)^3[(x^2+2) + 8x^2] = 10(9x^2+2)(x^2+2)^3$.

9. $g(t) = (2t^2-1)^2(3t^2)$, so

$g'(t) = 2(2t^2-1)(4t)(3t^2) - (2t^2-1)^2(6t) = 6t(2t^2-1)[4t^2 + (2t^2-1)] = 6t(2t^2-1)(6t^2-1)$

$= 6t(12t^4 - 8t^2 + 1) = 72t^5 - 48t^3 + 6t$

and $g''(t) = 360t^4 - 144t^2 + 6 = 6(60t^4 - 24t^2 + 1)$.

11. $f(x) = (2x^2+2)^{7/2}$, so $f'(x) = \frac{7}{2}(2x^2+2)^{5/2}(4x) = 14x(2x^2+2)^{5/2}$ and

$f''(x) = 14(2x^2+2)^{5/2} + 14x\left(\frac{5}{2}\right)(2x^2+2)^{3/2}(4x) = 14(2x^2+2)^{3/2}[(2x^2+2) + 10x^2]$

$= 28(6x^2+1)(2x^2+2)^{3/2}$.

13. $f(x) = x(x^2+1)^2$, so

$f'(x) = (x^2+1)^2 + x(2)(x^2+1)(2x) = (x^2+1)[(x^2+1) + 4x^2] = (x^2+1)(5x^2+1)$ and

$f''(x) = 2x(5x^2+1) + (x^2+1)(10x) = 2x(5x^2+1+5x^2+5) = 4x(5x^2+3)$.

15. $f(x) = \dfrac{x}{2x+1}$, so $f'(x) = \dfrac{(2x+1)(1) - x(2)}{(2x+1)^2} = \dfrac{1}{(2x+1)^2}$ and

$f''(x) = \dfrac{d}{dx}(2x+1)^{-2} = -2(2x+1)^{-3}(2) = -\dfrac{4}{(2x+1)^3}$.

17. $f(s) = \dfrac{s-1}{s+1}$, so $f'(s) = \dfrac{(s+1)(1) - (s-1)(1)}{(s+1)^2} = \dfrac{2}{(s+1)^2}$ and

$f''(s) = 2\dfrac{d}{ds}(s+1)^{-2} = -4(s+1)^{-3} = -\dfrac{4}{(s+1)^3}$.

19. $f(u) = \sqrt{4-3u} = (4-3u)^{1/2}$, so $f'(u) = \frac{1}{2}(4-3u)^{-1/2}(-3) = -\dfrac{3}{2\sqrt{4-3u}}$ and

$f''(u) = -\dfrac{3}{2} \cdot \dfrac{d}{du}(4-3u)^{-1/2} = -\dfrac{3}{2}\left(-\dfrac{1}{2}\right)(4-3u)^{-3/2}(-3) = -\dfrac{9}{4(4-3u)^{3/2}}$.

21. $f(x) = 3x^4 - 4x^3$, so $f'(x) = 12x^3 - 12x^2$, $f''(x) = 36x^2 - 24x$, and $f'''(x) = 72x - 24$.

23. $f(x) = \dfrac{1}{x}$, so $f'(x) = \dfrac{d}{dx}\left(x^{-1}\right) = -x^{-2} = -\dfrac{1}{x^2}$, $f''(x) = 2x^{-3} = \dfrac{2}{x^3}$, and $f'''(x) = -6x^{-4} = -\dfrac{6}{x^4}$.

25. $g(s) = (3s-2)^{1/2}$, so $g'(s) = \frac{1}{2}(3s-2)^{-1/2}(3) = \dfrac{3}{2(3s-2)^{1/2}}$,

$g''(s) = \frac{3}{2}\left(-\frac{1}{2}\right)(3s-2)^{-3/2}(3) = -\frac{9}{4}(3s-2)^{-3/2} = -\dfrac{9}{4(3s-2)^{3/2}}$, and

$g'''(s) = \frac{27}{8}(3s-2)^{-5/2}(3) = \frac{81}{8}(3s-2)^{-5/2} = \dfrac{81}{8(3s-2)^{5/2}}$.

27. $f(x) = (2x-3)^4$, so $f'(x) = 4(2x-3)^3(2) = 8(2x-3)^3$, $f''(x) = 24(2x-3)^2(2) = 48(2x-3)^2$, and $f'''(x) = 96(2x-3)(2) = 192(2x-3)$.

29. Its velocity at any time t is $v(t) = \frac{d}{dt}\left(16t^2\right) = 32t$. The hammer strikes the ground when $16t^2 = 256$ or $t = 4$ (we reject the negative root). Therefore, its velocity at the instant it strikes the ground is $v(4) = 32(4) = 128$ ft/sec. Its acceleration at time t is $a(t) = \frac{d}{dt}(32t) = 32$. In particular, its acceleration at $t = 4$ is 32 ft/sec^2.

31. $P(t) = 0.38t^2 + 1.3t + 3$.

a. The projected percentage is $P(5) = 0.38(5)^2 + 1.3(5) + 3 = 19$, or 19%.

b. $P'(t) = 0.76t + 1.3$, so the percentage of vehicles with transmissions that have 7 or more speeds is projected to be changing at the rate of $P'(5) = 0.76(5) + 1.3 = 5.1$, or 5.1% per year (in 2015).

c. $P''(15) = 0.76$, so the rate of increase in vehicles with such transmissions is itself increasing at the rate of 0.76% per year per year in 2025.

33. $N(t) = -0.1t^3 + 1.5t^2 + 100$.

a. $N'(t) = -0.3t^2 + 3t = 0.3t(10-t)$. Because $N'(t) > 0$ for $t = 0, 1, 2, \ldots, 8$, it is evident that $N(t)$ (and therefore the crime rate) was increasing from 2006 through 2014.

b. $N''(t) = -0.6t + 3 = 0.6(5-t)$. Now $N''(4) = 0.6 > 0$, $N''(5) = 0$, $N''(6) = -0.6 < 0$, $N''(7) = -1.2 < 0$, and $N''(8) = -1.8 < 0$. This shows that the rate of the rate of change was decreasing beyond $t = 5$ (in the year 2011). This indicates that the program was working.

35. $S(t) = 4t^3 + 2t^2 + 300t$, so $S(6) = 4(6)^3 + 2(6)^2 + 300(6) = 2736$. This says that 6 months after the grand opening of the store, monthly LP sales are projected to be 2736 units.

$S'(t) = 12t^2 + 4t + 300$, so $S'(6) = 12(6)^2 + 4(6) + 300 = 756$. Thus, monthly sales are projected to be increasing by 756 units per month.

$S''(t) = 24t + 4$, so $S''(6) = 24(6) + 4 = 148$. This says that the rate of increase of monthly sales is itself increasing at the rate of 148 units per month per month.

37. $P(t) = 0.0004t^3 + 0.0036t^2 + 0.8t + 12$, so $P'(t) = 0.00012t^2 + 0.0072t + 0.8$. Thus, $P'(t) \geq 0.8$ for $0 \leq t \leq 13$. $P''(t) = 0.00024t + 0.0072$, and for $0 \leq t \leq 13$, $P''(t) > 0$. This means that the proportion of the U.S. population that was obese was increasing at an increasing rate from 1991 through 2004.

39. $A(t) = -0.00006t^5 + 0.00468t^4 - 0.1316t^3 + 1.915t^2 - 17.63t + 100$, so $A'(t) = -0.0003t^4 + 0.01872t^3 - 0.3948t^2 + 3.83t - 17.63$ and $A''(t) = -0.0012t^3 + 0.05616t^2 - 0.7896t + 3.83$. Thus, $A'(10) = -3.09$ and $A''(10) = 0.35$. Our calcutations show that 10 minutes after the start of the test, the smoke remaining is decreasing at a rate of 3.09% per minute, but the rate at which the rate of smoke is decreasing is decreasing at the rate of 0.35 percent per minute per minute.

41. $f(t) = 10.72(0.9t + 10)^{0.3}$, so $f'(t) = 10.72(0.3)(0.9t + 10)^{-0.7}(0.9) = 2.8944(0.9t + 10)^{-0.7}$ and $f''(t) = 2.8944(-0.7)(0.9t + 10)^{-1.7}(0.9) = -1.823472(0.9t + 10)^{-1.7}$. Thus, $f''(10) = -1.823472(19)^{-1.7} \approx -0.01222$, which says that the rate of the rate of change of the population is decreasing at the rate of $0.01\%/\text{yr}^2$.

43. True. If $h = fg$ where f and g have derivatives of order 2, then $h''(x) = f''(x)g(x) + 2f'(x)g'(x) + f(x)g''(x)$.

45. True. Suppose $P(t)$ represents the population of bacteria at time t and suppose $P'(t) > 0$ and $P''(t) < 0$. Then the population is increasing at time t, but at a decreasing rate.

47. $\overline{C}(x) = \dfrac{C(x)}{x}$, so $\overline{C}'(x) = \dfrac{xC'(x) - C(x) \cdot 1}{x^2} = \dfrac{xC'(x) - C(x)}{x^2}$ and

$$\overline{C}''(x) = \frac{x^2\left[xC''(x) + C'(x) - C'(x)\right] - \left[xC'(x) - C(x)\right]2x}{x^4} = \frac{x^3C''(x) - 2x^2C'(x) + 2xC(x)}{x^4}$$

$$= \frac{C''(x)}{x} - \frac{2C'(x)}{x^2} + \frac{2C(x)}{x^3}.$$

49. Consider the function $f(x) = x^{(2n+1)/2} = x^{n+(1/2)}$. We calculate $f'(x) = \left(n + \frac{1}{2}\right)x^{n-(1/2)}$, $f''(x) = \left(n + \frac{1}{2}\right)\left(n - \frac{1}{2}\right)x^{n-(3/2)}$, ..., $f^{(n)}(x) = \left(n + \frac{1}{2}\right)\left(n - \frac{1}{2}\right)\cdots\frac{3}{2}x^{1/2}$, and $f^{(n+1)}(x) = \left(n + \frac{1}{2}\right)\left(n - \frac{1}{2}\right)\cdots\frac{1}{2}x^{-1/2}$. The first n derivatives exist at $x = 0$, but the $(n+1)$st derivative fails to be defined there.

Using Technology page 221

1. -18 **3.** 15.2762 **5.** -0.6255 **7.** 0.1973

3.6 Implicit Differentiation and Related Rates

Problem-Solving Tips

1. If an equation expresses y implicitly as a function of x, then we can use implicit differentiation to find its derivative. We apply the chain rule to find the derivative of any term involving y. (Note that the derivative of any term involving y will include the factor dy/dx.) The terms involving only x are differentiated in the usual manner.

2. Guideline 3 for solving related rates problems, on page 221 of the text, asks you to find an equation giving the relationship between the variables in the related rates problem.. Make sure that you differentiate this equation implicitly with respect to t before substituting the values of the variables (Step 5).

Concept Questions page 230

1. a. We differentiate both sides of $F(x, y) = 0$ with respect to x, then solve for dy/dx.

b. The Chain Rule is used to differentiate any expression involving the dependent variable y.

3. Suppose x and y are two variables that are related by an equation. Furthermore, suppose x and y are both functions of a third variable t. (Normally t represents time.) Then a related rates problem involves finding dx/dt or dy/dt.

Exercises page 231

1. a. Solving for y in terms of x, we have $y = -\frac{1}{2}x + \frac{5}{2}$. Therefore, $y' = -\frac{1}{2}$.

b. Next, differentiating $x + 2y = 5$ implicitly, we have $1 + 2y' = 0$, or $y' = -\frac{1}{2}$.

3. a. $xy = 1$, $y = \dfrac{1}{x}$, and $\dfrac{dy}{dx} = -\dfrac{1}{x^2}$.

b. $x\dfrac{dy}{dx} + y = 0$, so $x\dfrac{dy}{dx} = -y$ and $\dfrac{dy}{dx} = -\dfrac{y}{x} = -\dfrac{-1/x}{x} = -\dfrac{1}{x^2}$.

5. $x^3 - x^2 - xy = 4$.

a. $-xy = 4 - x^3 + x^2$, so $y = -\dfrac{4}{x} + x^2 - x$ and $\dfrac{dy}{dx} = \dfrac{4}{x^2} + 2x - 1$.

b. $x^3 - x^2 - xy = 4$, so $-x\dfrac{dy}{dx} = -3x^2 + 2x + y$, and therefore

$$\frac{dy}{dx} = 3x - 2 - \frac{y}{x} = 3x - 2 - \frac{1}{x}\left(-\frac{4}{x} + x^2 - x\right) = 3x - 2 + \frac{4}{x^2} - x + 1 = \frac{4}{x^2} + 2x - 1.$$

7. a. $\dfrac{x}{y} - x^2 = 1$ is equivalent to $\dfrac{x}{y} = x^2 + 1$, or $y = \dfrac{x}{x^2+1}$. Therefore, $y' = \dfrac{(x^2+1) - x(2x)}{(x^2+1)^2} = \dfrac{1-x^2}{(x^2+1)^2}$.

b. Next, differentiating the equation $x - x^2y = y$ implicitly, we obtain $1 - 2xy - x^2y' = y'$, $y'(1+x^2) = 1 - 2xy$, and thus $y' = \dfrac{1-2xy}{(1+x^2)}$. This may also be written in the form $-2y^2 + \dfrac{y}{x}$. To show that this is equivalent to the results obtained earlier, use the earlier value of y to get $y' = \dfrac{1 - 2x\left(\dfrac{x}{x^2+1}\right)}{1+x^2} = \dfrac{x^2+1-2x^2}{(1+x^2)^2} = \dfrac{1-x^2}{(1+x^2)^2}$.

9. $x^2 + y^2 = 16$. Differentiating both sides of the equation implicitly, we obtain $2x + 2yy' = 0$, and so $y' = -x/y$.

11. $x^2 - 2y^2 = 16$. Differentiating implicitly with respect to x, we have $2x - 4y\dfrac{dy}{dx} = 0$, and so $\dfrac{dy}{dx} = \dfrac{x}{2y}$.

13. $x^2 - 2xy = 6$. Differentiating both sides of the equation implicitly, we obtain $2x - 2y - 2xy' = 0$ and so $y' = \dfrac{x-y}{x} = 1 - \dfrac{y}{x}$.

15. $x^2y^2 - xy = 8$. Differentiating both sides of the equation implicitly, we obtain $2xy^2 + 2x^2yy' - y - xy' = 0$, $2xy^2 - y + y'\left(2x^2y - x\right) = 0$, and so $y' = \dfrac{y\,(1-2xy)}{x\,(2xy-1)} = -\dfrac{y}{x}$.

17. $x^{1/2} + y^{1/2} = 1$. Differentiating implicitly with respect to x, we have $\frac{1}{2}x^{-1/2} + \frac{1}{2}y^{-1/2}\dfrac{dy}{dx} = 0$. Therefore, $\dfrac{dy}{dx} = -\dfrac{x^{-1/2}}{y^{-1/2}} = -\dfrac{\sqrt{y}}{\sqrt{x}}$.

19. $\sqrt{x+y} = x$. Differentiating both sides of the equation implicitly, we obtain $\frac{1}{2}\,(x+y)^{-1/2}\left(1+y'\right) = 1$, $1 + y' = 2\,(x+y)^{1/2}$, and so $y' = 2\sqrt{x+y} - 1$.

21. $\dfrac{1}{x^2} + \dfrac{1}{y^2} = 1$. Differentiating both sides of the equation implicitly, we obtain $-\dfrac{2}{x^3} - \dfrac{2}{y^3}y' = 0$, or $y' = -\dfrac{y^3}{x^3}$.

23. $\sqrt{xy} = x + y$. Differentiating both sides of the equation implicitly, we obtain $\frac{1}{2}\,(xy)^{-1/2}\left(xy' + y\right) = 1 + y'$, so $xy' + y = 2\sqrt{xy}\left(1+y'\right)$, $y'\left(x - 2\sqrt{xy}\right) = 2\sqrt{xy} - y$, and so $y' = -\dfrac{2\sqrt{xy} - y}{2\sqrt{xy} - x} = \dfrac{2\sqrt{xy} - y}{x - 2\sqrt{xy}}$.

25. $\dfrac{x+y}{x-y} = 3x$, or $x + y = 3x^2 - 3xy$. Differentiating both sides of the equation implicitly, we obtain $1 + y' = 6x - 3xy' - 3y$, so $y' + 3xy' = 6x - 3y - 1$ and $y' = \dfrac{6x - 3y - 1}{3x + 1}$.

27. $xy^{3/2} = x^2 + y^2$. Differentiating implicitly with respect to x, we obtain $y^{3/2} + x\left(\frac{3}{2}\right)y^{1/2}\dfrac{dy}{dx} = 2x + 2y\dfrac{dy}{dx}$. Multiply both sides by 2 to get $2y^{3/2} + 3xy^{1/2}\dfrac{dy}{dx} = 4x + 4y\dfrac{dy}{dx}$. Then $\left(3xy^{1/2} - 4y\right)\dfrac{dy}{dx} = 4x - 2y^{3/2}$, so $\dfrac{dy}{dx} = \dfrac{2\left(2x - y^{3/2}\right)}{3xy^{1/2} - 4y}$.

29. $(x+y)^3 + x^3 + y^3 = 0$. Differentiating implicitly with respect to x, we obtain $3\,(x+y)^2\left(1 + \dfrac{dy}{dx}\right) + 3x^2 + 3y^2\dfrac{dy}{dx} = 0$, $(x+y)^2 + (x+y)^2\dfrac{dy}{dx} + x^2 + y^2\dfrac{dy}{dx} = 0$, $\left[(x+y)^2 + y^2\right]\dfrac{dy}{dx} = -\left[(x+y)^2 + x^2\right]$, and thus $\dfrac{dy}{dx} = -\dfrac{2x^2 + 2xy + y^2}{x^2 + 2xy + 2y^2}$.

31. $4x^2 + 9y^2 = 36$. Differentiating the equation implicitly, we obtain $8x + 18yy' = 0$. At the point $(0, 2)$, we have $0 + 36y' = 0$, and the slope of the tangent line is 0. Therefore, an equation of the tangent line is $y = 2$.

33. $x^2y^3 - y^2 + xy - 1 = 0$. Differentiating implicitly with respect to x, we have $2xy^3 + 3x^2y^2\frac{dy}{dx} - 2y\frac{dy}{dx} + y + x\frac{dy}{dx} = 0$. At $(1, 1)$, $2 + 3\frac{dy}{dx} - 2\frac{dy}{dx} + 1 + \frac{dy}{dx} = 0$, and so $2\frac{dy}{dx} = -3$ and $\frac{dy}{dx} = -\frac{3}{2}$. Using the point-slope form of an equation of a line, we have $y - 1 = -\frac{3}{2}(x - 1)$, and the equation of the tangent line to the graph of the function f at $(1, 1)$ is $y = -\frac{3}{2}x + \frac{5}{2}$.

35. $xy = 1$. Differentiating implicitly, we have $xy' + y = 0$, or $y' = -\frac{y}{x}$. Differentiating implicitly once again, we have $xy'' + y' + y' = 0$. Therefore, $y'' = -\frac{2y'}{x} = \frac{2\left(\frac{y}{x}\right)}{x} = \frac{2y}{x^2}$.

37. $y^2 - xy = 8$. Differentiating implicitly we have $2yy' - y - xy' = 0$, and so $y' = \frac{y}{2y - x}$. Differentiating implicitly again, we have $2(y')^2 + 2yy'' - y' - y' - xy'' = 0$, so $y'' = \frac{2y' - 2(y')^2}{2y - x} = \frac{2y'(1 - y')}{2y - x}$. Then

$$y'' = \frac{2\left(\frac{y}{2y - x}\right)\left(1 - \frac{y}{2y - x}\right)}{2y - x} = \frac{2y(2y - x - y)}{(2y - x)^3} = \frac{2y(y - x)}{(2y - x)^3}.$$

39. a. Differentiating the given equation with respect to t, we obtain $\frac{dV}{dt} = \pi r^2\frac{dh}{dt} + 2\pi rh\frac{dr}{dt} = \pi r\left(r\frac{dh}{dt} + 2h\frac{dr}{dt}\right)$.

b. Substituting $r = 2$, $h = 6$, $\frac{dr}{dt} = 0.1$, and $\frac{dh}{dt} = 0.3$ into the expression for $\frac{dV}{dt}$, we obtain $\frac{dV}{dt} = \pi(2)[2(0.3) + 2(6)(0.1)] = 3.6\pi$, and so the volume is increasing at the rate of 3.6π in^3/sec.

41. We are given $\frac{dp}{dt} = 2$ and wish to find $\frac{dx}{dt}$ when $x = 9$ and $p = 63$. Differentiating the equation $p + x^2 = 144$ with respect to t, we obtain $\frac{dp}{dt} + 2x\frac{dx}{dt} = 0$. When $x = 9$, $p = 63$, and $\frac{dp}{dt} = 2$, we have $2 + 2(9)\frac{dx}{dt} = 0$, and so and $\frac{dx}{dt} = -\frac{1}{9} \approx -0.111$. Thus, the quantity demanded is decreasing at the rate of approximately 111 tires per week.

43. $100x^2 + 9p^2 = 3600$. Differentiating the given equation implicitly with respect to t, we have $200x\frac{dx}{dt} + 18p\frac{dp}{dt} = 0$. Next, when $p = 14$, the given equation yields $100x^2 + 9(14)^2 = 3600$, so $100x^2 = 1836$, or $x \approx 4.2849$. When $p = 14$, $\frac{dp}{dt} = -0.15$, and $x \approx 4.2849$, we have $200(4.2849)\frac{dx}{dt} + 18(14)(-0.15) = 0$, and so $\frac{dx}{dt} \approx 0.0441$. Thus, the quantity demanded is increasing at the rate of approximately 44 headphones per week.

45. Differentiating $625p^2 - x^2 = 100$ implicitly, we have $1250p\frac{dp}{dt} - 2x\frac{dx}{dt} = 0$. When $p = 1.0770$, $x = 25$, and $\frac{dx}{dt} = -1$, we find that $1250(1.077)\frac{dp}{dt} - 2(25)(-1) = 0$, and so $\frac{dp}{dt} = -\frac{50}{1250(1.077)} = -0.037$. We conclude that the price is decreasing at the rate of 3.7 cents per carton.

47. $p = -0.01x^2 - 0.2x + 8$. Differentiating the given equation implicitly with respect to p, we have $1 = -0.02x\frac{dx}{dp} - 0.2\frac{dx}{dp} = -[0.02x + 0.2]\frac{dx}{dp}$, so $\frac{dx}{dp} = -\frac{1}{0.02x + 0.2}$. When $x = 15$, $p = -0.01(15)^2 - 0.2(15) + 8 = 2.75$, and so and $\frac{dx}{dp} = -\frac{1}{0.02(15) + 0.2} = -2$. Therefore,

$$E(p) = -\frac{pf'(p)}{f(p)} = -\frac{(2.75)(-2)}{15} \approx 0.37 < 1,$$ and the demand is inelastic.

49. a. The required output is $Q(32, 243) = 20\left(32^{3/5}\right)\left(243^{2/5}\right) = 1440$, or \$1440 billion.

b. Differentiating $20x^{3/5}y^{2/5} = 1440$ implicitly with respect to x, we have $20\left(\frac{3}{5}x^{-2/5}y^{2/5}\right) + 20\left(x^{3/5}\right)\left(\frac{2}{5}y^{-3/5}\frac{dy}{dx}\right) = 0$, so $\frac{dy}{dx} = -\frac{3}{5}x^{-2/5}y^{2/5}\left(\frac{5}{2}x^{-3/5}y^{3/5}\right) = -\frac{3y}{2x}$. If $x = 32$ and $y = 243$, then $\frac{dy}{dx} = -\frac{3 \cdot 243}{2 \cdot 32} \approx -11.39$, so the amount spent on capital should decrease by approximately \$11.4 billion. The MRTS is \$11.4 billion per billion dollars.

51. $A = \pi r^2$. Differentiating with respect to t, we obtain $\frac{dA}{dt} = 2\pi r \frac{dr}{dt}$. When the radius of the circle is 60 ft and increasing at the rate of $\frac{1}{2}$ ft/sec, $\frac{dA}{dt} = 2\pi(60)\left(\frac{1}{2}\right) = 60\pi$ ft^2/sec. Thus, the area is increasing at a rate of approximately 188.5 ft^2/sec.

53. $A = \pi r^2$, so $r = \left(\frac{A}{\pi}\right)^{1/2}$. Differentiating with respect to t, we obtain $\frac{dr}{dt} = \frac{1}{2}\left(\frac{A}{\pi}\right)^{-1/2}\frac{dA}{dt}$. When the area of the spill is 1600π ft^2 and increasing at the rate of 80π ft^2/sec, $\frac{dr}{dt} = \frac{1}{2}\left(\frac{1600\pi}{\pi}\right)^{-1/2}(80\pi) = \pi$ ft/sec. Thus, the radius is increasing at the rate of approximately 3.14 ft/sec.

55. Let $(x, 0)$ and $(0, y)$ denote the position of the two cars at time t. Then $y = t^2 + 2t$. Now $D^2 = x^2 + y^2$ so $2D\frac{dD}{dt} = 2x\frac{dx}{dt} + 2y\frac{dy}{dt}$ and thus $D\frac{dD}{dt} = x\frac{dx}{dt} + \left(t^2 + 2t\right)(2t + 2)$. When $t = 4$, we have $x = -20$, $\frac{dx}{dt} = -9$, and $y = 24$, so $\sqrt{(-20)^2 + (24)^2}\frac{dD}{dt} = (-20)(-9) + (24)(10)$, and therefore $\frac{dD}{dt} = \frac{420}{\sqrt{976}} \approx 13.44$. That is, the distance is changing at approximately 13.44 ft/sec.

57. Referring to the diagram, we see that $D^2 = 120^2 + x^2$. Differentiating this last equation with respect to t, we have $2D\frac{dD}{dt} = 2x\frac{dx}{dt}$, and so $\frac{dD}{dt} = \frac{x\frac{dx}{dt}}{D}$. When $x = 50$ and $\frac{dx}{dt} = 20$, $D = \sqrt{120^2 + 50^2} = 130$ and $\frac{dD}{dt} = \frac{(20)(50)}{130} \approx 7.69$, or 7.69 ft/sec.

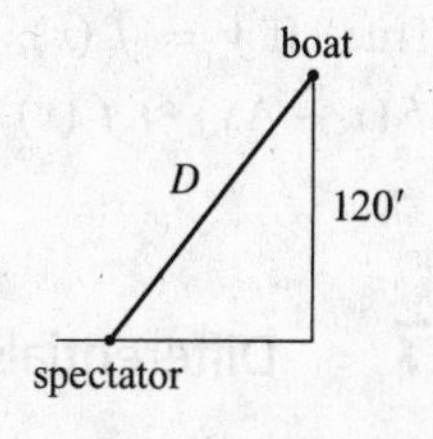

59. Let V and S denote its volume and surface area. Then we are given that $\frac{dV}{dt} = kS$, where k is the constant of proportionality. But from $V = \frac{4}{3}\pi r^3$, we find, upon differentiating both sides with respect to t, that $\frac{dV}{dt} = \frac{d}{dt}\left(\frac{4}{3}\pi r^3\right) = 4\pi r^2 \frac{dr}{dt} = kS = k\left(4\pi r^2\right)$. Therefore, $\frac{dr}{dt} = k$ a constant.

61. We are given that $\frac{dx}{dt} = 264$. Using the Pythagorean Theorem, $s^2 = x^2 + 1000^2 = x^2 + 1{,}000{,}000$. We want to find $\frac{ds}{dt}$ when $s = 1500$. Differentiating both sides of the equation with respect to t, we have $2s\frac{ds}{dt} = 2x\frac{dx}{dt}$ and so $\frac{ds}{dt} = \frac{x\frac{dx}{dt}}{s}$. When $s = 1500$, we have $1500^2 = x^2 + 1{,}000{,}000$, or $x = \sqrt{1{,}250{,}000}$. Therefore, $\frac{ds}{dt} = \frac{\sqrt{1{,}250{,}000} \cdot (264)}{1500} \approx 196.8$, that is, the aircraft is receding from the trawler at the speed of approximately 196.8 ft/sec.

x
1000′
s
trawler

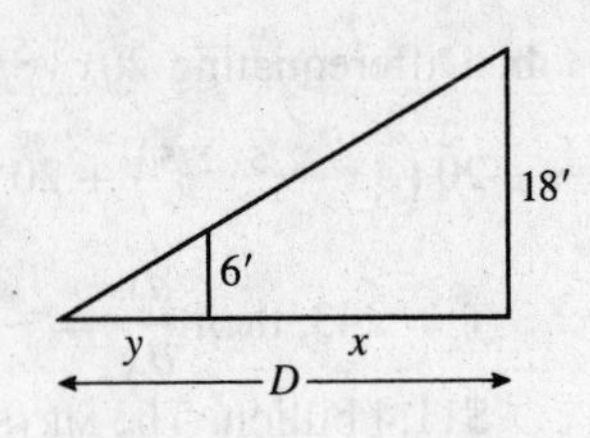

63. $\dfrac{y}{6} = \dfrac{y+x}{18}$, $18y = 6(y+x)$, so $3y = y + x$, $2y = x$, and $y = \frac{1}{2}x$.

Thus, $D = y + x = \frac{3}{2}x$. Differentiating implicitly, we have

$\frac{dD}{dt} = \frac{3}{2} \cdot \frac{dx}{dt}$, and when $\frac{dx}{dt} = 6$, $\frac{dD}{dt} = \frac{3}{2}(6) = 9$, or 9 ft/sec.

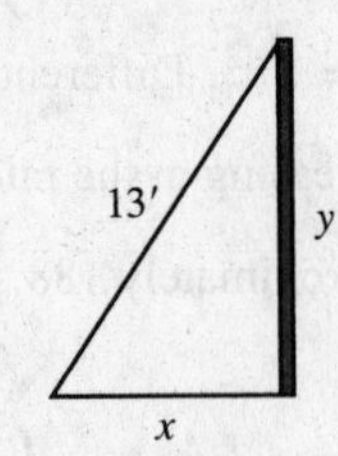

65. Differentiating $x^2 + y^2 = 13^2 = 169$ with respect to t gives

$2x\frac{dx}{dt} + 2y\frac{dy}{dt} = 0$. When $x = 12$, we have $144 + y^2 = 169$, or $y = 5$.

Therefore, with $x = 12$, $y = 5$, and $\frac{dx}{dt} = 8$, we find

$2(12)(8) + 2(5)\frac{dy}{dt} = 0$, or $\frac{dy}{dt} = -19.2$. Thus, the top of the ladder is

sliding down the wall at the rate of 19.2 ft/sec.

67. $P^5V^7 = C$, so $V^7 = CP^{-5}$ and $7V^6\dfrac{dV}{dt} = -5CP^{-6}\dfrac{dP}{dt}$. Therefore,

$\dfrac{dV}{dt} = -\dfrac{5C}{7P^6V^6}\dfrac{dP}{dt} = -\dfrac{5P^5V^7}{7P^6V^6}\dfrac{dP}{dt} = -\dfrac{5}{7}\dfrac{V}{P}\dfrac{dP}{dt}$. When $V = 4$ L, $P = 100$ kPa, and $\dfrac{dP}{dt} = -5\,\dfrac{\text{kPa}}{\text{sec}}$, we have

$\dfrac{dV}{dt} = -\dfrac{5}{7} \cdot \dfrac{4}{100}(-5) = \dfrac{1}{7}\left(\dfrac{\text{L}}{\text{kPa}} \cdot \dfrac{\text{kPa}}{\text{s}}\right) = \dfrac{1}{7}\dfrac{\text{L}}{\text{s}}$.

69. False. There are no real numbers x and y such that $x^2 + y^2 = -1$.

71. True. Differentiating both sides of the equation with respect to x, we have $\frac{d}{dx}[f(x)g(y)] = \frac{d}{dx}(0)$, so

$f(x)g'(y)\dfrac{dy}{dx} + f'(x)g(y) = 0$, and therefore $\frac{dy}{dx} = -\frac{f'(x)g(y)}{f(x)g'(y)}$, provided $f(x) \neq 0$ and $g'(y) \neq 0$.

73. True. If $y = f(x)$, then $\Delta y = f(x + \Delta x) - f(x) \approx f'(x)\Delta x$, from which it follows that

$f(x + \Delta x) \approx f(x) + f'(x)\Delta x$.

3.7 Differentials

Concept Questions page 240

1. The differential of x is dx. The differential of y is $dy = f'(x)\,dx$.

3. Because $\Delta P = P(t_0 + \Delta t) - P(t_0) \approx P'(t_0)\Delta t$, we see that $P'(t_0)\Delta t$ is an approximation of the change in the population from time t_0 to time $t_0 + \Delta t$.

Exercises page 240

1. $f(x) = 2x^2$ and $dy = 4x\,dx$.

3. $f(x) = x^3 - x$ and $dy = (3x^2 - 1)\,dx$.

5. $f(x) = \sqrt{x+1} = (x+1)^{1/2}$ and $dy = \frac{1}{2}(x+1)^{-1/2}\,dx = \dfrac{dx}{2\sqrt{x+1}}$.

7. $f(x) = 2x^{3/2} + x^{1/2}$ and $dy = \left(3x^{1/2} + \frac{1}{2}x^{-1/2}\right)dx = \frac{1}{2}x^{-1/2}(6x+1)\,dx = \dfrac{6x+1}{2\sqrt{x}}\,dx$.

9. $f(x) = x + \frac{2}{x}$ and $dy = \left(1 - \frac{2}{x^2}\right) dx = \frac{x^2 - 2}{x^2} dx$.

11. $f(x) = \frac{x-1}{x^2+1}$ and $dy = \frac{x^2 + 1 - (x-1)\,2x}{(x^2+1)^2}\,dx = \frac{-x^2 + 2x + 1}{(x^2+1)^2}\,dx$.

13. $f(x) = \sqrt{3x^2 - x} = (3x^2 - x)^{1/2}$ and $dy = \frac{1}{2}(3x^2 - x)^{-1/2}(6x - 1)\,dx = \frac{6x - 1}{2\sqrt{3x^2 - x}}\,dx$.

15. $f(x) = x^2 - 1$.

a. $dy = 2x\,dx$.

b. $dy \approx 2(1)(0.02) = 0.04$.

c. $\Delta y = \left[(1.02)^2 - 1\right] - (1 - 1) = 0.0404$.

17. $f(x) = \frac{1}{x}$.

a. $dy = -\frac{dx}{x^2}$.

b. $dy \approx -0.05$.

c. $\Delta y = \frac{1}{-0.95} - \frac{1}{-1} \approx -0.05263$.

19. $y = \sqrt{x}$ and $dy = \frac{dx}{2\sqrt{x}}$. Therefore, $\sqrt{10} \approx 3 + \frac{1}{2 \cdot \sqrt{9}} \approx 3.167$.

21. $y = \sqrt{x}$ and $dy = \frac{dx}{2\sqrt{x}}$. Therefore, $\sqrt{49.5} \approx 7 + \frac{0.5}{2 \cdot 7} \approx 7.0357$.

23. $y = x^{1/3}$ and $dy = \frac{1}{3}x^{-2/3}\,dx$. Therefore, $\sqrt[3]{7.8} \approx 2 - \frac{0.2}{3 \cdot 4} \approx 1.983$.

25. $y = \sqrt{x}$ and $dy = \frac{dx}{2\sqrt{x}}$. Therefore, $\sqrt{0.089} = \frac{1}{10}\sqrt{8.9} \approx \frac{1}{10}\left(3 - \frac{0.1}{2 \cdot 3}\right) \approx 0.298$.

27. $y = f(x) = \sqrt{x} + \frac{1}{\sqrt{x}} = x^{1/2} + x^{-1/2}$. Therefore, $\frac{dy}{dx} = \frac{1}{2}x^{-1/2} - \frac{1}{2}x^{-3/2}$, so $dy = \left(\frac{1}{2x^{1/2}} - \frac{1}{2x^{3/2}}\right) dx$.

Letting $x = 4$ and $dx = 0.02$, we find $\sqrt{4.02} + \frac{1}{\sqrt{4.02}} - f(4) = f(4.02) - f(4) = \Delta y \approx dy$, so

$\sqrt{4.02} + \frac{1}{\sqrt{4.02}} \approx f(4) + dy \approx 2 + \frac{1}{2} + \left(\frac{1}{2 \cdot 2} - \frac{1}{16}\right)(0.02) = 2.50375$.

29. The volume of the cube is given by $V = x^3$. Then $dV = 3x^2\,dx$, and when $x = 12$ and $dx = 0.02$, $dV = 3(144)(\pm 0.02) = \pm 8.64$. The possible error that might occur in calculating the volume is $\pm 8.64\ \text{cm}^3$.

31. The volume of the hemisphere is given by $V = \frac{2}{3}\pi r^3$. The amount of rust-proofer needed is

$\Delta V = \frac{2}{3}\pi(r + \Delta r)^3 - \frac{2}{3}\pi r^3 \approx dV = \frac{2}{3}(3\pi r^2)\,dr$. Thus, with $r = 60$ and $dr = \frac{1}{12}(0.01)$, we have

$\Delta V \approx 2\pi(60^2)\left(\frac{1}{12}\right)(0.01) \approx 18.85$. So we need approximately 18.85 ft^3 of rust-proofer.

33. $dR = \frac{d}{dr}(k\ell r^{-4})\,dr = -4k\ell r^{-5}\,dr$. With $\frac{dr}{r} = 0.1$, we find $\frac{dR}{R} = -\frac{4k\ell r^{-5}}{k\ell r^{-4}}\,dr = -4\frac{dr}{r} = -4(0.1) = -0.4$.
In other words, the resistance will drop by 40%.

35. $f(n) = 4n\sqrt{n-4} = 4n(n-4)^{1/2}$, so $df = 4\left[(n-4)^{1/2} + \frac{1}{2}n(n-4)^{-1/2}\right]dn$. When $n = 85$ and $dn = 5$,

$df = 4\left(9 + \frac{85}{2\cdot 9}\right)5 \approx 274$ seconds.

37. $N(r) = \frac{7}{1+0.02r^2}$ and $dN = -\frac{0.28r}{\left(1+0.02r^2\right)^2}\,dr$. To estimate the decrease in the number of housing starts when the mortgage rate is increased from 6% to 6.5%, we set $r = 6$ and $dr = 0.5$ and compute

$dN = -\frac{(0.28)(6)(0.5)}{(1.72)^2} \approx -0.283937$, or 283,937 fewer housing starts.

39. $p = \frac{30}{0.02x^2+1}$ and $dp = -\frac{1.2x}{\left(0.02x^2+1\right)^2}\,dx$. To estimate the change in the price p when the quantity demanded changed from 5000 to 5500 units per week (that is, x changes from 5 to 5.5), we compute

$dp = \frac{(-1.2)(5)(0.5)}{[0.02(25)+1]^2} \approx -1.33$, a decrease of \$1.33.

41. $P(x) = -0.000032x^3 + 6x - 100$ and $dP = \left(-0.000096x^2 + 6\right)dx$. To determine the error in the estimate of Trappee's profits corresponding to a maximum error in the forecast of 15 percent [that is, $dx = \pm 0.15(200)$], we compute $dP = \left[(-0.000096)(200)^2 + 6\right](\pm 30) = (2.16)(30) = \pm 64.80$, or \$64,800.

43. The approximate change in the quantity demanded is given by

$\Delta x \approx dx = f'(p)\,\Delta p = \frac{d}{dp}(144-p)^{1/2}\,\Delta p = -\frac{1}{2}\cdot\frac{1}{\sqrt{144-p}}\cdot\Delta p$. When $\Delta p = 110 - 108 = 2$, we find

$\Delta x = -\frac{1}{2}\cdot\frac{1}{\sqrt{144-108}}(2) = -\frac{1}{6} \approx -0.1667$. Thus, the quantity demanded decreases by approximately 167 tires/week.

45. $N(x) = \frac{500(400+20x)^{1/2}}{(5+0.2x)^2}$ and

$N'(x) = \frac{(5+0.2x)^2\,250(400+20x)^{-1/2}(20) - 500(400+20x)^{1/2}(2)(5+0.2x)(0.2)}{(5+0.2x)^4}\,dx$. To estimate the change in the number of crimes if the level of reinvestment changes from 20 cents to 22 cents per dollar deposited, we compute

$dN = \frac{(5+4)^2(250)(800)^{-1/2}(20) - 500(400+400)^{1/2}(2)(9)(0.2)}{(5+4)^4}(2) \approx \frac{(14318.91 - 50911.69)}{9^4}(2)$

≈ -11, a decrease of approximately 11 crimes per year.

47. $A = 10{,}000\left(1+\frac{r}{12}\right)^{120}$.

a. $dA = 10{,}000(120)\left(1+\frac{r}{12}\right)^{119}\left(\frac{1}{12}\right)dr = 100{,}000\left(1+\frac{r}{12}\right)^{119}dr$.

b. At 3.1%, it will be worth $100{,}000\left(1+\frac{0.03}{12}\right)^{119}(0.001)$, or approximately \$134.60 more. At 3.2%, it will be worth $100{,}000\left(1+\frac{0.03}{12}\right)^{119}(0.002)$, or approximately \$269.20 more. At 3.3%, it will be worth $100{,}000\left(1+\frac{0.03}{12}\right)^{119}(0.003)$, or approximately \$403.80 more.

49. True. $dy = f'(x)\,dx = \frac{d}{dx}(ax+b)\,dx = a\,dx$. On the other hand, $\Delta y = f(x+\Delta x) - f(x) = [a(x+\Delta x)+b] - (ax+b) = a\,\Delta x = a\,dx$.

Using Technology page 244

1. $dy = f'(3)\,dx = 757.87\,(0.01) \approx 7.5787$.

3. $dy = f'(1)\,dx = 1.04067285926\,(0.03) \approx 0.031220$.

5. $dy = f'(4)\,(0.1) \approx -0.198761598\,(0.1) = -0.01988$.

7. If the interest rate changes from 5% to 5.3% per year, the monthly payment will increase by $dP = f'(0.05)\,(0.003) \approx 44.00$, or approximately \$44.00 per month. If the rate changes from 5% to 5.4% per year, the payment will increase by \$58.67 per month, and if it changes from 5% to 5.5% per year, the payment will increase by \$73.34 per month.

9. $dx = f'(40)\,(2) \approx -0.625$. That is, the quantity demanded will decrease by 625 watches per week.

CHAPTER 3 Concept Review Questions page 246

1. a. 0 **b.** nx^{n-1} **c.** $cf'(x)$. **d.** $f'(x) \pm g'(x)$

3. a. $g'(f(x))\,f'(x)$ **b.** $n\left[f(x)\right]^{n-1} f'(x)$

5. a. $-\dfrac{pf'(p)}{f(p)}$ **b.** Elastic, unitary, inelastic

7. $y, dy/dt, a$

9. a. $x_2 - x_1$ **b.** $f(x+\Delta x) - f(x)$

CHAPTER 3 Review Exercises page 247

1. $f'(x) = \dfrac{d}{dx}\left(3x^5 - 2x^4 + 3x^2 - 2x + 1\right) = 15x^4 - 8x^3 + 6x - 2$.

3. $g'(x) = \dfrac{d}{dx}\left(-2x^{-3} + 3x^{-1} + 2\right) = 6x^{-4} - 3x^{-2}$.

5. $g'(t) = \dfrac{d}{dt}\left(2t^{-1/2} + 4t^{-3/2} + 2\right) = -t^{-3/2} - 6t^{-5/2}$.

7. $f'(t) = \dfrac{d}{dt}\left(t + 2t^{-1} + 3t^{-2}\right) = 1 - 2t^{-2} - 6t^{-3} = 1 - \dfrac{2}{t^2} - \dfrac{6}{t^3}$.

9. $h'(x) = \dfrac{d}{dx}\left(x^2 - 2x^{-3/2}\right) = 2x + 3x^{-5/2} = 2x + \dfrac{3}{x^{5/2}}$.

11. $g(t) = \dfrac{t^2}{2t^2+1}$, so $g'(t) = \dfrac{(2t^2+1)\dfrac{d}{dt}(t^2) - t^2\dfrac{d}{dt}(2t^2+1)}{(2t^2+1)^2} = \dfrac{(2t^2+1)(2t) - t^2(4t)}{(2t^2+1)^2} = \dfrac{2t}{(2t^2+1)^2}$.

13. $f(x) = \dfrac{\sqrt{x}-1}{\sqrt{x}+1} = \dfrac{x^{1/2}-1}{x^{1/2}+1}$, so

$$f'(x) = \frac{(x^{1/2}+1)\left(\frac{1}{2}x^{-1/2}\right) - (x^{1/2}-1)\left(\frac{1}{2}x^{-1/2}\right)}{(x^{1/2}+1)^2} = \frac{\frac{1}{2}+\frac{1}{2}x^{-1/2}-\frac{1}{2}+\frac{1}{2}x^{-1/2}}{(x^{1/2}+1)^2} = \frac{x^{-1/2}}{(x^{1/2}+1)^2}$$

$$= \frac{1}{\sqrt{x}\left(\sqrt{x}+1\right)^2}.$$

15. $f(x) = \dfrac{x^2(x^2+1)}{x^2-1}$, so

$$f'(x) = \frac{(x^2-1)\dfrac{d}{dx}(x^4+x^2) - (x^4+x^2)\dfrac{d}{dx}(x^2-1)}{(x^2-1)^2} = \frac{(x^2-1)(4x^3+2x) - (x^4+x^2)(2x)}{(x^2-1)^2}$$

$$= \frac{4x^5+2x^3-4x^3-2x-2x^5-2x^3}{(x^2-1)^2} = \frac{2x^5-4x^3-2x}{(x^2-1)^2} = \frac{2x(x^4-2x^2-1)}{(x^2-1)^2}.$$

17. $f(x) = (3x^3-2)^8$, so $f'(x) = 8(3x^3-2)^7(9x^2) = 72x^2(3x^3-2)^7$.

19. $f'(t) = \dfrac{d}{dt}(2t^2+1)^{1/2} = \frac{1}{2}(2t^2+1)^{-1/2}\dfrac{d}{dt}(2t^2+1) = \frac{1}{2}(2t^2+1)^{-1/2}(4t) = \dfrac{2t}{\sqrt{2t^2+1}}$.

21. $s(t) = (3t^2-2t+5)^{-2}$, so

$$s'(t) = -2(3t^2-2t+5)^{-3}(6t-2) = -4(3t^2-2t+5)^{-3}(3t-1) = -\frac{4(3t-1)}{(3t^2-2t+5)^3}.$$

23. $h(x) = \left(x+\dfrac{1}{x}\right)^2 = (x+x^{-1})^2$, so

$$h'(x) = 2(x+x^{-1})(1-x^{-2}) = 2\left(x+\frac{1}{x}\right)\left(1-\frac{1}{x^2}\right) = 2\left(\frac{x^2+1}{x}\right)\left(\frac{x^2-1}{x^2}\right) = \frac{2(x^2+1)(x^2-1)}{x^3}.$$

25. $h(t) = (t^2+t)^4(2t^2)$, so

$$h'(t) = (t^2+t)^4\frac{d}{dt}(2t^2) + 2t^2\frac{d}{dt}(t^2+t)^4 = (t^2+t)^4(4t) + 2t^2\cdot 4(t^2+t)^3(2t+1)$$

$$= 4t(t^2+t)^3\left[(t^2+t)+4t^2+2t\right] = 4t^2(5t+3)(t^2+t)^3.$$

27. $g(x) = x^{1/2}(x^2-1)^3$, so

$$g'(x) = \frac{d}{dx}\left[x^{1/2}(x^2-1)^3\right] = x^{1/2}\cdot 3(x^2-1)^2(2x) + (x^2-1)^3\cdot\tfrac{1}{2}x^{-1/2}$$

$$= \tfrac{1}{2}x^{-1/2}(x^2-1)^2\left[12x^2+(x^2-1)\right] = \frac{(13x^2-1)(x^2-1)^2}{2\sqrt{x}}.$$

29. $h(x) = \dfrac{(3x+2)^{1/2}}{4x-3}$, so

$$h'(x) = \frac{(4x-3)\frac{1}{2}(3x+2)^{-1/2}(3) - (3x+2)^{1/2}(4)}{(4x-3)^2} = \frac{\frac{1}{2}(3x+2)^{-1/2}[3(4x-3) - 8(3x+2)]}{(4x-3)^2}$$
$$= -\frac{12x+25}{2\sqrt{3x+2}\,(4x-3)^2}.$$

31. $f(x) = 2x^4 - 3x^3 + 2x^2 + x + 4$, so $f'(x) = \dfrac{d}{dx}(2x^4 - 3x^3 + 2x^2 + x + 4) = 8x^3 - 9x^2 + 4x + 1$ and $f''(x) = \dfrac{d}{dx}(8x^3 - 9x^2 + 4x + 1) = 24x^2 - 18x + 4 = 2(12x^2 - 9x + 2)$.

33. $h(t) = \dfrac{t}{t^2+4}$, so $h'(t) = \dfrac{(t^2+4)(1) - t(2t)}{(t^2+4)^2} = \dfrac{4-t^2}{(t^2+4)^2}$ and

$$h''(t) = \frac{(t^2+4)^2(-2t) - (4-t^2)\,2(t^2+4)(2t)}{(t^2+4)^4} = \frac{-2t(t^2+4)\left[(t^2+4) + 2(4-t^2)\right]}{(t^2+4)^4} = \frac{2t(t^2-12)}{(t^2+4)^3}.$$

35. $f'(x) = \frac{d}{dx}(2x^2+1)^{1/2} = \frac{1}{2}(2x^2+1)^{-1/2}(4x) = 2x(2x^2+1)^{-1/2}$, so

$$f''(x) = 2(2x^2+1)^{-1/2} + 2x\cdot\left(-\tfrac{1}{2}\right)(2x^2+1)^{-3/2}(4x) = 2(2x^2+1)^{-3/2}\left[(2x^2+1) - 2x^2\right] = \frac{2}{(2x^2+1)^{3/2}}.$$

37. $6x^2 - 3y^2 = 9$. Differentiating this equation implicitly, we have $12x - 6y\dfrac{dy}{dx} = 0$ and $-6y\dfrac{dy}{dx} = -12x$. Therefore, $\dfrac{dy}{dx} = \dfrac{-12x}{-6y} = \dfrac{2x}{y}$.

39. $y^3 + 3x^2 = 3y$. Differentiating this equation implicitly, we have $3y^2y' + 6x = 3y'$, $3y^2y' - 3y' = -6x$, and $y'(3y^2 - 3) = -6x$. Therefore, $y' = -\dfrac{6x}{3(y^2-1)} = -\dfrac{2x}{y^2-1}$.

41. $x^2 - 4xy - y^2 = 12$. Differentiating this equation implicitly, we have $2x - 4xy' - 4y - 2yy' = 0$ and $y'(-4x - 2y) = -2x + 4y$. Therefore, $y' = \dfrac{-2(x-2y)}{-2(2x+y)} = \dfrac{x-2y}{2x+y}$.

43. $f(x) = x^2 + \dfrac{1}{x^2}$, so $df = f'(x)\,dx = (2x - 2x^{-3})\,dx = \left(2x - \dfrac{2}{x^3}\right)dx = \dfrac{2(x^4-1)}{x^3}\,dx$.

45. a. $df = f'(x)\,dx = \dfrac{d}{dx}(2x^2+4)^{1/2}\,dx = \frac{1}{2}(2x^2+4)^{-1/2}(4x) = \dfrac{2x}{\sqrt{2x^2+4}}\,dx$.

b. Setting $x = 4$ and $dx = 0.1$, we find $\Delta f \approx df = \dfrac{2(4)(0.1)}{\sqrt{2(16)+4}} = \dfrac{0.8}{6} = \dfrac{8}{60} = \dfrac{2}{15}$.

c. $\Delta f = f(4.1) - f(4) = \sqrt{2(4.1)^2+4} - \sqrt{2(16)+4} \approx 0.1335$. From part (b), $\Delta f \approx \dfrac{2}{15} \approx 0.1333$.

47. $f(x) = 2x^3 - 3x^2 - 16x + 3$ and $f'(x) = 6x^2 - 6x - 16$.

a. To find the point(s) on the graph of f where the slope of the tangent line is equal to -4, we solve $6x^2 - 6x - 16 = -4$, obtaining $6x^2 - 6x - 12 = 0$, $6(x^2 - x - 2) = 0$, and $6(x-2)(x+1) = 0$. Thus, $x = 2$ or $x = -1$. Now $f(2) = 2(2)^3 - 3(2)^2 - 16(2) + 3 = -25$ and $f(-1) = 2(-1)^3 - 3(-1)^2 - 16(-1) + 3 = 14$, so the points are $(2, -25)$ and $(-1, 14)$.

b. Using the point-slope form of the equation of a line, we find that the equation of the tangent line at $(2, -25)$ is $y - (-25) = -4(x-2)$, $y + 25 = -4x + 8$, or $y = -4x - 17$, and the equation of the tangent line at $(-1, 14)$ is $y - 14 = -4(x+1)$, or $y = -4x + 10$.

49. $y = (4 - x^2)^{1/2}$, so $y' = \frac{1}{2}(4 - x^2)^{-1/2}(-2x) = -\dfrac{x}{\sqrt{4 - x^2}}$. The slope of the tangent line is obtained by letting $x = 1$, giving $m = -\frac{1}{\sqrt{3}} = -\frac{\sqrt{3}}{3}$. Therefore, an equation of the tangent line at $x = 1$ is $y - \sqrt{3} = -\frac{\sqrt{3}}{3}(x - 1)$, or $y = -\frac{\sqrt{3}}{3}x + \frac{4\sqrt{3}}{3}$.

51. $f(x) = (2x-1)^{-1}$, so $f'(x) = -2(2x-1)^{-2}$, $f''(x) = 8(2x-1)^{-3} = \dfrac{8}{(2x-1)^3}$, and $f'''(x) = -48(2x-1)^4 = -\dfrac{48}{(2x-1)^4}$. Because $(2x-1)^4 = 0$ when $x = \frac{1}{2}$, we see that the domain of f''' is $\left(-\infty, \frac{1}{2}\right) \cup \left(\frac{1}{2}, \infty\right)$.

53. a. The number of UK digital viewers in 2015 is projected to be $N(t) = 65.71(5)^{0.085} \approx 75.3$, or 75.3 million.

b. $N'(t) = 65.71(0.085)t^{-0.915}$, so $N'(5) = 65.71(0.085)(5)^{-0.915} \approx 1.28$. Thus, the number of viewers is expected to be increasing at the rate of approximately 1.3 million per year.

55. a. The number of cameras that will be shipped after 2 years is given by $N(2) = 6(2^2) + 200(2) + 4\sqrt{2} + 20{,}000 \approx 20{,}429.7$, or approximately 20,430 cameras.

b. The rate of change in the number of cameras shipped after 2 years is given by $N'(2) = \left(12t + 200 + 2t^{-1/2}\right)\big|_2 = 12(2) + 200 + \frac{2}{\sqrt{2}} \approx 225.4$, or approximately 225 cameras/yr.

57. a. The population after 3 years is given by $P(3) = 30 - \dfrac{20}{2(3)+3} \approx 27.7778$, or approximately 27,778. The current population is $P(0) = 30 - \frac{20}{3} \approx 23.333$, or approximately 23,333. So the population will have changed by $27{,}778 - 23{,}333 = 4445$; that is, it would have increased by 4445.

b. $P'(t) = \dfrac{d}{dt}\left[30 - 20(2t+3)^{-1}\right] = \dfrac{40}{(2t+3)^2}$, so the rate of change after 3 years is $P'(3) = \dfrac{40}{[2(3)+3]^2} \approx 0.4938$; that is, it will be increasing at the rate of approximately 494 people/yr.

59. $N(x) = 1000(1+2x)^{1/2}$, so $N'(x) = 1000\left(\frac{1}{2}\right)(1+2x)^{-1/2}(2) = \dfrac{1000}{\sqrt{1+2x}}$. The rate of increase at the end of the twelfth week is $N'(12) = \frac{1000}{\sqrt{25}} = 200$, or 200 subscribers/week.

61. He can expect to live $f(100) = 46.9[1 + 1.09(100)]^{0.1} \approx 75.0433$, or approximately 75.04 years.
$f'(t) = 46.9(0.1)(1 + 1.09t)^{-0.9}(1.09) = 5.1121(1 + 1.09t)^{-0.9}$; so the required rate of change is
$f'(100) = 5.1121[1 + 1.09(100)]^{-0.9} \approx 0.074$, or approximately 0.07 yr/yr.

63. $p'(x) = \frac{d}{dt}\left[\frac{1}{10}x^{3/2} + 10\right] = \frac{3}{20}x^{1/2} = \frac{3}{20}\sqrt{x}$, so $p'(40) = \frac{3}{20}\sqrt{40} \approx 0.9487$, or \$0.9487. When the number of units is 40,000, the price will increase \$0.9487 for each 1000 radios demanded.

65. a. The actual cost incurred in the manufacturing of the 301st MP3 player is

$$\begin{aligned} C(301) - C(300) &= \left[0.0001(301)^3 - 0.02(301)^2 + 24(301) + 2000\right] \\ &\quad - \left[0.0001(300)^3 - 0.02(300)^2 + 24(300) + 2000\right] \\ &\approx 39.07, \text{ or approximately \$39.07.} \end{aligned}$$

b. The marginal cost is $C'(300) = \left(0.0003x^2 - 0.04x + 24\right)\big|_{x=300} \approx 39$, or approximately \$39.

67. a. $R(x) = px = (2000 - 0.04x)x = 2000x - 0.04x^2$, so

$$\begin{aligned} P(x) &= R(x) - C(x) = \left(2000x - 0.04x^2\right) - \left(0.000002x^3 - 0.02x^2 + 1000x + 120{,}000\right) \\ &= -0.000002x^3 - 0.02x^2 + 1000x - 120{,}000. \end{aligned}$$

Therefore,

$$\overline{C}(x) = \frac{C(x)}{x} = \frac{0.000002x^3 - 0.02x^2 + 1000x + 120{,}000}{x} = 0.000002x^2 - 0.02x + 1000 + \frac{120{,}000}{x}.$$

b.

$$\begin{aligned} C'(x) &= \frac{d}{dx}\left(0.000002x^3 - 0.02x^2 + 1000x + 120{,}000\right) = 0.000006x^2 - 0.04x + 1000, \\ R'(x) &= \frac{d}{dx}\left(2000x - 0.04x^2\right) = 2000 - 0.08x, \\ P'(x) &= \frac{d}{dx}\left(-0.000002x^3 - 0.02x^2 + 1000x - 120{,}000\right) = -0.000006x^2 - 0.04x + 1000, \text{ and} \\ \overline{C}'(x) &= \frac{d}{dx}\left(0.000002x^2 - 0.02x + 1000 + 120{,}000x^{-1}\right) = 0.000004x - 0.02 - 120{,}000x^{-2}. \end{aligned}$$

c. $C'(3000) = 0.000006(3000)^2 - 0.04(3000) + 1000 = 934$, $R'(3000) = 2000 - 0.08(3000) = 1760$, and
$P'(3000) = -0.000006(3000)^2 - 0.04(3000) + 1000 = 826$.

d. $\overline{C}'(5000) = 0.000004(5000) - 0.02 - 120{,}000(5000)^{-2} = -0.0048$, and
$\overline{C}'(8000) = 0.000004(8000) - 0.02 - 120{,}000(8000)^{-2} \approx 0.0101$. At a production level of 5000 machines, the average cost of each additional unit is decreasing at a rate of 0.48 cents. At a production level of 8000 machines, the average cost of each additional unit is increasing at a rate of approximately 1 cent per unit.

69. $x = f(p) = -\frac{5}{2}p + 30$, so $f'(p) = -\frac{5}{2}$ and $E(p) = -\frac{pf'(p)}{f(p)} = -\frac{p\left(-\frac{5}{2}\right)}{-\frac{5}{2}p + 30} = \frac{p}{12 - p}$.

a. $E(3) = \frac{3}{9} = \frac{1}{3}$, so demand is inelastic.

b. $E(6) = \frac{6}{12-6} = 1$, so demand is unitary.

c. $E(9) = \frac{9}{12-9} = 3$, so demand is elastic.

71. $x = 100 - 0.01p^2$, so $f'(p) = -0.02p$ and $E(p) = -\dfrac{p(-0.02p)}{100 - 0.01p^2} = \dfrac{p^2}{5000 - \frac{1}{2}p^2}$.

a. $E(40) = \dfrac{1600}{5000 - \frac{1}{2}(1600)} = \dfrac{1600}{4200} = \dfrac{8}{21} < 1$ and so demand is inelastic.

b. Because demand is inelastic, raising the unit price slightly causes revenue to increase.

73. $G'(t) = \dfrac{d}{dt}(-0.3t^3 + 1.2t^2 + 500) = -0.9t^2 + 2.4t$, so $G'(2) = -0.9(4) + 2.4(2) = 1.2$. Thus, the GDP is growing at the rate of \$1.2 billion/year. $G''(2) = (-1.8t + 2.4)|_{t=2} = -1.2$, so the rate of rate of change of the GDP is decreasing at the rate of \$1.2 billion/yr/yr.

CHAPTER 3 Before Moving On... page 250

1. $f(x) = 2x^3 - 3x^{1/3} + 5x^{-2/3}$, so $f'(x) = 2(3x^2) - 3\left(\frac{1}{3}x^{-2/3}\right) + 5\left(-\frac{2}{3}x^{-5/3}\right) = 6x^2 - x^{-2/3} - \frac{10}{3}x^{-5/3}$.

2.
$$g'(x) = \frac{d}{dx}\left[x(2x^2-1)^{1/2}\right] = (2x^2-1)^{1/2} + x\left(\tfrac{1}{2}\right)(2x^2-1)^{-1/2}\frac{d}{dx}(2x^2-1)$$
$$= (2x^2-1)^{1/2} + \tfrac{1}{2}x(2x^2-1)^{-1/2}(4x) = (2x^2-1)^{-1/2}\left[(2x^2-1) + 2x^2\right] = \frac{4x^2-1}{\sqrt{2x^2-1}}.$$

3. $y = f(x) = \dfrac{2x+1}{x^2+x+1}$, so
$$\frac{dy}{dx} = \frac{(x^2+x+1)(2) - (2x+1)(2x+1)}{(x^2+x+1)^2} = \frac{2x^2+2x+2-(4x^2+4x+1)}{(x^2+x+1)^2} = -\frac{2x^2+2x-1}{(x^2+x+1)^2}.$$

4. $f(x) = \dfrac{1}{\sqrt{x+1}} = (x+1)^{-1/2}$, so $f'(x) = \dfrac{d}{dx}(x+1)^{-1/2} = -\frac{1}{2}(x+1)^{-3/2} = -\dfrac{1}{2(x+1)^{3/2}}$.

Thus, $f''(x) = -\frac{1}{2}\left(-\frac{3}{2}\right)(x+1)^{-5/2} = \frac{3}{4}(x+1)^{-5/2} = \dfrac{3}{4(x+1)^{5/2}}$ and

$f'''(x) = \frac{3}{4}\left(-\frac{5}{2}\right)(x+1)^{-7/2} = -\frac{15}{8}(x+1)^{-7/2} = -\dfrac{15}{8(x+1)^{7/2}}$.

5. $xy^2 - x^2y + x^3 = 4$. Differentiating both sides of the equation implicitly with respect to x gives $y^2 + x(2yy') - 2xy - x^2y' + 3x^2 = 0$, so $(2xy - x^2)y' + (y^2 - 2xy + 3x^2) = 0$ and $y' = \dfrac{-y^2 + 2xy - 3x^2}{2xy - x^2} = \dfrac{-y^2 + 2xy - 3x^2}{x(2y - x)}$.

6. a. $y = x\sqrt{x^2+5}$, so
$$dy = \frac{d}{dx}\left[x(x^2+5)^{1/2}\right]dx = \left[x\left(\tfrac{1}{2}\right)(x^2+5)^{-1/2}(2x)\right]dx + \left[(x^2+5)^{1/2}(1)\right]dx$$
$$= (x^2+5)^{-1/2}\left[(x^2+5) + x^2\right]dx = \frac{2x^2+5}{\sqrt{x^2+5}}dx.$$

b. Here $dx = \Delta x = 2.01 - 2 = 0.01$. Therefore, $\Delta y \approx dy = \dfrac{2(4)+5}{\sqrt{4+5}}(0.01) = \dfrac{0.13}{3} \approx 0.043$.

4 APPLICATIONS OF THE DERIVATIVE

4.1 Applications of the First Derivative

Problem-Solving Tips

1. The critical number of a function f is any number x in the domain of f such that $f'(x) = 0$ or $f'(x)$ does not exist. Note that the definition requires that x be in the domain of f. For example, consider the function $f(x) = x + \frac{1}{x}$ in Example 8 on page 262 of the text. Even though f' is discontinuous at $x = 0$, this value does not qualify as a critical number because it does not lie in the domain of f.

2. Note that when you use test values to find the sign of a derivative over an interval, you don't need to evaluate the derivative at a test value. You need only find the **sign** of the derivative at that test value. For example, to find the sign of $f'(x) = \dfrac{(x+1)(x-1)}{x^2}$ in the interval $(0, 1)$ using the test value $x = \frac{1}{2}$, we simply note that the numerator is the product of a positive number and a negative number, so it is negative. Because x^2 is always positive, the denominator is always positive. The quotient of a negative number and a positive number is negative, so f' is negative over the interval $(0, 1)$.

Concept Questions page 263

1. **a.** f is increasing on I if whenever x_1 and x_2 are in I with $x_1 < x_2$, then $f(x_1) < f(x_2)$.

   b. f is decreasing on I if whenever x_1 and x_2 are in I with $x_1 < x_2$, then $f(x_1) > f(x_2)$.

3. **a.** f has a relative maximum at $x = a$ if there is an open interval I containing a such that $f(x) \le f(a)$ for all x in I.

   b. f has a relative minimum at $x = a$ if there is an open interval I containing a such that $f(x) \ge f(a)$ for all x in I.

5. See page 260 of the text.

Exercises page 264

1. f is decreasing on $(-\infty, 0)$ and increasing on $(0, \infty)$.

3. f is increasing on $(-\infty, -1)$ and $(1, \infty)$, and decreasing on $(-1, 1)$.

5. f is increasing on $(0, 2)$ and decreasing on $(-\infty, 0)$ and $(2, \infty)$.

7. f is decreasing on $(-\infty, -1)$ and $(1, \infty)$ and increasing on $(-1, 1)$.

9. Increasing on $(20.2, 20.6)$ and $(21.7, 21.8)$, constant on $(19.6, 20.2)$ and $(20.6, 21.1)$, and decreasing on $(21.1, 21.7)$ and $(21.8, 22.7)$.

11. a. f is decreasing on $(0, 4)$. **b.** f is constant on $(4, 12)$. **c.** f is increasing on $(12, 24)$.

13. a. 3, 5, and 7 are critical numbers because $f'(3) = f'(5) = f'(7) = 0$ and 9 is a critical number because $f'(9)$ is not defined.

b. f' not defined

+ + + 0 − 0 + 0 + | − − − sign of f'; x: 0, 3, 5, 7, 9

c. f has relative maxima at $(3, 3)$ and $(9, 6)$ and a relative minimum at $(5, 1)$.

15. $f(x) = 3x + 5$, so $f'(x) = 3 > 0$ for all x. Thus, f is increasing on $(-\infty, \infty)$.

17. $f(x) = 2x^2 + x + 1$, so $f'(x) = 4x + 1 = 0$ if $x = -\frac{1}{4}$. From the sign diagram of f', we see that f is decreasing on $\left(-\infty, -\frac{1}{4}\right)$ and increasing on $\left(-\frac{1}{4}, \infty\right)$.

− − − − 0 + + + + + sign of f'; x: $-\frac{1}{4}$, 0

19. $g(x) = x - x^3$, so $g'(x) = 1 - 3x^2$ is continuous everywhere and is equal to zero when $1 - 3x^2 = 0$, or $x = \pm\frac{\sqrt{3}}{3}$. From the sign diagram, we see that f is decreasing on $\left(-\infty, -\frac{\sqrt{3}}{3}\right)$ and $\left(\frac{\sqrt{3}}{3}, \infty\right)$ and increasing on $\left(-\frac{\sqrt{3}}{3}, \frac{\sqrt{3}}{3}\right)$.

− − 0 + + + + + 0 − − sign of g'; x: $-\frac{\sqrt{3}}{3}$, 0, $\frac{\sqrt{3}}{3}$

21. $g(x) = x^3 + 3x^2 + 1$, so $g'(x) = 3x^2 + 6x = 3x(x + 2)$. From the sign diagram, we see that g is increasing on $(-\infty, -2)$ and $(0, \infty)$ and decreasing on $(-2, 0)$.

+ + + 0 − − 0 + + + sign of f'; x: −2, 0

23. $f(x) = \frac{2}{3}x^3 - 2x^2 - 6x - 2$, so $f'(x) = 2x^2 - 4x - 6 = 2(x^2 - 2x - 3) = 2(x - 3)(x + 1) = 0$ if $x = -1$ or 3. From the sign diagram of f', we see that f is increasing on $(-\infty, -1)$ and $(3, \infty)$ and decreasing on $(-1, 3)$.

+ + + 0 − − − 0 + + + sign of f'; x: −1, 0, 3

25. $h(x) = x^4 - 4x^3 + 10$, so $h'(x) = 4x^3 - 12x^2 = 4x^2(x - 3) = 0$ if $x = 0$ or 3. From the sign diagram of h', we see that h is increasing on $(3, \infty)$ and decreasing on $(-\infty, 3)$.

− − − 0 − − − 0 + + + sign of h'; x: 0, 3

27. $h(x) = \dfrac{1}{2x + 3}$, so $h'(x) = \dfrac{-2}{(2x + 3)^2}$ and we see that h' is not defined at $x = -\frac{3}{2}$. But $h'(x) < 0$ for all x except $x = -\frac{3}{2}$. Therefore, h is decreasing on $\left(-\infty, -\frac{3}{2}\right)$ and $\left(-\frac{3}{2}, \infty\right)$.

29. $g(t) = \dfrac{2t}{t^2 + 1}$, so $g'(t) = \dfrac{(t^2 + 1)(2) - (2t)(2t)}{(t^2 + 1)^2} = \dfrac{2t^2 + 2 - 4t^2}{(t^2 + 1)^2} = -\dfrac{2(t^2 - 1)}{(t^2 + 1)^2}$. Next, $g'(t) = 0$ if $t = \pm 1$.

From the sign diagram of g', we see that g is increasing on $(-1, 1)$ and decreasing on $(-\infty, -1)$ and $(1, \infty)$.

− − − 0 + + + + + 0 − − − sign of g'; t: −1, 0, 1

31. $f(x) = x^{2/3} + 5$, so $f'(x) = \frac{2}{3}x^{-1/3} = \dfrac{2}{3x^{1/3}}$, and so f' is not defined at $x = 0$. Now $f'(x) < 0$ if $x < 0$ and $f'(x) > 0$ if $x > 0$, and so f is decreasing on $(-\infty, 0)$ and increasing on $(0, \infty)$.

33. $f(x) = (x-5)^{2/3}$, so

$f'(x) = \frac{2}{3}(x-5)^{-1/3} = \dfrac{2}{3(x-5)^{1/3}}$. From the sign diagram, we see that f is decreasing on $(-\infty, 5)$ and increasing on $(5, \infty)$.

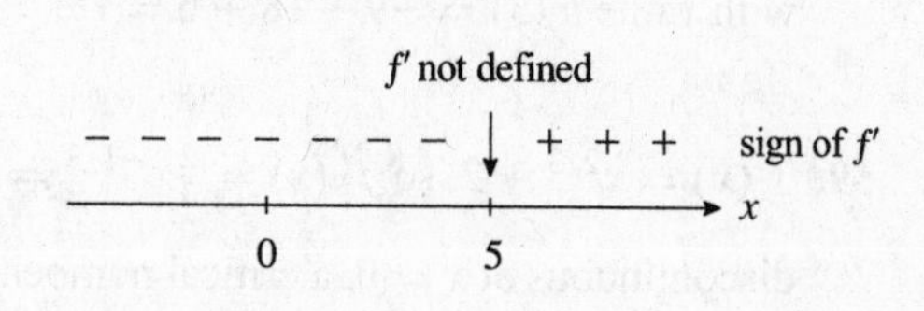

35. $g(x) = x(x+1)^{1/2}$, so

$$g'(x) = (x+1)^{1/2} + x\left(\tfrac{1}{2}\right)(x+1)^{-1/2} = (x+1)^{-1/2}\left(x+1+\tfrac{1}{2}x\right) = (x+1)^{-1/2}\left(\tfrac{3}{2}x+1\right) = \frac{3x+2}{2\sqrt{x+1}}.$$

Thus, g' is continuous on $(-1, \infty)$ and has a zero at $x = -\frac{2}{3}$. From the sign diagram, we see that g is decreasing on $\left(-1, -\frac{2}{3}\right)$ and increasing on $\left(-\frac{2}{3}, \infty\right)$.

– 0 + + + + + + + + sign of g' ; x ; -1, $-\frac{2}{3}$, 0

37. $h(x) = \dfrac{x^2}{x-1}$, so $h'(x) = \dfrac{(x-1)(2x) - x^2}{(x-1)^2} = \dfrac{x^2 - 2x}{(x-1)^2} = \dfrac{x(x-2)}{(x-1)^2}$. Thus, h' is continuous everywhere except at $x = 1$ and has zeros at $x = 0$ and $x = 2$. From the sign diagram, we see that h is increasing on $(-\infty, 0)$ and $(2, \infty)$ and decreasing on $(0, 1)$ and $(1, 2)$.

h' not defined ; + + + 0 – – – – 0 + + + sign of h' ; t ; 0, 1, 2

39. f has a relative maximum of $f(0) = 1$ and relative minima of $f(-1) = 0$ and $f(1) = 0$.

41. f has a relative maximum of $f(-1) = 2$ and a relative minimum of $f(1) = -2$.

43. f has a relative maximum of $f(1) = 3$ and a relative minimum of $f(2) = 2$.

45. f has a relative maximum at $\left(-3, -\frac{9}{2}\right)$ and a relative minimum at $\left(3, \frac{9}{2}\right)$.

47. f is decreasing on the interval (a, c), where $f'(x) < 0$; f is increasing on (c, d), where $f'(x) > 0$; f is constant on (d, e), where $f'(x) = 0$; and finally f is decreasing on (e, b), where $f'(x) < 0$.

49. The profit is increasing at a level of production between 0 units and c units, corresponding to the interval $(0, c)$ on which $P'(t) > 0$. The profit is neither increasing nor decreasing when the level of production is c units; this corresponds to the number c at which $P'(t) = 0$. Finally, the profit is decreasing when the level of production is between c units and b units.

51. a

53. d

55. $g(x) = x^2 + 3x + 8$, so $g'(x) = 2x + 3$ has a critical point at $x = -\frac{3}{2}$. From the sign diagram, we see that $g\left(-\frac{3}{2}\right) = \frac{23}{4}$ is a relative minimum by the First Derivative Test.

– – – – 0 + + + + + + sign of g' ; x ; $-\frac{3}{2}$, 0

57. $h(t) = -t^2 + 6t + 6$, so $h'(t) = -2t + 6 = -2(t - 3) = 0$ if $t = 3$, a critical number. The sign diagram and the First Derivative Test imply that h has a relative maximum at 3 with value $h(3) = -9 + 18 + 6 = 15$.

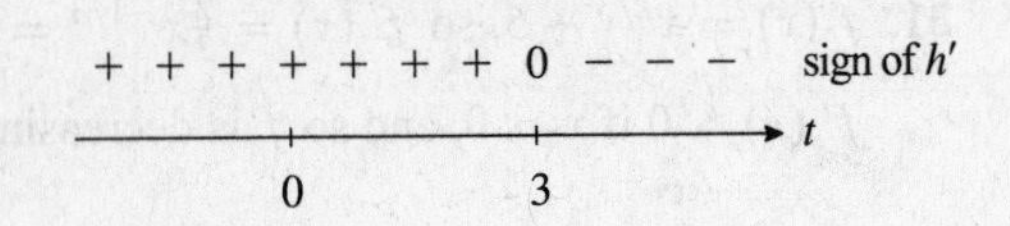

59. $f(x) = x^{2/3} + 2$, so $f'(x) = \frac{2}{3}x^{-1/3} = \dfrac{2}{3x^{1/3}}$ is discontinuous at $x = 0$, a critical number. From the sign diagram and the First Derivative Test, we see that f has a relative minimum at $(0, 2)$.

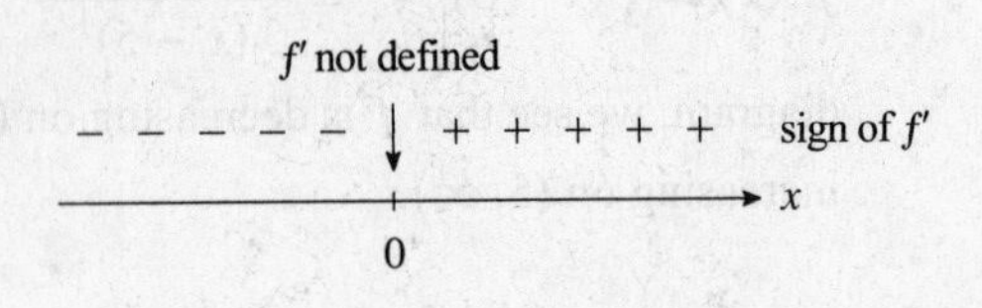

61. $f(x) = x^3 - 3x + 6$. Setting $f'(x) = 3x^2 - 3 = 3(x^2 - 1) = 3(x + 1)(x - 1) = 0$ gives $x = -1$ and $x = 1$ as critical numbers. The sign diagram of f' shows that $(-1, 8)$ is a relative maximum and $(1, 4)$ is a relative minimum.

63. $f(x) = \frac{1}{2}x^4 - x^2$, so $f'(x) = 2x^3 - 2x = 2x(x^2 - 1) = 2x(x + 1)(x - 1)$ is continuous everywhere and has zeros at $x = -1$, 0, and 1, the critical numbers of f. Using the First Derivative Test and the sign diagram of f', we see that $f(-1) = -\frac{1}{2}$ and $f(1) = -\frac{1}{2}$ are relative minima of f and $f(0) = 0$ is a relative maximum of f.

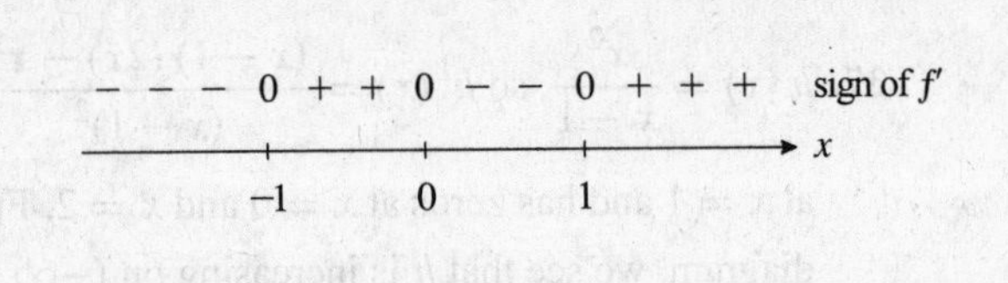

65. $g(x) = x^4 - 4x^3 + 12$. Setting $g'(x) = 4x^3 - 12x^2 = 4x^2(x - 3) = 0$ gives $x = 0$ and $x = 3$ as critical numbers. From the sign diagram, we see that $x = 3$ gives a relative minimum. Its value is $g(3) = 3^4 - 4(3)^3 + 12 = -15$.

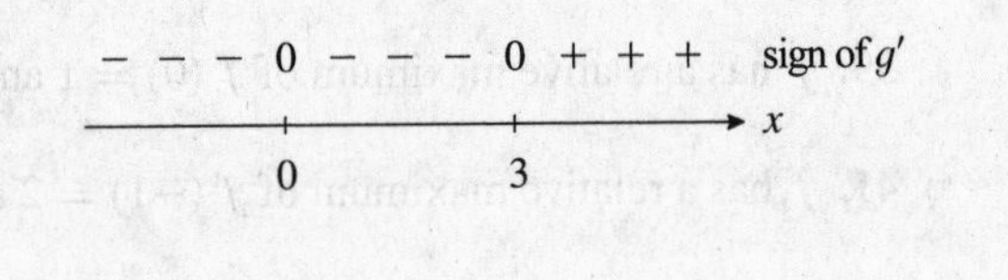

67. $F(t) = 3t^5 - 20t^3 + 20$. Setting $F'(t) = 15t^4 - 60t^2 = 15t^2(t^2 - 4) = 15t^2(t + 2)(t - 2) = 0$ gives $t = -2$, 0, and 2 as critical numbers. From the sign diagram, we see that $t = -2$ gives a relative maximum and $t = 2$ gives a relative minimum. The values are $F(-2) = 3(-32) - 20(-8) + 20 = 84$ and $F(2) = 3(32) - 20(8) + 20 = -44$.

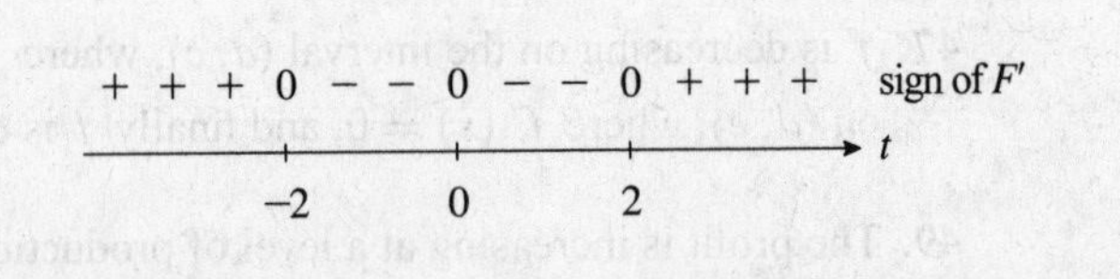

69. $g'(x) = \dfrac{d}{dx}\left(1 + \dfrac{1}{x}\right) = -\dfrac{1}{x^2}$. Observe that g' is nonzero for all values of x. Furthermore, g' is undefined at $x = 0$, but $x = 0$ is not in the domain of g. Therefore, g has no critical number and hence no relative extremum.

71. $g(x) = 2x^2 + \dfrac{4000}{x} + 10$, so $g'(x) = 4x - \dfrac{4000}{x^2} = \dfrac{4(x^3 - 1000)}{x^2}$. The only critical number of g is $x = 10$; $x = 0$ is not a critical number of g since $g(x)$ is not defined there. Using the sign diagram of g' and the First Derivative Test, we conclude that $(10, 610)$ is a relative minimum of g.

73. $g(x) = \frac{x}{x^2 - 1}$, so $g'(x) = \frac{(x^2 - 1) - x(2x)}{(x^2 - 1)^2} = -\frac{1 + x^2}{(x^2 - 1)^2}$ is never zero. Furthermore, $x \pm 1$ are not critical numbers since they are not in the domain of g. Therefore, g has no relative extremum.

75. $g(x) = x\sqrt{x - 4} = x(x - 4)^{1/2}$, so
$g'(x) = (x - 4)^{1/2} + x\left(\frac{1}{2}\right)(x - 4)^{-1/2} = \frac{1}{2}(x - 4)^{-1/2}[2(x - 4) + x] = \frac{3x - 8}{2\sqrt{x - 4}}$ is continuous everywhere except at $x = 4$ and has a zero at $x = \frac{8}{3}$. But both $\frac{8}{3}$ and 4 lie outside the interval $(4, \infty)$, so there is no critical number, and accordingly g has no relative minimum.

g' not defined
+ + + + + + + sign of g'
4 x

77. $N(t) = 1.1375t^2 + 0.25t + 4.6$, so $N'(t) = 2.275t + 0.25$. Because $N'(t) > 0$ for all t in the interval under consideration, we see that N is increasing on the interval $(0, 4)$. We conclude that the percentage of U.S. homes and businesses with digital meters was always increasing between 2008 and 2012.

79. a. $f'(t) = \frac{d}{dt}(0.469t^2 + 0.758t + 0.44) = 0.938t + 0.758$. Because $f'(t) > 0$ for t in $(0, 10)$, we conclude that f is increasing on that interval.

b. The result of part (a) tells us that sales in managed services grew from 1999 through 2009.

81. $h(t) = -\frac{1}{3}t^3 + 16t^2 + 33t + 10$, so $h'(t) = -t^2 + 32t + 33 = -(t + 1)(t - 33)$. The sign diagram for h' shows that the rocket is ascending on the time interval $(0, 33)$ and descending on $(33, T)$ for some positive number T.

+ + + + 0 − − − sign of h'
0 33 t

83. $f(t) = 20t - 40\sqrt{t} + 50 = 20t - 40t^{1/2} + 50$, so $f'(t) = 20 - 40\left(\frac{1}{2}t^{-1/2}\right) = 20\left(1 - \frac{1}{\sqrt{t}}\right) = \frac{20(\sqrt{t} - 1)}{\sqrt{t}}$.
Thus, f' is continuous on $(0, 4)$ and is equal to zero at $t = 1$. From the sign diagram, we see that f is decreasing on $(0, 1)$ and increasing on $(1, 4)$. We conclude that the average speed decreases from 6 a.m. to 7 a.m. and then picks up from 7 a.m. to 10 a.m.

− 0 + + + + + + sign of f'
0 1 4 t

85. $D'(t) = \frac{d}{dt}(0.0032t^3 - 0.0698t^2 + 0.6048t + 3.22) = 0.0096t^2 - 0.1396t + 0.6048$. The discriminant of the quadratic equation $D'(t) = 0$ is $b^2 - 4ac = (-0.1396)^2 - 4(0.0096)(0.6048) = -0.00373616$. Since this is negative, we conclude that the function D has no zeros on $(-\infty, \infty)$ and, in particular, on $(0, 21)$. Also, $D'(0) = 0.6048 > 0$. These two results tell us that $D'(t) > 0$ on $(0, 21)$. Thus, the U.S. public debt outstanding was increasing throughout the period under consideration.

87. a. $f'(t) = \frac{d}{dt}\left(-0.05t^3 + 0.56t^2 + 5.47t + 7.5\right) = -0.15t^2 + 1.12t + 5.47$. Setting $f'(t) = 0$ gives $-0.15t^2 + 1.12t + 5.47 = 0$. Using the quadratic formula, we find $t = \frac{-1.12 \pm \sqrt{(1.12)^2 - 4(-0.15)(5.47)}}{-0.3}$; that is, $t = -3.37$ or 10.83. Because f' is continuous, the only critical numbers of f are $t \approx -3.4$ and $t \approx 10.8$, both of which lie outside the interval of interest. Nevertheless, this result can be used to tell us that f' does not change sign in the interval $(-3.4, 10.8)$. Using $t = 0$ as the test number, we see that $f'(0) = 5.47 > 0$ and so we see that f is increasing on $(-3.4, 10.8)$ and, in particular, in the interval $(0, 10)$. Thus, we conclude that f is increasing on $(0, 10)$.

b. The result of part (a) tells us that sales in the web hosting industry will be increasing from 1999 through 2009.

89. $f'(t) = \frac{d}{dt}\left[90.7\left(0.01t^2 + 0.01t + 1\right)^{-1}\right] = -90.7\left(0.01t^2 + 0.01t + 1\right)^{-2}(0.02t + 0.01)$

$$= -\frac{90.7(0.02t + 0.01)}{\left(0.01t^2 + 0.01t + 1\right)^2}.$$

We see that $f'(t)$ is negative for $t \geq 0$, and thus for $0 \leq t \leq 4$, showing that the volume of first class mail deliveries has been declining throughout the period under consideration.

91. $A(t) = -96.6t^4 + 403.6t^3 + 660.9t^2 + 250$, so

$A'(t) = -386.4t^3 + 1210.8t^2 + 1321.8t = t\left(-386.4t^2 + 1210.8t + 1321.8\right)$. Solving $A'(t) = 0$, we find $t = 0$ and $t = \frac{-1210.8 \pm \sqrt{(1210.8)^2 - 4(-386.4)(1321.8)}}{-2(386.4)} = \frac{-1210.8 \pm 1873.2}{-2(386.4)} \approx 4$. Because t lies in the interval $(0, 5)$, we see that the continuous function A' has zeros at $t = 0$ and $t = 4$.

From the sign diagram, we see that A is increasing on $(0, 4)$ and decreasing on $(4, 5)$. We conclude that the cash in the Central Provident Trust Funds will be increasing from 2005 to 2045 and decreasing from 2045 to 2055.

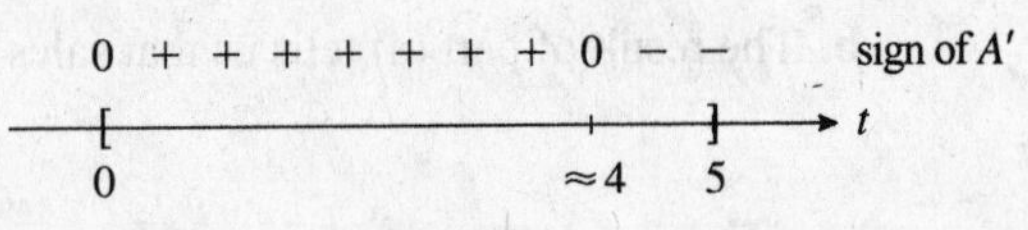

93. a. In 2005, the percentage was $f(0) = \frac{5.3\sqrt{0} - 300}{\sqrt{0} - 10} = 30$ (%). In 2015, it will be

$$f(10) = \frac{5.3\sqrt{10} - 300}{\sqrt{10} - 10} \approx 41.4 \text{ (\%)}.$$

b. $f'(t) = \frac{\left(t^{1/2} - 10\right)(5.3)\left(\frac{1}{2}t^{-1/2}\right) - \left(5.3t^{1/2} - 300\right)\left(\frac{1}{2}t^{-1/2}\right)}{\left(t^{1/2} - 10\right)^2} = \frac{247}{2\sqrt{t}\left(\sqrt{t} - 10\right)^2} > 0$. Thus, f is increasing on $(0, 10)$, indicating that the percentage of small and lower-midsize vehicles is increasing from 2005 through 2015.

95. a. $R(x) = px = (a - bx)x = ax - bx^2$.

b. $R'(x) = a - 2bx$. Solving $R'(x) = a - bx = 0$ gives $x = \frac{a}{2b}$. From the sign diagram for R', we see that the revenue is increasing when the level of production is between $x = 0$ and $x = \frac{a}{2b}$, revenue is stable at a production level of $x = \frac{a}{2b}$, and revenue is decreasing when the production level is between $x = \frac{a}{2b}$ and $x = \frac{a}{b}$.

+ + + + 0 − − − − sign of R' ; x: 0, $\frac{a}{2b}$, $\frac{a}{b}$

c. From the results of part (b), we conclude that a production level of $x = \frac{a}{2b}$ would yield a maximum revenue of $R\left(\frac{a}{2b}\right) = \frac{a^2}{4b}$.

97. False. The function $f(x) = \begin{cases} -x+1 & \text{if } x < 0 \\ -\frac{1}{2}x+1 & \text{if } x \geq 0 \end{cases}$ is decreasing on $(-1, 1)$, but $f'(0)$ does not exist.

99. True. Let $a < x_1 < x_2 < b$. Then $f(x_2) > f(x_1)$ and $g(x_2) > g(x_1)$. Therefore, $(f+g)(x_2) = f(x_2) + g(x_2) > f(x_1) + g(x_1) = (f+g)(x_1)$, and so $f+g$ is increasing on (a, b).

101. False. Consider $f(x) = g(x) = -x$ on $(-1, 1)$. Then both f and g are decreasing on $(-1, 1)$, but $(fg)(x) = f(x)g(x) = (-x)(-x) = x^2$ is not decreasing on $(-1, 1)$.

103. True. The derivative of a polynomial function of degree n $(n \geq 2)$ is a polynomial function of degree $n-1$. Since a polynomial of degree $n-1$ has at most $n-1$ zeros, and therefore at most $n-1$ critical numbers, it follows that the original function of degree n has at most $n-1$ relative extrema.

105. We compute $f'(x) = m$. If $m > 0$, then $f'(x) > 0$ for all x and f is increasing; if $m < 0$, then $f'(x) < 0$ for all x and f is decreasing; if $m = 0$, then $f'(x) = 0$ for all x and f is a constant function.

107. We require that $f'(-1) = 0$; that is, $f'(-1) = \left(3ax^2 + 12x + b\right)\big|_{x=-1} = 3a - 12 + b = 0$, and therefore $f'(2) = 0$, or $f'(2) = \left(3ax^2 + 12x + b\right)\big|_{x=2} = 12a + 24 + b = 0$. Solving the system $\begin{cases} 3a + b = 12 \\ 12a + b = -24 \end{cases}$ simultaneously gives $a = -4$ and $b = 24$.

109. a. $f'(x) = \begin{cases} 2x & \text{if } x \leq 0 \\ -\frac{2}{x^3} & \text{if } x > 0 \end{cases}$ The sign diagram of f' shows that f' does not change sign as we move across $x = 0$.

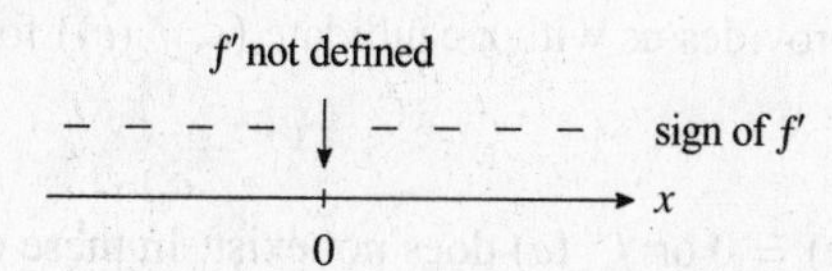

b. From the graph of f, we see that f has a relative minimum at $x = 0$. This does not contradict the First Derivative Test since f is not a continuous function.

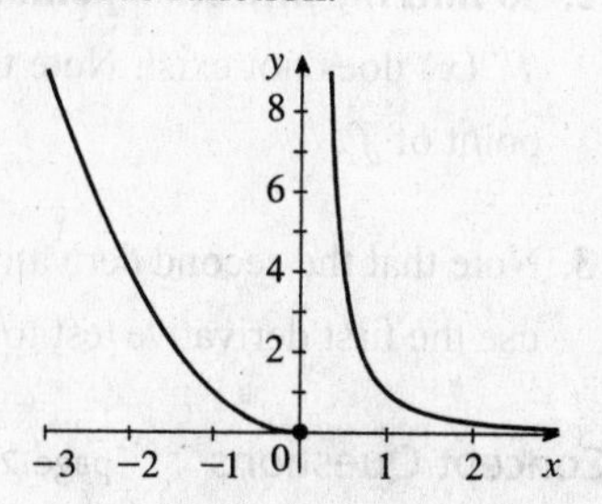

Using Technology page 271

1. a. f is decreasing on $(-\infty, -0.2934)$ and increasing on $(-0.2934, \infty)$.

b. Relative minimum $f(-0.2934) = -2.5435$.

3. a. f is increasing on $(-\infty, -1.6144)$ and $(0.2390, \infty)$ and decreasing on $(-1.6144, 0.2390)$.

b. Relative maximum $f(-1.6144) = 26.7991$, relative minimum $f(0.2390) = 1.6733$.

5. a. f is decreasing on $(-\infty, -1)$ and $(0.33, \infty)$ and increasing on $(-1, 0.33)$.

b. Relative maximum $f(0.33) = 1.11$, relative minimum $f(-1) = -0.63$.

7. a. f is decreasing on $(-1, -0.71)$ and increasing on $(-0.71, 1)$.

b. f has a relative minimum at $(-0.71, -1.41)$.

9. a.

b. $f(t) = 0.004545t^3 - 0.1113t^2 + 1.385t + 11$, so $f'(t) = 0.013635t^2 - 0.2226t + 1.385$. The discriminant of $f'(t) = 0$ is $(-0.2226)^2 - 4(0.013635)(1.385) \approx -0.026$ and $f'(0) = 1.385 > 0$, so f is increasing everywhere.

11. The PSI is increasing on the interval $(0, 4.5)$ and decreasing on $(4.5, 11)$. It is highest when $t = 4.5$ (at 11:30 a.m.) and has value 164.

4.2 Applications of the Second Derivative

Problem-Solving Tips

1. If $f''(x) > 0$, then the graph of f "holds water" and we say the graph of f is concave upward.

If $f''(x) < 0$, then the graph of f "loses water" and we say the graph of f is concave downward.

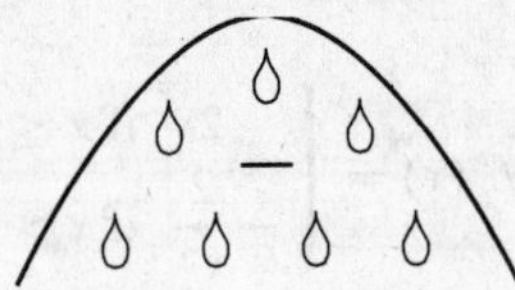

2. To find the inflection points of a function f, determine the number(s) in the domain of f for which $f''(x) = 0$ or $f''(x)$ does not exist. Note that each of these numbers c provides us with a candidate $(c, f(c))$ for an inflection point of f.

3. Note that the second derivative test is not valid when $f''(c) = 0$ or $f''(c)$ does not exist. In these cases you need to use the first derivative test to determine the relative extrema.

Concept Questions page 282

1. a. f is concave upward on (a, b) if f' is increasing on (a, b). f is concave downward on (a, b) if f' is decreasing on (a, b).

b. For the procedure for determining where f is concave upward and where f is concave downward, see page 274 of the text.

3. The Second Derivative Test is stated on page 280 of the text. In general, if f'' is easy to compute, then use the Second Derivative Test. However, keep in mind that (1) in order to use this test f'' must exist, (2) the test is inconclusive if $f''(c) = 0$, and (3) the test is inconvenient to use if f'' is difficult to compute.

Exercises page 282

1. f is concave downward on $(-\infty, 0)$ and concave upward on $(0, \infty)$. f has an inflection point at $(0, 0)$.

3. f is concave downward on $(-\infty, 0)$ and $(0, \infty)$.

5. f is concave upward on $(-\infty, 0)$ and $(1, \infty)$ and concave downward on $(0, 1)$. $(0, 0)$ and $(1, -1)$ are inflection points of f.

7. f is concave downward on $(-\infty, -2)$ and $(-2, 2)$ and $(2, \infty)$.

9. a. f is concave upward on $(0, 2)$, $(4, 6)$, $(7, 9)$, and $(9, 12)$ and concave downward on $(2, 4)$ and $(6, 7)$.

b. f has inflection points at $\left(2, \frac{5}{2}\right)$, $(4, 2)$, $(6, 2)$, and $(7, 3)$.

11. (a)

13. (b)

15. a. $D_1'(t) > 0$, $D_2'(t) > 0$, $D_1''(t) > 0$, and $D_2''(t) < 0$ on $(0, 12)$.

b. With or without the proposed promotional campaign, the deposits will increase, but with the promotion, the deposits will increase at an increasing rate whereas without the promotion, the deposits will increase at a decreasing rate.

17. (c)

19. (d)

21. a. Between 8 a.m. and 10 a.m. the rate of change of the rate of smartphone assembly is increasing; between 10 a.m. and 12 noon, that rate is decreasing.

b. If you look at the tangent lines to the graph of N, you will see that the tangent line at P has the greatest slope. This means that the rate at which the average worker is assembling smartphones is greatest—that is, the worker is most efficient—when $t = 2$, at 10 a.m.

23. The significance of the inflection point Q is that the restoration process is working at its peak at the time t_0 corresponding to the t-coordinate of Q.

25. $f(x) = 4x^2 - 12x + 7$, so $f'(x) = 8x - 12$ and $f''(x) = 8$. Thus, $f''(x) > 0$ everywhere, and so f is concave upward everywhere.

27. $f(x) = \frac{1}{x^4} = x^{-4}$, so $f'(x) = -\frac{4}{x^5}$ and $f''(x) = \frac{20}{x^6} > 0$ for all values of x in $(-\infty, 0)$ and $(0, \infty)$, and so f is concave upward on its domain.

29. $f(x) = 2x^2 - 3x + 4$, so $f'(x) = 4x - 3$ and $f''(x) = 4x > 0$ for all values of x. Thus, f is concave upward on $(-\infty, \infty)$.

31. $f(x) = 1 - x^3$, so $f'(x) = -3x^2$ and $f''(x) = -6x$. From the sign diagram of f'', we see that f is concave upward on $(-\infty, 0)$ and concave downward on $(0, \infty)$.

+ + + + + 0 − − − − − sign of f''
x
0

33. $f(x) = x^4 - 6x^3 + 2x + 8$, so $f'(x) = 4x^3 - 18x^2 + 2$ and $f''(x) = 12x^2 - 36x = 12x(x-3)$.

The sign diagram of f'' shows that f is concave upward on $(-\infty, 0)$ and $(3, \infty)$ and concave downward on $(0, 3)$.

+ + + 0 − − − 0 + + + sign of f''; x: 0, 3

35. $f(x) = x^{4/7}$, so $f'(x) = \frac{4}{7}x^{-3/7}$ and $f''(x) = -\frac{12}{49}x^{-10/7} = -\dfrac{12}{49x^{10/7}}$. Observe that $f''(x) < 0$ for all $x \neq 0$, so f is concave downward on $(-\infty, 0)$ and $(0, \infty)$.

37. $f(x) = (4-x)^{1/2}$, so $f'(x) = \frac{1}{2}(4-x)^{-1/2}(-1) = -\frac{1}{2}(4-x)^{-1/2}$ and $f''(x) = \frac{1}{4}(4-x)^{-3/2}(-1) = -\dfrac{1}{4(4-x)^{3/2}} < 0$ whenever it is defined, so f is concave downward on $(-\infty, 4)$.

39. $f'(x) = \dfrac{d}{dx}(x-2)^{-1} = -(x-2)^{-2}$ and

$f''(x) = 2(x-2)^{-3} = \dfrac{2}{(x-2)^3}$. The sign diagram of f''shows that f is concave downward on $(-\infty, 2)$ and concave upward on $(2, \infty)$.

f'' not defined; − − − − − − | + + + sign of f''; x: 0, 2

41. $f'(x) = \dfrac{d}{dx}(2+x^2)^{-1} = -(2+x^2)^{-2}(2x) = -2x(2+x^2)^{-2}$ and

$f''(x) = -2(2+x^2)^{-2} - 2x(-2)(2+x^2)^{-3}(2x) = 2(2+x^2)^{-3}\left[-(2+x^2) + 4x^2\right] = \dfrac{2(3x^2-2)}{(2+x^2)^3} = 0$ if $x = \pm\sqrt{\frac{2}{3}} = \frac{\sqrt{6}}{3}$. From the sign diagram of f'', we see that f is concave upward on $\left(-\infty, -\frac{\sqrt{6}}{3}\right)$ and $\left(\frac{\sqrt{6}}{3}, \infty\right)$ and concave downward on $\left(-\frac{\sqrt{6}}{3}, \frac{\sqrt{6}}{3}\right)$.

+ + + 0 − − − − − − 0 + + + sign of f''; x: $-\frac{\sqrt{6}}{3}$, 0, $\frac{\sqrt{6}}{3}$

43. $h(t) = \dfrac{t^2}{t-1}$, so $h'(t) = \dfrac{(t-1)(2t) - t^2(1)}{(t-1)^2} = \dfrac{t^2-2t}{(t-1)^2}$ and

$h''(t) = \dfrac{(t-1)^2(2t-2) - (t^2-2t)2(t-1)}{(t-1)^4} = \dfrac{(t-1)(2t^2-4t+2-2t^2+4t)}{(t-1)^4} = \dfrac{2}{(t-1)^3}$.

The sign diagram of h'' shows that h is concave downward on $(-\infty, 1)$ and concave upward on $(1, \infty)$.

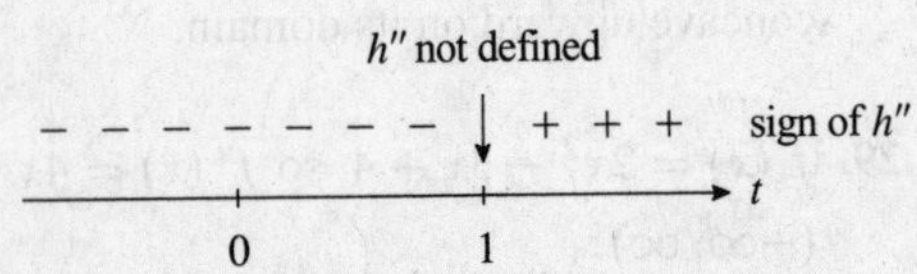

45. $g(x) = x + \dfrac{1}{x^2}$, so $g'(x) = 1 - 2x^{-3}$ and $g''(x) = 6x^{-4} = \dfrac{6}{x^4} > 0$ whenever $x \neq 0$. Therefore, g is concave upward on $(-\infty, 0)$ and $(0, \infty)$.

47. $g(t) = (2t-4)^{1/3}$, so $g'(t) = \frac{1}{3}(2t-4)^{-2/3}(2) = \frac{2}{3}(2t-4)^{-2/3}$ and $g''(t) = -\frac{4}{9}(2t-4)^{-5/3} = -\dfrac{4}{9(2t-4)^{5/3}}$. The sign diagram of g'' shows that g is concave upward on $(-\infty, 2)$ and concave downward on $(2, \infty)$.

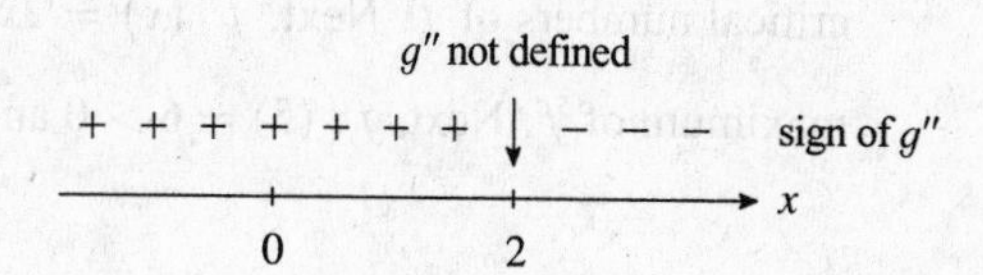

49. $f(x) = x^3 - 2$, so $f'(x) = 3x^2$ and $f''(x) = 6x$. $f''(x)$ is continuous everywhere and has a zero at $x = 0$. From the sign diagram of f'', we conclude that $(0, -2)$ is an inflection point of f.

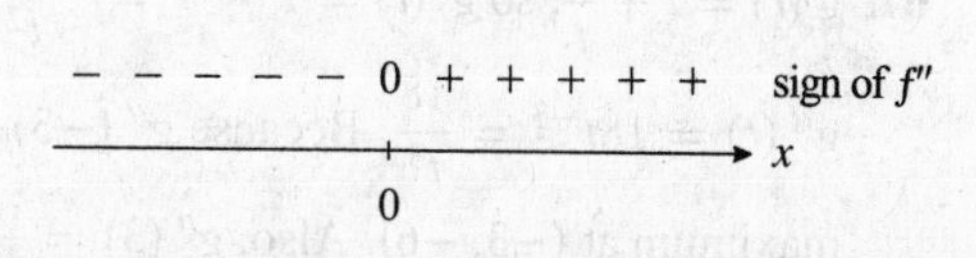

51. $f(x) = 6x^3 - 18x^2 + 12x - 20$, so $f'(x) = 18x^2 - 36x + 12$ and $f''(x) = 36x - 36 = 36(x-1) = 0$ if $x = 1$. The sign diagram of f'' shows that f has an inflection point at $(1, -20)$.

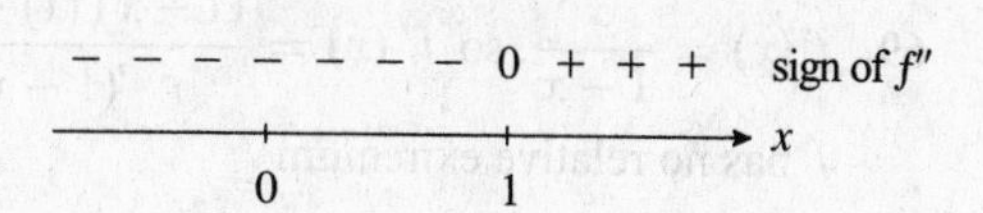

53. $f(x) = 3x^4 - 4x^3 + 1$, so $f'(x) = 12x^3 - 12x^2$ and $f''(x) = 36x^2 - 24x = 12x(3x-2) = 0$ if $x = 0$ or $\frac{2}{3}$. These are candidates for inflection points. The sign diagram of f'' shows that $(0, 1)$ and $\left(\frac{2}{3}, \frac{11}{27}\right)$ are inflection points of f.

+ + + 0 − − − 0 + + + sign of f''; x: 0, $\frac{2}{3}$

55. $g(t) = t^{1/3}$, so $g'(t) = \frac{1}{3}t^{-2/3}$ and $g''(t) = -\frac{2}{9}t^{-5/3} = -\dfrac{2}{9t^{5/3}}$. Observe that $t = 0$ is in the domain of g. Next, since $g''(t) > 0$ if $t < 0$ and $g''(t) < 0$ if $t > 0$, we see that $(0, 0)$ is an inflection point of g.

57. $f(x) = (x-1)^3 + 2$, so $f'(x) = 3(x-1)^2$ and $f''(x) = 6(x-1)$. Observe that $f''(x) < 0$ if $x < 1$ and $f''(x) > 0$ if $x > 1$ and so $(1, 2)$ is an inflection point of f.

59. $f(x) = \dfrac{2}{1+x^2} = 2(1+x^2)^{-1}$, so $f'(x) = -2(1+x^2)^{-2}(2x) = -4x(1+x^2)^{-2}$ and $f''(x) = -4(1+x^2)^{-2} - 4x(-2)(1+x^2)^{-3}(2x) = 4(1+x^2)^{-3}\left[-(1+x^2) + 4x^2\right] = \dfrac{4(3x^2-1)}{(1+x^2)^3}$, which is continuous everywhere and has zeros at $x = \pm\frac{\sqrt{3}}{3}$. From the sign diagram of f'', we conclude that $\left(-\frac{\sqrt{3}}{3}, \frac{3}{2}\right)$ and $\left(\frac{\sqrt{3}}{3}, \frac{3}{2}\right)$ are inflection points of f.

+ + + 0 − − − − − 0 + + + sign of f''; x: $-\frac{\sqrt{3}}{3}$, 0, $\frac{\sqrt{3}}{3}$

61. $f(x) = -x^2 + 2x + 4$, so $f'(x) = -2x + 2$. The critical number of f is $x = 1$. Because $f''(x) = -2$ and $f''(1) = -2 < 0$, we conclude that $f(1) = 5$ is a relative maximum of f.

63. $f(x) = 2x^3 + 1$, so $f'(x) = 6x^2 = 0$ if $x = 0$ and this is a critical number of f. Next, $f''(x) = 12x$, and so $f''(0) = 0$. Thus, the Second Derivative Test fails. But the First Derivative Test shows that $(0, 0)$ is not a relative extremum.

65. $f(x) = \frac{1}{3}x^3 - 2x^2 - 5x - 5$, so $f'(x) = x^2 - 4x - 5 = (x-5)(x+1)$ and this gives $x = -1$ and $x = 5$ as critical numbers of f. Next, $f''(x) = 2x - 4$. Because $f''(-1) = -6 < 0$, we see that $\left(-1, -\frac{7}{3}\right)$ is a relative maximum of f. Next, $f''(5) = 6 > 0$ and this shows that $\left(5, -\frac{115}{3}\right)$ is a relative minimum of f.

67. $g(t) = t + \frac{9}{t}$, so $g'(t) = 1 - \frac{9}{t^2} = \frac{t^2 - 9}{t^2} = \frac{(t+3)(t-3)}{t^2}$, showing that $t = \pm 3$ are critical numbers of g. Now, $g''(t) = 18t^{-3} = \frac{18}{t^3}$. Because $g''(-3) = -\frac{18}{27} < 0$, the Second Derivative Test implies that g has a relative maximum at $(-3, -6)$. Also, $g''(3) = \frac{18}{27} > 0$ and so g has a relative minimum at $(3, 6)$.

69. $f(x) = \frac{x}{1-x}$, so $f'(x) = \frac{(1-x)(1) - x(-1)}{(1-x)^2} = \frac{1}{(1-x)^2}$ is never zero. Thus, there is no critical number and f has no relative extremum.

71. $f(t) = t^2 - \frac{16}{t}$, so $f'(t) = 2t + \frac{16}{t^2} = \frac{2t^3 + 16}{t^2} = \frac{2(t^3 + 8)}{t^2}$. Setting $f'(t) = 0$ gives $t = -2$ as a critical number. Next, we compute $f''(t) = \frac{d}{dt}\left(2t + 16t^{-2}\right) = 2 - 32t^{-3} = 2 - \frac{32}{t^3}$. Because $f''(-2) = 2 - \frac{32}{(-8)} = 6 > 0$, we see that $(-2, 12)$ is a relative minimum.

73. $g(s) = \frac{s}{1+s^2}$, so $g'(s) = \frac{(1+s^2)(1) - s(2s)}{(1+s^2)^2} = \frac{1-s^2}{(1+s^2)^2} = 0$ gives $s = -1$ and $s = 1$ as critical numbers of g. Next, we compute

$$g''(s) = \frac{(1+s^2)^2(-2s) - (1-s^2)\,2(1+s^2)(2s)}{(1+s^2)^4} = \frac{2s(1+s^2)(-1-s^2-2+2s^2)}{(1+s^2)^4} = \frac{2s(s^2-3)}{(1+s^2)^3}.$$

Now $g''(-1) = \frac{1}{2} > 0$, and so $g(-1) = -\frac{1}{2}$ is a relative minimum of g. Next, $g''(1) = -\frac{1}{2} < 0$ and so $g(1) = \frac{1}{2}$ is a relative maximum of g.

75. $f(x) = \frac{x^4}{x-1}$, so $f'(x) = \frac{(x-1)(4x^3) - x^4(1)}{(x-1)^2} = \frac{4x^4 - 4x^3 - x^4}{(x-1)^2} = \frac{3x^4 - 4x^3}{(x-1)^2} = \frac{x^3(3x-4)}{(x-1)^2}$. Thus, $x = 0$ and $x = \frac{4}{3}$ are critical numbers of f. Next,

$$\begin{aligned} f''(x) &= \frac{(x-1)^2(12x^3 - 12x^2) - (3x^4 - 4x^3)(2)(x-1)}{(x-1)^4} \\ &= \frac{(x-1)(12x^4 - 12x^3 - 12x^3 + 12x^2 - 6x^4 + 8x^3)}{(x-1)^4} \\ &= \frac{6x^4 - 16x^3 + 12x^2}{(x-1)^3} = \frac{2x^2(3x^2 - 8x + 6)}{(x-1)^3}. \end{aligned}$$

Because $f''\left(\frac{4}{3}\right) > 0$, we see that $f\left(\frac{4}{3}\right) = \frac{256}{27}$ is a relative minimum. Because $f''(0) = 0$, the Second Derivative Test fails. Using the sign diagram for f' and the First Derivative Test, we see that $f(0) = 0$ is a relative maximum.

f' not defined

+ + 0 − − − − − ↓ − 0 + + sign of f'

0, 1, 4/3 (x-axis)

77.

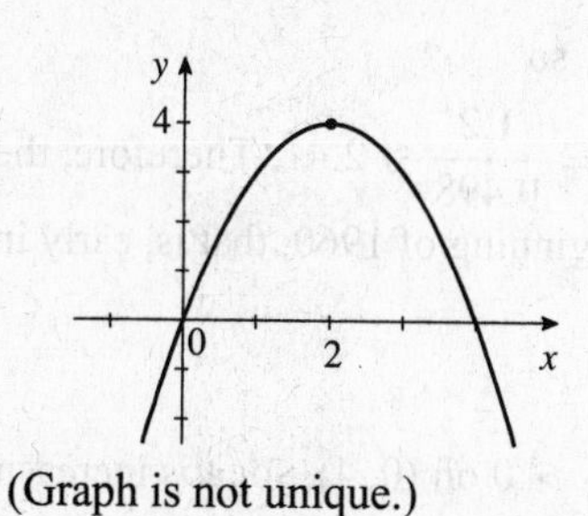

(Graph is not unique.)

79.

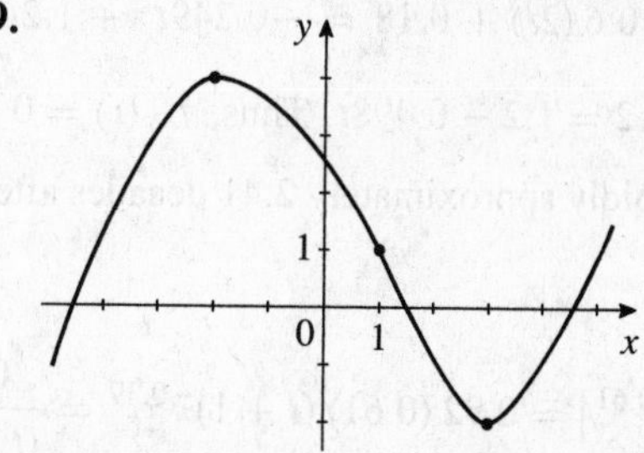

81.

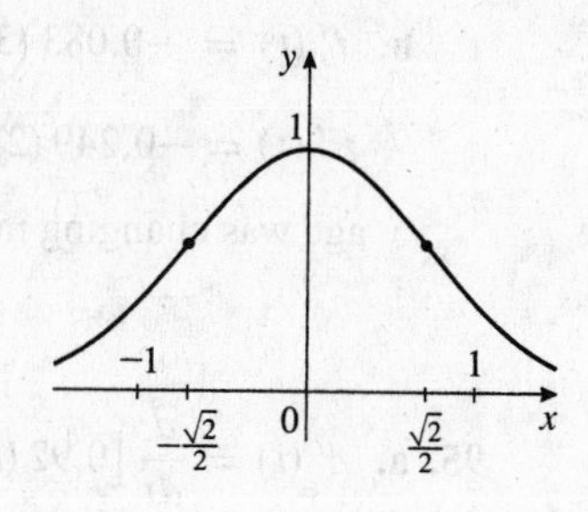

83. a. $N'(t)$ is positive because N is increasing on $(0, 12)$.

b. $N''(t) < 0$ on $(0, 6)$ and $N''(t) > 0$ on $(6, 12)$.

c. The rate of growth of the number of help-wanted advertisements was decreasing over the first six months of the year and increasing over the last six months.

85. $f(t)$ increases at an increasing rate until the water level reaches the middle of the vase (this corresponds to the inflection point of f). At this point, $f(t)$ is increasing at the fastest rate. Though $f(t)$ still increases until the vase is filled, it does so at a decreasing rate.

87. a. $f'(t) = \frac{d}{dt}\left(0.43t^{0.43}\right) = \left(0.43^2\right)t^{-0.57} = \frac{0.1849}{t^{0.57}}$ is positive if $t \geq 1$. This shows that f is increasing for $t \geq 1$, and this implies that the average state cigarette tax was increasing during the period in question.

b. $f''(t) = \frac{d}{dt}\left(0.1849t^{-0.57}\right) = (0.1849)(-0.57)t^{-1.57} = -\frac{0.105393}{t^{1.57}}$ is negative if $t \geq 1$. Thus, the rate of the increase of the cigarette tax is decreasing over the period in question.

89. a. $A'(t) = \frac{d}{dt}\left(0.012414t^2 + 0.7485t + 313.9\right) = 0.024828t + 0.7485 \geq 0.024828(1) + 0.7485 = 0.773328$, so $A'(t)$ is positive for $t \geq 1$. Therefore, A is increasing on $(1, 56)$, showing that the average amount of atmospheric carbon dioxide is increasing from 1958 through 2013.

b. $A''(t) = \frac{d}{dt}A'(t) = \frac{d}{dt}(0.024828t + 0.7485) = 0.024828 > 0$, showing that the rate of increase of the amount of atmospheric carbon dioxide is increasing from 1958 through 2013.

91. a. $D(t) = 0.0032t^3 - 0.0698t^2 + 0.6048t + 3.22$, $0 \leq t \leq 21$, so $D'(t) = 0.0096t^2 - 0.1396t + 0.6048$ and $D''(t) = 0.0192t - 0.1396$. Setting $D''(t) = 0$ gives $t \approx 7.27$. From the sign diagram for D'', we see that the graph of D is concave downward on approximately $(0, 7.27)$ and concave upward on approximately $(7.27, 21)$.

− − 0 + + + + sign of D''
0 ≈7.27 21 t

b. The inflection point of the graph of D is approximately $(7.27, 5.16)$. The U.S. public debt was increasing at a decreasing rate from 1990 through the first quarter of 1997 (approximately), and then continued to increase, but at an increasing rate, from that point onward.

Note: In Exercise 4.1.85, we showed that D is increasing on $(0, 21)$.

93. a. $f(t) = -0.083t^3 + 0.6t^2 + 0.18t + 20.1$, so at the beginning of 1960, the median age of women at first marriage was $f(0) = 20.1$. In 2000, it was $f(4) = -0.083(4)^3 + 0.6(4)^2 + 0.18(4) + 20.1 \approx 25.1$, and in 2001, it was $f(5) = -0.083(5)^3 + 0.6(5)^2 + 0.18(5) + 20.1 \approx 25.6$.

b. $f'(t) = -0.083\left(3t^2\right) + 0.6(2t) + 0.18 = -0.249t^2 + 1.2t + 0.18$, so

$f''(t) = -0.249(2)t + 1.2 = 1.2 - 0.498t$. Thus, $f''(t) = 0$ when $t = \dfrac{1.2}{0.498} \approx 2.41$. Therefore, the median age was changing most rapidly approximately 2.41 decades after the beginning of 1960; that is, early in 1984.

95. a. $A'(t) = \dfrac{d}{dt}\left[0.92(t+1)^{0.61}\right] = 0.92(0.61)(t+1)^{-0.39} = \dfrac{0.5612}{(t+1)^{0.39}} > 0$ on $(0,4)$, so A is increasing on $(0,4)$. This tells us that the spending is increasing over the years in question.

b. $A''(t) = (0.5612)(-0.39)(t+1)^{-1.39} = -\dfrac{0.218868}{(t+1)^{1.39}} < 0$ on $(0,4)$, so A'' is concave downward on $(0,4)$. This tells us that the spending is increasing but at a decreasing rate.

97. a. $P(t) = -0.007333t^3 + 0.91343t^2 + 8.507t + 439$, so $P'(t) = -0.021999t^2 + 1.82686t + 8.507$. Setting $P'(t) = 0$ and using the quadratic formula, we find

$t = \dfrac{-1.82686 \pm \sqrt{(1.82686)^2 - 4(-0.021999)(8.507)}}{2(-0.021999)} \approx -4.42$ or 87.46. Both roots lie outside $(0,31)$, so P has no critical number on that interval. Since $P'(1) \approx 10.31 > 0$, we see that $P'(t) > 0$ on $(0,31)$. We conclude that the number of people aged 80 and over in Canada was increasing from 1981 through 2011.

b. $P''(t) = -0.043998t + 1.82686$. Setting $P''(t) = 0$ and solving, we find $t \approx 41.52$, which lies outside the interval $(0,31)$. Because $P''(1) = 1.782862 > 0$, we see that $P''(t) > 0$ for all t in $(0,31)$. Therefore, P' is increasing on $(0,31)$. We conclude that the population of Canadians aged 80 and over was increasing at an increasing rate from 1981 through 2011.

99. a. $P'(t) = \dfrac{d}{dt}\left(44560t^3 - 89394t^2 + 234633t + 273288\right) = 133680t^2 - 178788 + 234633$. Observe that P' is continuous everywhere and $P'(t) = 0$ has no real solution since the discriminant $b^2 - 4ac = (-178788)^2 - 4(133680)(234633) = -93497808816 < 0$. Because $P'(0) = 234633 > 0$, we may conclude that $P'(t) > 0$ for all t in $(0,4)$, so the population is always increasing.

b. $P''(t) = 267360t - 178788 = 0$ implies that $t = 0.67$. The sign diagram of P'' shows that $t = 0.67$ is an inflection point of the graph of P, so the population was increasing at the slowest pace sometime during August of 1976.

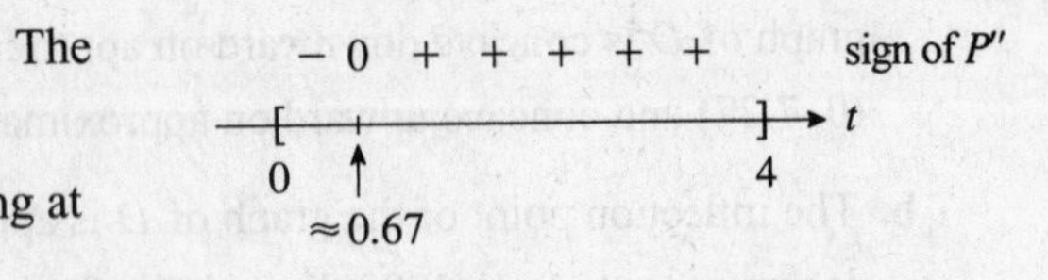

101. $A(t) = 1.0974t^3 - 0.0915t^4$, so $A'(t) = 3.2922t^2 - 0.366t^3$ and $A''(t) = 6.5844t - 1.098t^2$. Setting $A'(t) = 0$, we obtain $t^2(3.2922 - 0.366t) = 0$, and this gives $t = 0$ or $t \approx 8.995 \approx 9$. Using the Second Derivative Test, we find $A''(9) = 6.5844(9) - 1.098(81) = -29.6784 < 0$, and this tells us that $t \approx 9$ gives rise to a relative maximum of A. Our analysis tells us that on that May day, the level of ozone peaked at approximately 4 p.m.

103. a. $R'(t) = \frac{d}{dt}(0.00731t^4 - 0.174t^3 + 1.528t^2 + 0.48t + 19.3) = 0.02924t^3 - 0.522t^2 + 3.056t + 0.48$ and $R''(t) = 0.08772t^2 - 1.044t + 3.056$. Solving the equation $R''(t) = 0$, we obtain $t = \frac{1.044 \pm \sqrt{(-1.044)^2 - 4(0.08772)(3.056)}}{2(0.08772)} \approx 5.19$ or 6.71.

From the sign diagram of R'', we see that the inflection points are approximately $(5.19, 43.95)$ and $(6.71, 53.56)$. We see that the dependency ratio will be increasing at the greatest pace around $t = 5.2$, that is, around 2052.

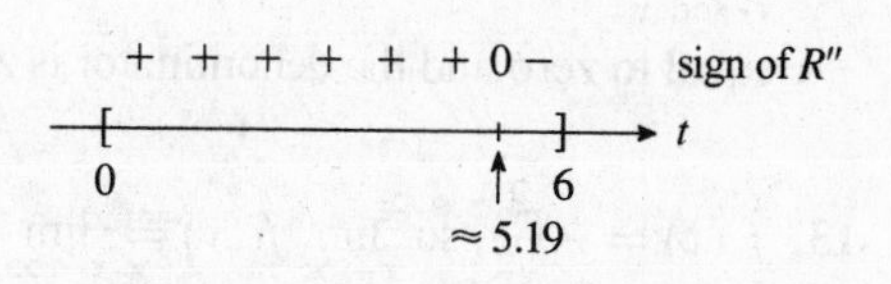

b. The dependency ratio will be $R(5.2) \approx 43.99$, or approximately 44.

105. False. Let $f(x) = x + \frac{1}{x}$ (see Example 2). Then f is concave downward on $(-\infty, 0)$ and concave upward on $(0, \infty)$, but f does not have an inflection point at 0.

107. False. Take $f(x) = x^{1/3}$ on $(-1, 1)$. Then f is defined on $(-1, 1)$ and f has an inflection point at $(0, 0)$, but $f'(x) = \frac{1}{3}x^{-2/3}$ and $f''(x) = -\frac{2}{9}x^{-5/3} = -\frac{2}{9x^{5/3}}$, so $f''(0)$ is undefined.

109. $f(x) = ax^2 + bx + c$, so $f'(x) = 2ax + b$ and $f''(x) = 2a$. Thus, $f''(x) > 0$ if $a > 0$, and the parabola opens upward. If $a < 0$, then $f''(x) < 0$ and the parabola opens downward.

4.3 Curve Sketching

Problem-Solving Tips

1. **To find the horizontal asymptotes of a function** f, find the limit of f as $x \to \infty$ and as $x \to -\infty$. If the limit is equal to a real number b, then $y = b$ is a horizontal asymptote of f.

2. **To find the vertical asymptotes of a rational function** $f(x) = P(x)/Q(x)$, determine the values a for which $Q(a) = 0$. If $Q(a) = 0$ but $P(a) \neq 0$, then the line $x = a$ is a vertical asymptote of f.

3. If a line $x = a$ is a vertical asymptote of the graph of a rational function f, then the denominator of $f(x)$ is equal to zero at $x = a$. However, if both numerator and denominator of $f(x)$ are equal to zero, then $x = a$ is not necessarily a vertical asymptote.

Concept Questions page 298

1. a. See the definition on page 293 of the text. **b.** See the definition on page 295 of the text.

3. See the procedure given on page 292 of the text.

Exercises page 298

1. $y = 0$ is a horizontal asymptote.

3. $y = 0$ is a horizontal asymptote and $x = 0$ is a vertical asymptote.

5. $y = 0$ is a horizontal asymptote and $x = -1$ and $x = 1$ are vertical asymptotes.

7. $y = 3$ is a horizontal asymptote and $x = 0$ is a vertical asymptote.

9. $y = 1$ and $y = -1$ are horizontal asymptotes.

11. $\lim_{x\to\infty} \frac{1}{x} = 0$, and so $y = 0$ is a horizontal asymptote. Next, since the numerator of the rational expression is not equal to zero and the denominator is zero at $x = 0$, we see that $x = 0$ is a vertical asymptote.

13. $f(x) = -\frac{2}{x^2}$, so $\lim_{x\to\infty} f(x) = \lim_{x\to\infty}\left(-\frac{2}{x^2}\right) = 0$. Thus, $y = 0$ is a horizontal asymptote. Next, the denominator of $f(x)$ is equal to zero at $x = 0$. Because the numerator of $f(x)$ is not equal to zero at $x = 0$, we see that $x = 0$ is a vertical asymptote.

15. $\lim_{x\to\infty} \frac{x-2}{x+2} = \lim_{x\to\infty} \frac{1-\frac{2}{x}}{1+\frac{2}{x}} = 1$, and so $y = 1$ is a horizontal asymptote. Next, the denominator is equal to zero at $x = -2$ and the numerator is not equal to zero at this number, so $x = -2$ is a vertical asymptote.

17. $h(x) = x^3 - 3x^2 + x + 1$. $h(x)$ is a polynomial function, and therefore it does not have any horizontal or vertical asymptotes.

19. $\lim_{t\to\infty} \frac{t^2}{t^2-16} = \lim_{t\to\infty} \frac{1}{1-\frac{16}{t^2}} = 1$, and so $y = 1$ is a horizontal asymptote. Next, observe that the denominator of the rational expression $t^2 - 16 = (t+4)(t-4) = 0$ if $t = -4$ or $t = 4$. But the numerator is not equal to zero at these numbers, so $t = -4$ and $t = 4$ are vertical asymptotes.

21. $\lim_{x\to\infty} \frac{3x}{x^2-x-6} = \lim_{x\to\infty} \frac{\frac{3}{x}}{1-\frac{1}{x}-\frac{6}{x^2}} = 0$ and so $y = 0$ is a horizontal asymptote. Next, observe that the denominator $x^2 - x - 6 = (x-3)(x+2) = 0$ if $x = -2$ or $x = 3$. But the numerator $3x$ is not equal to zero at these numbers, so $x = -2$ and $x = 3$ are vertical asymptotes.

23. $\lim_{t\to\infty}\left[2 + \frac{5}{(t-2)^2}\right] = 2$, and so $y = 2$ is a horizontal asymptote. Next observe that $\lim_{t\to 2^+} g(t) = \lim_{t\to 2^-}\left[2 + \frac{5}{(t-2)^2}\right] = \infty$, and so $t = 2$ is a vertical asymptote.

25. $\lim_{x\to\infty} \frac{x^2-2}{x^2-4} = \lim_{x\to\infty} \frac{1-\frac{2}{x^2}}{1-\frac{4}{x^2}} = 1$, and so $y = 1$ is a horizontal asymptote. Next, observe that the denominator $x^2 - 4 = (x+2)(x-2) = 0$ if $x = -2$ or 2. Because the numerator $x^2 - 2$ is not equal to zero at these numbers, the lines $x = -2$ and $x = 2$ are vertical asymptotes.

27. $g(x) = \frac{x^3 - x}{x(x+1)}$. Rewrite $g(x)$ as $\frac{x^2-1}{x+1}$ for $x \neq 0$, and note that $\lim_{x\to-\infty} g(x) = \lim_{x\to-\infty} \frac{x-\frac{1}{x}}{1+\frac{1}{x}} = -\infty$ and $\lim_{x\to\infty} g(x) = \infty$. Therefore, there is no horizontal asymptote. Next, note that the denominator of $g(x)$ is equal to zero at $x = 0$ and $x = -1$. However, since the numerator of $g(x)$ is also equal to zero when $x = 0$, we see that $x = 0$ is not a vertical asymptote. Also, the numerator of $g(x)$ is equal to zero when $x = -1$, so $x = -1$ is not a vertical asymptote.

29. f is the derivative function of the function g. Observe that at a relative maximum or minimum of g, $f(x) = 0$.

31.

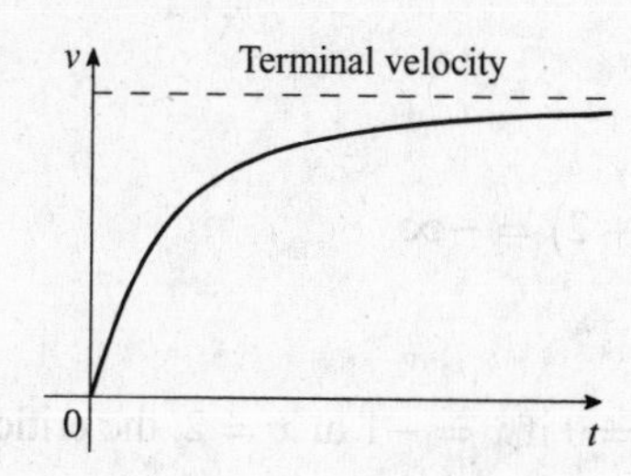

33.

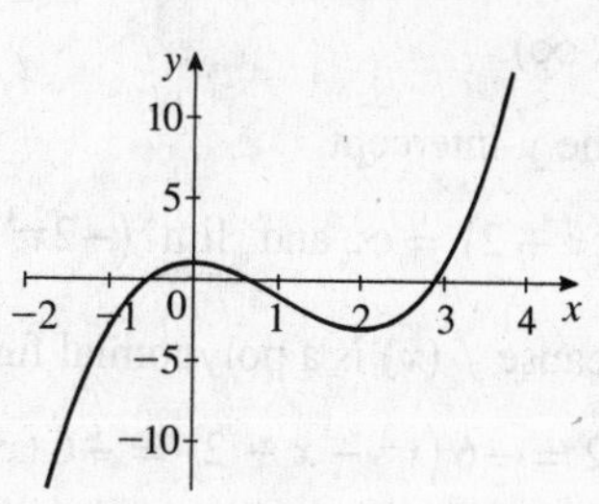

35.

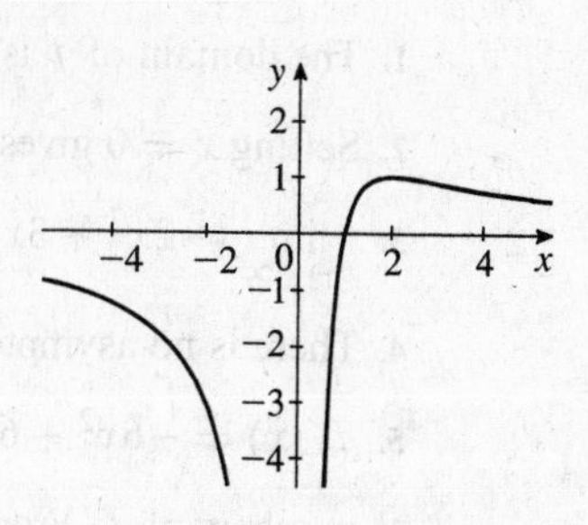

37. $g(x) = 4 - 3x - 2x^3$. We first gather the following information on f.

1. The domain of f is $(-\infty, \infty)$.
2. Setting $x = 0$ gives $y = 4$ as the y-intercept. Setting $y = g(x) = 0$ gives a cubic equation which is not easily solved, and we will not attempt to find the x-intercepts.
3. $\lim_{x \to -\infty} g(x) = \infty$ and $\lim_{x \to \infty} g(x) = -\infty$.
4. The graph of g has no asymptote.
5. $g'(x) = -3 - 6x^2 = -3\left(2x^2 + 1\right) < 0$ for all values of x and so g is decreasing on $(-\infty, \infty)$.
6. The results of step 5 show that g has no critical number and hence no relative extremum.
7. $g''(x) = -12x$. Because $g''(x) > 0$ for $x < 0$ and $g''(x) < 0$ for $x > 0$, we see that g is concave upward on $(-\infty, 0)$ and concave downward on $(0, \infty)$.
8. From the results of step 7, we see that $(0, 4)$ is an inflection point of g.

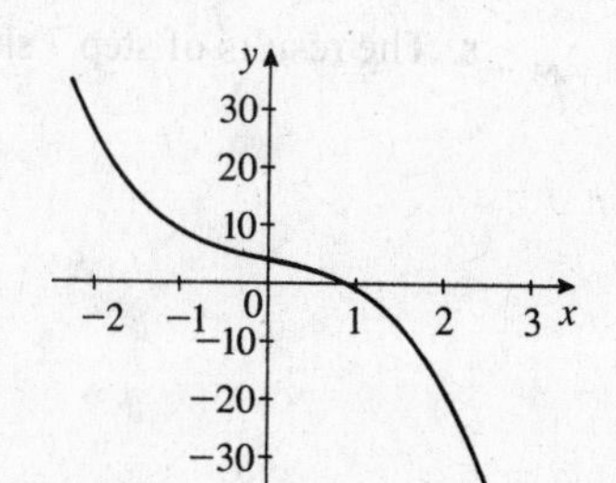

39. $h(x) = x^3 - 3x + 1$. We first gather the following information on h.

1. The domain of h is $(-\infty, \infty)$.
2. Setting $x = 0$ gives 1 as the y-intercept. We will not find the x-intercept.
3. $\lim_{x \to -\infty} \left(x^3 - 3x + 1\right) = -\infty$ and $\lim_{x \to \infty} \left(x^3 - 3x + 1\right) = \infty$.
4. There is no asymptote because $h(x)$ is a polynomial.
5. $h'(x) = 3x^2 - 3 = 3(x + 1)(x - 1)$, and we see that $x = -1$ and $x = 1$ are critical numbers. From the sign diagram, we see that h is increasing on $(-\infty, -1)$ and $(1, \infty)$ and decreasing on $(-1, 1)$.

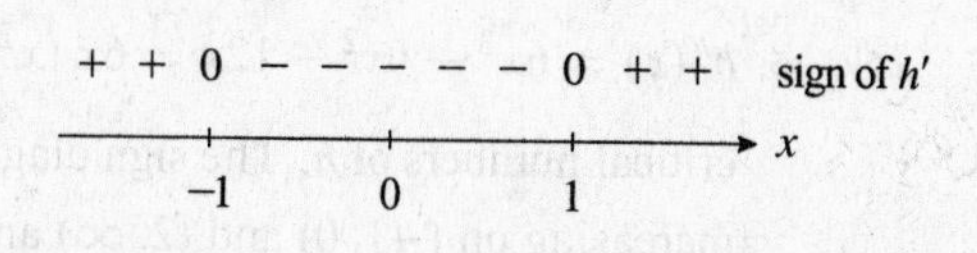

6. The results of step 5 show that $(-1, 3)$ is a relative maximum and $(1, -1)$ is a relative minimum.
7. $h''(x) = 6x$, so $h''(x) < 0$ if $x < 0$ and $h''(x) > 0$ if $x > 0$. Thus, the graph of h is concave downward on $(-\infty, 0)$ and concave upward on $(0, \infty)$.

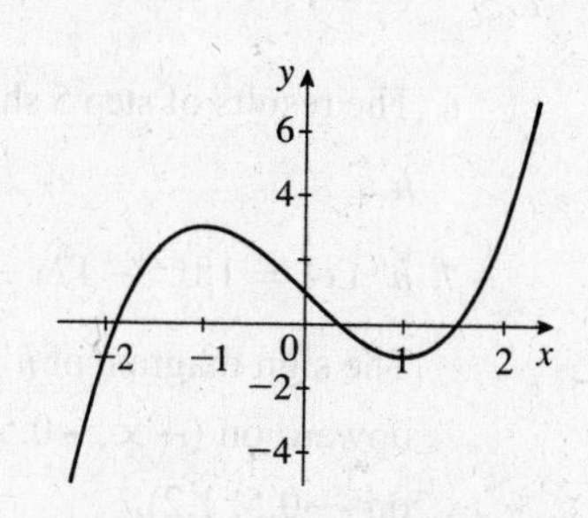

8. The results of step 7 show that $(0, 1)$ is an inflection point of h.

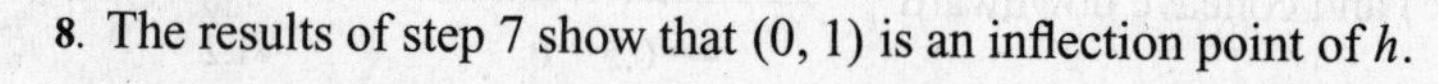

41. $f(x) = -2x^3 + 3x^2 + 12x + 2$. We first gather the following information on f.

1. The domain of f is $(-\infty, \infty)$.
2. Setting $x = 0$ gives 2 as the y-intercept.
3. $\lim_{x\to-\infty} (-2x^3 + 3x^2 + 12x + 2) = \infty$ and $\lim_{x\to\infty} (-2x^3 + 3x^2 + 12x + 2) = -\infty$
4. There is no asymptote because $f(x)$ is a polynomial function.
5. $f'(x) = -6x^2 + 6x + 12 = -6(x^2 - x - 2) = -6(x-2)(x+1) = 0$ if $x = -1$ or $x = 2$, the critical numbers of f. From the sign diagram, we see that f is decreasing on $(-\infty, -1)$ and $(2, \infty)$ and increasing on $(-1, 2)$.

   − − 0 + + + + + 0 − − sign of f'
   x: −1 0 2

6. The results of step 5 show that $(-1, -5)$ is a relative minimum and $(2, 22)$ is a relative maximum.
7. $f''(x) = -12x + 6 = 0$ if $x = \frac{1}{2}$. The sign diagram of f'' shows that the graph of f is concave upward on $\left(-\infty, \frac{1}{2}\right)$ and concave downward on $\left(\frac{1}{2}, \infty\right)$.

   + + + + + + 0 − − − − sign of f''
   x: 0 $\frac{1}{2}$

8. The results of step 7 show that $\left(\frac{1}{2}, \frac{17}{2}\right)$ is an inflection point.

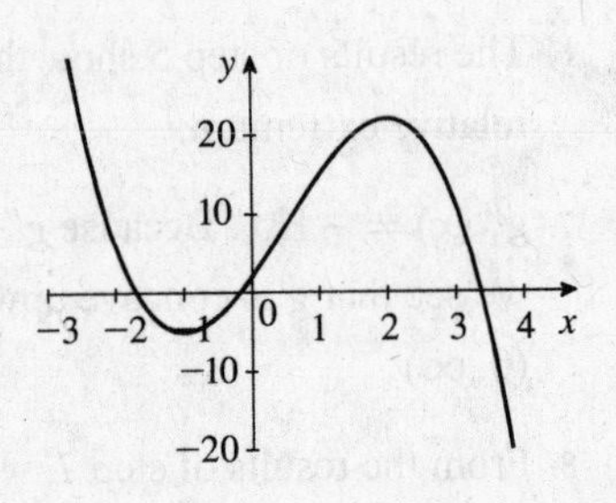

43. $h(x) = \frac{3}{2}x^4 - 2x^3 - 6x^2 + 8$. We first gather the following information on h.

1. The domain of h is $(-\infty, \infty)$.
2. Setting $x = 0$ gives 8 as the y-intercept.
3. $\lim_{x\to-\infty} h(x) = \lim_{x\to\infty} h(x) = \infty$
4. There is no asymptote.
5. $h'(x) = 6x^3 - 6x^2 - 12x = 6x(x^2 - x - 2) = 6x(x-2)(x+1) = 0$ if $x = -1$, 0, or 2, and these are the critical numbers of h. The sign diagram of h' shows that h is increasing on $(-1, 0)$ and $(2, \infty)$ and decreasing on $(-\infty, -1)$ and $(0, 2)$.

   − − 0 + 0 − − − 0 + + sign of h'
   x: −1 0 2

6. The results of step 5 show that $\left(-1, \frac{11}{2}\right)$ and $(2, -8)$ are relative minima of h and $(0, 8)$ is a relative maximum of h.
7. $h''(x) = 18x^2 - 12x - 12 = 6(3x^2 - 2x - 2)$. The zeros of h'' are $x = \frac{2\pm\sqrt{4+24}}{6} \approx -0.5$ or 1.2. The sign diagram of h'' shows that the graph of h is concave upward on $(-\infty, -0.5)$ and $(1.2, \infty)$ and concave downward on $(-0.5, 1.2)$.

   + + 0 − − − − − − − 0 + + sign of h''
   x: ≈−0.5 0 ≈1.2

8. The results of step 7 also show that $(-0.5, 6.8)$ and $(1.2, -1)$ are inflection points.

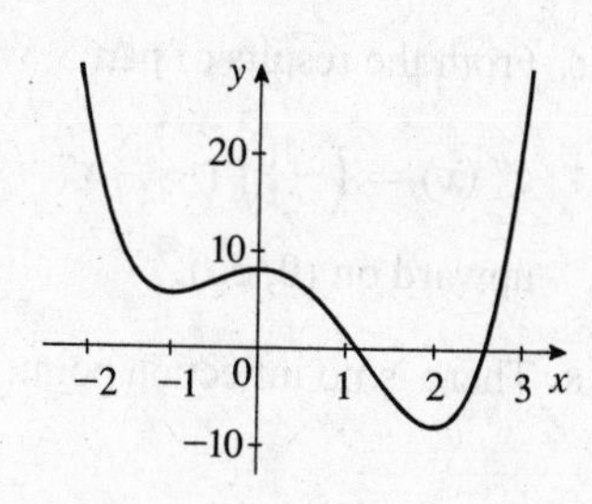

45. $f(t) = \sqrt{t^2 - 4}$. We first gather the following information on f.

1. The domain of f is found by solving $t^2 - 4 \geq 0$ to obtain $(-\infty, -2] \cup [2, \infty)$.
2. Because $t \neq 0$, there is no y-intercept. Next, setting $y = f(t) = 0$ gives the t-intercepts as -2 and 2.
3. $\lim_{t\to-\infty} f(t) = \lim_{t\to\infty} f(t) = \infty$.
4. There is no asymptote.
5. $f'(t) = \frac{1}{2}(t^2 - 4)^{-1/2}(2t) = t(t^2 - 4)^{-1/2} = \dfrac{t}{\sqrt{t^2 - 4}}$. Setting $f'(t) = 0$ gives $t = 0$. But $t = 0$ is not in the domain of f and so there is no critical number. From the sign diagram for f', we see that f is increasing on $(2, \infty)$ and decreasing on $(-\infty, -2)$.

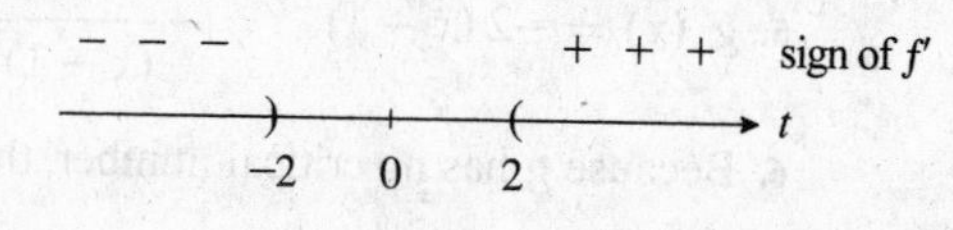

6. From the results of step 5 we see that there is no relative extremum.
7. $f''(t) = (t^2 - 4)^{-1/2} + t\left(-\frac{1}{2}\right)(t^2 - 4)^{-3/2}(2t)$
$$= (t^2 - 4)^{-3/2}(t^2 - 4 - t^2) = -\frac{4}{(t^2 - 4)^{3/2}}.$$
8. Because $f''(t) < 0$ for all t in the domain of f, we see that f is concave downward everywhere. From the results of step 7, we see that there is no inflection point.

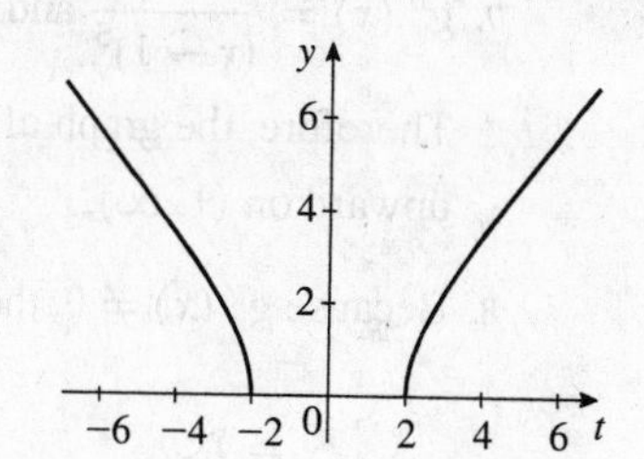

47. $g(x) = \frac{1}{2}x - \sqrt{x}$. We first gather the following information on g.

1. The domain of g is $[0, \infty)$.
2. The y-intercept is 0. To find the x-intercept(s), set $y = 0$, giving $\frac{1}{2}x - \sqrt{x} = 0$, $x = 2\sqrt{x}$, $x^2 = 4x$, $x(x - 4) = 0$, and so $x = 0$ or $x = 4$.
3. $\lim_{x\to\infty} \left(\frac{1}{2}x - \sqrt{x}\right) = \lim_{x\to\infty} \frac{1}{2}x\left(1 - \frac{2}{\sqrt{x}}\right) = \infty$.
4. There is no asymptote.
5. $g'(x) = \frac{1}{2} - \frac{1}{2}x^{-1/2} = \frac{1}{2}x^{-1/2}(x^{1/2} - 1) = \dfrac{\sqrt{x} - 1}{2\sqrt{x}}$, which is zero when $x = 1$. From the sign diagram for g', we see that g is decreasing on $(0, 1)$ and increasing on $(1, \infty)$.

− − 0 + + + sign of g'

0 1 x

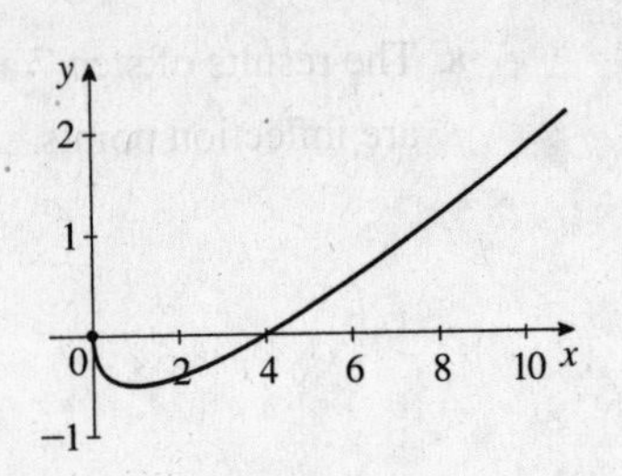

6. From the results of part 5, we see that $g(1) = -\frac{1}{2}$ is a relative minimum.

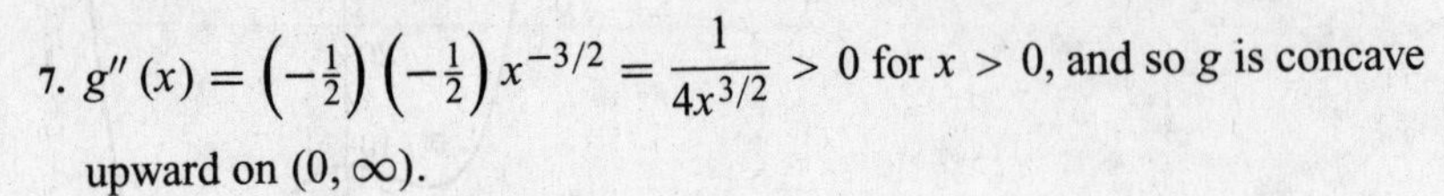

7. $g''(x) = \left(-\frac{1}{2}\right)\left(-\frac{1}{2}\right)x^{-3/2} = \frac{1}{4x^{3/2}} > 0$ for $x > 0$, and so g is concave upward on $(0, \infty)$.

8. There is no inflection point.

49. $g(x) = \frac{2}{x-1}$. We first gather the following information on g.

1. The domain of g is $(-\infty, 1) \cup (1, \infty)$.

2. Setting $x = 0$ gives -2 as the y-intercept. There is no x-intercept because $\frac{2}{x-1} \neq 0$ for all x.

3. $\lim_{x\to-\infty} \frac{2}{x-1} = 0$ and $\lim_{x\to\infty} \frac{2}{x-1} = 0$.

4. The results of step 3 show that $y = 0$ is a horizontal asymptote. Furthermore, the denominator of $g(x)$ is equal to zero at $x = 1$ but the numerator is not equal to zero there. Therefore, $x = 1$ is a vertical asymptote.

5. $g'(x) = -2(x-1)^{-2} = -\frac{2}{(x-1)^2} < 0$ for all $x \neq 1$ and so g is decreasing on $(-\infty, 1)$ and $(1, \infty)$.

6. Because g has no critical number, there is no relative extremum.

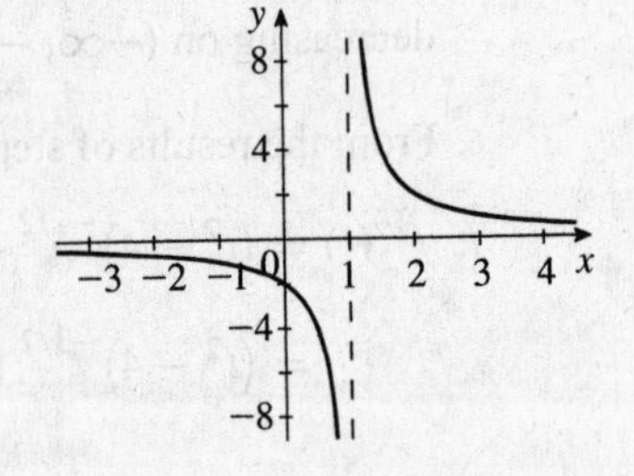

7. $g''(x) = \frac{4}{(x-1)^3}$ and so $g''(x) < 0$ if $x < 1$ and $g''(x) > 0$ if $x > 1$.

   Therefore, the graph of g is concave downward on $(-\infty, 1)$ and concave upward on $(1, \infty)$.

8. Because $g''(x) \neq 0$, there is no inflection point.

51. $h(x) = \frac{x+2}{x-2}$. We first gather the following information on h.

1. The domain of h is $(-\infty, 2) \cup (2, \infty)$.

2. Setting $x = 0$ gives $y = -1$ as the y-intercept. Next, setting $y = 0$ gives $x = -2$ as the x-intercept.

3. $\lim_{x\to\infty} h(x) = \lim_{x\to-\infty} \frac{1+\frac{2}{x}}{1-\frac{2}{x}} = \lim_{x\to-\infty} h(x) = 1$.

4. Setting $x - 2 = 0$ gives $x = 2$. Furthermore, $\lim_{x\to2^+} \frac{x+2}{x-2} = \infty$, so $x = 2$ is a vertical asymptote of h. Also, from the results of step 3, we see that $y = 1$ is a horizontal asymptote of h.

5. $h'(x) = \frac{(x-2)(1)-(x+2)(1)}{(x-2)^2} = -\frac{4}{(x-2)^2}$. We see that h has no critical number. (Note that $x = 2$ is not in the domain of h.) The sign diagram of h' shows that h is decreasing on $(-\infty, 2)$ and $(2, \infty)$.

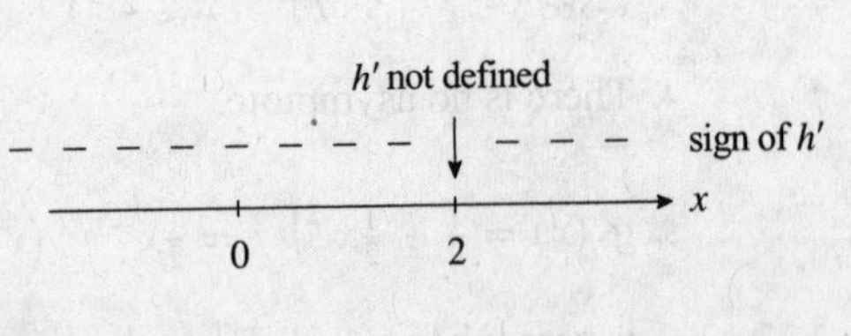

6. From the results of step 5, we see that there is no relative extremum.

7. $h''(x) = \frac{8}{(x-2)^3}$. Note that $x = 2$ is not a candidate for an inflection point because $h(2)$ is not defined. Because $h''(x) < 0$ for $x < 2$ and $h''(x) > 0$ for $x > 2$, we see that h is concave downward on $(-\infty, 2)$ and concave upward on $(2, \infty)$.

8. From the results of step 7, we see that there is no inflection point.

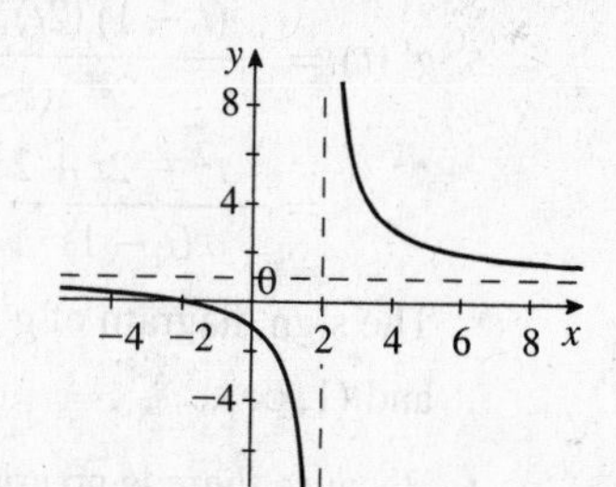

53. $f(t) = \frac{t^2}{1+t^2}$. We first gather the following information on f.

1. The domain of f is $(-\infty, \infty)$.
2. Setting $t = 0$ gives the y-intercept as 0. Similarly, setting $y = 0$ gives the t-intercept as 0.
3. $\lim_{t\to-\infty} \frac{t^2}{1+t^2} = \lim_{t\to\infty} \frac{t^2}{1+t^2} = 1$.
4. The results of step 3 show that $y = 1$ is a horizontal asymptote. There is no vertical asymptote since the denominator is never zero.
5. $f'(t) = \frac{(1+t^2)(2t) - t^2(2t)}{(1+t^2)^2} = \frac{2t}{(1+t^2)^2} = 0$, if $t = 0$, the only critical number of f. Because $f'(t) < 0$ if $t < 0$ and $f'(t) > 0$ if $t > 0$, we see that f is decreasing on $(-\infty, 0)$ and increasing on $(0, \infty)$.
6. The results of step 5 show that $(0, 0)$ is a relative minimum.
7. $f''(t) = \frac{(1+t^2)^2(2) - 2t(2)(1+t^2)(2t)}{(1+t^2)^4} = \frac{2(1+t^2)\left[(1+t^2) - 4t^2\right]}{(1+t^2)^4} = \frac{2(1-3t^2)}{(1+t^2)^3} = 0$ if $t = \pm\frac{\sqrt{3}}{3}$.

   The sign diagram of f'' shows that f is concave downward on $\left(-\infty, -\frac{\sqrt{3}}{3}\right)$ and $\left(\frac{\sqrt{3}}{3}, \infty\right)$ and concave upward on $\left(-\frac{\sqrt{3}}{3}, \frac{\sqrt{3}}{3}\right)$.

   $- \; - \; - \; 0 \; + \; + \; + \; + \; + \; 0 \; - \; - \; -$ sign of f''

   x: $-\frac{\sqrt{3}}{3}$, 0, $\frac{\sqrt{3}}{3}$

8. The results of step 7 show that $\left(-\frac{\sqrt{3}}{3}, \frac{1}{4}\right)$ and $\left(\frac{\sqrt{3}}{3}, \frac{1}{4}\right)$ are inflection points.

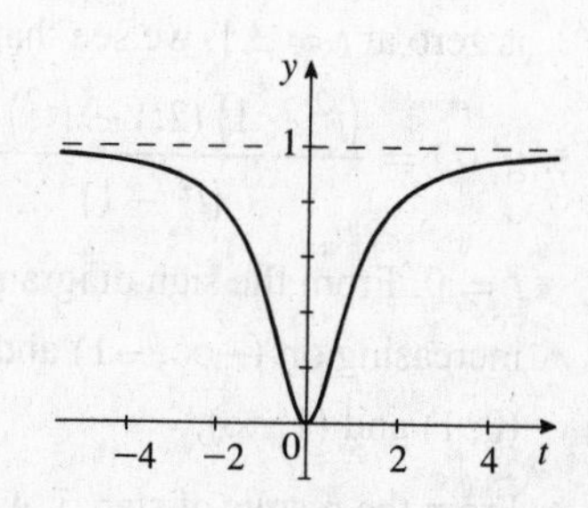

55. $g(t) = -\frac{t^2-2}{t-1}$. We first gather the following information on g.

1. The domain of g is $(-\infty, 1) \cup (1, \infty)$
2. Setting $t = 0$ gives -2 as the y-intercept.
3. $\lim_{t\to-\infty}\left(-\frac{t^2-2}{t-1}\right) = \infty$ and $\lim_{t\to\infty}\left(-\frac{t^2-2}{t-1}\right) = -\infty$.
4. There is no horizontal asymptotes. The denominator is equal to zero at $t = 1$ at which number the numerator is not equal to zero. Therefore $t = 1$ is a vertical asymptote.

5. $g'(t) = -\dfrac{(t-1)(2t) - (t^2-2)(1)}{(t-1)^2}$

$= -\dfrac{t^2-2t+2}{(t-1)^2} \neq 0$ for all values of t.

g' not defined
– – – – – – – – – sign of g'
0 1 t

The sign diagram of g' shows that g is decreasing on $(-\infty, 1)$ and $(1, \infty)$.

6. Because there is no critical number, g has no relative extremum.

7. $g''(t) = -\dfrac{(t-1)^2(2t-2) - (t^2-2t+2)(2)(t-1)}{(t-1)^4}$

$= \dfrac{-2(t-1)(t^2-2t+1-t^2+2t-2)}{(t-1)^4} = \dfrac{2}{(t-1)^3}$.

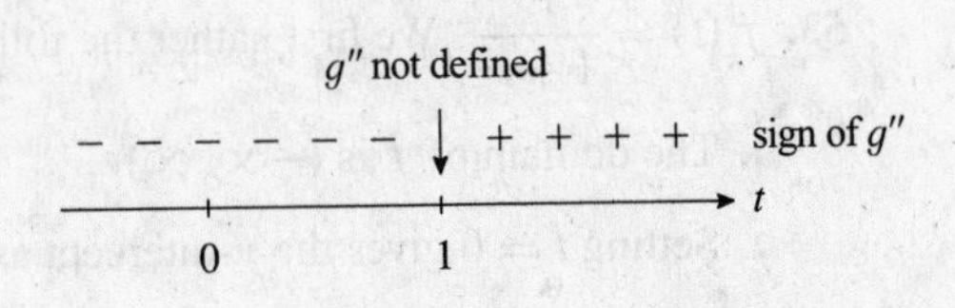

The sign diagram of g''shows that the graph of g is concave upward on $(1, \infty)$ and concave downward on $(-\infty, 1)$.

8. There is no inflection point because $g''(x) \neq 0$ for all x.

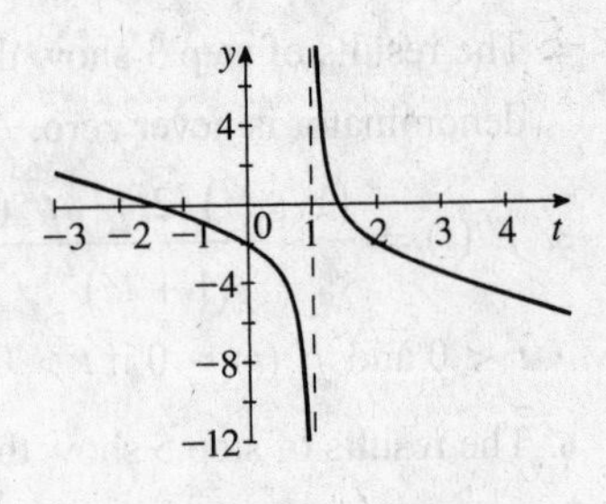

57. $g(t) = \dfrac{t^2}{t^2-1}$. We first gather the following information on g.

1. Because $t^2 - 1 = 0$ if $t = \pm 1$, we see that the domain of g is $(-\infty, -1) \cup (-1, 1) \cup (1, \infty)$.
2. Setting $t = 0$ gives 0 as the y-intercept. Setting $y = 0$ gives 0 as the t-intercept.
3. $\lim_{t \to -\infty} g(t) = \lim_{t \to \infty} g(t) = 1$.
4. The results of step 3 show that $y = 1$ is a horizontal asymptote. Because the denominator (but not the numerator) is zero at $t = \pm 1$, we see that $t = \pm 1$ are vertical asymptotes.
5. $g'(t) = \dfrac{(t^2-1)(2t) - (t^2)(2t)}{(t^2-1)^2} = -\dfrac{2t}{(t^2-1)^2} = 0$ if $t = 0$. From the sign diagram of g', we see that g is increasing on $(-\infty, -1)$ and $(-1, 0)$ and decreasing on $(0, 1)$ and $(1, \infty)$.

g' not defined
+ + + + – – – – – – sign of g'
-1 0 1 t

6. From the results of step 5, we see that g has a relative maximum at $t = 0$.
7. $g''(t) = \dfrac{(t^2-1)^2(-2) - (-2t)(2)(t^2-1)(2t)}{(t^2-1)^4} = \dfrac{2(t^2-1)\left[-(t^2-1)+4t^2\right]}{(t^2-1)^4} = \dfrac{2(-t^2+1+4t^2)}{(t^2-1)^3}$

$= \dfrac{2(3t^2+1)}{(t^2-1)^3}$.

From the sign diagram, we see that the graph of g is concave upward on $(-\infty, -1)$ and $(1, \infty)$ and concave downward on $(-1, 1)$.

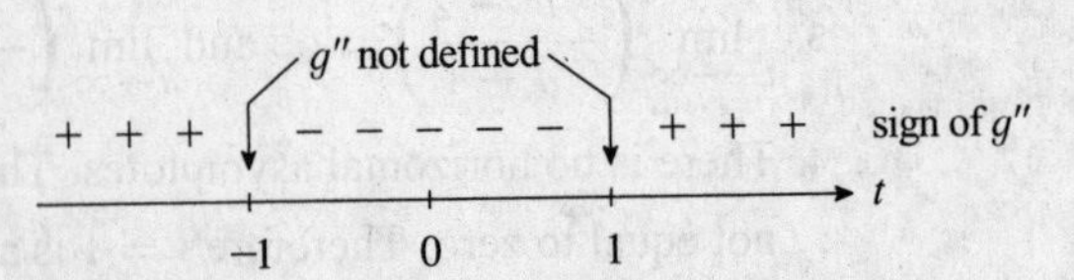

8. Because g is undefined at ± 1, the graph of g has no inflection point.

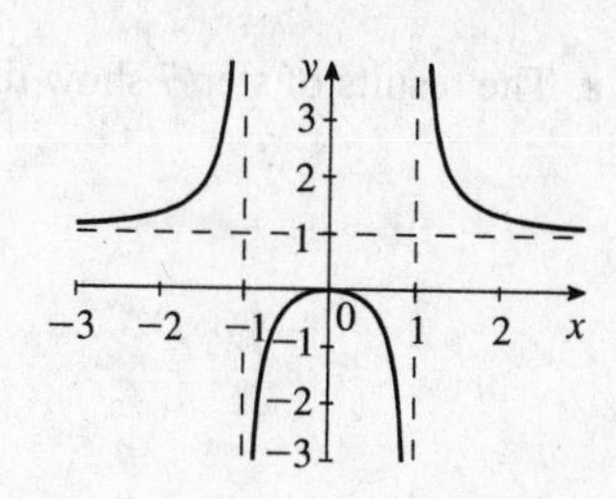

59. $h(x) = (x-1)^{2/3} + 1$. We begin by obtaining the following information on h.

1. The domain of h is $(-\infty, \infty)$.
2. Setting $x = 0$ gives 2 as the y-intercept; since $h(x) \neq 0$ there is no x-intercept.
3. $\lim_{x\to\infty} \left[(x-1)^{2/3}+1\right] = \infty$ and $\lim_{x\to-\infty} \left[(x-1)^{2/3}+1\right] = \infty$.
4. There is no asymptote.
5. $h'(x) = \frac{2}{3}(x-1)^{-1/3}$ and is positive if $x > 1$ and negative if $x < 1$. Thus, h is increasing on $(1, \infty)$ and decreasing on $(-\infty, 1)$.
6. From step 5, we see that h has a relative minimum at $(1, 1)$.
7. 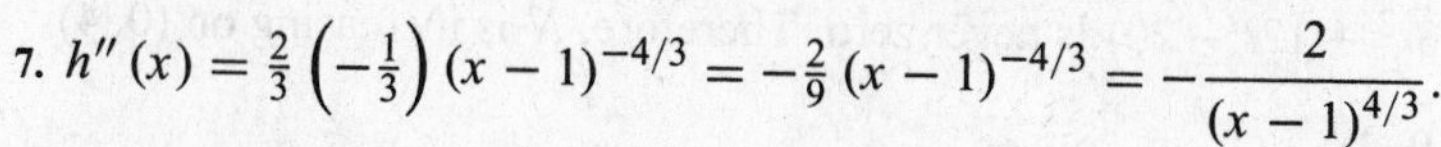
$h''(x) = \frac{2}{3}\left(-\frac{1}{3}\right)(x-1)^{-4/3} = -\frac{2}{9}(x-1)^{-4/3} = -\frac{2}{(x-1)^{4/3}}$.

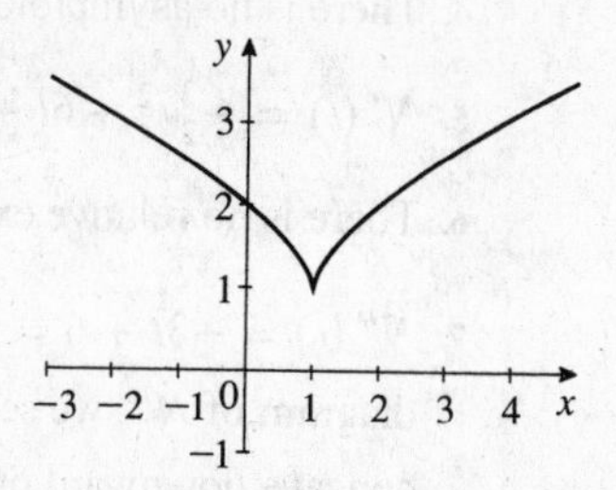

Because $h''(x) < 0$ on $(-\infty, 1)$ and $(1, \infty)$, we see that h is concave downward on $(-\infty, 1)$ and $(1, \infty)$. Note that $h''(x)$ is not defined at $x = 1$.

8. From the results of step 7, we see h has no inflection point.

61. a. The denominator of $C(x)$ is equal to zero if $x = 100$. Also, $\lim_{x\to 100^-} \frac{0.5x}{100-x} = \infty$ and $\lim_{x\to 100^+} \frac{0.5x}{100-x} = -\infty$. Therefore, $x = 100$ is a vertical asymptote of C.

b. No, because the denominator is equal to zero in that case.

63. a. Because $\lim_{t\to\infty} C(t) = \lim_{t\to\infty} \frac{0.2t}{t^2+1} = \lim_{t\to\infty} \left(\frac{0.2}{t+\frac{1}{t^2}}\right) = 0$, $y = 0$ is a horizontal asymptote.

b. Our results reveal that as time passes, the concentration of the drug decreases and approaches zero.

65. $G(t) = -0.2t^3 + 2.4t^2 + 60$. We first gather the following information on G.

1. The domain of G is restricted to $[0, 8]$.
2. Setting $t = 0$ gives 60 as the y-intercept.

Step 3 is unnecessary in this case because of the restricted domain.

4. There is no asymptote because G is a polynomial function.
5. $G'(t) = -0.6t^2 + 4.8t = -0.6t(t-8) = 0$ if $t = 0$ or $t = 8$, critical numbers of G. But $G'(t) > 0$ on $(0, 8)$, so G is increasing on its domain.
6. The results of step 5 tell us that there is no relative extremum.
7. $G''(t) = -1.2t + 4.8 = -1.2(t-4)$. The sign diagram of G'' shows that G is concave upward on $(0, 4)$ and concave downward on $(4, 8)$.

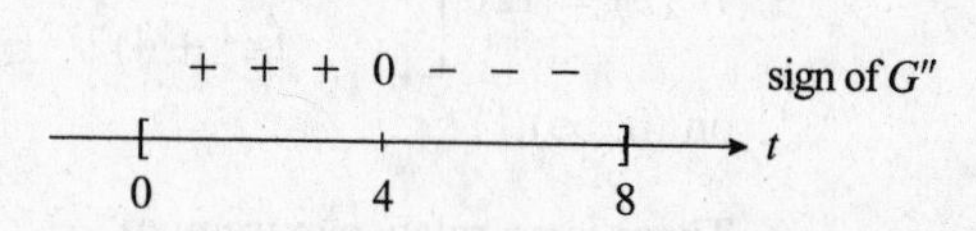

8. The results of step 7 show that (4, 85.6) is an inflection point.

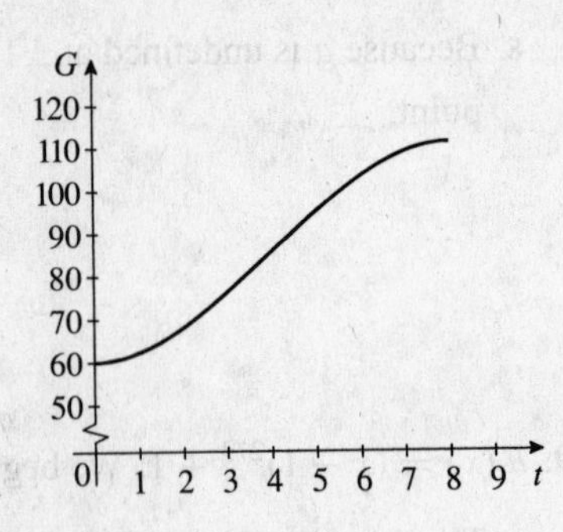

67. $N(t) = -\frac{1}{2}t^3 + 3t^2 + 10t$, $0 \le t \le 4$. We first gather the following information on N.

1. The domain of N is restricted to $[0, 4]$.

2. The y-intercept is 0.
Step 3 does not apply because the domain of $N(t)$ is $[0, 4]$.

4. There is no asymptote.

5. $N'(t) = -\frac{3}{2}t^2 + 6t + 10 = -\frac{1}{2}(3t^2 - 12t - 20)$ is never zero. Therefore, N is increasing on $(0, 4)$.

6. There is no relative extremum in $(0, 4)$.

7. $N''(t) = -3t + 6 = -3(t - 2) = 0$ at $t = 2$. From the sign diagram of N'', we see that N is concave upward on $(0, 2)$ and concave downward on $(2, 4)$.

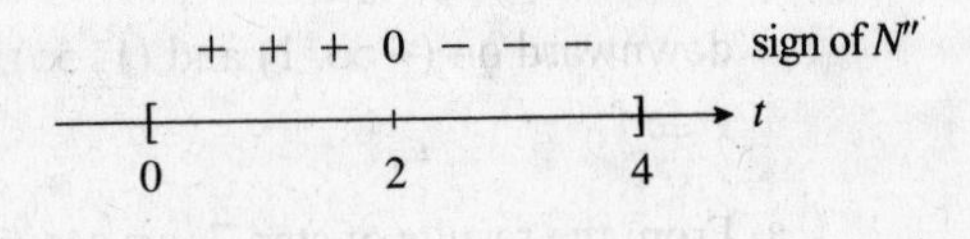

8. The point (2, 28) is an inflection point.

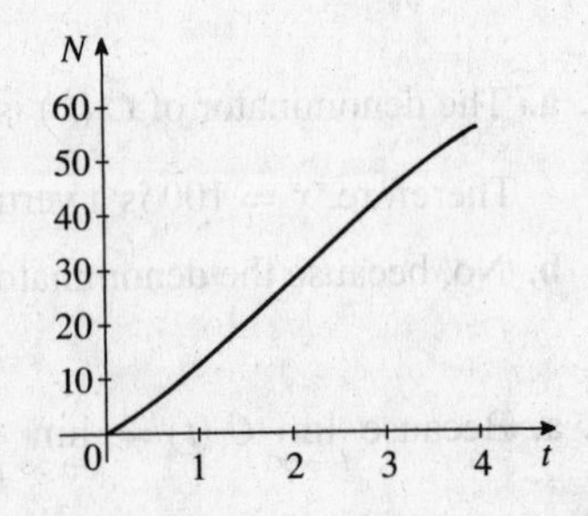

69. $T(x) = \dfrac{120x^2}{x^2 + 4}$. We first gather the following information on T.

1. The domain of T is $[0, \infty)$.

2. Setting $x = 0$ gives 0 as the y-intercept.

3. $\displaystyle\lim_{x\to\infty} \frac{120x^2}{x^2 + 4} = 120$.

4. The results of step 3 show that $y = 120$ is a horizontal asymptote.

5. $T'(x) = 120\left[\dfrac{(x^2 + 4)2x - x^2(2x)}{(x^2 + 4)^2}\right] = \dfrac{960x}{(x^2 + 4)^2}$. Because $T'(x) > 0$ if $x > 0$, we see that T is increasing on $(0, \infty)$.

6. There is no relative extremum.

7. $T''(x) = 960\left[\frac{(x^2+4)^2 - x(2)(x^2+4)(2x)}{(x^2+4)^4}\right] = \frac{960(x^2+4)\left[(x^2+4)-4x^2\right]}{(x^2+4)^4} = \frac{960(4-3x^2)}{(x^2+4)^3}$.

The sign diagram for T'' shows that T is concave downward on $\left(\frac{2\sqrt{3}}{3}, \infty\right)$ and concave upward on $\left(0, \frac{2\sqrt{3}}{3}\right)$.

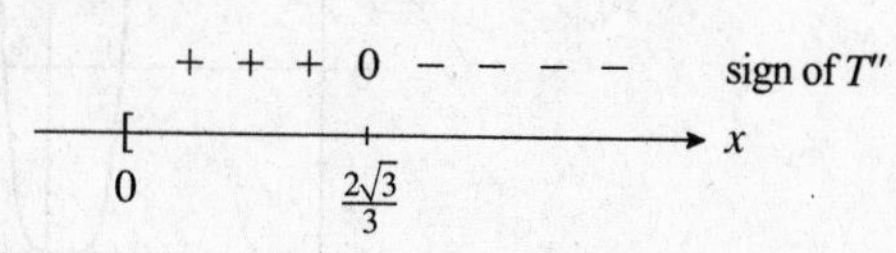

8. We see from the results of step 7 that $\left(\frac{2\sqrt{3}}{3}, 30\right)$ is an inflection point.

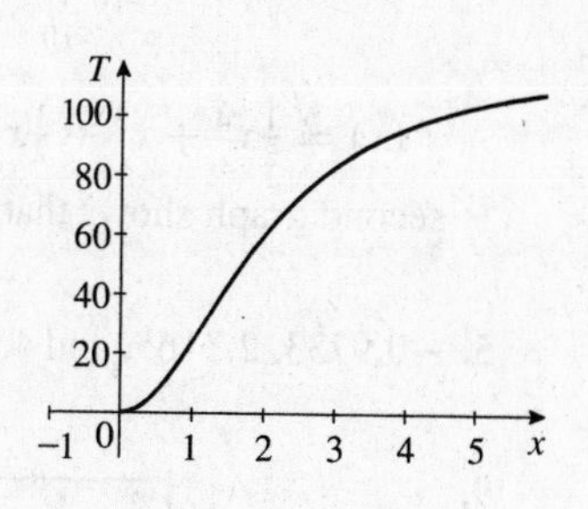

71. $C(x) = \frac{0.5x}{100-x}$. We first gather the following information on C.

1. The domain of C is $[0, 100)$.
2. Setting $x = 0$ gives the y-intercept as 0.

Because of the restricted domain, we omit steps 3 and 4.

5. $C'(x) = 0.5\left[\frac{(100-x)(1)-x(-1)}{(100-x)^2}\right] = \frac{50}{(100-x)^2} > 0$ for all $x \neq 100$. Therefore C is increasing on $(0, 100)$.
6. There is no relative extremum.
7. $C''(x) = -\frac{100}{(100-x)^3}$, so $C''(x) > 0$ if $x < 100$ and the graph of C is concave upward on $(0, 100)$.
8. There is no inflection point.

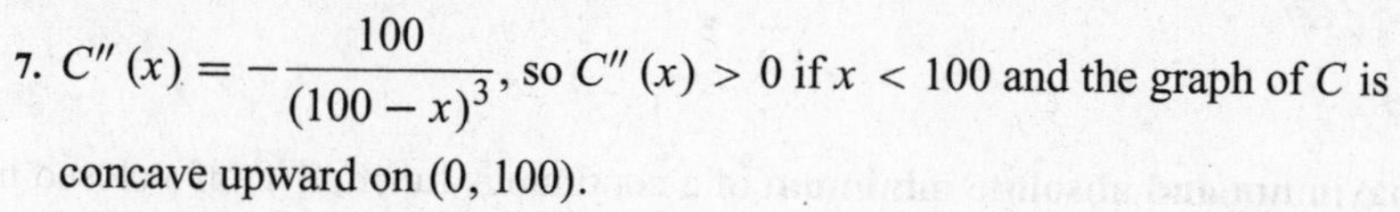

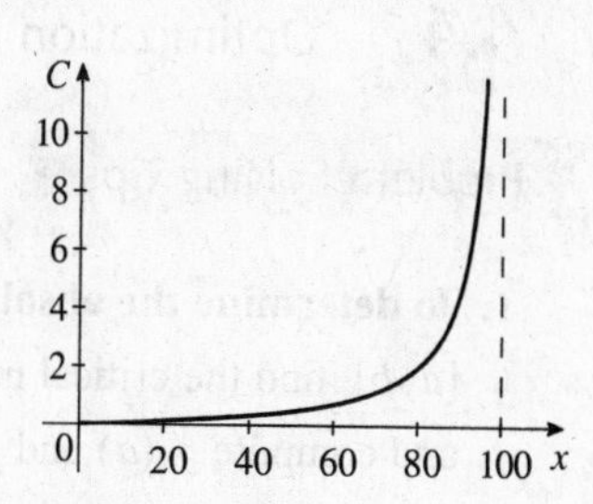

73. False. Consider $f(x) = \begin{cases} 0 & x \leq 0 \\ 1/x & x > 0 \end{cases}$ The graph of f intersects its vertical asymptote at the point $(0, 0)$.

Using Technology page 304

1.

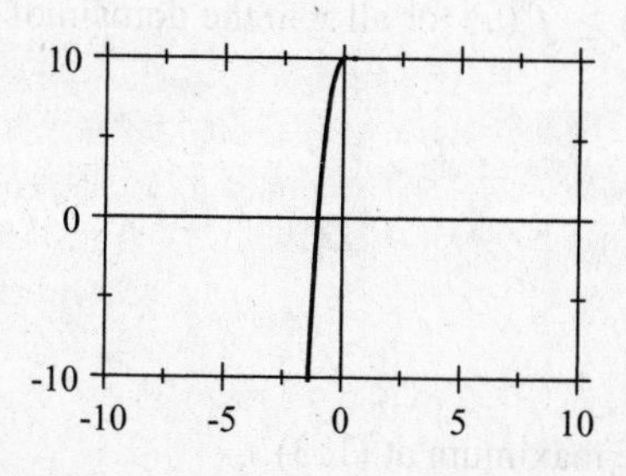

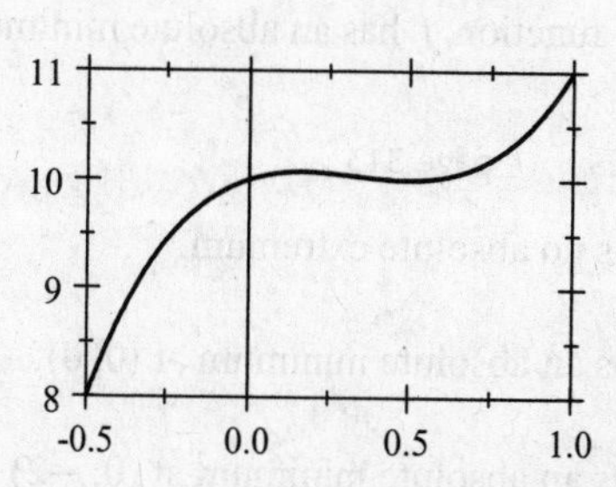

$f(x) = 4x^3 - 4x^2 + x + 10$, so $f'(x) = 12x^2 - 8x + 1 = (6x-1)(2x-1) = 0$ if $x = \frac{1}{6}$ or $x = \frac{1}{2}$. The second graph shows that f has a maximum at $x = \frac{1}{6}$ and a minimum at $x = \frac{1}{2}$.

3. 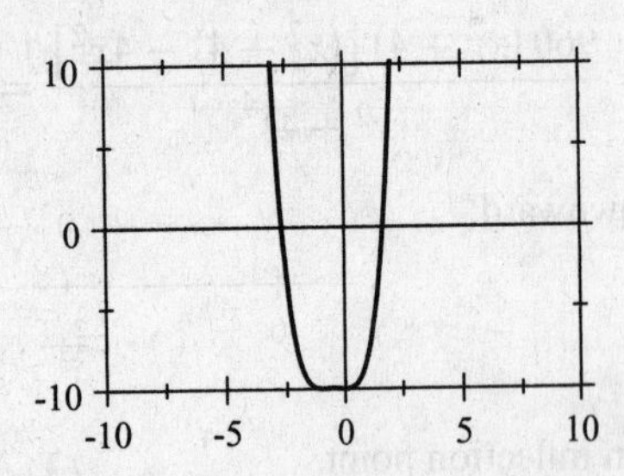

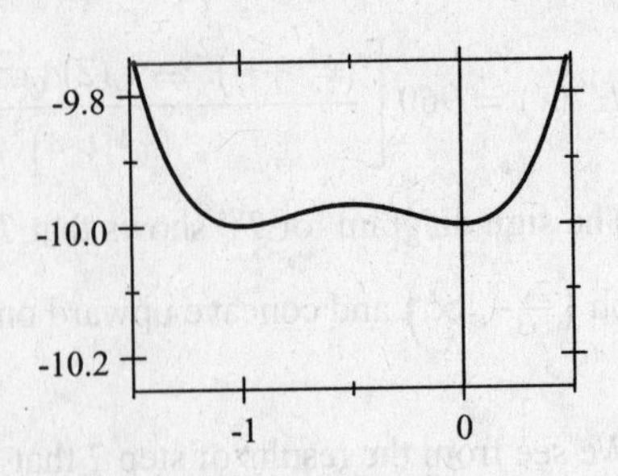

$f(x) = \frac{1}{2}x^4 + x^3 + \frac{1}{2}x^2 - 10$, so $f'(x) = 2x^3 + 3x^2 + x = x(x+1)(2x+1) = 0$ if $x = -1, -\frac{1}{2}$, or 0. The second graph shows that x has minima at $x = -1$ and $x = 0$ and a maximum at $x = -\frac{1}{2}$.

5. -0.9733, 2.3165, and 4.6569.

7. 1.5142

9. 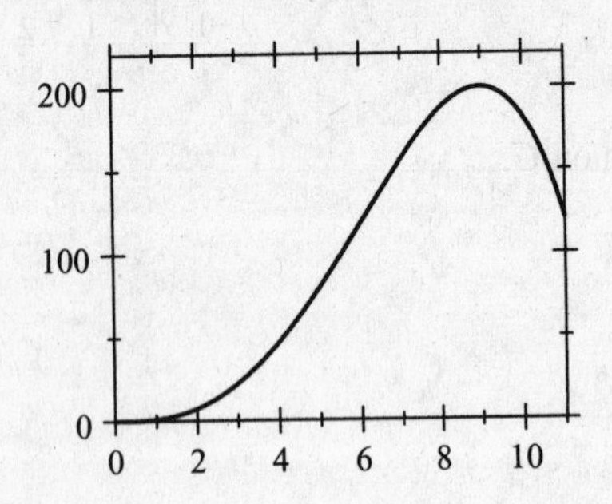

4.4 Optimization I

Problem-Solving Tips

1. **To determine the absolute maximum and absolute minimum** of a continuous function f on a closed interval $[a, b]$, find the critical numbers of f that lie in (a, b). Then compute the value of f at each critical number of f and compute $f(a)$ and $f(b)$. The largest and smallest of these values are the absolute maximum and absolute minimum values of f, respectively.

2. Note that the procedure in Tip 1 holds only for a *continuous* function f over a *closed* interval $[a, b]$.

Concept Questions page 313

1. a. A function f has an absolute maximum at a if $f(x) \leq f(a)$ for all x in the domain of f.

b. A function f has an absolute minimum at a if $f(x) \geq f(a)$ for all x in the domain of f.

Exercises page 313

1. f has no absolute extremum.

3. f has an absolute minimum at $(0, 0)$.

5. f has an absolute minimum at $(0, -2)$ and an absolute maximum at $(1, 3)$.

7. f has an absolute minimum at $\left(\frac{3}{2}, -\frac{27}{16}\right)$ and an absolute maximum at $(-1, 3)$.

9. The graph of $f(x) = 2x^2 + 3x - 4$ is a parabola that opens upward. Therefore, the vertex of the parabola is the absolute minimum of f. To find the vertex, we solve the equation $f'(x) = 4x + 3 = 0$, finding $x = -\frac{3}{4}$. We conclude that the absolute minimum value is $f\left(-\frac{3}{4}\right) = -\frac{41}{8}$.

11. Because $\lim_{x \to -\infty} x^{1/3} = -\infty$ and $\lim_{x \to \infty} x^{1/3} = \infty$, we see that h is unbounded. Therefore, it has no absolute extremum.

13. $f(x) = \dfrac{1}{1 + x^2}$. Using the techniques of graphing, we sketch the graph of f (see Figure 40 on page 278 of the text). The absolute maximum of f is $f(0) = 1$. Alternatively, observe that $1 + x^2 \geq 1$ for all real values of x. Therefore, $f(x) \leq 1$ for all x, and we see that the absolute maximum is attained when $x = 0$.

15. $f(x) = x^2 - 2x - 3$ and $f'(x) = 2x - 2 = 0$, so $x = 1$ is a critical number. From the table, we conclude that the absolute maximum value is $f(-2) = 5$ and the absolute minimum value is $f(1) = -4$.

x	-2	1	3
$f(x)$	5	-4	0

17. $f(x) = -x^2 + 4x + 6$; The function f is continuous and defined on the closed interval $[0, 5]$. $f'(x) = -2x + 4$, and so $x = 2$ is a critical number. From the table, we conclude that $f(2) = 10$ is the absolute maximum value and $f(5) = 1$ is the absolute minimum value.

x	0	2	5
$f(x)$	6	10	1

19. The function $f(x) = x^3 + 3x^2 - 1$ is continuous and defined on the closed interval $[-3, 2]$ and differentiable in $(-3, 2)$. The critical numbers of f are found by solving $f'(x) = 3x^2 + 6x = 3x(x + 2) = 0$, giving $x = -2$ and $x = 0$. From the table, we see that the absolute maximum value of f is $f(2) = 19$ and the absolute minimum value is $f(-3) = f(0) = -1$.

x	-3	-2	0	2
$f(x)$	-1	3	-1	19

21. The function $g(x) = 3x^4 + 4x^3$ is continuous on the closed interval $[-2, 1]$ and differentiable in $(-2, 1)$. The critical numbers of g are found by solving $g'(x) = 12x^3 + 12x^2 = 12x^2(x + 1) = 0$, giving $x = 0$ and $x = -1$. From the table, we see that $g(-2) = 16$ is the absolute maximum value of g and $g(-1) = -1$ is the absolute minimum value of g.

x	-2	-1	0	1
$g(x)$	16	-1	0	7

23. $f(x) = \dfrac{x + 1}{x - 1}$ on $[2, 4]$. Next, we compute $f'(x) = \dfrac{(x - 1)(1) - (x + 1)(1)}{(x - 1)^2} = -\dfrac{2}{(x - 1)^2}$. Because there is no critical number ($x = 1$ is not in the domain of f), we need only test the endpoints. We conclude that $f(4) = \frac{5}{3}$ is the absolute minimum value and $f(2) = 3$ is the absolute maximum value.

25. $f(x) = 4x + \dfrac{1}{x}$ is continuous on $[1, 4]$ and differentiable in $(1, 4)$. To find the critical numbers of f, we solve $f'(x) = 4 - \dfrac{1}{x^2} = 0$, obtaining $x = \pm\frac{1}{2}$. Because these critical numbers lie outside the interval $[1, 4]$, they are not candidates for the absolute extrema of f. Evaluating f at the endpoints of the interval $[1, 4]$, we find that the absolute maximum value of f is $f(4) = \frac{65}{4}$, and the absolute minimum value of f is $f(1) = 5$.

27. $f(x) = \frac{1}{2}x^2 - 2\sqrt{x} = \frac{1}{2}x^2 - 2x^{1/2}$. To find the critical numbers of f, we solve $f'(x) = x - x^{-1/2} = 0$, or $x^{3/2} - 1 = 0$, obtaining $x = 1$. From the table, we conclude that $f(3) \approx 1.04$ is the absolute maximum value and $f(1) = -\frac{3}{2}$ is the absolute minimum value.

x	0	1	3
$f(x)$	0	$-\frac{3}{2}$	$\frac{9}{2} - 2\sqrt{3} \approx 1.04$

29.

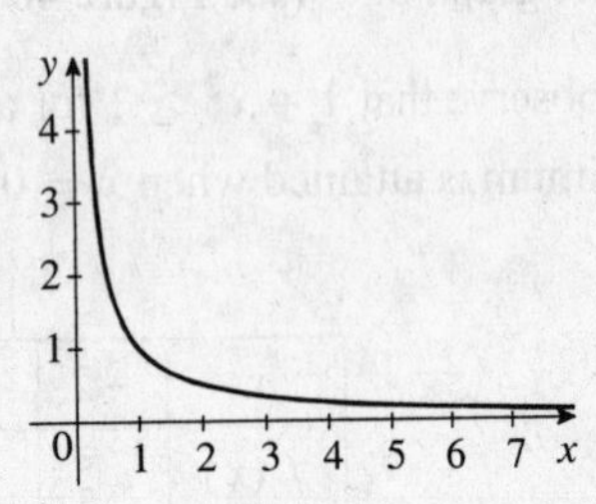

From the graph of $f(x) = \dfrac{1}{x}$ for $x > 0$, we conclude that f has no absolute extremum.

31. $f(x) = 3x^{2/3} - 2x$. The function f is continuous on $[0, 3]$ and differentiable on $(0, 3)$. To find the critical numbers of f, we solve $f'(x) = 2x^{-1/3} - 2 = 0$, obtaining $x = 1$ as the critical number. From the table, we conclude that the absolute maximum value is $f(1) = 1$ and the absolute minimum value is $f(0) = 0$.

x	0	1	3
$f(x)$	0	1	$3^{5/3} - 6 \approx 0.24$

33. $f(x) = x^{2/3}(x^2 - 4)$, so $f'(x) = x^{2/3}(2x) + \frac{2}{3}x^{-1/3}(x^2 - 4) = \frac{2}{3}x^{-1/3}\left[3x^2 + (x^2 - 4)\right] = \dfrac{8(x^2 - 1)}{3x^{1/3}} = 0$. Observe that f' is not defined at $x = 0$. Furthermore, $f'(x) = 0$ at $x \pm 1$. So the critical numbers of f are -1 and 0, and 1. From the table, we see that f has absolute minima at $(-1, -3)$ and $(1, -3)$ and absolute maxima at $(0, 0)$ and $(2, 0)$.

x	-1	0	1	2
$f(x)$	-3	0	-3	0

35. $f(x) = \dfrac{x}{x^2 + 2}$. To find the critical numbers of f, we solve $f'(x) = \dfrac{(x^2 + 2) - x(2x)}{(x^2 + 2)^2} = \dfrac{2 - x^2}{(x^2 + 2)^2} = 0$, obtaining $x = \pm\sqrt{2}$. Because $x = -\sqrt{2}$ lies outside $[-1, 2]$, $x = \sqrt{2}$ is the only critical number in the given interval.

From the table, we conclude that $f\left(\sqrt{2}\right) = \frac{\sqrt{2}}{4} \approx 0.35$ is the absolute maximum value and $f(-1) = -\frac{1}{3}$ is the absolute minimum value.

x	-1	$\sqrt{2}$	2
$f(x)$	$-\frac{1}{3}$	$\frac{\sqrt{2}}{4} \approx 0.35$	$\frac{1}{3}$

37. The function $f(x) = \dfrac{x}{\sqrt{x^2+1}} = \dfrac{x}{(x^2+1)^{1/2}}$ is continuous on the closed interval $[-1, 1]$ and differentiable on $(-1, 1)$. To find the critical numbers of f, we first compute

$$f'(x) = \frac{(x^2+1)^{1/2}(1) - x\left(\frac{1}{2}\right)(x^2+1)^{-1/2}(2x)}{\left[(x^2+1)^{1/2}\right]^2} = \frac{(x^2+1)^{-1/2}\left[x^2+1-x^2\right]}{x^2+1} = \frac{1}{(x^2+1)^{3/2}},$$ which is never equal to zero. We compute $f(x)$ at the endpoints, and conclude that $f(-1) = -\frac{\sqrt{2}}{2}$ is the absolute minimum value and $f(1) = \frac{\sqrt{2}}{2}$ is the absolute maximum value.

39. $h(t) = -16t^2 + 64t + 80$. To find the maximum value of h, we solve $h'(t) = -32t + 64 = -32(t-2) = 0$, giving $t = 2$ as the critical number of h. Furthermore, this value of t gives rise to the absolute maximum value of h since the graph of h is a parabola that opens downward. The maximum height is given by $h(2) = -16(4) + 64(2) + 80 = 144$, or 144 feet.

41. $f(t) = 0.136t^2 + 0.127t + 18.1$, so $f'(t) = 0.272t + 0.127$. Setting $f'(t) = 0$ gives $0.272t = -0.127$, so $t \approx -0.467$. This value of t lies outside the interval $[0, 4]$, so f has no critical number on that interval. From the table, we conclude that the lowest and highest strikeout rates are 18.1% and 20.8%, occurring in 2009 and 2013 respectively.

t	0	4
$f(t)$	18.1	20.784

43. Observe that f is continuous on $[0, 4]$. Next, we compute

$$f'(t) = \frac{d}{dt}\left(20t - 40t^{1/2} + 50\right) = 20 - 20t^{-1/2} = 20t^{-1/2}\left(t^{1/2} - 1\right) = 20\frac{\sqrt{t}-1}{\sqrt{t}}.$$ Observe that $t = 1$ is the only critical number of f in $(0, 4)$. Because $f(0) = 50$, $f(1) = 30$, and $f(4) = 50$, we conclude that f attains its minimum value of 30 at $t = 1$. This tells us that the traffic is moving at the slowest rate at 7 a.m. and the average speed of a vehicle at that time is 30 mph.

45. $h(t) = -\frac{1}{3}t^3 + 4t^2 + 20t + 2$, so $h'(t) = -t^2 + 8t + 20 = -\left(t^2 - 8t - 20\right) = -(t-10)(t+2) = 0$ if $t = -2$ or $t = 10$. Rejecting the negative root, we take $t = 10$. Next, we compute $h''(t) = -2t + 8$. Because $h''(10) = -20 + 8 = -12 < 0$, the Second Derivative Test indicates that the point $t = 10$ gives a relative maximum. From physical considerations, or from a sketch of the graph of h, we conclude that the rocket attains its maximum altitude at $t = 10$ with a maximum height of $h(10) = -\frac{1}{3}(10)^3 + 4(10)^2 + 20(10) + 2$, or approximately 268.7 ft.

47. The revenue is $R(x) = px = -0.00042x^2 + 6x$. Therefore, the profit is $P(x) = R(x) - C(x) = -0.00042x^2 + 6x - \left(600 + 2x - 0.00002x^2\right) = -0.0004x^2 + 4x - 600$. $P'(x) = -0.0008x + 4 = 0$ if $x = 5000$, a critical number of P. From the table, we see that Phonola should produce 5000 discs/month.

x	0	5000	12,000
$P(x)$	-600	9400	$-10{,}200$

49. The cost function is $C(x) = V(x) + 20{,}000 = 0.000001x^3 - 0.01x^2 + 50x + 20{,}000$, so the profit function is

$$P(x) = R(x) - C(x) = -0.02x^2 + 150x - 0.000001x^3 + 0.01x^2 - 50x + 20{,}000$$
$$= -0.000001x^3 - 0.01x^2 + 100x - 20{,}000.$$

We want to maximize P on $[0, 7000]$. $P'(x) = -0.000003x^2 - 0.02x + 100$. Setting $P'(x) = 0$ gives $3x^2 + 20{,}000x - 100{,}000{,}000 = 0$, so or $x = \dfrac{-20{,}000 \pm \sqrt{20{,}000^2 + 1{,}200{,}000{,}000}}{6} = -10{,}000$ or $3{,}333.33$.

Thus, $x = 3333.33$ is a critical number in the interval $[0, 7500]$. From the table, we see that a level of production of 3,333 pagers per week will yield a maximum profit of \$165,185.20 per week.

x	0	3333.33	7500
$P(x)$	$-20{,}000$	165,185.2	$-254{,}375$

51. The cost function is $C(x) = 0.2\left(0.01x^2 + 120\right)$ and the average cost function is $\overline{C}(x) = \dfrac{C(x)}{x} = 0.2\left(0.01x + \dfrac{120}{x}\right) = 0.002x + \frac{24}{x}$. To find the minimum average cost, we first compute $\overline{C}'(x) = 0.002 - \dfrac{24}{x^2}$. Setting $\overline{C}'(x) = 0$ gives $0.002 - \dfrac{24}{x^2} = 0$, so $x^2 = \dfrac{24}{0.002} = 12{,}000$, and thus $x \approx \pm 110$. We reject the negative root, leaving $x = 110$ as the only critical number of $\overline{C}(x)$. Because $\overline{C}''(x) = 48x^{-3} > 0$ for all $x > 0$, we see that $\overline{C}(x)$ is concave upward on $(0, \infty)$. We conclude that $\overline{C}(110) \approx 0.44$ is the absolute minimum value of $\overline{C}(x)$ and that the average cost is minimized when $x = 110$ units.

53. a. $C(x) = 0.000002x^3 + 5x + 400$, so $\overline{C}(x) = \dfrac{C(x)}{x} = 0.000002x^2 + 5 + \dfrac{400}{x}$.

b. $\overline{C}'(x) = 0.000004x - \dfrac{400}{x^2} = \dfrac{0.000004x^3 - 400}{x^2} = \dfrac{0.000004\left(x^3 - 100{,}000{,}000\right)}{x^2}$. Setting $\overline{C}'(x) = 0$ gives $x = 464$, the only critical number of $\overline{C}$. Next, $\overline{C}''(x) = 0.000004 + \dfrac{800}{x^3}$, so $\overline{C}''(464) > 0$ and by the Second Derivative Test, the point $x = 464$ gives rise to a relative minimum. Because $\overline{C}''(x) > 0$ for all $x > 0$, $\overline{C}$ is concave upward on $(0, \infty)$ and $x = 464$ gives rise to an absolute minimum of $\overline{C}$. Thus, the smallest average product cost occurs when the level of production is 464 cases per day.

c. We want to solve the equation $\overline{C}(x) = C'(x)$, that is, $0.000002x^2 + 5 + \dfrac{400}{x} = 0.000006x^2 + 5$, so $0.000004x^3 = 400$, $x^3 = 100{,}000{,}000$, and $x = 464$.

d. The results are as expected.

55. a. $\overline{C}(x) = \dfrac{C(x)}{x} = 0.0025x + 80 + \dfrac{10{,}000}{x}$.

b. Using the result of Exercise 54, we set $\dfrac{C(x)}{x} = \overline{C}(x) = C'(x)$, obtaining $0.0025x + 80 + \dfrac{10{,}000}{x} = 0.005x + 80$. This is the same equation obtained in Exercise 52(b). The lowest average production cost occurs when the production level is 2000 cases per day.

c. The average cost is equal to the marginal cost when the production level is 2000 cases per day.

d. They are the same, as expected.

57. The revenue function is $R(x) = xp = \frac{50x}{0.01x^2+1}$. To find the maximum value of R, we compute $R'(x) = \frac{(0.01x^2+1)\,50 - 50x\,(0.02x)}{(0.01x^2+1)^2} = -\frac{0.5\,(x^2-100)}{(0.01x^2+1)^2}$. Now $R'(x) = 0$ implies $x = -10$, or $x = 10$. The first root is rejected since x must be greater than or equal to zero. Thus, $x = 10$ is the only critical number. From the table, we conclude that $R(10) = 250$ is the absolute maximum value of R. Thus, the revenue is maximized by selling 10,000 watches.

x	0	10	20
$R(x)$	0	250	200

59. $f(t) = 100\left(\frac{t^2-4t+4}{t^2+4}\right)$.

a. $f'(t) = 100\left[\frac{(t^2+4)(2t-4) - (t^2-4t+4)(2t)}{(t^2+4)^2}\right] = \frac{400\,(t^2-4)}{(t^2+4)^2} = \frac{400\,(t-2)(t+2)}{(t^2+4)^2}$.

From the sign diagram for f', we see that $t = 2$ gives a relative minimum, and we conclude that the oxygen content is the lowest 2 days after the organic waste has been dumped into the pond.

$- - - \ 0 \ + + +$ sign of f', on t from 0, with 0 at $t = 2$.

b. $f''(t) = 400\left[\frac{(t^2+4)^2(2t) - (t^2-4)\,2\,(t^2+4)(2t)}{(t+4)^4}\right] = 400\left[\frac{(2t)(t^2+4)(t^2+4-2t^2+8)}{(t^2+4)^4}\right]$

$= -\frac{800t\,(t^2-12)}{(t^2+4)^3}$.

$f''(t) = 0$ when $t = 0$ and $t = \pm 2\sqrt{3}$. We reject $t = 0$ and $t = -2\sqrt{3}$. From the sign diagram for f'', we see that $t = 2\sqrt{3}$ gives an inflection point of f and we conclude that this is an absolute maximum. Therefore, the rate of oxygen regeneration is greatest 3.5 days after the organic waste has been dumped into the pond.

$0 \ + + + \ 0 \ - - -$ sign of f'', on t from 0, with 0 at $t = 2\sqrt{3}$.

61. We compute $\overline{R}'(x) = \frac{xR'(x) - R(x)}{x^2}$. Setting $\overline{R}'(x) = 0$ gives $xR'(x) - R(x) = 0$, or $R'(x) = \frac{R(x)}{x} = \overline{R}(x)$, so a critical number of $\overline{R}$ occurs when $\overline{R}(x) = R'(x)$. Next, we compute $\overline{R}''(x) = \frac{x^2\left[R'(x) + xR''(x) - R'(x)\right] - \left[xR'(x) - R(x)\right](2x)}{x^4} = \frac{R''(x)}{x} < 0$. Thus, by the Second Derivative Test, the critical number does give the maximum revenue.

63. $G(t) = -0.2t^3 + 2.4t^2 + 60$, so the growth rate is $G'(t) = -0.6t^2 + 4.8t$. To find the maximum growth rate, we compute $G''(t) = -1.2t + 4.8$. Setting $G''(t) = 0$ gives $t = 4$ as a critical number. From the table, we see that G is maximal at $t = 4$; that is, the growth rate is greatest in 2010.

t	0	4	8
$G'(t)$	0	9.6	0

65. $N(t) = -87.244444t^3 - 2482.35t^2 + 46009.26t + 579185$, so
$N'(t) = -261.733328t^2 - 4964.7t + 46009.26$. Setting $N'(t) = 0$ and using the quadratic formula gives
$t = \dfrac{-(-4964.7) \pm \sqrt{(-4964.7)^2 - 4(-261.733328)(46009.26)}}{2(-261.73332)} \approx -25.8$
or 6.82, so 6.82 is an approximate critical number of N. From the table, we see that the number of new prison admissions did indeed peak in 2006 ($t = 6$) at approximately 749,833.

t	2	6.82	11
$N(t)$	660,576	749,833	668,800

67. $S'(t) = \frac{d}{dt}(0.000989t^3 - 0.0486t^2 + 0.7116t + 1.46) = 0.002967t^2 - 0.0972t + 0.7116$. Using the quadratic formula to solve the equation $f'(t) = 0$ gives $t = \dfrac{0.0972 \pm \sqrt{(-0.0972)^2 - 4(0.002967)(0.7116)}}{2(0.002967)} \approx 11.0$
or 21.7. From the table, we see that S has an absolute maximum when $t \approx 11$. Thus, children with superior intelligence have a cortex that reaches maximum thickness around 11 years of age.

t	5	11.0	19
$S(t)$	3.9	4.7	4.2

69. **a.** $P(t) = 0.00074t^3 - 0.0704t^2 + 0.89t + 6.04$, so $P'(t) = 0.00222t^2 - 0.1408t + 0.89 = 0$ implies
$t = \dfrac{0.1408 \pm \sqrt{(0.1408)^2 - 4(0.00222)(0.89)}}{2(0.00222)} \approx 7.12$ or 56.3. The root 56.3 is rejected because it lies outside the interval $[0, 10]$. $P''(t) = 0.00444t - 0.1408$ and $P''(7.12) = -0.109 < 0$, and so $t \approx 7.12$ gives a relative maximum. This occurs around 2071.

b. The population will peak at $P(7.12) \approx 9.075$ billion.

71. We want to minimize the function $E(v) = \dfrac{aLv^3}{v-u}$. Because $v > u$, the function has no points of discontinuity. To find the critical numbers of $E(v)$, we solve the equation $E'(v) = \dfrac{(v-u)3aLv^2 - aLv^3}{(v-u)^2} = \dfrac{aLv^2(2v-3u)}{(v-u)^2} = 0$, obtaining $v = \frac{3}{2}u$ or $v = 0$. Now $v \neq 0$ since $u < v$, so $v = \frac{3}{2}u$ is the only critical number of interest. Because $E'(v) < 0$ if $v < \frac{3}{2}u$ and $E'(v) > 0$ if $v > \frac{3}{2}u$, we see that $v = \frac{3}{2}u$ gives a relative minimum. The nature of the problem suggests that $v = \frac{3}{2}u$ gives the absolute minimum of E (we can verify this by sketching the graph of E). Therefore, the fish must swim at $\frac{3}{2}u$ ft/sec in order to minimize the total energy expended.

73. $\dfrac{dR}{dD} = kD - D^2$ and $\dfrac{d^2R}{dD^2} = k - 2D$. Setting $\dfrac{d^2R}{dD^2} = 0$, we obtain $k = 2D$, or $D = \dfrac{k}{2}$. Because $\dfrac{d^2R}{dD^2} > 0$ for $k < 2D$ and $\dfrac{d^2R}{dD^2} < 0$ for $k > 2D$, we see that $k = 2D$ provides the relative (and absolute) maximum.

75. Setting $P' = 0$ gives $P' = \frac{d}{dR}\left[\frac{E^2 R}{(R+r)^2}\right] = E^2\left[\frac{(R+r)^2 - R(2)(R+r)}{(R+r)^4}\right] = \frac{E^2(r-R)}{(R+r)^3} = 0$. Therefore, $R = r$ is a critical number of P. Because $P'' = E^2\frac{(R+r)^3(-1) - (r-R)(3)(R+r)^2}{(R+r)^6} = \frac{2E^2(R-2r)}{(R+r)^4}$ and $P''(r) = \frac{-2E^2 r}{(2r)^4} = -\frac{E^2}{8r^3} < 0$, the Second Derivative Test and physical considerations both imply that $R = r$ gives a relative maximum value of P. The maximum power is $P = \frac{E^2 r}{(2r)^2} = \frac{E^2}{4r}$ watts.

77. $\frac{dx}{dt} = \frac{d}{dt}\left[1.5(10-t) - 0.0013(10-t)^4\right] = -1.5 - 0.0013(4)(10-t)^3(-1) = -1.5 + 0.0052(10-t)^3$ is continuous everywhere and has zeros where $0.0052\left(10 - t^3\right) = 1.5$; that is, $(10-t)^3 = \frac{1.5}{0.0052}$, or $t = 10 - \sqrt[3]{\frac{1.5}{0.0052}} \approx 3.4$, and so x has the critical number 3.4 in $(0, 10)$. Now $x(0) = 2$, $x(3.4) = 7.4$, and $x(10) = 0$, showing that after 3.4 minutes, the maximum amount of salt (roughly 7.4 lb) is in the tank.

79. False. Let $f(x) = \begin{cases} |x| & \text{if } x \neq 0 \\ 1 & \text{if } x = 0 \end{cases}$ on $[-1, 1]$.

81. True. The absolute extrema of f must occur for some x in (a, b) at which $f'(x) = 0$, or at an endpoint. Since $f'(x) \neq 0$ for all x in (a, b), the absolute extrema of f (and in particular its absolute maximum) must occur at $x = a$ or $x = b$, with value $f(a)$ or $f(b)$.

83. True. This follows from the Second Derivative Test applied to the function $P = R - C$.

85. Because $f(x) = c$ for all x, the function f satisfies $f(x) \leq c$ for all x and so f has absolute maxima at all values of x. Similarly, f has absolute minima at all values of x.

87. a. f is not continuous at $x = 0$ because $\lim_{x\to 0} f(x)$ does not exist.

b. $\lim_{x\to 0^-} f(x) = \lim_{x\to 0^-} \frac{1}{x} = -\infty$ and $\lim_{x\to 0^+} f(x) = \lim_{x\to 0^+} \frac{1}{x} = \infty$.

c.

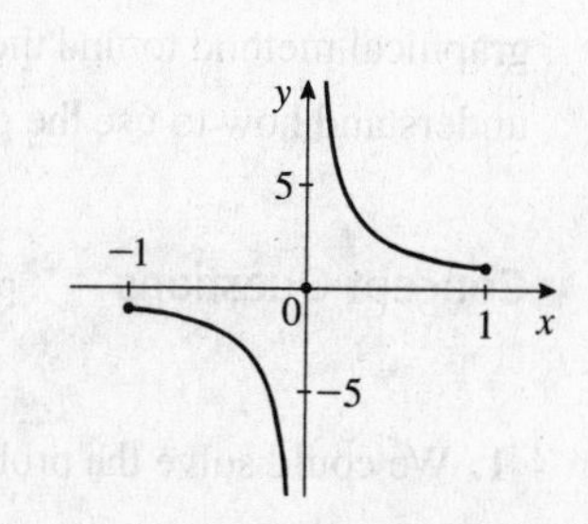

Using Technology page 320

1. Absolute maximum value 145.9, absolute minimum value −4.3834.

3. Absolute maximum value 16, absolute minimum value −0.1257.

5. Absolute maximum value 2.8889, absolute minimum value 0.

7. a.

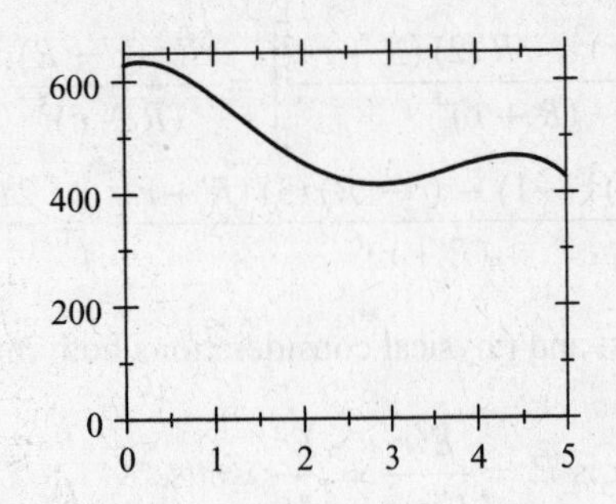

b. Using the function for finding the absolute minimum of f on $[0, 5]$, we see that the absolute minimum value of f is approximately 415.56, occurring when $x \approx 2.87$. This proves the assertion.

9. a. $N(t) = 1.2576t^4 - 26.357t^3 + 127.98t^2 + 82.3t + 43$, so $N'(t) = 5.0304t^3 - 79.071t^2 + 255.96t + 82.3$.

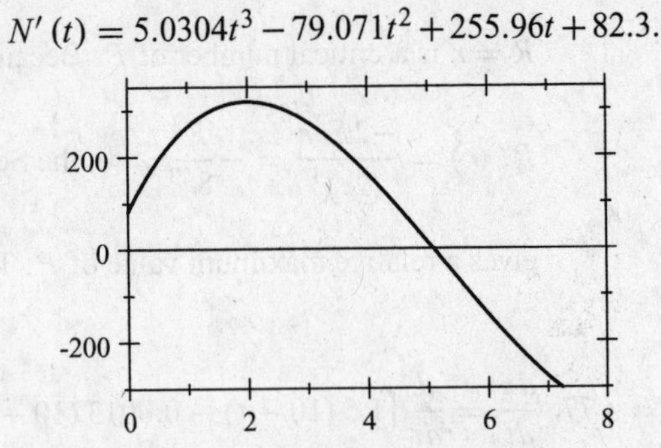

From the graph, we see that $N'(t)$ has a maximum when $t \approx 2$, on February 8.

b. From the graph in part (a), the maximum number of sickouts occurred when $N'(t) = 0$, that is, when $t = 5$. We calculate $N(5) \approx 1145$ canceled flights.

4.5 Optimization II

Problem-Solving Tips

Follow the guidelines given on page 321 of the text to solve the optimization problems that follow. Remember, Theorem 3 in Section 4.4 provides us with a method of computing the absolute extrema of a continuous function over a closed interval $[a, b]$. If the problem involves a function that is to be optimized over an interval that is not closed, then use the graphical method to find the optimal values of f. You might review Example 4 on page 324 of the text to make sure you understand how to use the graphical method.

Concept Questions page 327

1. We could solve the problem by sketching the graph of f and checking to see if there is an absolute extremum.

Exercises page 327

1. Let x and y denote the lengths of two adjacent sides of the rectangle. We want to maximize $A = xy$. But the perimeter is $2x + 2y$ and this is equal to 100, so $2x + 2y = 100$, and therefore $y = 50 - x$. Thus, $A = f(x) = x(50 - x) = -x^2 + 50x$, $0 \le x \le 50$. We allow the "degenerate"cases $x = 0$ and $x = 50$. $A' = -2x + 50 = 0$ implies that $x = 25$ is a critical number of f. $A(0) = 0$, $A(25) = 625$, and $A(50) = 0$, so we see that A is maximized for $x = 25$. The required dimensions are 25 ft by 25 ft.

3. We have $2x + y = 3000$ and we want to maximize the function $A = f(x) = xy = x(3000 - 2x) = 3000x - 2x^2$ on the interval $[0, 1500]$. The critical number of A is obtained by solving $f'(x) = 3000 - 4x = 0$, giving $x = 750$. From the table of values, we conclude that $x = 750$ yields the absolute maximum value of A. Thus, the required dimensions are 750×1500 yards. The maximum area is 1,125,000 yd^2.

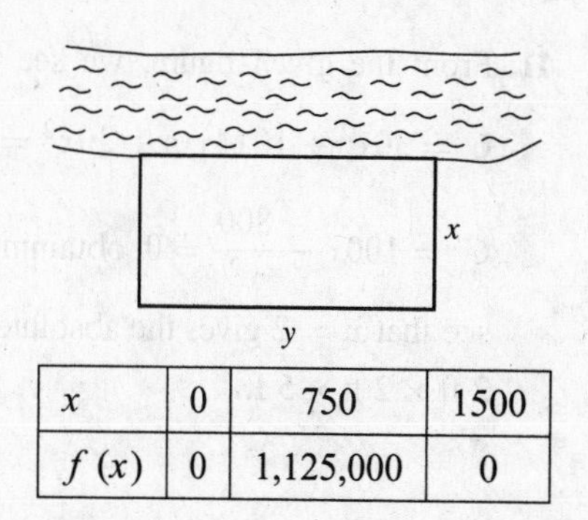

x	0	750	1500
$f(x)$	0	1,125,000	0

5. Let x denote the length of the side made of wood and y the length of the side made of steel. The cost of construction is $C = 6(2x) + 3y$, but $xy = 800$, so $y = \dfrac{800}{x}$. Therefore, $C = f(x) = 12x + 3\left(\dfrac{800}{x}\right) = 12x + \dfrac{2400}{x}$. To minimize C, we compute $f'(x) = 12 - \dfrac{2400}{x^2} = \dfrac{12x^2 - 2400}{x^2} = \dfrac{12(x^2 - 200)}{x^2}$. Setting $f'(x) = 0$ gives $x = \pm\sqrt{200}$ as critical numbers of f. The sign diagram of f' shows that $x = \pm\sqrt{200}$ gives a relative minimum of f.

– – – 0 + + + sign of f'

0 $\sqrt{200}$ x

$f''(x) = \dfrac{4800}{x^3} > 0$ if $x > 0$, and so f is concave upward for $x > 0$. Therefore, $x = \sqrt{200} = 10\sqrt{2}$ yields the absolute minimum. Thus, the dimensions of the enclosure should be $10\sqrt{2}$ ft $\times$ $40\sqrt{2}$ ft, or 14.1 ft $\times$ 56.6 ft.

7. Let the dimensions of each square that is cut out be $x'' \times x''$. Then the dimensions of the box are $(8 - 2x)''$ by $(8 - 2x)''$ by x'', and its volume is be $V = f(x) = x(8 - 2x)^2$. We want to maximize f on $[0, 4]$.

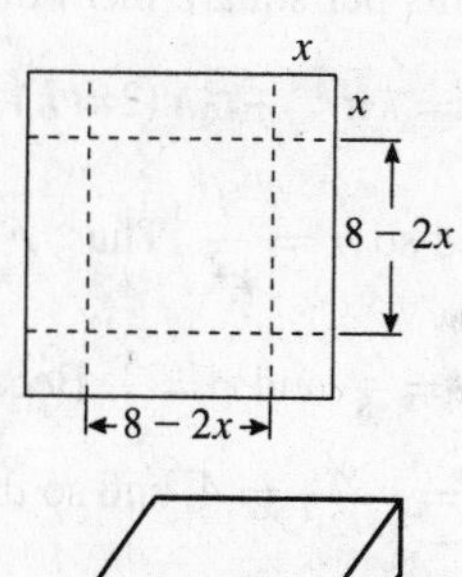

$$\begin{aligned} f'(x) &= (8 - 2x)^2 + x(2)(8 - 2x)(-2) \quad \text{(by the Product Rule)} \\ &= (8 - 2x)[(8 - 2x) - 4x] \\ &= (8 - 2x)(8 - 6x) \\ &= 0 \text{ if } x = 4 \text{ or } \tfrac{4}{3}. \end{aligned}$$

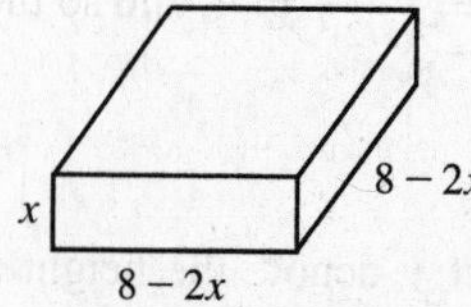

The latter is a critical number in $(0, 4)$. From the table, we see that $x = \frac{4}{3}$ yields an absolute maximum for f, so the dimensions of the box should be $\frac{16}{3}'' \times \frac{16}{3}'' \times \frac{4}{3}''$.

x	0	$\frac{4}{3}$	4
$f(x)$	0	$\frac{1024}{27}$	0

9. Let x denote the length of a side of the base and y the height of the cup, both measured in inches. Then the cost of constructing the cup is $C = 40x^2 + 15(4xy) = 40x^2 + 60xy$ (cents). The volume of the cup is 36 cubic inches, and so $x^2y = 36$ and $y = \dfrac{36}{x^2}$. Therefore, $C(x) = 40x^2 + 60x \cdot \dfrac{36}{x^2} = 40x^2 + \dfrac{2160}{x}$ and $C'(x) = 80x - \dfrac{2160}{x^2}$. Setting $C'(x) = 0$ gives $x^3 = 27$, so $x = 3$. Because $C''(3) = \left[80 + \dfrac{4320}{x^3}\right]_{x=3} > 0$, we see that $x = 3$ gives a relative (and absolute) minimum of C. Also, $y = \dfrac{36}{3^2} = 4$, so the required dimensions are $3'' \times 3'' \times 4''$.

11. From the given figure, we see that $x^2y = 20$ and $y = 20/x^2$, and so $C = 30x^2 + 10(4xy) + 20x^2 = 50x^2 + 40x\left(\frac{20}{x^2}\right) = 50x^2 + \frac{800}{x}$. To find the critical numbers of C, we solve $C' = 100x - \frac{800}{x^2} = 0$, obtaining $100x^3 = 800$, $x^3 = 8$, and $x = 2$. Next, $C'' = \frac{1600}{x^3} > 0$ for all $x > 0$, so we see that $x = 2$ gives the absolute minimum value of C. Because $y = \frac{20}{4} = 5$, we see that the dimensions are 2 ft × 2 ft × 5 ft.

13. $xy = 50$ and so $y = 50/x$. The printed area is $A = (x-1)(y-2) = (x-1)\left(\frac{50}{x} - 2\right) = (x-1)\left(\frac{50-2x}{x}\right) = -2x + 52 - \frac{50}{x}$, so $A' = -2 + \frac{50}{x^2} = \frac{-2(x^2-25)}{x^2} = 0$ if $x = \pm 5$. From the sign diagram for A', we see that $x = 5$ yields a maximum. Because $A'' = -\frac{100}{x^3} < 0$ for $x > 0$, we see that the graph of A is concave downward on $(0, \infty)$ and so $x = 5$ yields an absolute maximum. The dimensions of the paper should therefore be 5″ × 10″.

+ + + 0 − − − sign of A'

0 5 x

15. Denote the radius and height of the cup (in inches) by r and h respectively. Let k denote the price (in cents per square inch) of the material for the base of the cup. Then the cost of constructing the cup is $C = k\pi r^2 + \frac{3}{8}k(2\pi rh) = k\pi\left(r^2 + \frac{3}{4}rh\right)$. It suffices to minimize $F(r) = \frac{C(r)}{k\pi} = r^2 + \frac{3rh}{4}$. But $\pi r^2 h = 9\pi$, and so $h = \frac{9}{r^2}$. Thus, $F(r) = r^2 + \frac{3r}{4}\left(\frac{9}{r^2}\right) = r^2 + \frac{27}{4r}$. Now $F'(r) = 2r - \frac{27}{4r^2} = 0$ gives $8r^3 = 27$, so $r^3 = \frac{27}{8}$ and $r = \frac{3}{2}$. Because $F''(r) = 2 + \frac{27}{2r^2}$, we see that $F''\left(\frac{3}{2}\right) > 0$, and so F has a minimum at $r = \frac{3}{2}$. Also, $h = \frac{9}{(3/2)^2} = 4$, and so the required dimensions are a radius of 1.5 inches and a height of 4 inches.

17. Let y denote the height and x the width of the cabinet. Then $y = \frac{3}{2}x$. Because the volume is to be 2.4 ft^3, we have $xyd = 2.4$, where d *is* the depth of the cabinet. Thus, $x\left(\frac{3}{2}x\right)d = 2.4$, so $d = \frac{2.4(2)}{3x^2} = \frac{1.6}{x^2}$. The cost for constructing the cabinet is $C = 40(2xd + 2yd) + 20(2xy) = 80\left[\frac{1.6}{x} + \left(\frac{3x}{2}\right)\left(\frac{1.6}{x^2}\right)\right] + 40x\left(\frac{3x}{2}\right) = \frac{320}{x} + 60x^2$, so $C'(x) = -\frac{320}{x^2} + 120x = \frac{120x^3 - 320}{x^2} = 0$ if $x = \sqrt[3]{\frac{8}{3}} = \frac{2}{\sqrt[3]{3}} = \frac{2}{3}\sqrt[3]{9}$. Therefore, $x = \frac{2}{3}\sqrt[3]{9}$ is a critical number of C. The sign diagram shows that $x = \frac{2}{3}\sqrt[3]{9}$ gives a relative minimum. Next, $C''(x) = \frac{640}{x^3} + 120 > 0$ for all $x > 0$, telling us that the graph of C is concave upward, so $x = \frac{2}{3}\sqrt[3]{9}$ yields an absolute minimum. The required dimensions are $\frac{2}{3}\sqrt[3]{9}' \times \sqrt[3]{9}' \times \frac{2}{5}\sqrt[3]{9}'$.

− − − 0 + + + + sign of C'

0 $\frac{2\sqrt[3]{9}}{3}$ x

19. Let x denote the number of passengers beyond the 200th. We want to maximize the function $R(x) = (200 + x)(300 - x) = -x^2 + 100x + 60{,}000$. Now $R'(x) = -2x + 100 = 0$ gives $x = 50$, and this is a critical number of R. Because $R''(x) = -2 < 0$, we see that $x = 50$ gives an absolute maximum of R. Therefore, the number of passengers should be 250. The fare will then be \$250/passenger and the revenue will be \$62,500.

21. Let x denote the number of people beyond 20 who sign up for the cruise. Then the revenue is $R(x) = (20 + x)(600 - 4x) = -4x^2 + 520x + 12{,}000$. We want to maximize R on the closed bounded interval $[0, 70]$. $R'(x) = -8x + 520 = 0$ implies $x = 65$, a critical number of R. Evaluating R at this critical number and the endpoints, we see that R is maximized if $x = 65$. Therefore, 85 passengers will result in a maximum revenue of \$28,900. The fare in this case is \$340/passenger.

x	0	65	70
$R(x)$	12,000	28,900	28,800

23. The fuel cost is $x/600$ dollars per mile and the labor cost is $18/x$ dollars per mile. Therefore, the total cost is $C(x) = \frac{18}{x} + \frac{3x}{600}$. We calculate $C'(x) = -\frac{18}{x^2} + \frac{3}{600} = 0$, giving $-\frac{18}{x^2} = -\frac{3}{600}$, $3x^2 = 18(600)$, $x^2 = 3600$, and so $x = 60$. Next, $C''(x) = \frac{48}{x^3} > 0$ for all $x > 0$ so C is concave upward. Therefore, $x = 60$ gives the absolute minimum. The most economical speed is 60 mph.

25. We want to maximize $S = kh^2w$. But $h^2 + w^2 = 24^2$, or $h^2 = 576 - w^2$, so $S = f(w) = kw(576 - w^2) = k(576w - w^3)$. Now, setting $f'(w) = k(576 - 3w^2) = 0$ gives $w = \pm\sqrt{192} \approx \pm 13.86$. Only the positive root is a critical number of interest. Next, we find $f''(w) = -6kw$, and in particular, $f''\left(\sqrt{192}\right) = -6\sqrt{192}k < 0$, so that $w \approx 13.86$ gives a relative maximum of f. Because $f''(w) < 0$ for $w > 0$, we see that the graph of f is concave downward on $(0, \infty)$, and so $w = \sqrt{192}$ gives an absolute maximum of f. We find $h^2 = 576 - 192 = 384$ and so $h \approx 19.60$, so the width and height of the log should be approximately 13.86 inches and 19.60 inches, respectively.

27. We want to minimize $C(x) = 1.50(10{,}000 - x) + 2.50\sqrt{3000^2 + x^2}$ subject to $0 \le x \le 10{,}000$. Now $C'(x) = -1.50 + 2.5\left(\frac{1}{2}\right)(9{,}000{,}000 + x^2)^{-1/2}(2x) = -1.50 + \frac{2.50x}{\sqrt{9{,}000{,}000 + x^2}} = 0$ if $2.5x = 1.50\sqrt{9{,}000{,}000 + x^2}$, or $6.25x^2 = 2.25(9{,}000{,}000 + x^2)$, or $4x^2 = 20{,}250{,}000$, giving $x = 2250$. From the table, we see that $x = 2250$, or 2250 ft, gives the absolute minimum.

x	0	2250	10,000
$C(x)$	22,500	21,000	26,101

29. The time of flight is $T = f(x) = \frac{12 - x}{6} + \frac{\sqrt{x^2 + 9}}{4}$, so

$f'(x) = -\frac{1}{6} + \frac{1}{4}\left(\frac{1}{2}\right)(x^2 + 9)^{-1/2}(2x) = -\frac{1}{6} + \frac{x}{4\sqrt{x^2 + 9}} = \frac{3x - 2\sqrt{x^2 + 9}}{12\sqrt{x^2 + 9}}$. Setting $f'(x) = 0$ gives $3x = 2\sqrt{x^2 + 9}$, $9x^2 = 4(x^2 + 9)$, and $5x^2 = 36$. Therefore, $x = \pm\frac{6}{\sqrt{5}} = \pm\frac{6\sqrt{5}}{5}$. Only the critical number $x = \frac{6\sqrt{5}}{5}$ is of interest. The nature of the problem suggests $x \approx 2.68$ gives an absolute minimum for T.

31. The area enclosed by the rectangular region of the racetrack is $A = (\ell)(2r) = 2r\ell$. The length of the racetrack is $2\pi r + 2\ell$, and is equal to 1760. That is, $2(\pi r + \ell) = 1760$, and $\pi r + \ell = 880$. Therefore, we want to maximize $A = f(r) = 2r(880 - \pi r) = 1760r - 2\pi r^2$. The restriction on r is $0 \le r \le \frac{880}{\pi}$. To maximize A, we compute $f'(r) = 1760 - 4\pi r$. Setting $f'(r) = 0$ gives $r = \frac{1760}{4\pi} = \frac{440}{\pi} \approx 140$. Because $f(0) = f\left(\frac{880}{\pi}\right) = 0$, we see that the maximum rectangular area is enclosed if we take $r = \frac{440}{\pi}$ and $\ell = 880 - \pi\left(\frac{440}{\pi}\right) = 440$. So $r = 140$ and $\ell = 440$. The total area enclosed is $2r\ell + \pi r^2 = 2\left(\frac{440}{\pi}\right)(440) + \pi\left(\frac{440}{\pi}\right)^2 = \frac{2(440)^2}{\pi} + \frac{440^2}{\pi} = \frac{580{,}800}{\pi} \approx 184{,}874\text{ ft}^2$.

33. Let x denote the number of bottles in each order. We want to minimize $C(x) = 200\left(\frac{2{,}000{,}000}{x}\right) + \frac{x}{2}(0.40) = \frac{400{,}000{,}000}{x} + 0.2x$. We compute $C'(x) = -\frac{400{,}000{,}000}{x^2} + 0.2$.

Setting $C'(x) = 0$ gives $x^2 = \frac{400{,}000{,}000}{0.2} = 2{,}000{,}000{,}000$, or $x = 44{,}721$, a critical number of C. $C''(x) = \frac{800{,}000{,}000}{x^3} > 0$ for all $x > 0$, and we see that the graph of C is concave upward and so $x = 44{,}721$ gives an absolute minimum of C. Therefore, there should be $2{,}000{,}000/x \approx 45$ orders per year (since we can not have fractions of an order.) Each order should be for $2{,}000{,}000/45 \approx 44{,}445$ bottles.

35. a. Because the sales are assumed to be steady and D units are expected to be sold per year, the number of orders per year is D/x. Because is costs \$$K$ per order, the ordering cost is KD/x. The purchasing cost is pD (cost per item times number purchased). Finally, the holding cost is $\frac{1}{2}xh$ (the average number on hand times holding cost per item). Therefore, $C(x) = \frac{KD}{x} + pD + \frac{hx}{2}$.

b. $C'(x) = -\frac{KD}{x^2} + \frac{h}{2} = 0$ implies $\frac{KD}{x^2} = \frac{h}{2}$, so $x^2 = \frac{2KD}{h}$ and $x = \pm\sqrt{\frac{2KD}{h}}$. We reject the negative root. So $x = \sqrt{\frac{2KD}{h}}$ is the only critical number. Next, $C''(x) = \frac{2KD}{x^3} > 0$ for $x > 0$, so $C''\left(\sqrt{\frac{2KD}{h}}\right) > 0$ and the Second Derivative Test shows that $x = \sqrt{\frac{2KD}{h}}$ does give a relative minimum. Because C is concave upward, this is also the absolute minimum.

CHAPTER 4 Concept Review Questions page 332

1. a. $f(x_1) < f(x_2)$ **b.** $f(x_1) > f(x_2)$

3. a. $f(x) \le f(c)$ **b.** $f(x) \ge f(c)$

5. a. $f'(x)$ **b.** > 0 **c.** concavity **d.** relative maximum; relative extremum

7. 0, 0

9. a. $f(x) \le f(c)$, absolute maximum value **b.** $f(x) \ge f(c)$, open interval

CHAPTER 4 Review Exercises page 333

1. a. $f(x) = \frac{1}{3}x^3 - x^2 + x - 6$, so $f'(x) = x^2 - 2x + 1 = (x-1)^2$. $f'(x) = 0$ gives $x = 1$, the critical number of f. Now $f'(x) > 0$ for all $x \neq 1$. Thus, f is increasing on $(-\infty, \infty)$.

b. Because $f'(x)$ does not change sign as we move across the critical number $x = 1$, the First Derivative Test implies that $x = 1$ does not give a relative extremum of f.

c. $f''(x) = 2(x-1)$. Setting $f''(x) = 0$ gives $x = 1$ as a candidate for an inflection point of f. Because $f''(x) < 0$ for $x < 1$, and $f''(x) > 0$ for $x > 1$, we see that f is concave downward on $(-\infty, 1)$ and concave upward on $(1, \infty)$.

d. The results of part (c) imply that $\left(1, -\frac{17}{3}\right)$ is an inflection point.

3. a. $f(x) = x^4 - 2x^2$, so $f'(x) = 4x^3 - 4x = 4x(x^2-1) = 4x(x+1)(x-1)$. The sign diagram of f' shows that f is decreasing on $(-\infty, -1)$ and $(0, 1)$ and increasing on $(-1, 0)$ and $(1, \infty)$.

− − 0 + + 0 − − 0 + + sign of f'
x: −1, 0, 1

b. The results of part (a) and the First Derivative Test show that $(-1, -1)$ and $(1, -1)$ are relative minima and $(0, 0)$ is a relative maximum.

c. $f''(x) = 12x^2 - 4 = 4(3x^2 - 1) = 0$ if $x = \pm\frac{\sqrt{3}}{3}$. The sign diagram shows that f is concave upward on $\left(-\infty, -\frac{\sqrt{3}}{3}\right)$ and $\left(\frac{\sqrt{3}}{3}, \infty\right)$ and concave downward on $\left(-\frac{\sqrt{3}}{3}, \frac{\sqrt{3}}{3}\right)$.

+ + 0 − − − − − − 0 + + sign of f''
x: $-\frac{\sqrt{3}}{3}$, 0, $\frac{\sqrt{3}}{3}$

d. The results of part (c) show that $\left(-\frac{\sqrt{3}}{3}, -\frac{5}{9}\right)$ and $\left(\frac{\sqrt{3}}{3}, -\frac{5}{9}\right)$ are inflection points.

5. a. $f(x) = \dfrac{x^2}{x-1}$, so $f'(x) = \dfrac{(x-1)(2x) - x^2(1)}{(x-1)^2} = \dfrac{x^2 - 2x}{(x-1)^2} = \dfrac{x(x-2)}{(x-1)^2}$.

The sign diagram of f' shows that f is increasing on $(-\infty, 0)$ and $(2, \infty)$ and decreasing on $(0, 1)$ and $(1, 2)$.

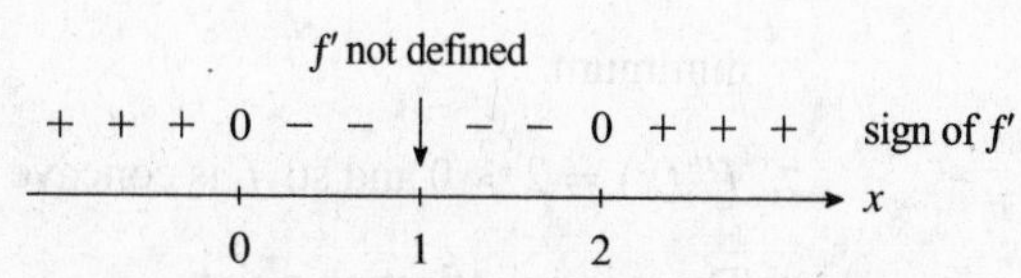

b. The results of part (a) show that $(0, 0)$ is a relative maximum and $(2, 4)$ is a relative minimum.

c. $f''(x) = \dfrac{(x-1)^2(2x-2) - x(x-2)2(x-1)}{(x-1)^4} = \dfrac{2(x-1)\left[(x-1)^2 - x(x-2)\right]}{(x-1)^4} = \dfrac{2}{(x-1)^3}$. Because $f''(x) < 0$ if $x < 1$ and $f''(x) > 0$ if $x > 1$, we see that f is concave downward on $(-\infty, 1)$ and concave upward on $(1, \infty)$.

d. Because $x = 1$ is not in the domain of f, there is no inflection point.

7. a. $f(x) = (1-x)^{1/3}$, so $f'(x) = -\frac{1}{3}(1-x)^{-2/3} = -\dfrac{1}{3(1-x)^{2/3}}$. The sign diagram for f' shows that f is decreasing on $(-\infty, \infty)$.

f' not defined
− − − − − − − − − − sign of f'
x: 0, 1

b. There is no relative extremum.

c. $f''(x) = -\frac{2}{9}(1-x)^{-5/3} = -\dfrac{2}{9(1-x)^{5/3}}$. The sign diagram for f'' shows that f is concave downward on $(-\infty, 1)$ and concave upward on $(1, \infty)$.

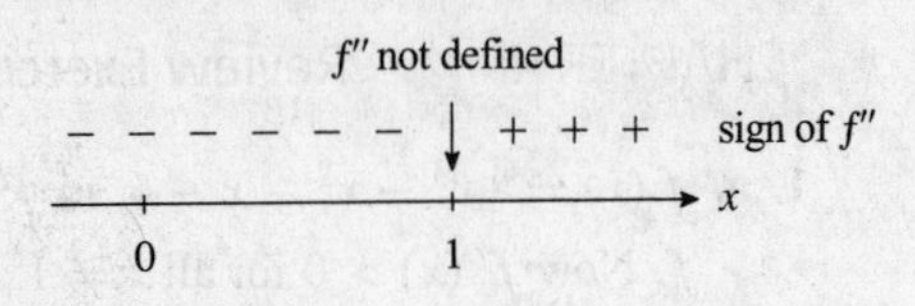

d. $x = 1$ is a candidate for an inflection point of f. Referring to the sign diagram for f'', we see that $(1, 0)$ is an inflection point.

9. a. $f(x) = \dfrac{2x}{x+1}$, so $f'(x) = \dfrac{(x+1)(2) - 2x(1)}{(x+1)^2} = \dfrac{2}{(x+1)^2} > 0$ if $x \neq -1$. Therefore f is increasing on $(-\infty, -1)$ and $(-1, \infty)$.

b. Because there is no critical number, f has no relative extremum.

c. $f''(x) = -4(x+1)^{-3} = -\dfrac{4}{(x+1)^3}$. Because $f''(x) > 0$ if $x < -1$ and $f''(x) < 0$ if $x > -1$, we see that f is concave upward on $(-\infty, -1)$ and concave downward on $(-1, \infty)$.

d. There is no inflection point because $f''(x) \neq 0$ for all x in the domain of f.

11. $f(x) = x^2 - 5x + 5$.

1. The domain of f is $(-\infty, \infty)$.
2. Setting $x = 0$ gives 5 as the y-intercept.
3. $\lim_{x\to-\infty}(x^2 - 5x + 5) = \lim_{x\to\infty}(x^2 - 5x + 5) = \infty$.
4. There is no asymptote because f is a quadratic function.
5. $f'(x) = 2x - 5 = 0$ if $x = \frac{5}{2}$. The sign diagram shows that f is increasing on $\left(\frac{5}{2}, \infty\right)$ and decreasing on $\left(-\infty, \frac{5}{2}\right)$.

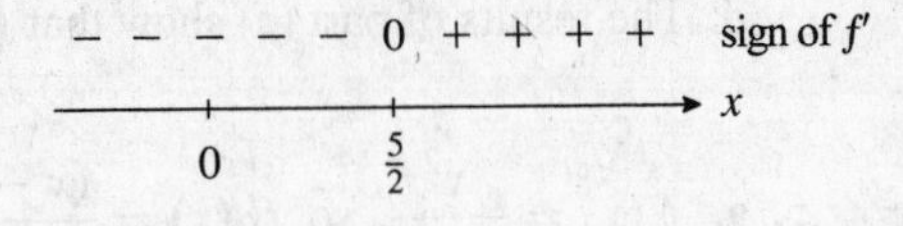

6. The First Derivative Test implies that $\left(\frac{5}{2}, -\frac{5}{4}\right)$ is a relative minimum.
7. $f''(x) = 2 > 0$ and so f is concave upward on $(-\infty, \infty)$.
8. There is no inflection point.

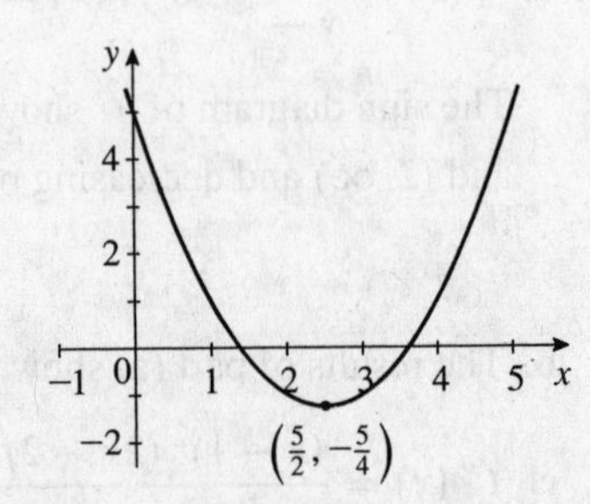

13. $g(x) = 2x^3 - 6x^2 + 6x + 1$.

1. The domain of g is $(-\infty, \infty)$.
2. Setting $x = 0$ gives 1 as the y-intercept.
3. $\lim_{x\to-\infty} g(x) = -\infty$ and $\lim_{x\to\infty} g(x) = \infty$.
4. There is no vertical or horizontal asymptote.
5. $g'(x) = 6x^2 - 12x + 6 = 6(x^2 - 2x + 1) = 6(x-1)^2$. Because $g'(x) > 0$ for all $x \neq 1$, we see that g is increasing on $(-\infty, 1)$ and $(1, \infty)$.

6. $g'(x)$ does not change sign as we move across the critical number $x = 1$, so there is no extremum.

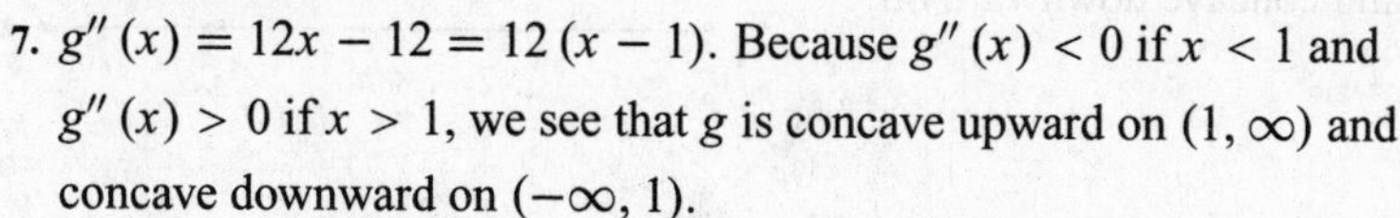

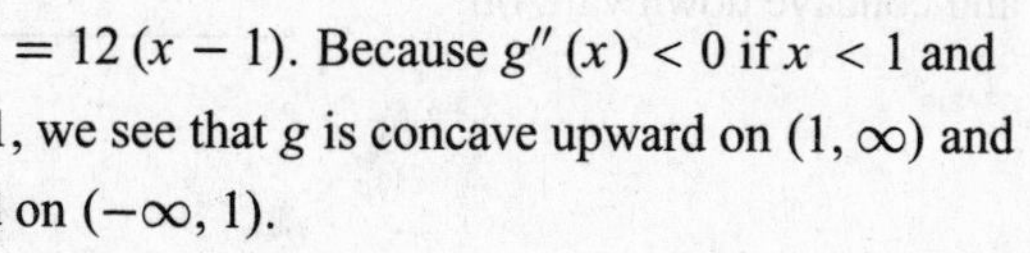

7. $g''(x) = 12x - 12 = 12(x - 1)$. Because $g''(x) < 0$ if $x < 1$ and $g''(x) > 0$ if $x > 1$, we see that g is concave upward on $(1, \infty)$ and concave downward on $(-\infty, 1)$.

8. The point $x = 1$ gives rise to the inflection point $(1, 3)$.

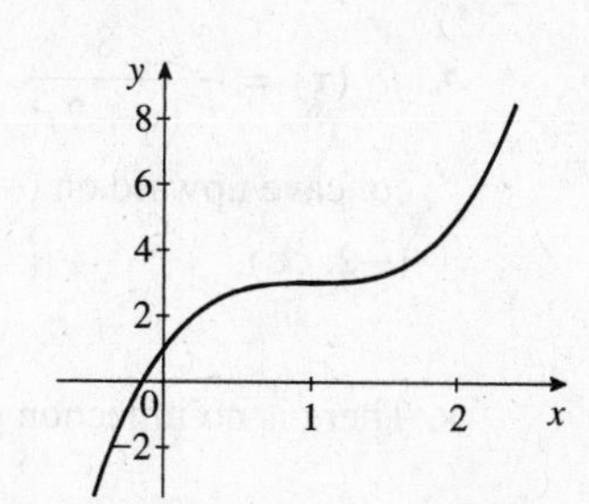

15. $h(x) = x\sqrt{x-2}$.

1. The domain of h is $[2, \infty)$.

2. There is no y-intercept. Setting $y = 0$ gives 2 as the x-intercept.

3. $\lim_{x\to\infty} x\sqrt{x-2} = \infty$.

4. There is no asymptote.

5. $h'(x) = (x-2)^{1/2} + x\left(\frac{1}{2}\right)(x-2)^{-1/2} = \frac{1}{2}(x-2)^{-1/2}[2(x-2) + x] = \dfrac{3x-4}{2\sqrt{x-2}} > 0$ on $[2, \infty)$, and so h is increasing on $[2, \infty)$.

6. Because h has no critical number in $(2, \infty)$, there is no relative extremum.

7. $h''(x) = \dfrac{1}{2}\left[\dfrac{(x-2)^{1/2}(3) - (3x-4)\frac{1}{2}(x-2)^{-1/2}}{x-2}\right] = \dfrac{(x-2)^{-1/2}[6(x-2) - (3x-4)]}{4(x-2)} = \dfrac{3x-8}{4(x-2)^{3/2}}$.

The sign diagram for h'' shows that h is concave downward on $\left(2, \frac{8}{3}\right)$ and concave upward on $\left(\frac{8}{3}, \infty\right)$.

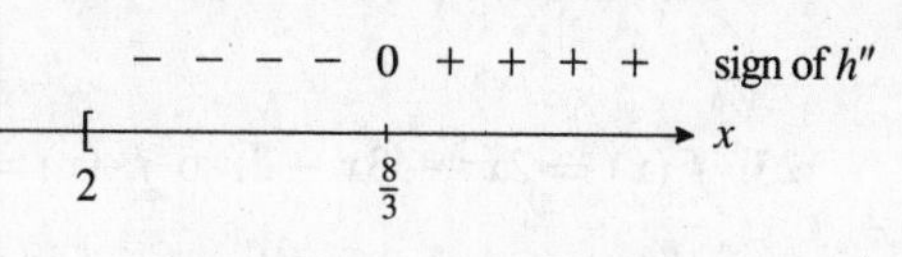

8. The results of step 7 tell us that $\left(\frac{8}{3}, \frac{8\sqrt{6}}{9}\right)$ is an inflection point.

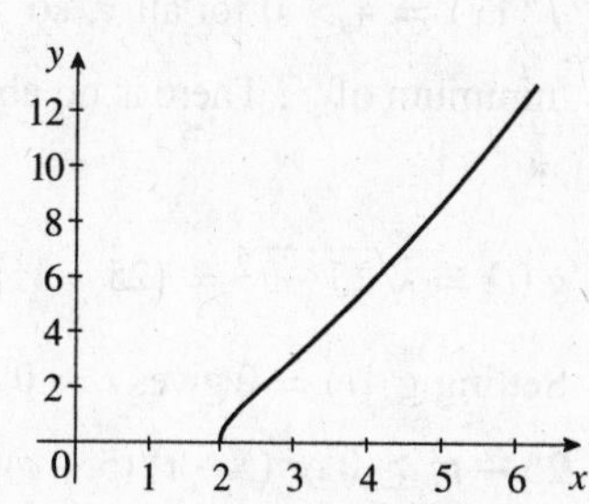

17. $f(x) = \dfrac{x-2}{x+2}$.

1. The domain of f is $(-\infty, -2) \cup (-2, \infty)$.

2. Setting $x = 0$ gives -1 as the y-intercept. Setting $y = 0$ gives 2 as the x-intercept.

3. $\lim_{x\to-\infty} \dfrac{x-2}{x+2} = \lim_{x\to\infty} \dfrac{x-2}{x+2} = 1$.

4. The results of step 3 tell us that $y = 1$ is a horizontal asymptote. Next, observe that the denominator of $f(x)$ is equal to zero at $x = -2$, but its numerator is not equal to zero there. Therefore, $x = -2$ is a vertical asymptote.

5. $f'(x) = \dfrac{(x+2)(1) - (x-2)(1)}{(x+2)^2} = \dfrac{4}{(x+2)^2}$. The sign diagram of f' tells us that f is increasing on $(-\infty, -2)$ and $(-2, \infty)$.

f' not defined

+ + + ↓ + + + + + + + sign of f'

x: -2, 0

6. The results of step 5 tell us that there is no relative extremum.

7. $f''(x) = -\frac{8}{(x+2)^3}$. The sign diagram of f'' shows that f is concave upward on $(-\infty, -2)$ and concave downward on $(-2, \infty)$.

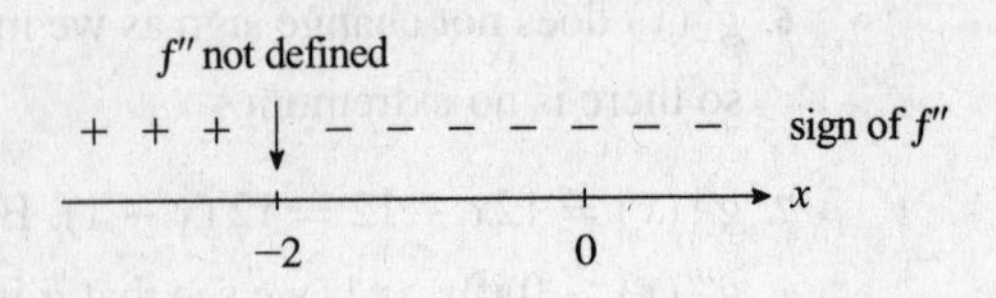

8. There is no inflection point.

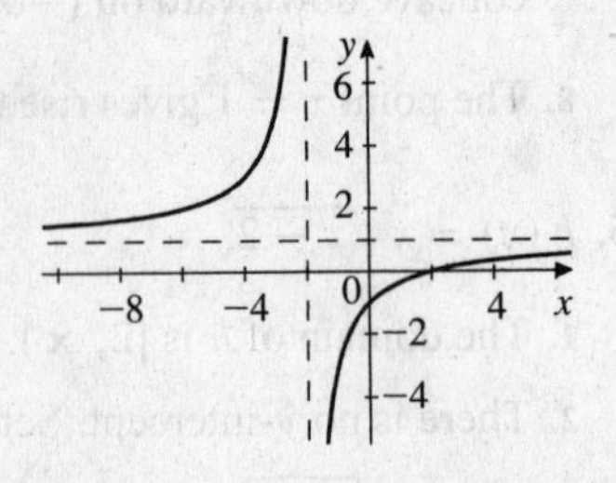

19. $\lim_{x\to-\infty} \frac{1}{2x+3} = \lim_{x\to\infty} \frac{1}{2x+3} = 0$ and so $y = 0$ is a horizontal asymptote. Because the denominator is equal to zero at $x = -\frac{3}{2}$ but the numerator is not equal to zero there, we see that $x = -\frac{3}{2}$ is a vertical asymptote.

21. $\lim_{x\to-\infty} \frac{5x}{x^2-2x-8} = \lim_{x\to\infty} \frac{5x}{x^2-2x-8} = 0$, so $y = 0$ is a horizontal asymptote. Next, note that the denominator is zero if $x^2 - 2x - 8 = (x-4)(x+2) = 0$, that is, if $x = -2$ or $x = 4$. Because the numerator is not equal to zero at these points, we see that $x = -2$ and $x = 4$ are vertical asymptotes.

23. $f(x) = 2x^2 + 3x - 2$, so $f'(x) = 4x + 3$. Setting $f'(x) = 0$ gives $x = -\frac{3}{4}$ as a critical number of f. Next, $f''(x) = 4 > 0$ for all x, so f is concave upward on $(-\infty, \infty)$. Therefore, $f\left(-\frac{3}{4}\right) = -\frac{25}{8}$ is an absolute minimum of f. There is no absolute maximum.

25. $g(t) = \sqrt{25-t^2} = \left(25-t^2\right)^{1/2}$. Differentiating $g(t)$, we have $g'(t) = \frac{1}{2}\left(25-t^2\right)^{-1/2}(-2t) = -\frac{t}{\sqrt{25-t^2}}$.
Setting $g'(t) = 0$ gives $t = 0$ as a critical number of g. The domain of g is given by solving the inequality $25 - t^2 \geq 0$ or $(5-t)(5+t) \geq 0$ which implies that $t \in [-5, 5]$. From the table, we conclude that $g(0) = 5$ is the absolute maximum of g and $g(-5) = 0$ and $g(5) = 0$ is the absolute minimum value of g.

t	−5	0	5
$g(t)$	0	5	0

27. $h(t) = t^3 - 6t^2$, so $h'(t) = 3t^2 - 12t = 3t(t-4) = 0$ if $t = 0$ or $t = 4$, critical numbers of h. But only $t = 4$ lies in $(2, 5)$. From the table, we see that h has an absolute minimum at $(4, -32)$ and an absolute maximum at $(2, -16)$.

t	2	4	5
$h(t)$	−16	−32	−25

29. $f(x) = x - \frac{1}{x}$ on $[1, 3]$, so $f'(x) = 1 + \frac{1}{x^2}$. Because $f'(x)$ is never zero, f has no critical number. Calculating $f(x)$ at the endpoints, we see that $f(1) = 0$ is the absolute minimum value and $f(3) = \frac{8}{3}$ is the absolute maximum value.

31. $f(s) = s\sqrt{1-s^2}$ on $[-1, 1]$. The function f is continuous on $[-1, 1]$ and differentiable on $(-1, 1)$. Next, $f'(s) = (1-s^2)^{1/2} + s\left(\frac{1}{2}\right)(1-s^2)^{-1/2}(-2s) = \dfrac{1-2s^2}{\sqrt{1-s^2}}$. Setting $f'(s) = 0$, we find that $s = \pm\frac{\sqrt{2}}{2}$ are critical numbers of f. From the table, we see that $f\left(-\frac{\sqrt{2}}{2}\right) = -\frac{1}{2}$ is the absolute minimum value and $f\left(\frac{\sqrt{2}}{2}\right) = \frac{1}{2}$ is the absolute maximum value of f.

x	-1	$-\frac{\sqrt{2}}{2}$	$\frac{\sqrt{2}}{2}$	1
$f(x)$	0	$-\frac{1}{2}$	$\frac{1}{2}$	0

33. a. The sign of R_1' is negative and the sign of R_2' is positive on $(0, T)$. The sign of R_1'' is negative and the sign of R_2'' is positive on $(0, T)$.

b. The revenue of the neighborhood bookstore is decreasing at an increasing rate, while the revenue of the new bookstore is increasing at an increasing rate.

35. We want to maximize $P(x) = -x^2 + 8x + 20$. Now, $P'(x) = -2x + 8 = 0$ if $x = 4$, a critical number of P. Because $P''(x) = -2 < 0$, the graph of P is concave downward. Therefore, the critical number $x = 4$ yields an absolute maximum. So, to maximize profit, the company should spend \$4000 per month on advertising.

37. a. $f'(t) = \dfrac{d}{dt}\left(0.0117t^3 + 0.0037t^2 + 0.7563t + 4.1\right) = 0.0351t^2 + 0.0074t + 0.7563 \geq 0.7563$ for all t in the interval $[0, 9]$. This shows that f is increasing on $(0, 9)$, which tells us that the projected amount of AMT will keep on increasing over the years in question.

b. $f''(t) = \dfrac{d}{dt}\left(0.0351t^2 + 0.0074t + 0.7563\right) = 0.0702t + 0.0074 \geq 0.0074$. This shows that f' is increasing on $(0, 9)$. Our result tells us that not only is the amount of AMT paid increasing over the period in question, but it is actually accelerating.

39. $S(x) = -0.002x^3 + 0.6x^2 + x + 500$, so $S'(x) = -0.006x^2 + 1.2x + 1$ and $S''(x) = -0.012x + 1.2$. $x = 100$ is a candidate for an inflection point of S. The sign diagram for S'' shows that $(100, 4600)$ is an inflection point of S.

+ + + 0 − − − sign of S''

0 100 200 x

41. $C(x) = 0.0001x^3 - 0.08x^2 + 40x + 5000$, so $C'(x) = 0.0003x^2 - 0.16x + 40$ and $C''(x) = 0.0006x - 0.16$. Thus, $x = 266.67$ is a candidate for an inflection point of C. The sign diagram for C'' shows that C has an inflection point at $(266.67, 11874.08)$.

− − − − 0 + + + + sign of C''

0 ≈ 266.67 x

43. a. $f(0) \approx 12.98$, so the proportion in 2005 was approximately 13.0%. The projected proportion in 2015 is given by $f(10) \approx 22.21$, or approximately 22.2%.

b. $f'(t) = \dfrac{\left(59 - t^{1/2}\right)(150)\left(\frac{1}{2}t^{-1/2}\right) - \left(150t^{1/2} + 766\right)\left(-\frac{1}{2}t^{-1/2}\right)}{\left(59 - t^{1/2}\right)^2} = \dfrac{4.808}{\sqrt{t}\left(59 - t^{1/2}\right)^2} > 0$ for $0 < t \leq 10$, so f is increasing on $(0, 10)$. This says that the percentage of small and lower-midsize vehicles will be growing over the period from 2005 to 2015.

45. The revenue is $R(x) = px = x(-0.0005x^2 + 60) = -0.0005x^3 + 60x$. Therefore, the total profit is $P(x) = R(x) - C(x) = -0.0005x^3 + 0.001x^2 + 42x - 4000$. $P'(x) = -0.0015x^2 + 0.002x + 42$, and setting $P'(x) = 0$ gives $3x^2 - 4x - 84{,}000 = 0$. Solving for x, we find $x = \frac{4 \pm \sqrt{16 - 4(3)(84{,}000)}}{2(3)} = \frac{4 \pm 1004}{6} = 168$ or -167. We reject the negative root. Next, $P''(x) = -0.003x + 0.002$ and $P''(168) = -0.003(168) + 0.002 = -0.502 < 0$. By the Second Derivative Test, $x = 168$ gives a relative maximum. Therefore, the required level of production is 168 DVDs.

47. a. $C(x) = 0.001x^2 + 100x + 4000$, so $\overline{C}(x) = \frac{C(x)}{x} = \frac{0.001x^2 + 100x + 4000}{x} = 0.001x + 100 + \frac{4000}{x}$.

b. $\overline{C}'(x) = 0.001 - \frac{4000}{x^2} = \frac{0.001x^2 - 4000}{x^2} = \frac{0.001(x^2 - 4{,}000{,}000)}{x^2}$. Setting $\overline{C}'(x) = 0$ gives $x = \pm 2000$. We reject the negative root. The sign diagram of $\overline{C}'$ shows that $x = 2000$ gives rise to a relative minimum of $\overline{C}$. Because $\overline{C}''(x) = \frac{8000}{x^3} > 0$ if $x > 0$, we see that $\overline{C}$ is concave upward on $(0, \infty)$, and so $x = 2000$ yields an absolute minimum. The required production level is 2000 units.

− − − 0 + + + sign of $\overline{C}'$

0 2000 x

49. a. $P(t) = -0.0002t^3 + 0.018t^2 - 0.36t + 10$, so $P'(t) = -0.0006t^2 + 0.036t - 0.36$. Setting $P'(t) = 0$ gives $-0.0006t^2 + 0.036t - 0.36 = 0$, or $t^2 - 60t + 600 = 0$. Thus, $t = \frac{60 \pm \sqrt{60^2 - 4(1)(600)}}{2} \approx 12.7$ or 47.3. We reject the root 47.3 because it lies outside $[0, 30]$. The sign diagram for P' shows that P is decreasing on $(0, 12.7)$ and increasing on $(12.7, 30)$.

− − − 0 + + + + + sign of P'

0 ≈12.68 30 t

b. The absolute minimum of P occurs at $t = 12.7$, and $P(12.7) \approx 7.9$.

c. The percentage of women 65 and older in the workforce was decreasing from 1970 to September 1982 and increasing from September 1982 to 2000. It reached a minimum value of 7.9% in September 1982.

51. $R'(x) = k\frac{d}{dx}x(M - x) = k[(M - x) + x(-1)] = k(M - 2x)$. Setting $R'(x) = 0$ gives $M - 2x = 0$, or $x = \frac{1}{2}M$, a critical number of R. Because $R''(x) = -2k < 0$, we see that $x = \frac{1}{2}M$ gives a maximum; that is, R is greatest when half the population is infected.

53. a. $f'(x) = -2x$ if $x \neq 0$, $f'(-1) = 2$, and $f'(1) = -2$, so $f'(x)$ changes sign from positive to negative as we move across $x = 0$.

b. f does not have a relative maximum at $x = 0$ because $f(0) = 2$ but a neighborhood of $x = 0$, for example $\left(-\frac{1}{2}, \frac{1}{2}\right)$, contains numbers with values larger than 2. This does not contradict the First Derivative Test because f is not continuous at $x = 0$.

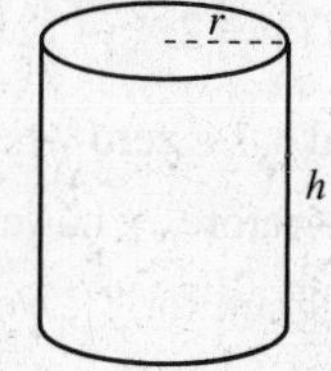

55. Suppose the radius is r and the height is h. Then the capacity is $\pi r^2 h$, and this must be equal to 32π ft^3; that is, $\pi r^2 h = 32\pi$. Let the cost per square foot for the sides be \$$c$. Then the cost of construction is $C = 2\pi rhc + 2(\pi r^2)(2c) = 2\pi crh + 4\pi cr^2$. But $h = \dfrac{32\pi}{\pi r^2} = \dfrac{32}{r^2}$, so $C = f(r) = \dfrac{64c\pi}{r} + 4\pi cr^2$, giving $C' = f'(r) = -\dfrac{64\pi c}{r^2} + 8\pi cr = \dfrac{-64\pi c + 8\pi cr^3}{r^2} = \dfrac{8\pi c(-8 + r^3)}{r^2}$. Setting $f'(r) = 0$ gives $r^3 = 8$, or $r = 2$. Next, $f''(r) = \dfrac{128\pi c}{r^3} + 8\pi c$, and so $f''(2) > 0$. Therefore, $r = 2$ minimizes f. The required dimensions are $r = 2$ and $h = \frac{32}{4} = 8$. That is, the radius is 2 ft and the height is 8 ft.

57. Let x denote the number of cases in each order. Then the average number of cases of beer in storage during the year is $\frac{1}{2}x$. The storage cost in dollars is $2\left(\frac{1}{2}x\right) = x$. Next, we see that the number of orders required is $\dfrac{800{,}000}{x}$, and so the ordering cost is $\dfrac{500(800{,}000)}{x} = \dfrac{400{,}000{,}000}{x}$. Thus, the total cost incurred by the company per year is given by $C(x) = x + \dfrac{400{,}000{,}000}{x}$. We want to minimize C in the interval $(0, \infty)$, so we calculate $C'(x) = 1 - \dfrac{400{,}000{,}000}{x^2}$. Setting $C'(x) = 0$ gives $x^2 = 400{,}000{,}000$, or $x = 20{,}000$ (we reject $x = -20{,}000$). Next, $C''(x) = \dfrac{800{,}000{,}000}{x^3} > 0$ for all x, so C is concave upward. Thus, $x = 20{,}000$ gives rise to the absolute minimum of C. The company should order 20,000 cases of beer per order.

59. $f(x) = x^2 + ax + b$, so $f'(x) = 2x + a$. We require that $f'(2) = 0$, so $(2)(2) + a = 0$, and $a = -4$. Next, $f(2) = 7$ implies that $f(2) = 2^2 + (-4)(2) + b = 7$, so $b = 11$. Thus, $f(x) = x^2 - 4x + 11$. Because the graph of f is a parabola that opens upward, $(2, 7)$ is a relative minimum.

61. Because $(a, f(a))$ is an inflection point of f, $f''(a) = 0$. This shows that a is a critical number of f'. Next, f changes concavity at $(a, f(a))$. If the concavity changes from concave downward to concave upward [that is, $f''(x) < 0$ for $x < a$ and $f''(x) > 0$ for $x > a$], then f' has a relative minimum at a. On the other hand, if the concavity changes from concave upward to concave downward, [$f''(x) > 0$ for $x < a$ and $f''(x) < 0$ for $x > a$], then f' has a relative maximum at a. In either case, f' has a relative extremum at a.

CHAPTER 4 Before Moving On... page 336

1. $f(x) = \dfrac{x^2}{1-x}$, so $f'(x) = \dfrac{(1-x)(2x) - x^2(-1)}{(1-x)^2} = \dfrac{2x - 2x^2 + x^2}{(1-x)^2} = \dfrac{x(2-x)}{(1-x)^2}$; f' is not defined at 1 and has zeros at 0 and 2. The sign diagram of f shows that f is decreasing on $(-\infty, 0)$ and $(2, \infty)$ and increasing on $(0, 1)$ and $(1, 2)$.

f′ not defined

− − − 0 + + ↓ + + 0 − − − sign of f′

x: 0, 1, 2

2. $f(x) = 2x^2 - 12x^{1/3}$, so $f'(x) = 4x - 4x^{-2/3} = 4x^{-2/3}\left(x^{5/3} - 1\right) = \dfrac{4\left(x^{5/3} - 1\right)}{x^{2/3}}$. f' is discontinuous at $x = 0$ and has a zero where $x^{5/3} = 1$ or $x = 1$.

Therefore, f has critical numbers at 0 and 1. From the sign diagram for f', we see that $x = 1$ gives a relative minimum. Because $f(1) = 2 - 12 = -10$, the relative minimum is $(1, -10)$. There is no relative maximum.

− − − − − − − − − − 0 + + + sign of f''
x: 0, 2

3. $f(x) = \frac{1}{3}x^3 - \frac{1}{4}x^2 - \frac{1}{2}x + 1$, so $f'(x) = x^2 - \frac{1}{2}x - \frac{1}{2}$ and $f''(x) = 2x - \frac{1}{2} = 0$ gives $x = \frac{1}{4}$.

The sign diagram of f'' shows that f is concave downward on $\left(-\infty, \frac{1}{4}\right)$ and concave upward on $\left(\frac{1}{4}, \infty\right)$. Because $f\left(\frac{1}{4}\right) = \frac{1}{3}\left(\frac{1}{4}\right)^3 - \frac{1}{4}\left(\frac{1}{4}\right)^2 - \frac{1}{2}\left(\frac{1}{4}\right) + 1 = \frac{83}{96}$, the inflection point is $\left(\frac{1}{4}, \frac{83}{96}\right)$.

− − − − − − − − − − 0 + + + sign of f''
x: 0, $\frac{1}{4}$

4. $f(x) = 2x^3 - 9x^2 + 12x - 1$.

1. The domain of f is $(-\infty, \infty)$.
2. Setting $y = f(x) = 0$ gives -1 as the y-intercept of f.
3. $\lim_{x \to -\infty} f(x) = -\infty$ and $\lim_{x \to \infty} f(x) = \infty$.
4. There is no asymptote.
5. $f'(x) = 6x^2 - 18x + 12 = 6\left(x^2 - 3x + 2\right) = 6(x - 2)(x - 1)$.

   The sign diagram of f' shows that f is increasing on $(-\infty, 1)$ and $(2, \infty)$ and decreasing on $(1, 2)$.

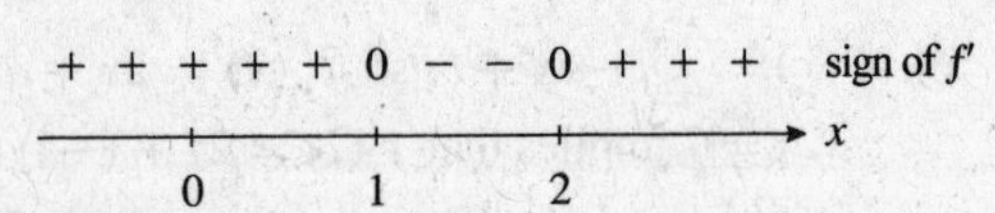

6. We see that $(1, 4)$ is a relative maximum and $(2, 3)$ is a relative minimum.
7. $f''(x) = 12x - 18 = 6(2x - 3)$. The sign diagram of f'' shows that f is concave downward on $\left(-\infty, \frac{3}{2}\right)$ and concave upward on $\left(\frac{3}{2}, \infty\right)$.

   − − − − − − − − − − 0 + + + sign of f''
   x: 0, $\frac{3}{2}$

8. $f\left(\frac{3}{2}\right) = 2\left(\frac{3}{2}\right)^3 - 9\left(\frac{3}{2}\right)^2 + 12\left(\frac{3}{2}\right) - 1 = \frac{7}{2}$, so $\left(\frac{3}{2}, \frac{7}{2}\right)$ is an inflection point of f.

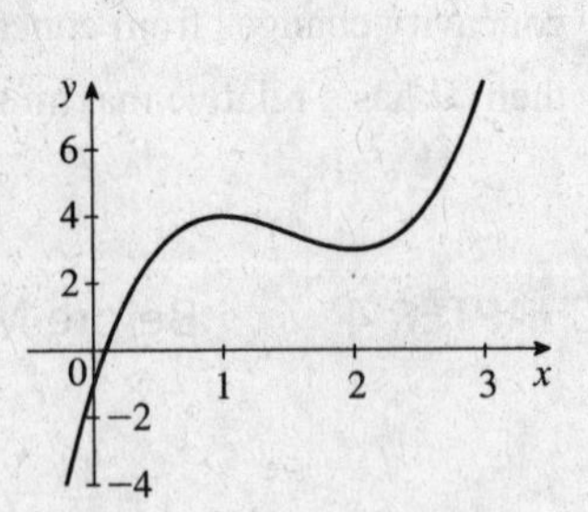

5. $f(x) = 2x^3 + 3x^2 - 1$ is continuous on the closed interval $[-2, 3]$. $f'(x) = 6x^2 + 6x = 6x(x + 1)$, so the critical numbers of f are -1 and 0. From the table, we see that the absolute maximum value of f is 80 and the absolute minimum value is -5.

x	-2	-1	0	3
y	-5	0	-1	80

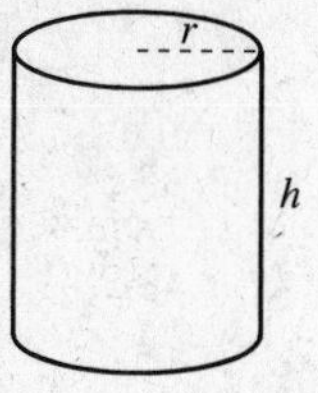

6. The amount of material used (the surface area) is $A = \pi r^2 + 2\pi rh$. But $V = \pi r^2 h = 1$, and so $h = \dfrac{1}{\pi r^2}$. Therefore, $A = \pi r^2 + 2\pi r\left(\dfrac{1}{\pi r^2}\right) = \pi r^2 + \dfrac{2}{r}$, so $A' = 2\pi r - \dfrac{2}{r^2} = 0$ implies $2\pi r = \dfrac{2}{r^2}$, $r^3 = \dfrac{2}{r^2}$, $r^3 = \dfrac{1}{\pi}$, and $r = \dfrac{1}{\sqrt[3]{\pi}}$. Because $A'' = 2\pi + \dfrac{4}{r^3} > 0$ for $r > 0$, we see that $r = \dfrac{1}{\sqrt[3]{\pi}}$ does give an absolute minimum.

Also, $h = \dfrac{1}{\pi r^2} = \dfrac{1}{\pi} \cdot \pi^{2/3} = \dfrac{1}{\pi^{1/3}} = \dfrac{1}{\sqrt[3]{\pi}}$. Therefore, the radius and height should each be $\dfrac{1}{\sqrt[3]{\pi}}$ ft.

r

h

5 EXPONENTIAL AND LOGARITHMIC FUNCTIONS

5.1 Exponential Functions

Problem-Solving Tips

1. Remember the order of operations when working with exponents. Note that $-5^2 \neq 25$, but rather $-5^2 = -(5)^2 = -25$. On the other hand, $(-5)^2 = 25$.

2. $b^{-x} = \dfrac{1}{b^x} = \left(\dfrac{1}{b}\right)^x$. If $b > 1$, then $0 < \dfrac{1}{b} < 1$, so the graph of b^{-x} for $b > 1$ is similar to the graph of $y = \left(\dfrac{1}{2}\right)^x$. (See Figure 3 on page 340 of the text.)

Concept Questions page 342

1. $f(x) = b^x$ with $b > 0$ and $b \neq 1$.

Exercises page 342

1. a. $4^{-3} \times 4^5 = 4^{-3+5} = 4^2 = 16$.

b. $3^{-3} \times 3^6 = 3^{6-3} = 3^3 = 27$.

3. a. $9(9)^{-1/2} = \dfrac{9}{9^{1/2}} = \dfrac{9}{3} = 3$.

b. $5(5)^{-1/2} = 5^{1/2} = \sqrt{5}$.

5. a. $\dfrac{(-3)^4(-3)^5}{(-3)^8} = (-3)^{4+5-8} = (-3)^1 = -3$.

b. $\dfrac{(2^{-4})(2^6)}{2^{-1}} = 2^{-4+6+1} = 2^3 = 8$.

7. a. $\dfrac{5^{3.3} \cdot 5^{-1.6}}{5^{-0.3}} = \dfrac{5^{3.3-1.6}}{5^{-0.3}} = 5^{1.7+(0.3)} = 5^2 = 25$.

b. $\dfrac{4^{2.7} \cdot 4^{-1.3}}{4^{-0.4}} = 4^{2.7-1.3+0.4} = 4^{1.8} \approx 12.126$.

9. a. $(64x^9)^{1/3} = 64^{1/3}(x^{9/3}) = 4x^3$.

b. $(25x^3y^4)^{1/2} = (25^{1/2})(x^{3/2})(y^{4/2}) = 5x^{3/2}y^2 = 5xy^2\sqrt{x}$.

11. a. $\dfrac{6a^{-4}}{3a^{-3}} = 2a^{-4+3} = 2a^{-1} = \dfrac{2}{a}$.

b. $\dfrac{4b^{-4}}{12b^{-6}} = \frac{1}{3}b^{-4+6} = \frac{1}{3}b^2$.

13. a. $(2x^3y^2)^3 = 2^3 \times x^{3(3)} \times y^{2(3)} = 8x^9y^6$.

b. $(4x^2y^2z^3)^2 = 4^2 \cdot x^{2(2)} \cdot y^{2(2)} \cdot z^{3(2)} = 16x^4y^4z^6$.

15. a. $\dfrac{5^0}{(2^{-3}x^{-3}y^2)^2} = \dfrac{1}{2^{-3(2)}x^{-3(2)}y^{2(2)}} = \dfrac{2^6x^6}{y^4} = \dfrac{64x^6}{y^4}$.

b. $\dfrac{(x+y)(x-y)}{(x-y)^0} = (x+y)(x-y)$.

17. $6^{2x} = 6^6$ if and only if $2x = 6$ or $x = 3$.

19. $3^{3x-4} = 3^5$ if and only if $3x - 4 = 5$, $3x = 9$, or $x = 3$.

21. $(2.1)^{x+2} = (2.1)^5$ if and only if $x + 2 = 5$, or $x = 3$.

23. $8^x = \left(\frac{1}{32}\right)^{x-2}$, $\left(2^3\right)^x = (32)^{2-x} = \left(2^5\right)^{2-x}$, so $2^{3x} = 2^{5(2-x)}$, $3x = 10 - 5x$, $8x = 10$, or $x = \frac{5}{4}$.

25. Let $y = 3^x$. Then the given equation is equivalent to $y^2 - 12y + 27 = 0$, or $(y - 9)(y - 3) = 0$, giving $y = 3$ or 9. So $3^x = 3$ or $3^x = 9$, and therefore, $x = 1$ or $x = 2$.

27. $y = 2^x$, $y = 3^x$, and $y = 4^x$.

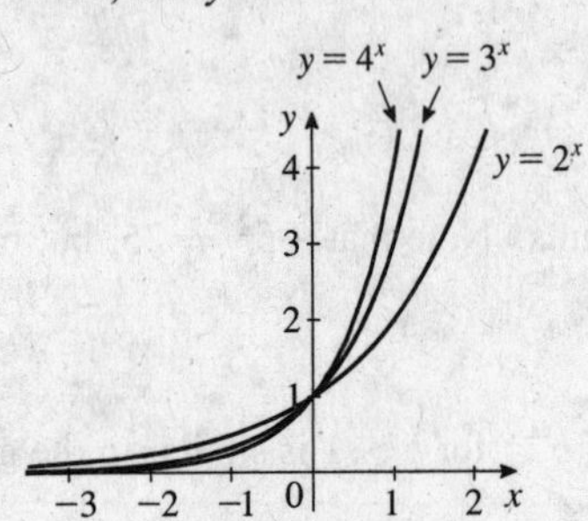

29. $y = 2^{-x}$, $y = 3^{-x}$, and $y = 4^{-x}$.

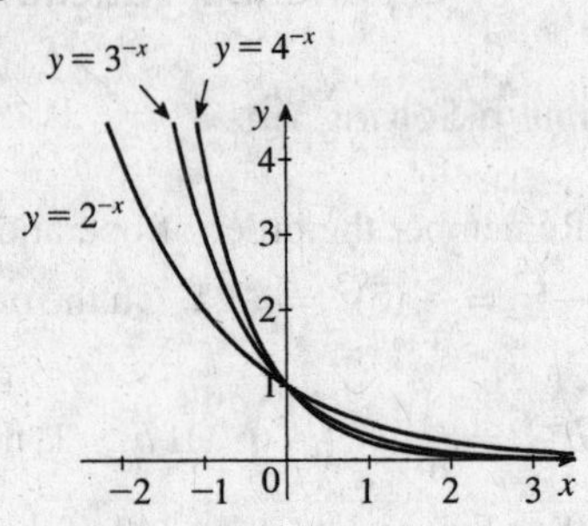

31. $y = 4^{0.5x}$, $y = 4^x$, and $y = 4^{2x}$.

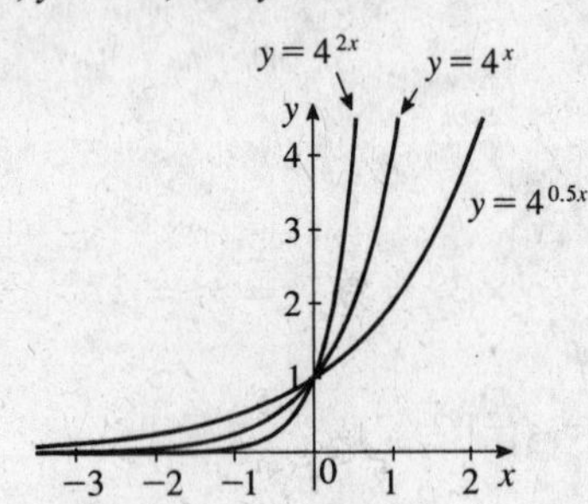

33. $y = e^{0.5x}$, $y = e^x$, $y = e^{1.5x}$.

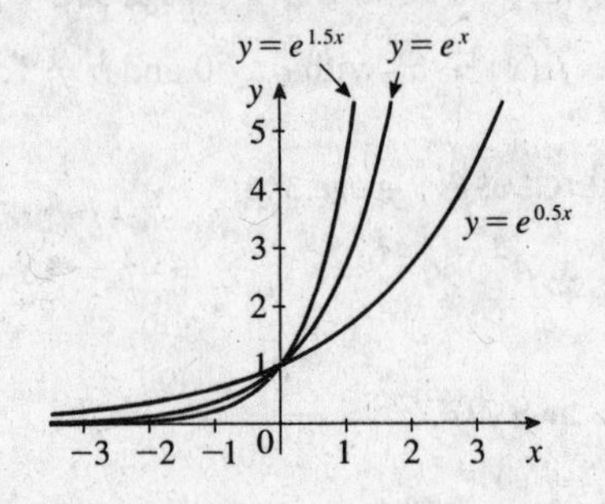

35. $y = 0.5e^{-x}$, $y = e^{-x}$, and $y = 2e^{-x}$.

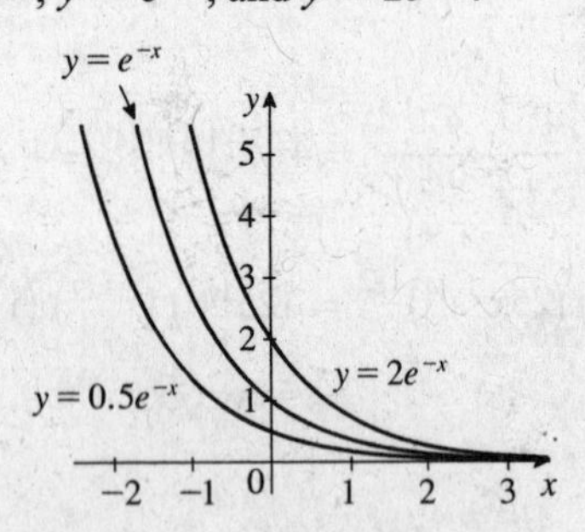

37. Because $f(0) = A = 100$ and $f(1) = 120$, we have $100e^k = 120$, and so $e^k = \frac{12}{10} = \frac{6}{5}$. Therefore, $f(x) = 100e^{kx} = 100\left(e^k\right) = 100\left(\frac{6}{5}\right)^x$.

39. $f(0) = 20$ implies that $\dfrac{1000}{1+B} = 20$, so $1000 = 20 + 20B$, or $B = \dfrac{980}{20} = 49$. Therefore, $f(t) = \dfrac{1000}{1+49e^{-kt}}$. Next, $f(2) = 30$, so $\dfrac{1000}{1+49e^{-2t}} = 30$. We have $1 + 49e^{-2k} = \dfrac{1000}{30} = \dfrac{100}{3}$, $49e^{-2k} = \dfrac{100}{3} - 1 = \dfrac{97}{3}$, $e^{-2k} = \dfrac{97}{147}$, and finally $e^{-k} = \left(\dfrac{97}{147}\right)^{1/2}$. Therefore, $f(t) = \dfrac{1000}{1+49\left(\frac{97}{147}\right)^{t/2}}$, so

$f(5) = \dfrac{1000}{1+49\left(\frac{97}{147}\right)^{5/2}} \approx 54.6.$

41. **a.** $f(t) = 64e^{0.188t}$.

t	0	1	2	3
$f(t)$	64	77.2	93.2	112.5

b.

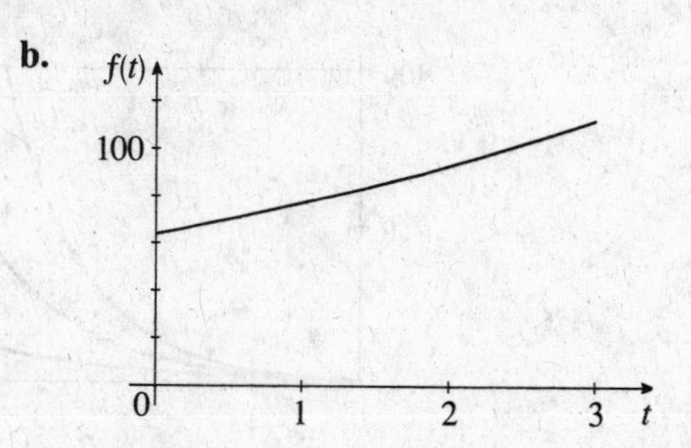

43. **a.** The number of internet users in 2005 was $f(1) = 115.423$, or 115,423,000. In 2006, it was $f(2) = 94.5e^{0.2(2)} \approx 140.977$, or 140,977,000. The number of internet users in 2010 was $f(6) = 94.5e^{1.2} \approx 313.751$, or 313,751,000.

b.

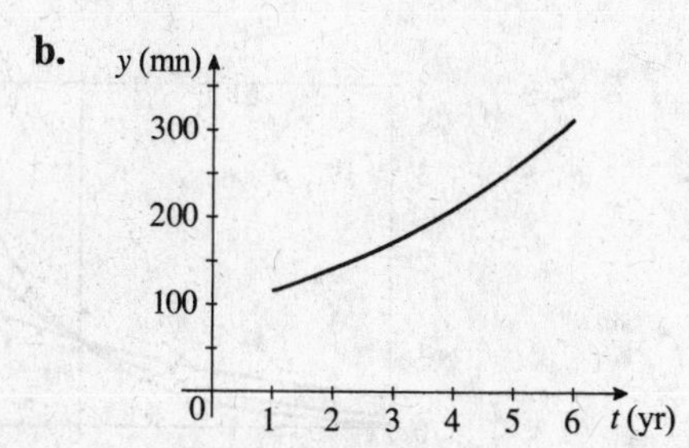

45. $N(t) = \dfrac{35.5}{1 + 6.89e^{-0.8674t}}$, so $N(6) = \dfrac{35.5}{1 + 6.89e^{-0.8674(6)}} \approx 34.2056$, or 34.21 million.

47. **a.** The initial concentration is given by $C(0) = 0.3(0) - 18\left(1 - e^{-0/60}\right)$, or 0 g/cm^3.

b. The concentration after 10 seconds is given by $C(10) = 0.3(10) - 18\left(1 - e^{-10/60}\right) = 0.23667$, or 0.2367 g/cm^3.

c. The concentration after 30 seconds is given by $C(30) = 18e^{-30/60} - 12e^{-(30-20)/60} = 0.75977$, or 0.7598 g/cm^3.

d. The concentration of the drug in the long run is given by $\lim_{t\to\infty} C(t) = \lim_{t\to\infty} \left(18e^{-t/60} - 12e^{-(t-20)/60}\right) = 0$.

49. False. Take $a = b = x = 2$. Then the left-hand side is $(2+2)^2 = 16$, but the right-hand side is $2^2 + 2^2 = 8$.

51. True. If $0 < b < 1$, then $f(x) = b^x$ is a decreasing function of x and so if $x < y$, then $f(x) > f(y)$; that is, $b^x > b^y$.

53. True. If $k > 0$, then $f(x) = e^{kx} = \left(e^k\right)^x = b^x$ (where $b = e^k > 1$) and so f is increasing. If $k < 0$, then $f(x) = \left(e^k\right)^x = b^x$ (where $b = e^k < 1$) and so f is decreasing.

Using Technology page 345

1.

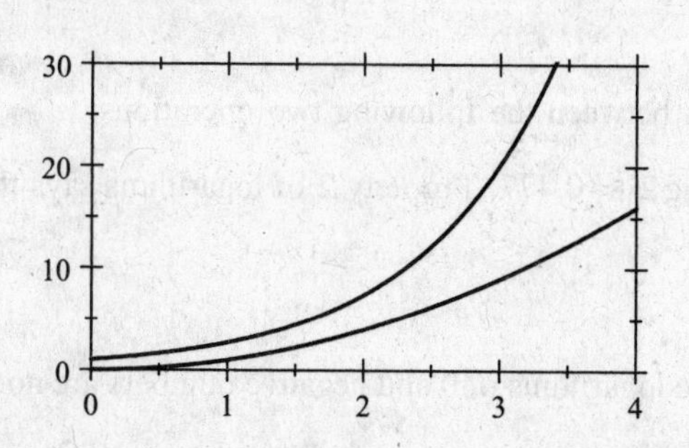

3.

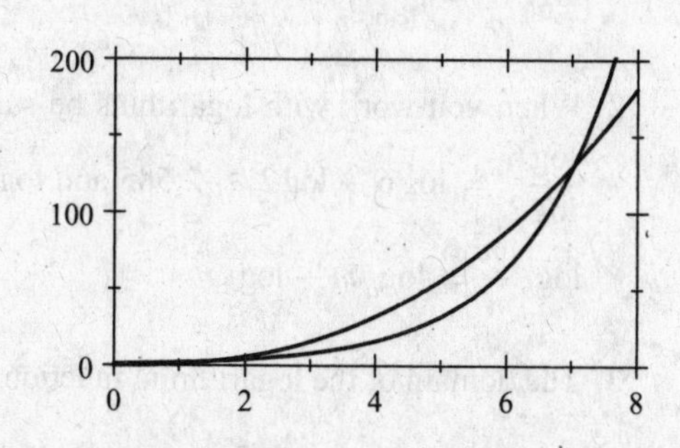

5.

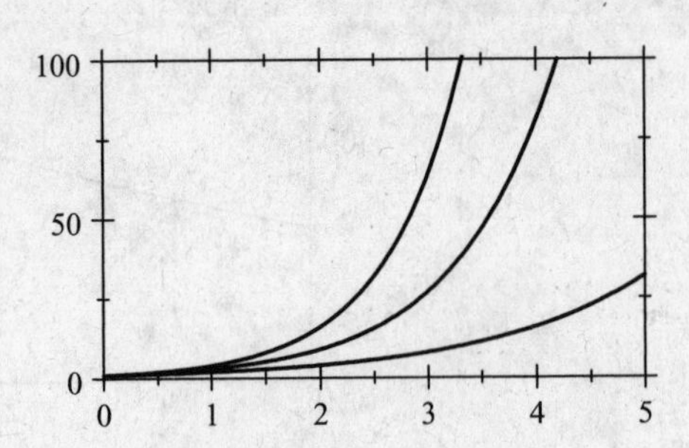

7.

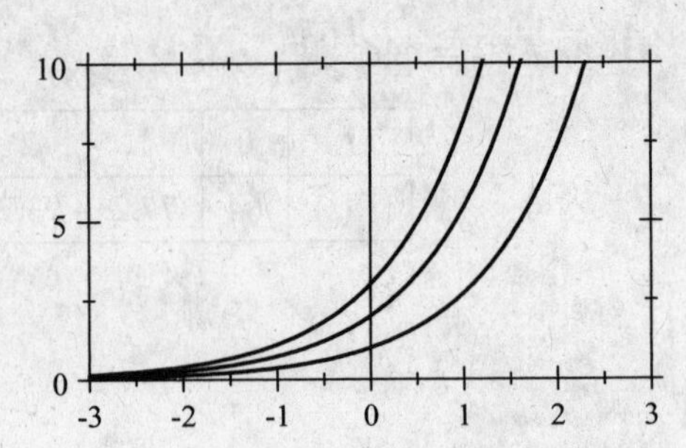

9.

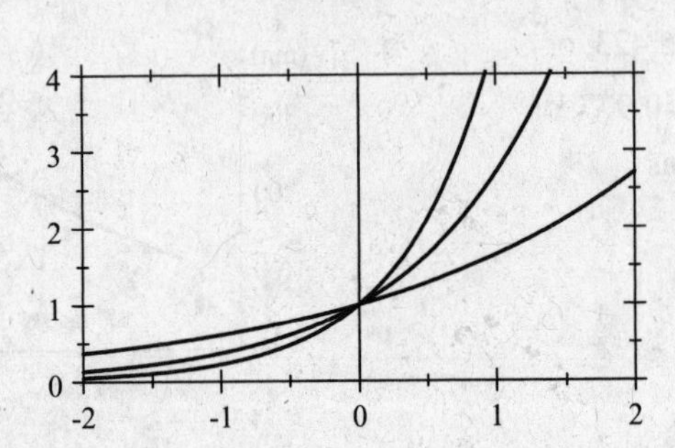

11. a.

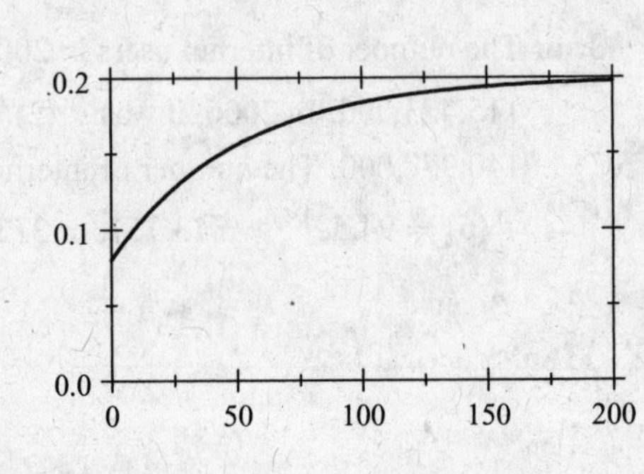

b. 0.08 g/cm^3. **c.** 0.12 g/cm^3.

d. 0.2 g/cm^3

13. a.

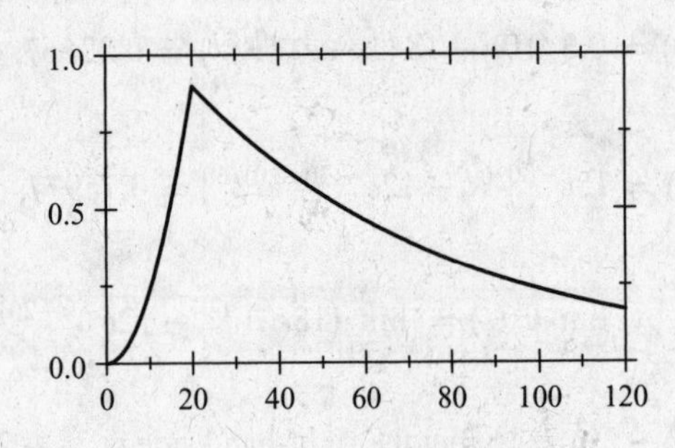

b. 20 seconds. **c.** 35.1 seconds.

5.2 Logarithmic Functions

Problem-Solving Tips

1. Property 1 of logarithms says that $\log_b mn = \log_b m + \log_b n$. However, $\log_b (m + n) \neq \log_b m + \log_b n$ and $\log_b \frac{m}{n} \neq \frac{\log_b m}{\log_b n}$.

2. When you work with logarithms be sure to distinguish between the following two operations: $\frac{\log 6}{\log 2} = \log 6 \div \log 2 \approx 2.585$ and $\log \frac{6}{2} = \log 6 - \log 2 \approx 0.477$. Property 2 of logarithms says that $\log_b \frac{m}{n} = \log_b m - \log_b n$.

3. The domain of the logarithmic function is $(0, \infty)$, so the logarithms of 0 and negative numbers are not defined.

Concept Questions page 351

1. a. $y = \log_b x$ if and only if $x = b^y$.

b. $f(x) = \log_b x$, $b > 0$, $b \neq 1$. Its domain is $(0, \infty)$.

3. a. $e^{\ln x} = x$.

b. $\ln e^x = x$.

Exercises page 351

1. $\log_2 64 = 6$.

3. $\log_4 \frac{1}{16} = -2$.

5. $\log_{1/3} \frac{1}{3} = 1$.

7. $\log_{32} 16 = \frac{4}{5}$.

9. $\log_{10} 0.001 = -3$.

11. $\log 12 = \log 4 \times 3 = \log 4 + \log 3 = 0.6021 + 0.4771 = 1.0792$.

13. $\log 16 = \log 4^2 = 2\log 4 = 2(0.6021) = 1.2042$.

15. $\log 48 = \log\left(3 \cdot 4^2\right) = \log 3 + 2\log 4 = 0.4771 + 2(0.6021) = 1.6813$.

17. $2\ln a + 3\ln b = \ln a^2 b^3$.

19. $\ln 3 + \frac{1}{2}\ln x + \ln y - \frac{1}{3}\ln z = \ln \dfrac{3\sqrt{x}y}{\sqrt[3]{z}}$.

21. $\log x(x+1)^4 = \log x + \log(x+1)^4 = \log x + 4\log(x+1)$.

23. $\log \dfrac{\sqrt{x+1}}{x^2+1} = \log(x+1)^{1/2} - \log\left(x^2+1\right) = \frac{1}{2}\log(x+1) - \log\left(x^2+1\right)$.

25. $\ln xe^{-x^2} = \ln x - x^2$.

27. $\ln\left(\dfrac{x^{1/2}}{x^2\sqrt{1+x^2}}\right) = \ln x^{1/2} - \ln x^2 - \ln\left(1+x^2\right)^{1/2} = \frac{1}{2}\ln x - 2\ln x - \frac{1}{2}\ln\left(1+x^2\right) = -\frac{3}{2}\ln x - \frac{1}{2}\ln\left(1+x^2\right)$.

29. $y = \log_3 x$.

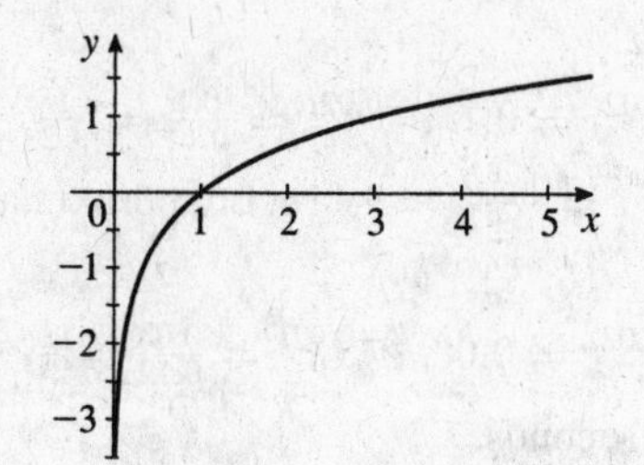

31. $y = \ln 2x$.

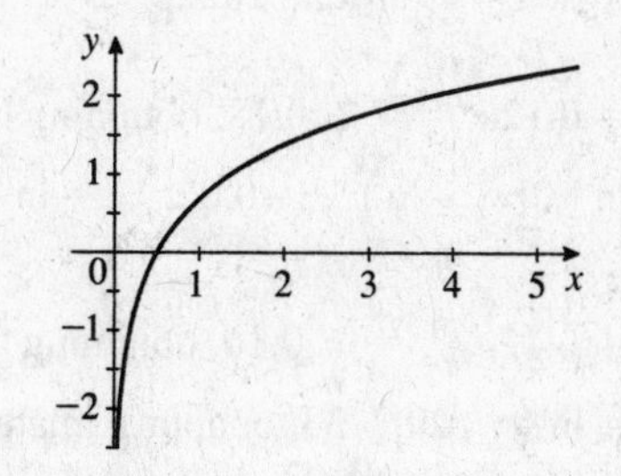

33. $y = 2^x$ and $y = \log_2 x$.

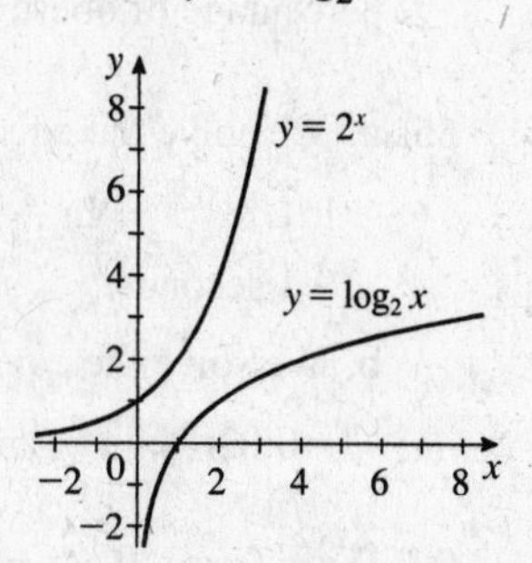

35. $e^{0.4t} = 8$, so $0.4t \ln e = \ln 8$ and thus $0.4t = \ln 8$ because $\ln e = 1$. Therefore, $t = \dfrac{\ln 8}{0.4} \approx 5.1986$.

37. $5e^{-2t} = 6$, so $e^{-2t} = \frac{6}{5} = 1.2$. Taking the logarithm, we have $-2t \ln e = \ln 1.2$, so $t = -\dfrac{\ln 1.2}{2} \approx -0.0912$.

39. $2e^{-0.2t} - 4 = 6$, so $2e^{-0.2t} = 10$. Taking the logarithm on both sides of this last equation, we have $\ln e^{-0.2t} = \ln 5$, $-0.2t \ln e = \ln 5$, $-0.2t = \ln 5$, and and $t = -\frac{\ln 5}{0.2} \approx -8.0472$.

41. $\dfrac{50}{1 + 4e^{0.2t}} = 20$, so $1 + 4e^{0.2t} = \dfrac{50}{20} = 2.5$, $4e^{0.2t} = 1.5$, $e^{0.2t} = \dfrac{1.5}{4} = 0.375$, $\ln e^{0.2t} = \ln 0.375$, and $0.2t = \ln 0.375$. Thus, $t = \dfrac{\ln 0.375}{0.2} \approx -4.9041$.

43. Taking logarithms of both sides, we obtain $\ln A = \ln Be^{-t/2}$, $\ln A = \ln B + \ln e^{-t/2}$, and $\ln A - \ln B = -\dfrac{t}{2} \ln e$, so $\ln \dfrac{A}{B} = -\dfrac{t}{2}$ and $t = -2 \ln \dfrac{A}{B} = 2 \ln \dfrac{B}{A}$.

45. $f(1) = 2$, so $a + b(0) = 2$. Thus, $a = 2$. Therefore, $f(x) = 2 + b \ln x$. We calculate $f(2) = 4$, so $2 + b \ln 2 = 4$. Solving for b, we obtain $b = \dfrac{2}{\ln 2} \approx 2.8854$, so $f(x) = 2 + 2.8854 \ln x$.

47. $p(x) = 19.4 \ln x + 18$. For a child weighing 92 lb, we find $p(92) = 19.4 \ln 92 + 18 \approx 105.7$, or approximately 106 millimeters of mercury.

49. a. $30 = 10 \log \dfrac{I}{I_0}$, so $3 = \log \dfrac{I}{I_0}$, and $\dfrac{I}{I_0} = 10^3 = 1000$. Thus, $I = 1000 I_0$.

b. When $D = 80$, $I = 10^8 I_0$ and when $D = 30$, $I = 10^3 I_0$. Therefore, an 80-decibel sound is $10^8/10^3 = 10^5 = 100{,}000$ times louder than a 30-decibel sound.

c. It is $10^{15}/10^8 = 10^7 = 10{,}000{,}000$ times louder.

51. a. The temperature when it was first poured is given by $T(0) = 70 + 100e^0 = 170$, or 170°F.

b. We solve the equation $70 + 100e^{-0.0446t} = 120$; $100e^{-0.0446t} = 50$, obtaining $e^{-0.0446t} = \frac{50}{100} = \frac{1}{2}$, $\ln e^{-0.0446t} = \ln \frac{1}{2} = \ln 1 - \ln 2 = -\ln 2$, $-0.0446t = -\ln 2$, and so $t = \dfrac{\ln 2}{0.0446} \approx 15.54$. Thus, it will take approximately 15.54 minutes.

53. When $f(t) = 40$, we have $\dfrac{46.5}{1 + 2.324e^{-0.05113t}} = 40$, so $1 + 2.324e^{-0.05113t} = \dfrac{46.5}{40}$, $2.324e^{-0.05113t} = \dfrac{46.5}{40} - 1 = 0.1625$, $e^{-0.05113t} = \dfrac{0.1625}{2.324}$, $-0.05113t = \ln\left(\dfrac{0.1625}{2.324}\right)$, and $t \approx 52.03$. Thus, the percentage of obese adults will reach 40% around 2022.

55. a. We solve the equation $0.08 + 0.12e^{-0.02t} = 0.18$, obtaining $0.12e^{-0.02t} = 0.1$, $e^{-0.02t} = \frac{0.1}{0.12} = \frac{1}{1.2}$, $\ln e^{-0.02t} = \ln \frac{1}{1.2} = \ln 1 - \ln 1.2 = -\ln 1.2$, $-0.02t = -\ln 1.2$, and $t = \frac{\ln 1.2}{0.02} \approx 9.116$, or approximately 9.1 seconds.

b. We solve the equation $0.08 + 0.12e^{-0.02t} = 0.16$, obtaining $0.12e^{-0.02t} = 0.08$, $e^{-0.02t} = \frac{0.08}{0.12} = \frac{2}{3}$, $-0.02t = \ln \frac{2}{3}$, and $t = -\frac{1}{0.02} \ln \frac{2}{3} \approx 20.2733$, or approximately 20.3 seconds.

57. With $T_0 = 70$, $T_1 = 98.6$, and $T = 80$, we have $80 = 70 + (98.6 - 70)(0.97)^t$, so $28.6(0.97)^t = 10$ and $(0.97)^t = 0.34965$. Taking logarithms, we have $\ln (0.97)^t = \ln 0.34965$, or $t = \dfrac{\ln 0.34965}{\ln 0.97} \approx 34.50$. Thus, he was killed $34\frac{1}{2}$ hours earlier, at 1:30 p.m.

59. False. Take $x = e$. Then $(\ln e)^3 = 1^3 = 1 \neq 3 \ln e = 3$.

61. True. $e^{\ln b} = b$ and $\ln e^b = b$ as well.

63. True. $g(x) = \ln x$ is continuous and greater than zero on $(1, \infty)$. Therefore, $f(x) = \dfrac{1}{\ln x}$ is continuous on $(1, \infty)$.

65. a. Taking logarithms of both sides gives $\ln 2^x = \ln e^{kx}$, so $x \ln 2 = kx (\ln e) = kx$. Thus, $x(\ln 2 - k) = 0$ for all x, and this implies that $k = \ln 2$.

b. Proceeding as in part (a), we find that $k = \ln b$.

67. Let $\log_b m = p$. Then $m = b^p$. Therefore, $m^n = (b^p)^n = b^{np}$, and so $\log_b m^n = \log_b b^{np} = np \log_b b = np$ (since $\log_b b = 1$) $= n \log_b m$.

5.3 Compound Interest

Problem-Solving Tips

In applied problems involving interest rates, it is important to choose the correct interest formula to solve the problem. If the problems asks for the *future value* of an investment, then use the compound interest formula. If the problem asks for the *amount of money that needs to be invested now* to accumulate a certain sum in the future, then use the present value formula for compound interest. If the interest in the applied problem is compounded continuously then use the corresponding formulas for continuous compound interest.

Concept Questions page 365

1. a. When simple interest is computed, the interest earned is based on the original principal. When compound interest is computed, the interest earned is periodically added to the principal and thereafter earns interest at the same rate.

b. The simple interest formula is $A = P(1 + rt)$ and the compound interest formula is $A = P\left(1 + \dfrac{r}{m}\right)^{mt}$.

3. $P = A\left(1 + \dfrac{r}{m}\right)^{-mt}$.

Exercises page 365

1. $A = 2500\left(1 + \dfrac{0.04}{2}\right)^{20} = 3714.87$, or \$3714.87.

3. $A = 150{,}000\left(1 + \dfrac{0.06}{12}\right)^{48} = 190{,}573.37$, or \$190,573.37.

5. a. Using the formula $r_{\text{eff}} = \left(1 + \dfrac{r}{m}\right)^m - 1$ with $r = 0.06$ and $m = 2$, we have $r_{\text{eff}} = \left(1 + \dfrac{0.06}{2}\right)^2 - 1 = 0.0609$, or 6.09%/yr.

b. Using the formula $r_{\text{eff}} = \left(1 + \dfrac{r}{m}\right)^m - 1$ with $r = 0.05$ and $m = 4$, we have $r_{\text{eff}} = \left(1 + \dfrac{0.05}{4}\right)^4 - 1 = 0.05095$, or 5.095%/yr.

7. a. The present value is given by $P = 40{,}000\left(1 + \dfrac{0.05}{2}\right)^{-8} = 32{,}829.86$, or \$32,829.86.

b. The present value is given by $P = 40{,}000\left(1+\dfrac{0.05}{4}\right)^{-16} = 32{,}789.85$, or \$32,789.85.

9. $A = 5000e^{0.05(4)} \approx 6107.01$, or \$6107.01.

11. We use Formula (8) with $A = 10{,}000$, $m = 365$, $r = 0.04$, and $t = 2$. The required deposit is $P = 10{,}000\left(1+\dfrac{0.04}{365}\right)^{-365(2)} \approx 9231.20$, or \$9231.20.

13. We use Formula (11) with $A = 20{,}000$, $r = 0.06$, and $t = 3$. Jack should deposit $P = 20{,}000e^{-(0.06)(3)} \approx 16{,}705.404$, or \$16,705.40.

15. $P = Ae^{-rt} = 59{,}673e^{-(0.06)5} \approx 44{,}206.85$, or approximately \$44,206.85.

17. We use Formula (6) with $A = 7500$, $P = 5000$, $m = 12$, and $t = 3$. Thus, $7500 = 5000\left(1+\frac{r}{12}\right)^{36}$, $\left(1+\frac{r}{12}\right)^{36} = \frac{7500}{5000} = \frac{3}{2}$, $\ln\left(1+\frac{r}{12}\right)^{36} = \ln 1.5$, $36\ln\left(1+\frac{r}{12}\right) = \ln 1.5$, $\ln\left(1+\frac{r}{12}\right) = \frac{\ln 1.5}{36} = 0.0112629$, $1+\frac{r}{12} = e^{0.0112629} = 1.011327$, $\frac{r}{12} = 0.011327$, and $r = 0.13592$. The annual interest rate is 13.59%.

19. We use Formula (6) with $A = 5500$, $P = 5000$, $m = 12$, and $t = \frac{1}{2}$. Thus, $5500 = 5000\left(1+\frac{r}{12}\right)^{6}$, and so $\left(1+\frac{r}{12}\right)^{6} = \frac{5500}{5000} = 1.1$. Proceeding as in the previous exercise, we find $r = 0.1921$, so the required annual interest rate is 19.21%.

21. We use Formula (6) with $A = 6000$, $P = 2000$, $m = 12$, and $t = 5$. Thus, $6000 = 2000\left(1+\frac{r}{12}\right)^{60}$. Thus, $\left(1+\frac{r}{12}\right)^{60} = 3$, $60\ln\left(1+\frac{r}{12}\right) = \ln 3$, $\ln\left(1+\frac{r}{12}\right) = \frac{\ln 3}{60}$, $1+\frac{r}{12} = e^{(\ln 3)/60}$, $\frac{r}{12} = e^{(\ln 3)/60} - 1$, and $r = 12\left(e^{(\ln 3)/60} - 1\right) \approx 0.2217$, so the required interest rate is 22.17% per year.

23. We use Formula (6) with $A = 6500$, $P = 5000$, $m = 12$, and $r = 0.06$. Thus, $6500 = 5000\left(1+\frac{0.06}{12}\right)^{12t}$, $(1.005)^{12t} = \frac{6500}{5000} = 1.3$, $12t\ln(1.005) = \ln 1.3$, and so $t = \dfrac{\ln 1.3}{12\ln 1.005} \approx 4.384$. It will take approximately 4.4 years.

25. We use Formula (6) with $A = 15000$, $P = 5000$, $m = 365$, and $r = 0.04$. Thus, $15{,}000 = 5000\left(1+\dfrac{0.04}{365}\right)^{365t}$, from which we find $t = \dfrac{\ln\left(\frac{15{,}000}{5000}\right)}{365\ln\left(1+\frac{0.04}{365}\right)} \approx 27.47$. Thus, it will take approximately 27.5 years.

27. We use Formula (10) with $A = 8000$, $P = 4000$, and $t = 5$. Thus, $8000 = 4000e^{5r}$, $e^{5r} = \frac{8000}{4000} = 2$, $5r = \ln 2$, and $r = \dfrac{\ln 2}{5} \approx 0.13863$. The annual interest rate is 13.86%.

29. We use Formula (10) with $A = 16{,}000$, $P = 8000$, and $r = 0.05$. Thus, $16{,}000 = 8000e^{0.05t}$, and we find that $t = \frac{\ln 2}{0.05} \approx 13.863$. It will take 13.9 years.

31. The utility company will have to increase its generating capacity by a factor of $(1.08)^{10} \approx 2.16$.

33. After 1 year, Maria's investment is worth $(1.2)(10{,}000)$ dollars and after 2 years, it is worth $(1.1)(1.2)(10{,}000)$ dollars.
After 1 year, Laura's investment is worth $(1.1)(10{,}000)$ dollars and after 2 years, it is worth $(1.2)(1.1)(10{,}000)$ dollars.
So after 2 years, both investments are worth the same amount, namely \$13,200.

35. Suppose Jack's portfolio is worth $\$P$ initially. After 1 year, it is worth $(0.8)P$ dollars, and after 2 years, it is worth $(1.2)(0.8)P$ or $0.96P$ dollars. This shows that after the second year, Jack's investment has not recouped all of its losses from the first year.

37. We use Formula 3 with $P = 15{,}000$, $r = 0.078$, $m = 12$, and $t = 4$, giving the worth of Jodie's account as $A = 15{,}000\left(1 + \frac{0.078}{12}\right)^{(12)(4)} \approx 20{,}471.641$, or approximately \$20,471.64.

39. We use Formula 3 with $P = 10{,}000$, $r = 0.0682$, $m = 4$, and $t = \frac{11}{2}$, giving the worth of Chris' account as $A = 10{,}000\left(1 + \frac{0.0682}{4}\right)^{(4)(11/2)} \approx 14{,}505.433$, or approximately \$14,505.43.

41. He can expect the minimum revenue for 2016 to be $240{,}000(1.2)(1.3)(1.25)^3 \approx 731{,}250$, or \$731,250.

43. We want the value of a 2013 dollar at the beginning of 2009. Denoting this value by x, we have $(1.027)(1.015)(1.030)(1.017)x = 1$, so $x \approx 0.916$. Thus, the purchasing power is approximately 92 cents.

45. He needs $65{,}000e^{0.03(10)} \approx 87{,}740.82$ or approximately \$87,740.82 annually.

47. The present value of the \$8000 loan due in 3 years is given by $P = 8000\left(1 + \frac{0.08}{2}\right)^{-6} \approx 6322.52$, or \$6322.52.

The present value of the \$15,000 loan due in 6 years is given by $P = 15{,}000\left(1 + \frac{0.08}{2}\right)^{-12} \approx 9368.96$, or \$9368.96.
Therefore, the amount the proprietors of the inn will be required to pay at the end of 5 years is given by $A = 15{,}691.48\left(1 + \frac{0.08}{2}\right)^{10} \approx 23{,}227.22$, or \$23,227.22.

49. $P(t) = V(t)e^{-rt} = 80{,}000e^{\sqrt{t}/2}e^{-rt} = 80{,}000e^{(\sqrt{t}/2)-0.09t}$. Thus, $P(4) = 80{,}000e^{1-0.09(4)} \approx 151{,}718.47$, or approximately \$151,718.

51. Suppose \$1 is invested in each investment. The accumulated amount in investment A is $\left(1 + \frac{0.08}{2}\right)^8 \approx 1.36857$ and the accumulated amount in investment B is $e^{0.0775(4)} \approx 1.36343$. Thus, investment A has a higher rate of return.

53. The effective annual rate of return on his investment is found by solving the equation $(1+r)^2 = \frac{32{,}100}{25{,}250}$. We find $1 + r = \left(\frac{32{,}100}{25{,}250}\right)^{1/2}$, so $1 + r \approx 1.1275$, and $r \approx 0.1275$, or 12.75%.

55. $r_{\text{eff}} = \lim_{m\to\infty}\left(1 + \frac{r}{m}\right)^m - 1 = e^r - 1$.

57. The effective rate of interest at Bank A is given by $R = \left(1 + \frac{0.07}{4}\right)^4 - 1 = 0.07186$, or 7.186%. The effective rate at Bank B is given by $R = e^r - 1 = e^{0.07125} - 1 = 0.07385$, or 7.385%. We conclude that Bank B has the higher effective rate of interest.

59. By definition, $A = P(1 + r_{\text{eff}})^t$, so $(1 + r_{\text{eff}})^t = \frac{A}{P}$, $1 + r_{\text{eff}} = \left(\frac{A}{P}\right)^{1/t}$, and $r_{\text{eff}} = \left(\frac{A}{P}\right)^{1/t} - 1$.

61. Using the formula $r_{\text{eff}} = \left(\frac{A}{P}\right)^{1/t} - 1$ with $A = 5070.42$, $P = 5000$, and $t = \frac{245}{365}$, we have $r_{\text{eff}} = \left(\frac{5070.42}{5000}\right)^{1/(245/365)} - 1 = \left(\frac{5070.42}{5000}\right)^{(365/245)} - 1 \approx 0.0211$, or 2.11%.

63. True. With $m = 1$, the effective rate is $r_{\text{eff}} = \left(1 + \frac{r}{1}\right)^1 - 1 = r$.

Using Technology page 370

1. \$5872.78 **3.** 8.95%/yr **5.** \$29,743.30

5.4 Differentiation of Exponential Functions

Problem-Solving Tips

1. The derivative of e^x is equal to e^x. By the chain rule, $\frac{d}{dx}\left(e^{3x}\right) = 3e^{3x}$ and $\frac{d}{dx}\left(e^{2x^2-1}\right) = 4xe^{2x^2-1}$. Note that the exponents in the original function and the derivative are the same.

2. Don't confuse functions of the type e^x with functions of the type x^r. The latter is a *power function* and its exponent is a *constant*; whereas the exponent in an *exponential function* such as e^x is a *variable*. A different rule is used to differentiate the two types of function. Thus, $\frac{d}{dx}\left(x^2e^x\right) = x^2\frac{d}{dx}\left(e^x\right) + e^x\frac{d}{dx}\left(x^2\right) = x^2e^x + e^x \cdot 2x = xe^x(x+2)$.

Concept Questions page 376

1. a. $f'(x) = e^x$

b. $g'(x) = e^{f(x)} \cdot f'(x)$

Exercises page 376

1. $f(x) = e^{3x}$, so $f'(x) = 3e^{3x}$

3. $g(t) = e^{-t}$, so $g'(t) = -e^{-t}$

5. $f(x) = e^x + x^2$, so $f'(x) = e^x + 2x$.

7. $f(x) = x^3e^x$, so $f'(x) = x^3e^x + e^x\left(3x^2\right) = x^2e^x(x+3)$.

9. $f(x) = \dfrac{e^x}{x}$, so $f'(x) = \dfrac{x\left(e^x\right) - e^x(1)}{x^2} = \dfrac{e^x(x-1)}{x^2}$.

11. $f(x) = 3\left(e^x + e^{-x}\right)$, so $f'(x) = 3\left(e^x - e^{-x}\right)$.

13. $f(w) = \dfrac{e^w + 2}{e^w} = 1 + \dfrac{2}{e^w} = 1 + 2e^{-w}$, so $f'(w) = -2e^{-w} = -\dfrac{2}{e^w}$.

15. $f(x) = 2e^{3x-1}$, so $f'(x) = 2e^{3x-1}(3) = 6e^{3x-1}$.

17. $h(x) = e^{-x^2}$, so $h'(x) = e^{-x^2}(-2x) = -2xe^{-x^2}$.

19. $f(x) = 3e^{-1/x}$, so $f'(x) = 3e^{-1/x} \cdot \dfrac{d}{dx}\left(-\dfrac{1}{x}\right) = 3e^{-1/x}\left(\dfrac{1}{x^2}\right) = \dfrac{3e^{-1/x}}{x^2}$.

21. $f(x) = (e^x+1)^{25}$, so $f'(x) = 25(e^x+1)^{24}e^x = 25e^x(e^x+1)^{24}$.

23. $f(x) = e^{\sqrt{x}}$, so $f'(x) = e^{\sqrt{x}}\dfrac{d}{dx}(x^{1/2}) = e^{\sqrt{x}}\frac{1}{2}x^{-1/2} = \dfrac{e^{\sqrt{x}}}{2\sqrt{x}}$.

25. $f(x) = (x-1)e^{3x+2}$, so $f'(x) = (x-1)(3)e^{3x+2} + e^{3x+2} = e^{3x+2}(3x-3+1) = e^{3x+2}(3x-2)$.

27. $f(x) = \dfrac{e^x-1}{e^x+1}$, so $f'(x) = \dfrac{(e^x+1)(e^x)-(e^x-1)(e^x)}{(e^x+1)^2} = \dfrac{e^x(e^x+1-e^x+1)}{(e^x+1)^2} = \dfrac{2e^x}{(e^x+1)^2}$.

29. $f(x) = e^{-4x} + e^{3x}$, so $f'(x) = -4e^{-4x} + 3e^{3x}$ and $f''(x) = 16e^{-4x} + 9e^{3x}$.

31. $f(x) = 2xe^{3x}$, so $f'(x) = 2e^{3x} + 2xe^{3x}(3) = 2(3x+1)e^{3x}$ and $f''(x) = 6e^{3x} + 2(3x+1)e^{3x}(3) = 6(3x+2)e^{3x}$.

33. $y = f(x) = e^{2x-3}$, so $f'(x) = 2e^{2x-3}$. To find the slope of the tangent line to the graph of f at $x = \frac{3}{2}$, we compute $f'\left(\frac{3}{2}\right) = 2e^{3-3} = 2$. Next, using the point-slope form of the equation of a line, we find that $y - 1 = 2\left(x - \frac{3}{2}\right) = 2x - 3$, or $y = 2x - 2$.

35. $f(x) = e^{-x^2/2}$, so $f'(x) = e^{-x^2/2}(-x) = -xe^{-x^2/2}$. Setting $f'(x) = 0$, gives $x = 0$ as the only critical point of f. From the sign diagram, we conclude that f is increasing on $(-\infty, 0)$ and decreasing on $(0, \infty)$.

+ + + + 0 − − − − sign of f'
→ x
0

37. $f(x) = \frac{1}{2}e^x - \frac{1}{2}e^{-x}$, so $f'(x) = \frac{1}{2}(e^x + e^{-x})$ and $f''(x) = \frac{1}{2}(e^x - e^{-x})$. Setting $f''(x) = 0$ gives $e^x = e^{-x}$ or $e^{2x} = 1$, and so $x = 0$. From the sign diagram for f'',we conclude that f is concave upward on $(0, \infty)$ and concave downward on $(-\infty, 0)$.

− − − − 0 + + + + sign of f''
→ x
0

39. $f(x) = xe^{-2x}$, so $f'(x) = e^{-2x} + xe^{-2x}(-2) = (1-2x)e^{-2x}$ and $f''(x) = -2e^{-2x} + (1-2x)e^{-2x}(-2) = 4(x-1)e^{-2x}$. Observe that $f''(x) = 0$ if $x = 1$. The sign diagram of f'' shows that $(1, e^{-2})$ is an inflection point.

− − − − 0 + + + + sign of f''
→ x
1

41. $f(x) = e^{-x^2}$, so $f'(x) = -2xe^{-x^2}$ and $f''(x) = -2e^{-x^2} - 2xe^{-x^2}(-2x) = -2e^{-x^2}(1 - 2x^2) = 0$ implies $x = \pm\frac{\sqrt{2}}{2}$. The sign diagram of f'' shows that the graph of f has inflection points at $\left(-\frac{\sqrt{2}}{2}, e^{-1/2}\right)$ and $\left(\frac{\sqrt{2}}{2}, e^{-1/2}\right)$. The slope of the tangent line at $\left(-\frac{\sqrt{2}}{2}, e^{-1/2}\right)$ is $f'\left(-\frac{\sqrt{2}}{2}\right) = \sqrt{2}e^{-1/2}$, and the tangent line has equation $y - e^{-1/2} = \sqrt{2}e^{-1/2}\left(x + \frac{\sqrt{2}}{2}\right)$, which can be simplified to $y = e^{-1/2}\left(\sqrt{2}x + 2\right)$. The slope of the tangent line at $\left(\frac{\sqrt{2}}{2}, e^{-1/2}\right)$ is $f'\left(\frac{\sqrt{2}}{2}\right) = -\sqrt{2}e^{-1/2}$, and this tangent line has equation $y - e^{-1/2} = -\sqrt{2}e^{-1/2}\left(x - \frac{\sqrt{2}}{2}\right)$ or $y = e^{-1/2}\left(-\sqrt{2}x + 2\right)$.

+ + + 0 − − − − − 0 + + + sign of f''

$-\frac{\sqrt{2}}{2}$ 0 $\frac{\sqrt{2}}{2}$ x

43. $f(x) = e^{-x^2}$, so $f'(x) = -2xe^{-x^2} = 0$ if $x = 0$, the only critical point of f. From the table, we see that f has an absolute minimum value of e^{-1} attained at $x = -1$ and $x = 1$. It has an absolute maximum at $(0, 1)$.

x	-1	0	1
$f(x)$	e^{-1}	1	e^{-1}

45. $g(x) = (2x - 1)e^{-x}$, so
$g'(x) = 2e^{-x} + (2x - 1)e^{-x}(-1) = (3 - 2x)e^{-x} = 0$ if $x = \frac{3}{2}$.
The graph of g shows that $\left(\frac{3}{2}, 2e^{-3/2}\right)$ is an absolute maximum, and $(0, -1)$ is an absolute minimum.

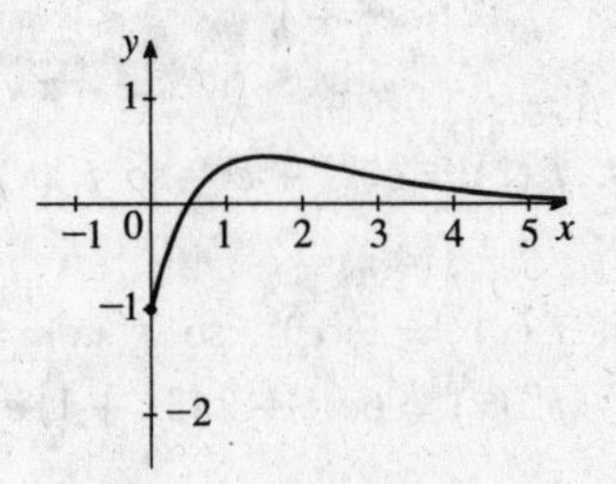

47. $f(t) = e^t - t$. We first gather the following information on f.

1. The domain of f is $(-\infty, \infty)$.
2. Setting $t = 0$ gives 1 as the y-intercept.
3. $\lim_{t\to-\infty}(e^t - t) = \infty$ and $\lim_{t\to\infty}(e^t - t) = \infty$.
4. There is no asymptote.
5. $f'(t) = e^t - 1$ if $t = 0$, a critical point of f. From the sign diagram for f', we see that f is decreasing on $(-\infty, 0)$ and increasing on $(0, \infty)$.

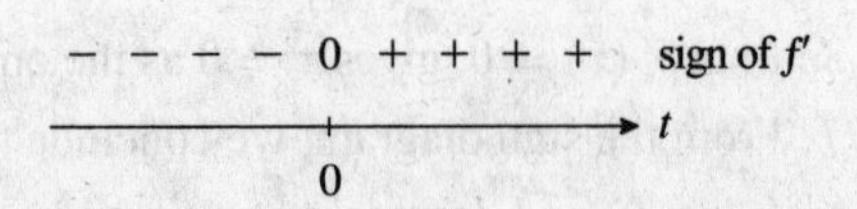

6. From the results of part 5, we see that $(0, 1)$ is a relative minimum of f.
7. $f''(t) = e^t > 0$ for all t, so the graph of f is concave upward on $(-\infty, \infty)$.
8. There is no inflection point.

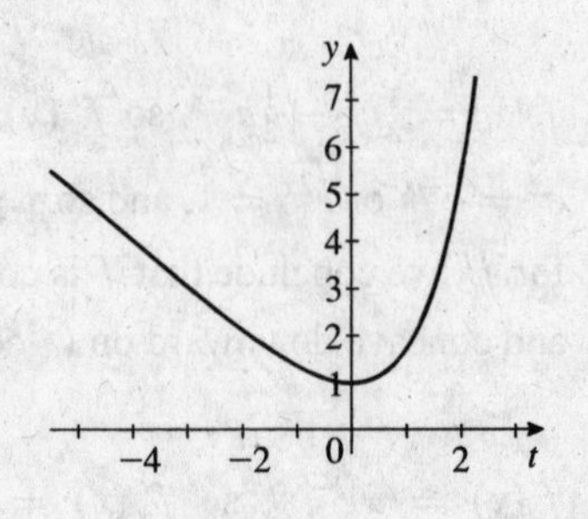

49. $f(x) = 2 - e^{-x}$. We first gather the following information on f.

1. The domain of f is $(-\infty, \infty)$.
2. Setting $x = 0$ gives 1 as the y-intercept.

3. $\lim_{x\to-\infty}(2-e^{-x}) = -\infty$ and $\lim_{x\to\infty}(2-e^{-x}) = 2$.

4. From the results of part 3, we see that $y = 2$ is a horizontal asymptote of f.

5. $f'(x) = e^{-x} > 0$ for all x in $(-\infty, \infty)$, so f is increasing on $(-\infty, \infty)$.

6. Because there is no critical point, f has no relative extremum.

7. $f''(x) = -e^{-x} < 0$ for all x in $(-\infty, \infty)$ and so the graph of f is concave downward on $(-\infty, \infty)$.

8. There is no inflection point.

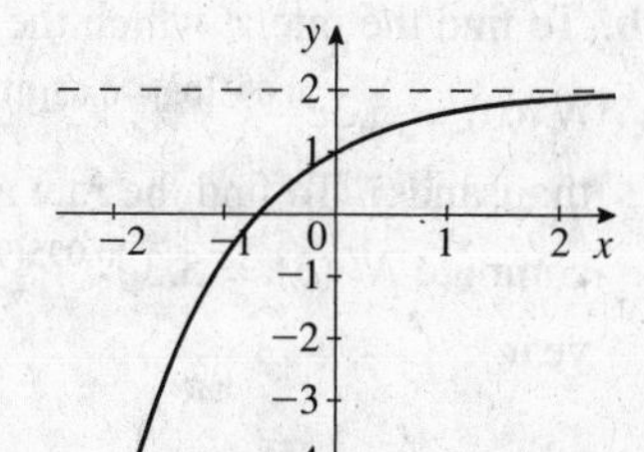

51. $x^2 + y^3 = 2e^{2y}$, so $2x + 3y^2y' = 4e^{2y}y'$ and $y'(4e^{2y} - 3y^2) = 2x$. Thus, $\dfrac{dy}{dx} = \dfrac{2x}{4e^{2y} - 3y^2}$.

53. $x = y + e^y$, so $1 = y' + e^y y' = (1 + e^y)y'$. Differentiating again, $0 = (e^y y')y' + (1 + e^y)y''$, and so $\dfrac{d^2y}{dx^2} = -\dfrac{e^y(dy/dx)^2}{1 + e^y}$. From our first differentiation, we have $\dfrac{dy}{dx} = \dfrac{1}{1 + e^y}$, so we can write $\dfrac{d^2y}{dx^2} = -\dfrac{e^y}{(1 + e^y)^3}$.

55. $xy + e^y = e$, so $y + x\dfrac{dy}{dx} + e^y\dfrac{dy}{dx} = 0$. When $x = 0$ and $y = 1$, we have $1 + 0 \cdot \dfrac{dy}{dx} + e^1\dfrac{dy}{dx} = 0$, so $e\dfrac{dy}{dx} = -1$ and $\dfrac{dy}{dx} = -\dfrac{1}{e}$.

57. a. The number of video viewers in 2012 was $N(5) = 135e^{0.067(5)}$ or approximately 188.7 million.

b. $N'(t) = 135(0.067)e^{0.067t} = 9.045e^{0.067t}$, so the number of viewers was changing at the rate of $N'(5) = 9.045e^{0.067(5)}$ or approximately 12.6 million viewers per year in 2012.

59. We find $f'(t) = \dfrac{d}{dt}(20.5e^{0.74t}) = 20.5(0.74e^{0.74t}) = 15.17e^{0.74t}$, so $f'(2) = 15.17e^{0.74(2)} \approx 66.64$. Thus, the value of stolen drugs is increasing at the rate of \$66.64 million per year at the beginning of 2008.

61. a. $S(t) = 20{,}000(1 + e^{-0.5t})$, so $S'(t) = 20{,}000(-0.5e^{-0.5t}) = -10{,}000e^{-0.5t}$. Thus, $S'(1) = -10{,}000e^{-0.5} \approx -6065$, or $-\$6065$/day/day; $S'(2) = -10{,}000e^{-1} \approx -3679$, or $-\$3679$/day/day; $S'(3) = -10{,}000(e^{-1.5}) \approx -2231$, or $-\$2231$/day/day; and $S'(4) = -10{,}000e^{-2} \approx -1353$, or $-\$1353$/day/day.

b. $S(t) = 20{,}000(1 + e^{-0.5t}) = 27{,}400$, so $1 + e^{-0.5t} = \frac{27{,}400}{20{,}000}$, $e^{-0.5t} = \frac{274}{200} - 1$, $-0.5t = \ln\left(\frac{274}{200} - 1\right)$, and so $t = \dfrac{\ln\left(\frac{274}{200} - 1\right)}{-0.5} \approx 2$, or 2 days.

63. $N(t) = 5.3e^{0.095t^2 - 0.85t}$.

a. $N'(t) = 5.3e^{0.095t^2 - 0.85t}(0.19t - 0.85)$. Because $N'(t)$ is negative for $0 \le t \le 4$, we see that $N(t)$ is decreasing over that interval.

b. To find the rate at which the number of polio cases was decreasing at the beginning of 1959, we compute $N'(0) = 5.3e^{0.095(0^2) - 0.85(0)}(0.85) \approx 5.3(-0.85) = -4.505$, or 4505 cases per year per year (t is measured in thousands). To find the rate at which the number of polio cases was decreasing at the beginning of 1962, we compute $N'(3) = 5.3e^{0.095(9) - 0.85(3)}(0.57 - 0.85) \approx (-0.28)(0.9731) \approx -0.273$, or 273 cases per year per year.

65. a. The frequency for a 70 year old is given by $f(1) = 0.71e^{0.71(1)} \approx 1.43$ (%), and for a 90 year old it is $f(5) = 0.71e^{0.71(5)} \approx 23.51$ (%).

b. $f'(t) = \dfrac{d}{dt}\left(0.71e^{0.7t}\right) = 0.7\left(0.71e^{0.7t}\right) = 0.497e^{0.7t}$, which is positive for $0 < t < 5$, and so f is increasing on $(0, 5)$. This says that the frequency of Alzheimer's disease increases with age in the age range under consideration.

c. $f''(t) = \dfrac{d}{dt}\left(0.497e^{0.7t}\right) = 0.3479e^{0.7t}$, which is positive for $0 < t < 5$, and so f is concave upward on $(0, 5)$. This says that the frequency of Alzheimer's disease is increasing at an increasing rate in the age range under consideration.

67. a. $R(x) = px = 100xe^{-0.0001x}$.

b. $R'(x) = 100e^{-0.0001x} + 100xe^{-0.0001x}(-0.0001) = 100(1 - 0.0001x)e^{-0.0001x}$.

c. $R'(10{,}000) = 100[1 - 0.0001(10{,}000)]e^{-0.001} = 0$, or \$0/pair.

69. The demand equation is $p(x) = 100e^{-0.0002x} + 150$. Next, $p'(x) = 100(-0.0002)e^{-0.0002x} = -0.02e^{-0.0002x}$.

a. To find the rate of change of the price per bottle when $x = 1000$, we compute $p'(1000) = -0.02e^{-0.0002(1000)} = -0.02e^{-0.2} \approx -0.0164$, or -1.64 cents per bottle. To find the rate of change of the price per bottle when $x = 2000$, we compute $p'(2000) = -0.02e^{-0.0002(2000)} = -0.02e^{-0.4} \approx -0.0134$, or -1.34 cents per bottle.

b. The price per bottle when $x = 1000$ is given by $p(1000) = 100e^{-0.0002(1000)} + 150 \approx 231.87$, or \$231.87/bottle. The price per bottle when $x = 2000$ is given by $p(2000) = 100e^{-0.0002(2000)} + 150 \approx 217.03$, or \$217.03/bottle.

71. a. $N(0) = \dfrac{3000}{1 + 99} = 30$.

b. $N'(x) = 3000\dfrac{d}{dx}\left(1 + 99e^{-x}\right)^{-1} = -3000\left(1 + 99e^{-x}\right)^{-2}\left(-99e^{-x}\right) = \dfrac{297{,}000e^{-x}}{\left(1 + 99e^{-x}\right)^2}$. Because $N'(x) > 0$ for all x in $(0, \infty)$, we see that N is increasing on $(0, \infty)$.

c. From the graph of N, we see that the total number of students who contracted influenza during that particular epidemic is approximately $\lim\limits_{x \to \infty} \dfrac{3000}{1 + 99e^{-x}} = 3000$.

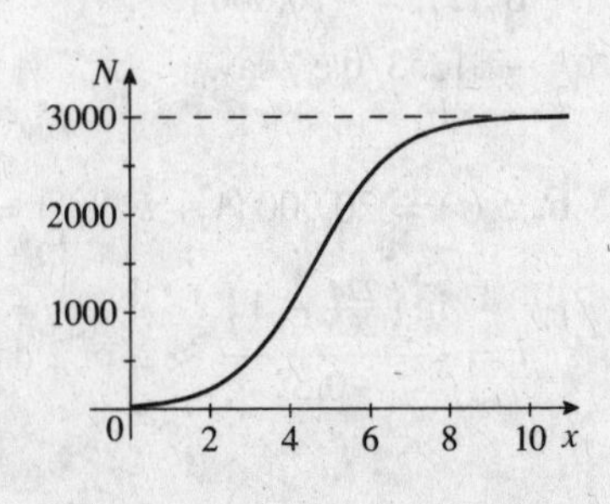

73. a. Here $f(p) = 50e^{-0.02p}$, so $E(p) = -\dfrac{pf'(p)}{f(p)} = -\dfrac{p\left(50e^{-0.02p}\right)(-0.02)}{50e^{-0.02p}} = 0.02p$.

b. Using the result from part (a), we see that $E(p) = 1$ if $0.02p = 1$, or $p = 50$; $E(p) < 1$ if $p < 50$; and $E(p) > 1$ if $p > 50$. Demand is inelastic if $0 < p < 50$, unitary if $p = 50$, and elastic if $p > 50$.

c. Because demand is inelastic when $p = 40$, decreasing the unit price slightly will cause revenue to decrease.

d. Because demand is elastic when $p = 60$, increasing the unit price slightly will cause revenue to decrease.

75. a. $W = 2.4e^{1.84h}$, so if $h = 1.6$, $W = 2.4e^{1.84(1.6)} \approx 45.58$, or approximately 45.6 kg.

b. $\Delta W \approx dW = (2.4)(1.84)e^{1.84h}\,dh$. With $h = 1.6$ and $dh = \Delta h = 1.65 - 1.6 = 0.05$, we find $\Delta W \approx (2.4)(1.84)e^{1.84(1.6)} \cdot (0.05) \approx 4.19$, or approximately 4.2 kg.

77. $P(t) = 80{,}000e^{\sqrt{t}/2-0.09t} = 80{,}000e^{(1/2)t^{1/2}-0.09t}$, so $P'(t) = 80{,}000\left(\frac{1}{4}t^{-1/2} - 0.09\right)e^{(1/2)t^{1/2}-0.09t}$. Setting $P'(t) = 0$, we have $\frac{1}{4}t^{-1/2} = 0.09$, so $t^{-1/2} = 0.36$, $\dfrac{1}{\sqrt{t}} = 0.36$, and $t = \left(\frac{1}{0.36}\right)^2 \approx 7.72$. Evaluating $P(t)$ at each of its endpoints and at the point $t = 7.72$, we find $P(0) = 80{,}000$, $P(7.72) \approx 160{,}207.69$, and $P(8) \approx 160{,}170.71$. We conclude that P is optimized at $t = 7.72$. The optimal price is approximately \$160,208.

79. $A(t) = 0.23te^{-0.4t}$, so $A'(t) = 0.23(1 - 0.4t)e^{-0.4t}$. Setting $A'(t) = 0$ gives $t = \frac{1}{0.4} = \frac{5}{2}$. From the graph of A, we see that the proportion of alcohol is highest $2\frac{1}{2}$ hours after drinking. The level is given by $A\left(\frac{5}{2}\right) \approx 0.2115$, or approximately 0.21%.

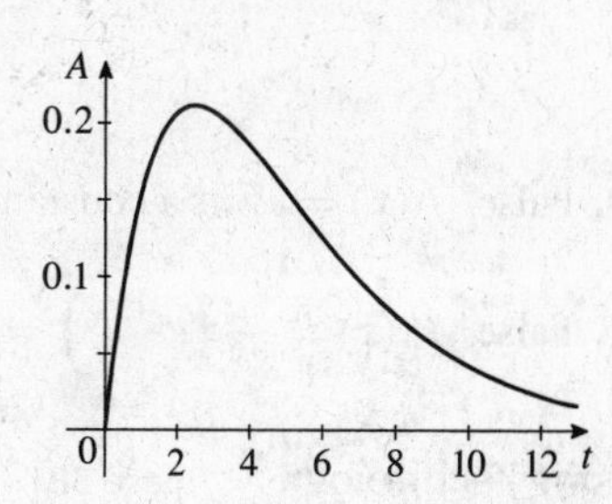

81. a. The temperature inside the house is given by $T(0) = 30 + 40e^0 = 70$, or 70°F.

b. The reading is changing at the rate of $T'(1) = 40(-0.98)e^{-0.98t}\big|_{t=1} \approx -14.7$. Thus, it is dropping at the rate of approximately 14.7°F/min.

c. The temperature outdoors is given by $\lim_{t\to\infty} T(t) = \lim_{t\to\infty}\left(30 + 40e^{-0.98t}\right) = 30 + 0 = 30$, or 30°F.

83. $A(t) = \dfrac{150\left(1 - e^{0.022662t}\right)}{1 - 2.5e^{0.022662t}}$.

a. Let $k = 0.022662$. Then $A'(t) = 150\dfrac{d}{dt}\left[\dfrac{1 - e^{kt}}{1 - 2.5e^{kt}}\right] = 150\dfrac{\left(1 - 2.5e^{kt}\right)\left(-ke^{kt}\right) - \left(1 - e^{kt}\right)\left(-2.5ke^{kt}\right)}{\left(1 - 2.5e^{kt}\right)^2}$.

Thus, the rate of formation of chemical C one minute after the interaction begins is

$A'(1) = \dfrac{5.09895e^{0.022662}}{\left(1 - 2.5e^{0.022662}\right)^2} \approx 2.15$, or 2.15 g/min.

b. The amount of chemical C that is eventually formed is

$$\lim_{t\to\infty} A(t) = 150 \lim_{t\to\infty} \frac{1 - e^{0.022662t}}{1 - 2.5e^{0.022662t}} = 150 \lim_{t\to\infty} \frac{e^{-0.022662t} - 1}{e^{-0.022662t} - 2.5} = \frac{150}{2.5} = 60 \text{ g}.$$

85. a. $x(t) = c\left(1 - e^{-at/V}\right)$, so $x'(t) = \frac{d}{dt}\left(c - ce^{-at/V}\right) = \frac{ac}{V}e^{-at/V}$. Because $a > 0$, $c > 0$, and $V > 0$, we see that $x'(t)$ is always positive and we conclude that $x(t)$ is always increasing.

b.

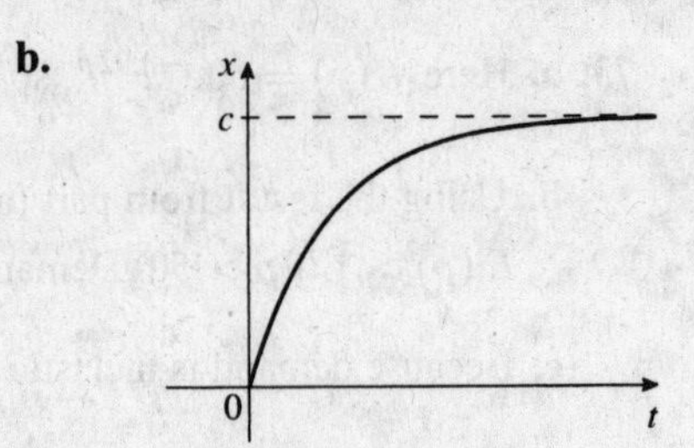

87. a. $A(t) = \begin{cases} 100e^{-1.4t} & \text{if } 0 \le t < 1 \\ 100\left(1 + e^{1.4}\right)e^{-1.4t} & \text{if } t \ge 1 \end{cases}$ so $A'(t) = \begin{cases} -140e^{-1.4t} & \text{if } 0 < t < 1 \\ -140\left(1 + e^{1.4}\right)e^{-1.4t} & \text{if } t > 1 \end{cases}$ Thus, after 12 hours the amount of drug is changing at the rate of $A'\left(\frac{1}{2}\right) = -140e^{-0.7} \approx -69.52$, or decreasing at the rate of 70 mg/day. After 2 days, it is changing at the rate of $A'(2) = -140\left(1 + e^{1.4}\right)e^{-2.8} \approx -43.04$, or decreasing at the rate of 43 mg/day.

b. From the graph of A, we see that the maximum occurs at $t = 1$, that is, at the time when she takes the second dose.

c. The maximum amount is $A(1) = 100\left(1 + e^{1.4}\right)e^{-1.4} \approx 124.66$, or 125 mg.

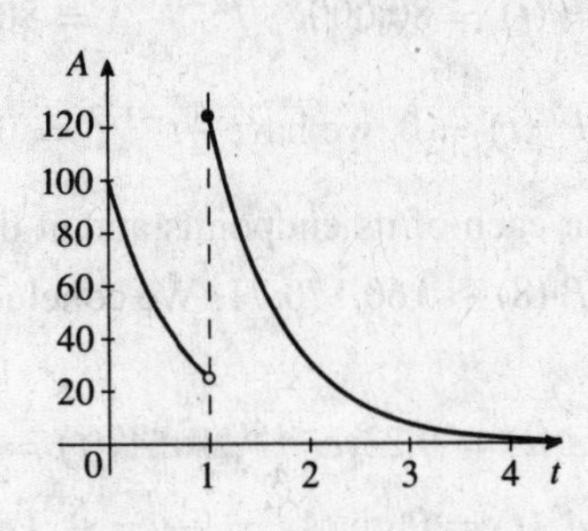

89. False. $f(x) = e^{\pi}$ is a constant function and so $f'(x) = 0$.

91. False. $f'(x) = \frac{d}{dx}\left(e^{x^2+x}\right) = (2x + 1)e^{x^2+x}$.

Using Technology page 381

1. 5.4366

3. 12.3929

5. 0.1861

7. a. The initial population of crocodiles is $P(0) = \frac{300}{6} = 50$.

b. $\lim_{t\to\infty} P(t) = \lim_{t\to\infty} \frac{300e^{-0.024t}}{5e^{-0.024t} + 1} = \frac{0}{0 + 1} = 0$.

c.

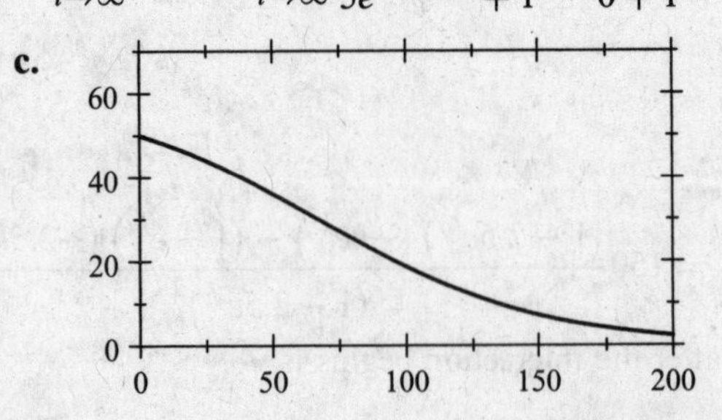

9. a.

1.5e+5
1e+5
50000
0
0 100 200 300

b. Initially, they owe $B(0) = \$160{,}000$, and their debt is decreasing at the rate of $B'(0) \approx \$87.07$ per month. After 180 payments, they owe $B(180) \approx \$126{,}928.78$ and their debt is decreasing at the rate of $B'(180) \approx \$334.18$ per month.

5.5 Differentiation of Logarithmic Functions

Problem-Solving Tips

1. If you trying to find the derivative of a complicated function involving products, quotients, or powers, check to see if you can use logarithmic differentiation to simplify the process. Look at Exercises 41–50 and familiarize yourself with the type of problem for which this method is especially suitable.

2. Example 7 provided us with a method for finding the derivative of the function $y = x^x$ $(x > 0)$. Note that we use the Power rule $\left(\frac{d}{dx}[x^n] = nx^{n-1}\right)$ to differentiate functions of the form $y = x^n$, where the base is a variable and the exponent is a constant, the rule for differentiating exponential functions to differentiate functions of the formy $= e^x$, where the base is the constant e and the exponent is a variable, and logarithmic differentiation to differentiate functions of the form $y = x^x$ where both the base and the exponent of the function are variables. Be sure that you can distinguish between these functions and the rule to be applied in each of these cases.

Concept Questions

page 386

1. a. $f'(x) = \frac{1}{x}$.

b. $g'(x) = \frac{f'(x)}{f(x)}$.

Exercises

page 387

1. $f(x) = 5\ln x$, so $f'(x) = 5\left(\frac{1}{x}\right) = \frac{5}{x}$.

3. $f(x) = \ln(x+1)$, so $f'(x) = \frac{1}{x+1}$.

5. $f(x) = \ln x^8$, so $f'(x) = \frac{8x^7}{x^8} = \frac{8}{x}$.

7. $f(x) = \ln x^{1/2}$, so $f'(x) = \frac{\frac{1}{2}x^{-1/2}}{x^{1/2}} = \frac{1}{2x}$.

9. $f(x) = \ln\left(\frac{1}{x^2}\right) = \ln x^{-2}$, so $f'(x) = -\frac{2x^{-3}}{x^{-2}} = -\frac{2}{x}$.

11. $f(x) = \ln\left(4x^2 - 5x + 3\right)$, so $f'(x) = \frac{8x-5}{4x^2-5x+3} = \frac{8x-5}{4x^2-5x+3}$.

13. $f(x) = \ln\left(\frac{2x}{x+1}\right) = \ln 2x - \ln(x+1)$, so

$$f'(x) = \frac{2}{2x} - \frac{1}{x+1} = \frac{1}{x} - \frac{1}{x+1} = \frac{(x+1)-x}{x(x+1)} = \frac{x+1-x}{x(x+1)} = \frac{1}{x(x+1)}.$$

15. $f(x) = x^2\ln x$, so $f'(x) = x^2\left(\frac{1}{x}\right) + (\ln x)(2x) = x + 2x\ln x = x(1 + 2\ln x)$.

17. $f(x) = \frac{2\ln x}{x}$, so $f'(x) = \frac{x\left(\frac{2}{x}\right) - 2\ln x}{x^2} = \frac{2(1-\ln x)}{x^2}$.

19. $f(u) = \ln(u-2)^3$, so $f'(u) = \frac{3(u-2)^2}{(u-2)^3} = \frac{3}{u-2}$.

21. $f(x) = (\ln x)^{1/2}$, so $f'(x) = \frac{1}{2}(\ln x)^{-1/2}\left(\frac{1}{x}\right) = \frac{1}{2x\sqrt{\ln x}}$.

23. $f(x) = (\ln x)^2$, so $f'(x) = 2(\ln x)\left(\dfrac{1}{x}\right) = \dfrac{2\ln x}{x}$.

25. $f(x) = \ln(x^3+1)$, so $f'(x) = \dfrac{3x^2}{x^3+1}$.

27. $f(x) = e^x \ln x$, so $f'(x) = e^x \ln x + e^x\left(\frac{1}{x}\right) = \dfrac{e^x(x\ln x + 1)}{x}$.

29. $f(t) = e^{2t}\ln(t+1)$, so $f'(t) = e^{2t}\left(\dfrac{1}{t+1}\right) + \ln(t+1)\cdot\left(2e^{2t}\right) = \dfrac{[2(t+1)\ln(t+1)+1]e^{2t}}{t+1}$.

31. $f(x)\dfrac{\ln x}{x^2}$, so $f'(x) = \dfrac{x^2\left(\frac{1}{x}\right) - \ln x\,(2x)}{x^4} = \dfrac{1-2\ln x}{x^3}$.

33. $f'(x) = \dfrac{d}{dx}[\ln(\ln x)] = \dfrac{\frac{d}{dx}(\ln x)}{\ln x} = \dfrac{\frac{1}{x}}{\ln x} = \dfrac{1}{x\ln x}$.

35. $f(x) = \ln 2 + \ln x$, so $f'(x) = \dfrac{1}{x}$ and $f''(x) = -\dfrac{1}{x^2}$.

37. $f(x) = \ln(x^2+2)$, so $f'(x) = \dfrac{2x}{(x^2+2)}$ and $f''(x) = \dfrac{(x^2+2)(2) - 2x(2x)}{(x^2+2)^2} = \dfrac{2(2-x^2)}{(x^2+2)^2}$.

39. $f'(x) = \dfrac{d}{dx}(x^2\ln x) = \dfrac{d}{dx}(x^2)\ln x + \dfrac{d}{dx}(\ln x)x^2 = 2x\ln x + \dfrac{1}{x}\cdot x^2 = 2x\ln x + x = x(2\ln x+1)$ and

$f''(x) = \dfrac{d}{dx}[x(2\ln x+1)] = \dfrac{d}{dx}(x)(2\ln x+1) + \dfrac{d}{dx}(2\ln x+1)x = 2\ln x + 1 + \dfrac{2}{x}\cdot x = 2\ln x + 3$.

41. $y = (x+1)^2(x+2)^3$, so

$\ln y = \ln(x+1)^2(x+2)^3 = \ln(x+1)^2 + \ln(x+2)^3 = 2\ln(x+1) + 3\ln(x+2)$.

Thus, $\dfrac{y'}{y} = \dfrac{2}{x+1} + \dfrac{3}{x+2} = \dfrac{2(x+2)+3(x+1)}{(x+1)(x+2)} = \dfrac{5x+7}{(x+1)(x+2)}$ and

$y' = \dfrac{(5x+7)(x+1)^2(x+2)^3}{(x+1)(x+2)} = (5x+7)(x+1)(x+2)^2$.

43. $y = (x-1)^2(x+1)^3(x+3)^4$, so $\ln y = 2\ln(x-1) + 3\ln(x+1) + 4\ln(x+3)$. Thus,

$$\frac{y'}{y} = \frac{2}{x-1} + \frac{3}{x+1} + \frac{4}{x+3} = \frac{2(x+1)(x+3) + 3(x-1)(x+3) + 4(x-1)(x+1)}{(x-1)(x+1)(x+3)}$$

$$= \frac{2x^2+8x+6+3x^2+6x-9+4x^2-4}{(x-1)(x+1)(x+3)} = \frac{9x^2+14x-7}{(x-1)(x+1)(x+3)}, \text{ and so}$$

$$y' = \frac{9x^2+14x-7}{(x-1)(x+1)(x+3)}\cdot y = \frac{(9x^2+14x-7)(x-1)^2(x+1)^3(x+3)^4}{(x-1)(x+1)(x+3)}$$

$$= (9x^2+14x-7)(x-1)(x+1)^2(x+3)^3.$$

45. $y = \dfrac{(2x^2-1)^5}{\sqrt{x+1}}$, so $\ln y = \ln \dfrac{(2x^2-1)^5}{(x+1)^{1/2}} = 5\ln(2x^2-1) - \frac{1}{2}\ln(x+1)$. Thus,

$$\frac{y'}{y} = \frac{20x}{2x^2-1} - \frac{1}{2(x+1)} = \frac{40x(x+1)-(2x^2-1)}{2(2x^2-1)(x+1)} = \frac{38x^2+40x+1}{2(2x^2-1)(x+1)}, \text{ and so}$$

$$y' = \frac{38x^2+40x+1}{2(2x^2-1)(x+1)} \cdot \frac{(2x^2-1)^5}{\sqrt{x+1}} = \frac{(38x^2+40x+1)(2x^2-1)^4}{2(x+1)^{3/2}}.$$

47. $y = 3^x$, so $\ln y = x\ln 3$, $\dfrac{1}{y}\cdot\dfrac{dy}{dx} = \ln 3$, and $\dfrac{dy}{dx} = y\ln 3 = 3^x \ln 3$.

49. $y = (x^2+1)^x$, so $\ln y = \ln(x^2+1)^x = x\ln(x^2+1)$,

$$\frac{y'}{y} = \ln(x^2+1) + x\left(\frac{2x}{x^2+1}\right) = \frac{(x^2+1)\ln(x^2+1)+2x^2}{x^2+1}, \text{ and}$$

$$y' = \frac{\left[(x^2+1)\ln(x^2+1)+2x^2\right](x^2+1)^x}{x^2+1} = (x^2+1)^{x-1}\left[(x^2+1)\ln(x^2+1)+2x^2\right].$$

51. $\dfrac{d}{dx}(\ln y - x\ln x) = \dfrac{d}{dx}(-1)$, so $\dfrac{d}{dx}\ln y - \dfrac{d}{dx}(x\ln x) = 0$, $\dfrac{y'}{y} = \left[\ln x + x\left(\dfrac{1}{x}\right)\right] = \ln x + 1$, and $y' = (\ln x + 1)\,y$.

53. $y = x\ln x$. The slope of the tangent line at any point is $y' = \ln x + x\left(\frac{1}{x}\right) = \ln x + 1$. In particular, the slope of the tangent line at $(1, 0)$ is $m = \ln 1 + 1 = 1$. Thus, an equation of the tangent line is $y - 0 = 1(x-1)$, or $y = x - 1$.

55. $f(x) = \ln x^2 = 2\ln x$ and so $f'(x) = 2/x$. Because $f'(x) < 0$ if $x < 0$ and $f'(x) > 0$ if $x > 0$, we see that f is decreasing on $(-\infty, 0)$ and increasing on $(0, \infty)$.

57. $f(x) = x^2 + \ln x^2$, so $f'(x) = 2x + \dfrac{2x}{x^2} = 2x + \dfrac{2}{x}$ and $f''(x) = 2 - \dfrac{2}{x^2}$. To find the intervals of concavity for f, we first set $f''(x) = 0$, giving $2 - \dfrac{2}{x^2} = 0$, $2 = \dfrac{2}{x^2}$, $2x^2 = 2$, $x^2 = 1$, and so $x = \pm 1$.

From the sign diagram for f'', we see that f is concave upward on $(-\infty, -1)$ and $(1, \infty)$ and concave downward on $(-1, 0)$ and $(0, 1)$.

f'' not defined

+ + + 0 − − | − − 0 + + + sign of f''

−1 0 1 x

59. $f(x) = \ln(x^2+1)$, so $f'(x) = \dfrac{2x}{x^2+1}$ and $f''(x) = \dfrac{(x^2+1)(2)-(2x)(2x)}{(x^2+1)^2} = -\dfrac{2(x^2-1)}{(x^2+1)^2}$. Setting $f''(x) = 0$ gives $x = \pm 1$ as candidates for inflection points of f. From the sign diagram of f'', we see that $(-1, \ln 2)$ and $(1, \ln 2)$ are inflection points of f.

− − − 0 + + 0 + + 0 − − − sign of f''

−1 0 1 x

61. $f(x) = x^2 + 2\ln x$, so $f'(x) = 2x + \frac{2}{x}$ and $f''(x) = 2 - \frac{2}{x^2} = 0$ implies $2 - \frac{2}{x^2} = 0$, $x^2 = 1$, and so $x = \pm 1$. We reject the negative root because the domain of f is $(0, \infty)$. The sign diagram of f'' shows that $(1, 1)$ is an inflection point of the graph of f. $f'(1) = 4$. So, an equation of the required tangent line is $y - 1 = 4(x - 1)$ or $y = 4x - 3$.

– – – – 0 + + + + sign of f''

0 1 x

63. $f(x) = x - \ln x$, so $f'(x) = 1 - \frac{1}{x} = \frac{x-1}{x} = 0$ if $x = 1$, a critical point of f. From the table, we see that f has an absolute minimum at $(1, 1)$ and an absolute maximum at $(3, 3 - \ln 3)$.

x	$\frac{1}{2}$	1	3
$f(x)$	$\frac{1}{2} + \ln 2$	1	$3 - \ln 3$

65. $\ln(xy) = x + y$, so $\ln x + \ln y = x + y$. Differentiating with respect to x, we obtain $\frac{1}{x} + \frac{1}{y}y' = 1 + y'$, $y'\left(\frac{1}{y} - 1\right) = 1 - 1/x$, and $y'\left(\frac{1-y}{y}\right) = \frac{x-1}{x}$. Thus, $y' = \frac{y(x-1)}{x(1-y)}$.

67. $\ln x + xy = 5$. Differentiating with respect to x, we obtain $\frac{1}{x} + y + xy' = 0$. Differentiating again, we have $-\frac{1}{x^2} + y' + y' + xy'' = 0$, so $xy'' = \frac{1}{x^2} - 2y' = \frac{1 - 2x^2y'}{x^2}$. But $y' = \frac{-\frac{1}{x} - y}{x} = -\frac{1 + xy}{x^2}$, so $y'' = \frac{1}{x^3}\left[1 + 2x^2\left(\frac{1+xy}{x^2}\right)\right] = \frac{3 + 2xy}{x^3}$.

69. $\ln y + xy = 1$. Differentiating with respect to x, we obtain $\frac{y'}{y} + y + xy' = 0$. Substituting $x = 1$ and $y = 1$ gives $y' + 1 + y' = 0$, so $y' = -\frac{1}{2}$.

71. $f(x) = 7.2956\ln\left(0.0645012x^{0.95} + 1\right)$, so

$$f'(x) = 7.2956 \cdot \frac{\frac{d}{dx}\left(0.0645012x^{0.95} + 1\right)}{0.0645012x^{0.95} + 1} = \frac{7.2956\,(0.0645012)\left(0.95x^{-0.05}\right)}{0.0645012x^{0.95} + 1} = \frac{0.4470462}{x^{0.05}\left(0.0645012x^{0.95} + 1\right)}.$$

Thus, $f'(100) = 0.05799$, or approximately 0.0580%/kg, and $f'(500) = 0.01330$, or approximately 0.0133%/kg.

73. a. $W'(t) = \frac{d}{dt}(49.9 + 17.1\ln t) = \frac{17.1}{t} > 0$ if $t > 0$, so $W'(t) > 0$ on $[1, 6]$ and W is increasing on $(1, 6)$.

b. $W''(t) = \frac{d}{dt}\left(\frac{17.1}{t}\right) = -\frac{17.1}{t^2} < 0$ on $(1, 6)$, so W is concave downward on $(1, 6)$.

75. a. $\ln V = \ln\left(C\left(1 - \frac{2}{N}\right)^n\right) = \ln C + n\ln\left(1 - \frac{2}{N}\right)$, so $\frac{d}{dn}\ln V = \frac{d}{dn}(\ln C) + \frac{d}{dn}\left[n\ln\left(1 - \frac{2}{N}\right)\right]$, $\frac{V'}{V} = \ln\left(1 - \frac{2}{N}\right)$, and $V' = V\ln\left(1 - \frac{2}{N}\right) = C\left(1 - \frac{2}{N}\right)^n\ln\left(1 - \frac{2}{N}\right)$.

b. The relative rate of change of $V(n)$ is given by $\frac{V'(n)}{V(n)} = \frac{C\left(1 - \frac{2}{N}\right)^n\ln\left(1 - \frac{2}{N}\right)}{C\left(1 - \frac{2}{N}\right)^n} = \ln\left(1 - \frac{2}{N}\right)$.

77. $\ln P(t) = \ln\left(\dfrac{40 + 80e^{0.06t}}{20 + e^{0.06t}}\right) = \ln\left(40 + 80e^{0.06t}\right) - \ln\left(20 + e^{0.06t}\right)$, so

$\dfrac{P'(t)}{P(t)} = \dfrac{d}{dt} P(t) = \dfrac{80\left(0.06e^{0.06t}\right)}{40 + 80e^{0.06t}} - \dfrac{0.06e^{0.06t}}{20 + e^{0.06t}}$ and $\left.\dfrac{P'(t)}{P(t)}\right|_{t=60} = \dfrac{80\left(0.06e^{3.6}\right)}{40 + 80e^{3.6}} - \dfrac{0.06e^{3.6}}{20 + e^{3.6}} \approx 0.0204$.

Therefore, the relative rate of growth of the population five years after the establishment of the biotech research center is approximately 2.04% per month.

79. $P(x) = 2\ln(2x + 1) + 2x - x^2 - 0.3$. We want to maximize the function P with respect to x. Setting $P'(x) = 0$ gives $P'(x) = \dfrac{2 \cdot 2}{2x + 1} + 2 - 2x = 0$, or $4 + (2x + 1)(2 - 2x) = 0$, $2x^2 - x - 3 = 0$, and $(2x - 3)(x + 1) = 0$. Therefore, $x = -1$ or $x = \frac{3}{2}$. We reject the negative root, so $\frac{3}{2}$ is the only critical number of P. Because $P''(x) = 4\dfrac{d}{dx}(2x + 1)^{-1} - 2 = -\dfrac{4 \cdot 2}{(2x + 1)^2} - 2 < 0$ for all $x > 0$, we see that the graph of P is concave downward on $(0, \infty)$, implying that $\frac{3}{2}$ gives an absolute maximum for P with value $P\left(\frac{3}{2}\right) \approx 3.22$. Thus, by employing 150 consultants, Seko makes an estimated annual profit of approximately \$3.22 million.

81. a. $100\dfrac{d}{dx}[\ln f(x)] = \dfrac{100 f'(x)}{f'(x)}$, and this is precisely the percentage rate of change of f.

b. The percentage rate of growth of the company t years from now is

$100\dfrac{d}{dt}[\ln R(t)] = 100\dfrac{d}{dt}\left[\ln\left(0.1t^{1.5}e^{0.2t}\right)\right] = 100\dfrac{d}{dt}(\ln 0.1 + 1.5\ln t + 0.2t) = 100\left(\dfrac{1.5}{t} + 0.2\right)$.

Thus, the percentage rate of growth 3 years from now is $100\left(\frac{1.5}{3} + 0.2\right) = 70$, or 70%/year.

83. The relative rate of change of P is

$$\frac{P'(t)}{P(t)} = \frac{d}{dt}[\ln P(t)] = \frac{d}{dt}\ln\left(Le^{-\ln(L/P_0)e^{-ct}}\right) = \frac{d}{dt}\left[\ln L - \ln\left(\frac{L}{P_0}\right)e^{-ct}\right] = 0 - \ln\left(\frac{L}{P_0}\right)\left(-ce^{-ct}\right)$$
$$= c\ln\left(\frac{L}{P_0}\right)e^{-ct}.$$

85. a. $R = \log\dfrac{10^6 I_0}{I_0} = \log 10^6 = 6$.

b. $I = I_0 10^R$ by definition. Taking the natural logarithm on both sides, we find $\ln I = \ln I_0 10^R = \ln I_0 + \ln 10^R = \ln I_0 + R\ln 10$. Differentiating implicitly with respect to R, we obtain $\dfrac{I'}{I} = \ln 10$. Therefore, $\Delta I \approx dI = \dfrac{dI}{dR}\Delta R = (\ln 10)\, I\, \Delta R$. If $|\Delta R| \leq (0.02)(6) = 0.12$ and $I = 1{,}000{,}000 I_0$, (see part (a)), then $|\Delta I| \leq (\ln 10)(1{,}000{,}000 I_0)(0.12) \approx 276{,}310.21 I_o$. Thus, the error is at most 276,310 times the standard reference intensity.

87. $-C\ln y + Dy = A\ln x - Bx + E$. Differentiating implicitly gives $-C\dfrac{y'}{y} + Dy' = \dfrac{Ax'}{x} - Bx'$, $\left(D - \dfrac{C}{y}\right)y' = \left(\dfrac{A}{x} - B\right)x'$, $\left(\dfrac{Dy - C}{y}\right)y' = \left(\dfrac{A - Bx}{x}\right)x'$, and so $y' = \dfrac{(A - Bx)\, yx'}{(Dy - C)\, x}$.

89. $f(x) = 2x - \ln x$. We first gather the following information on f.

1. The domain of f is $(0, \infty)$.
2. There is no y-intercept.

3. $\lim_{x\to\infty} (2x - \ln x) = \infty$.

4. There is no asymptote.

5. $f'(x) = 2 - \frac{1}{x} = \frac{2x-1}{x}$. Observe that $f'(x) = 0$ at $x = \frac{1}{2}$, a critical point of f. From the sign diagram of f', we conclude that f is decreasing on $\left(0, \frac{1}{2}\right)$ and increasing on $\left(\frac{1}{2}, \infty\right)$.

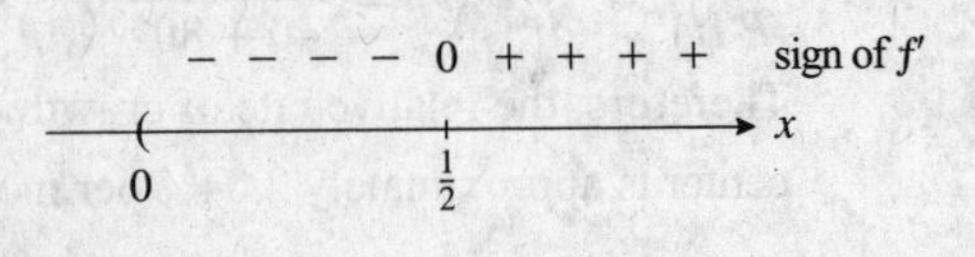

6. The results of part 5 show that $\left(\frac{1}{2}, 1 + \ln 2\right)$ is a relative minimum of f.

7. $f''(x) = \frac{1}{x^2}$ and is positive if $x > 0$, so the graph of f is concave upward on $(0, \infty)$.

8. The results of part 7 show that f has no inflection point.

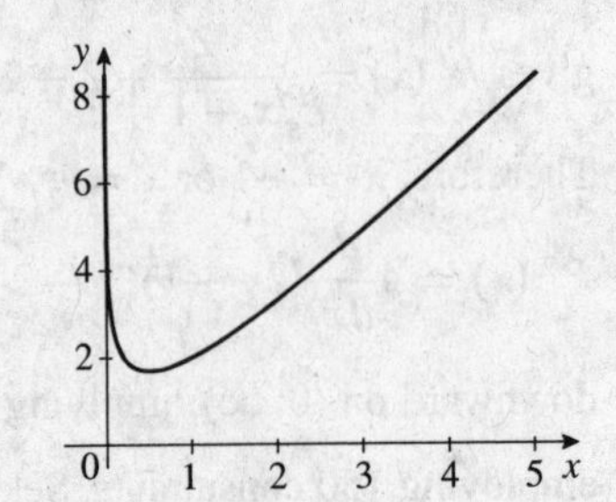

91. a. $f(x) = b^x$. Taking the logarithm of each side, we have $\ln f(x) = \ln b^x = x \ln b$, So $\frac{d}{dx}[\ln f(x)] = \frac{d}{dx}(x \ln b)$ and $\frac{f'(x)}{f(x)} = \ln b$. Therefore, $f'(x) = (\ln b) f(x) = (\ln b) b^x$.

b. $f'(x) = \frac{d}{dx}(3^x) = (\ln 3) 3^x$.

93. $f'(x) = \frac{d}{dx}(x^3 2^x) = x^3 \frac{d}{dx}(2^x) + 2^x \frac{d}{dx}(x^3) = (\ln 2) x^3 2^x + 3x^2 2^x = x^2 (x \ln 2 + 3) 2^x$.

95. $h'(x) = \frac{d}{dx}(x^2 \log_{10} x) = x^2 \frac{d}{dx} \log_{10} x + (\log_{10} x) \frac{d}{dx}(x^2) = \frac{x^2}{x \ln 10} + 2x \log_{10} x$
$= x\left(\frac{1}{\ln 10} + 2 \log_{10} x\right)$.

97. False. $\ln 5$ is a constant function and $f'(x) = 0$.

99. If $x \leq 0$, then $|x| = -x$. Therefore, $\ln|x| = \ln(-x)$. Writing $f(x) = \ln|x|$, we have $|x| = -x = e^{f(x)}$. Differentiating both sides with respect to x and using the Chain Rule,we have $-1 = e^{f(x)} \cdot f'(x)$, so $f'(x) = -\frac{1}{e^{f(x)}} = -\frac{1}{-x} = \frac{1}{x}$.

5.6 Exponential Functions as Mathematical Models

Problem-Solving Tips

Four mathematical models were introduced in this section:

1. **Exponential growth:** $Q(t) = Q_0 e^{kt}$ describes a quantity $Q(t)$ that is initially present in the amount $Q(0) = Q_0$ and whose rate of growth at any time t is directly proportional to the amount of the quantity present at time t.

2. **Exponential decay:** $Q(t) = Q_0 e^{-kt}$ describes a quantity $Q(t)$ that is initially present in the amount $Q(0) = Q_0$ and decreases at a rate that is directly proportional to its size.

3. **Learning curves:** $Q(t) = C - Ae^{-kt}$ describes a quantity $Q(t)$, where $Q(0) = C - A$, and $Q(t)$ increases and approaches the number C as t increases without bound.

4. **Logistic growth functions:** $Q(t) = \dfrac{A}{1 + Be^{-kt}}$ describes a quantity $Q(t)$, where $Q(0) = \dfrac{A}{1+B}$. Note that $Q(t)$ increases rapidly for small values of t but the rate of growth of $Q(t)$ decreases quickly as t increases. $Q(t)$ approaches the number A as t increases without bound.

Try to familiarize yourself with the examples and graphs for each of these models before you work through the applied problems in this section.

Concept Questions

page 398

1. $Q(t) = Q_0e^{kt}$ where $k > 0$ represents exponential growth and $k < 0$ represents exponential decay. The larger the magnitude of k, the more quickly the former grows and the more quickly the latter decays.

3. $Q(t) = \dfrac{A}{1 + Be^{-kt}}$, where A, B, and k are positive constants. Q increases rapidly for small values of t but the rate of increase slows down as Q (always increasing) approaches the number A.

Exercises

page 399

1. **a.** The growth constant is $k = 0.02$.

   b. Initially, there are 300 units present.

   c.

t	0	10	20	100	1000
Q	300	366	448	2217	1.46×10^{11}

3. **a.** $Q(t) = Q_0e^{kt}$. Here $Q_0 = 100$ and so $Q(t) = 100e^{kt}$. Because the number of cells doubles in 20 minutes, we have $Q(20) = 100e^{20k} = 200$, $e^{20k} = 2$, $20k = \ln 2$, and so $k = \frac{1}{20}\ln 2 \approx 0.03466$. Thus, $Q(t) = 100e^{0.03466t}$.

   b. We solve the equation $100e^{0.03466t} = 1{,}000{,}000$, obtaining $e^{0.03466t} = 10{,}000$, $0.03466t = \ln 10{,}000$, and so $t = \dfrac{\ln 10{,}000}{0.03466} \approx 266$, or 266 minutes.

   c. $Q(t) = 1000e^{0.03466t}$.

5. **a.** We solve the equation $5.3e^{0.02t} = 3(5.3)$, obtaining $e^{0.02t} = 3$, $0.02t = \ln 3$, and so $t = \dfrac{\ln 3}{0.02} \approx 54.93$. Thus, the world population will triple in approximately 54.93 years.

   b. If the growth rate is 1.8%, then proceeding as before, we find $N(t) = 5.3e^{0.018t}$. If $t = 54.93$, the population would be $N(54.93) = 5.3e^{0.018(54.93)} \approx 14.25$, or approximately 14.25 billion.

7. $P(h) = p_0e^{-kh}$, so $P(0) = p_0 = 15$. Thus, $P(4000) = 15e^{-4000k} = 12.5$, $e^{-4000k} = \frac{12.5}{15}$, $-4000k = \ln\left(\frac{12.5}{15}\right)$, and so $k = 0.00004558$. Therefore, $P(12{,}000) = 15e^{-0.00004558(12{,}000)} = 8.68$, or 8.7 lb/in^2. The rate of change of atmospheric pressure with respect to altitude is given by $P'(h) = \dfrac{d}{dh}\left(15e^{-0.00004558h}\right) = -0.0006837e^{-0.00004558h}$. Thus, the rate of change of atmospheric pressure with respect to altitude when the altitude is 12,000 feet is $P'(12{,}000) = -0.0006837e^{-0.00004558(12{,}000)} \approx -0.00039566$. That is, it is declining at the rate of approximately 0.0004 lb/in^2/ft.

9. Suppose the amount of P-32 at time t is given by $Q(t) = Q_0 e^{-kt}$, where Q_0 is the amount present initially and k is the decay constant. Because this element has a half-life of 14.2 days, we have $\frac{1}{2}Q_0 = Q_0 e^{-14.2k}$, so $e^{-14.2k} = \frac{1}{2}$, $-14.2k = \ln\frac{1}{2}$, and $k = -\frac{\ln(1/2)}{14.2} \approx 0.0488$. Therefore, the amount of P-32 present at any time t is given by $Q(t) = 100e^{-0.0488t}$. In particular, the amount left after 7.1 days is given by $Q(7.1) = 100e^{-0.0488(7.1)} = 100e^{-0.34648} \approx 70.717$, or 70.717 grams. The rate at which the element decays is $Q'(t) = \frac{d}{dt}\left(100e^{-0.0488t}\right) = 100(-0.0488)e^{-0.0488t} = -4.88e^{-0.0488t}$. Therefore, $Q'(7.1) = -4.88e^{-0.0488(7.1)} \approx -3.451$; that is, it is decreasing at the rate of 3.451 g/day.

11. We solve the equation $0.2Q_0 = Q_0 e^{-0.00012t}$, obtaining $\ln 0.2 = -0.00012t$ and $t = \frac{\ln 0.2}{-0.00012} \approx 13{,}412$, or approximately 13,412 years.

13. a. $f(t) = 157e^{-0.55t}$, so $f'(t) = 157(-0.55)e^{-0.55t} = -86.35e^{-0.55t}$. So the number of annual bank failures was changing at the rate of $f'(1) \approx -49.82$; that is, it was dropping at the rate of approximately 50/year in 2011.

b. The projected number of failures in 2013 is $f(3) = 157e^{-0.55(3)} \approx 30.15$, or approximately 30.

15. a. $S = S_0 e^{-kt}$, so $S(0) = S_0 = 100$. Thus, $S(t) = 100e^{kt}$. Next, $S(5) = 150$ gives $100e^{5k} = 150$, so $e^{5k} = \frac{150}{100} = 1.5$, $5k = \ln 1.5$, and $k \approx 0.0811$. Thus, the model is $S(t) = 100e^{0.0811t}$.

b. The sales of Garland Corporation in 2013 were $S(3) = 100e^{0.0811(3)} \approx 127.5$, or approximately \$127.5 million.

17.

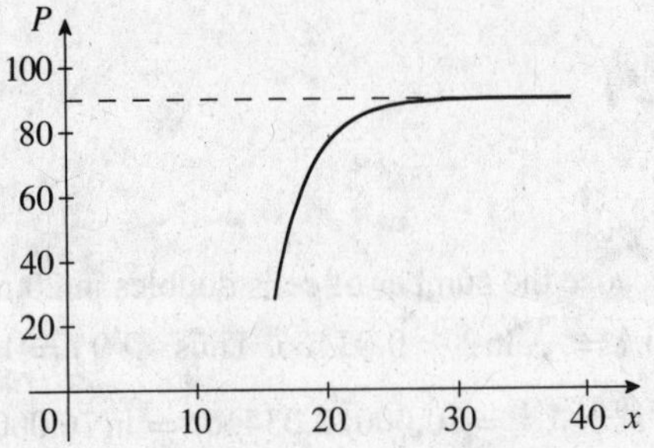

a. The percentage of 16-year-olds with a driver's license was $P(16) = 90\left[1 - e^{-0.37(16-15)}\right] \approx 27.8$, or approximately 27.8.

b. The percentage of 20-year-olds with a driver's license was $P(2) = 90\left[1 - e^{-0.37(20-15)}\right] \approx 75.8$, or approximately 75.8.

c. The percentage of 39-year-olds with a driver's license was $P(39) = 90\left[1 - e^{-0.37(39-15)}\right] \approx 90.0$, or approximately 90.0.

19.

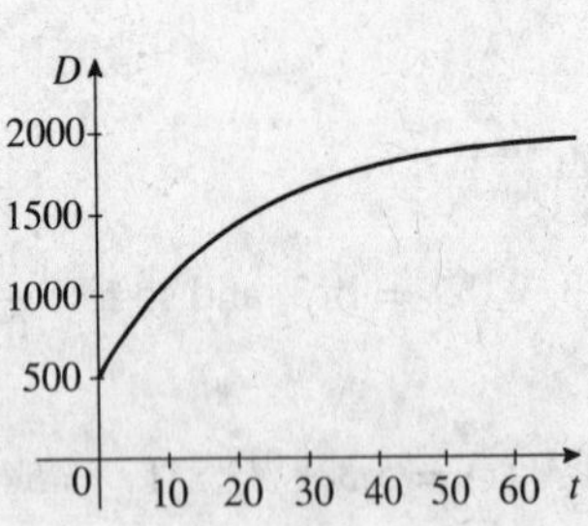

a. After 1 month, the demand is $D(1) = 2000 - 1500e^{-0.05} \approx 573$, after 12 months it is $D(12) = 2000 - 1500e^{-0.6} \approx 1177$, after 24 months it is $D(24) = 2000 - 1500e^{-1.2} \approx 1548$, and after 60 months, it is $D(60) = 2000 - 1500e^{-3} \approx 1925$.

b. $\lim_{t\to\infty} D(t) = \lim_{t\to\infty}\left(2000 - 1500e^{-0.05t}\right) = 2000$, and we conclude that the demand is expected to stabilize at 2000 computers per month.

c. $D'(t) = -1500e^{-0.05t}(-0.05) = 75e^{-0.05t}$. Therefore, the rate of growth after 10 months is $D'(10) = 75e^{-0.5} \approx 45.49$, or approximately 46 computers per month.

21. a. The length is given by $f(5) = 200\left(1 - 0.956e^{-0.18\cdot 5}\right) \approx 122.26$, or approximately 122.3 cm.

b. $f'(t) = 200(-0.956)e^{-0.18t}(-0.18) = 34.416e^{-0.18t}$, so a 5-year-old is growing at the rate of $f'(5) = 34.416e^{-0.18(5)} \approx 13.9925$, or approximately 14 cm/yr.

c. The maximum length is given by $\lim\limits_{t\to\infty} 200\left(1 - 0.956e^{-0.18t}\right) = 200$, or 200 cm.

23. a. $N(0) = \dfrac{400}{1+39} = 10$ flies.

b. $\lim\limits_{t\to\infty} \dfrac{400}{1+39e^{-0.16t}} = 400$ flies.

c. $N(20) = \dfrac{400}{1+39e^{-0.16(20)}} \approx 154.5$, or 154 flies.

d. $N'(t) = \dfrac{d}{dt}\left[400\left(1+39e^{-0.16t}\right)^{-1}\right] = -400\left(1+39e^{-0.16t}\right)^{-2}\dfrac{d}{dt}\left(39e^{-0.16t}\right) = \dfrac{2496e^{-0.16t}}{\left(1+39e^{-0.16t}\right)^2}$, so

$N'(20) = \dfrac{2496e^{-0.16\cdot 20}}{\left(1+39e^{-0.16\cdot 20}\right)^2} \approx 15.17$, or approximately 15 fruit flies per day.

25. $f(t) = \dfrac{40e^{-(t-1975)/20}}{\left[1+e^{-(t-1975)/20}\right]^2}$. Let $u = (t-1975)/20$. Then $\dfrac{du}{dt} = \dfrac{1}{20}$ and

$$f'(t) = f'(u)\frac{du}{dt} = 40\cdot\frac{\left(1+e^{-u}\right)^2(-1) - e^{-u}(2)\left(1+e^{-u}\right)e^{-u}(-1)}{\left(1+e^{-u}\right)^4}\left(\frac{1}{20}\right)$$

$$= 2\cdot\frac{\left(1+e^{-u}\right)e^{-u}\left[-\left(1+e^{-u}\right)+2e^{-u}\right]}{\left(1+e^{-u}\right)^4} = \frac{2e^{-u}\left(e^{-u}-1\right)}{\left(1+e^{-u}\right)^3} = \frac{2e^{-(t-1975)/20}\left[e^{-(t-1975)/20}-1\right]}{\left[1+e^{-(t-1975)/20}\right]^3}.$$

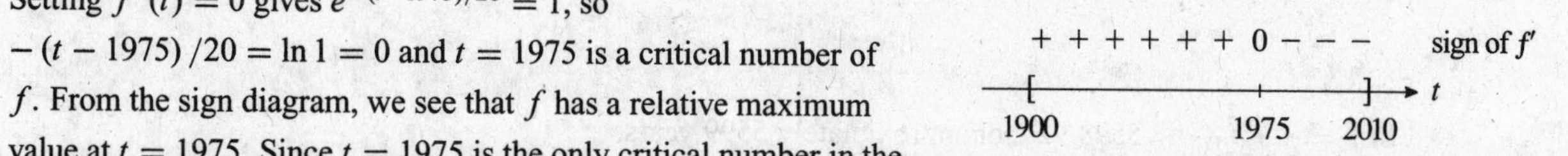

Setting $f'(t) = 0$ gives $e^{-(t-1975)/20} = 1$, so $-(t-1975)/20 = \ln 1 = 0$ and $t = 1975$ is a critical number of f. From the sign diagram, we see that f has a relative maximum value at $t = 1975$. Since $t = 1975$ is the only critical number in the interval (1900, 2010), we see that it gives an absolute maximum value of $f(1975) \approx 10$. We conclude that the maximum rate of production of crude oil in the U.S. occurred around 1975 and was approximately 10 million barrels per day.

27. The first of the given conditions implies that $f(0) = 300$, that is, $300 = \dfrac{3000}{1+Be^0} = \dfrac{3000}{1+B}$. Thus, $1+B = 10$, and $B = 9$. Therefore, $f(t) = \dfrac{3000}{1+9e^{-kt}}$. Next, the condition $f(2) = 600$ gives the equation $600 = \dfrac{3000}{1+9e^{-2k}}$, so $1+9e^{-2k} = 5$, $e^{-2k} = \frac{4}{9}$, and $k = -\frac{1}{2}\ln\frac{4}{9}$. Therefore, $f(t) = \dfrac{3000}{1+9e^{(1/2)t\cdot\ln(4/9)}} = \dfrac{3000}{1+9\left(\frac{4}{9}\right)^{t/2}}$. The number of students who had heard about the policy four hours later is given by $f(4) = \dfrac{3000}{1+9\left(\frac{4}{9}\right)^2} = 1080$, or 1080 students. To find the rate at which the rumor was spreading at any time time, we compute

$$f'(t) = \tfrac{d}{dt}\left[3000\left(1+9e^{-0.405465t}\right)^{-1}\right] = (3000)(-1)\left(1+9e^{-0.405465}\right)^{-2}\tfrac{d}{dt}\left(9e^{-0.405465t}\right)$$

$$= -3000(9)(-0.405465)e^{-0.405465t}\left(1+9e^{-0.405465t}\right)^{-2} = \frac{10947.555e^{-0.405465t}}{\left(1+9e^{-0.405465t}\right)^2}.$$

In particular, the rate at which the rumor was spreading 4 hours after the ceremony is given by

$f'(4) = \dfrac{10947.555e^{-0.405465\cdot 4}}{\left(1+9e^{-0.405465\cdot 4}\right)^2} \approx 280.26$. Thus, the rumor is spreading at the rate of 280 students per hour.

29. $x(t) = \dfrac{15\left(1 - \left(\frac{2}{3}\right)^{3t}\right)}{1 - \frac{1}{4}\left(\frac{2}{3}\right)^{3t}}$, so $\lim_{t\to\infty} x(t) = \lim_{t\to\infty} \dfrac{15\left[1 - \left(\frac{2}{3}\right)^{3t}\right]}{1 - \frac{1}{4}\left(\frac{2}{3}\right)^{3t}} = \dfrac{15(1-0)}{1-0} = 15$, or 15 lb.

31. a. $C(t) = \dfrac{k}{b-a}\left(e^{-at} - e^{-bt}\right)$, so

$C'(t) = \dfrac{k}{b-a}\left(-ae^{-at} + be^{-bt}\right) = \dfrac{kb}{b-a}\left[e^{-bt} - \left(\dfrac{a}{b}\right)e^{-at}\right] = \dfrac{kb}{b-a}e^{-bt}\left[1 - \dfrac{a}{b}e^{(b-a)t}\right]$.

$C'(t) = 0$ implies that $1 = \dfrac{a}{b}e^{(b-a)t}$, or $t = \dfrac{\ln\left(\frac{b}{a}\right)}{b-a}$. The sign diagram of C' shows that this value of t gives a maximum.

+ + + + + 0 − − − − sign of C'

t: 0, $\dfrac{\ln\frac{b}{a}}{b-a}$

b. $\lim_{t\to\infty} C(t) = 0$.

33. a. We solve $Q_0e^{-kt} = \frac{1}{2}Q_0$ for t, obtaining $e^{-kt} = \frac{1}{2}$, $\ln e^{-kt} = \ln\frac{1}{2} = \ln 1 - \ln 2 = -\ln 2$, $-kt = -\ln 2$, and so $\bar{t} = \frac{\ln 2}{k}$.

b. $\bar{t} = \frac{\ln 2}{0.0001238} \approx 5598.927$, or approximately 5599 years.

35. a. From the results of Exercise 34, we have $Q' = kQ\left(1 - \dfrac{Q}{A}\right)$, so

$Q'' = \dfrac{d}{dt}\left(kQ - \dfrac{k}{A}Q^2\right) = kQ' - \dfrac{2k}{A}QQ' = \dfrac{k}{A}Q'(A - 2Q)$. Setting $Q'' = 0$ gives $Q = \frac{A}{2}$ since $Q' > 0$ for all t. Furthermore, $Q'' > 0$ if $Q < \frac{A}{2}$ and $Q'' < 0$ if $Q > \frac{A}{2}$, So the graph of Q has an inflection point when $Q = \frac{A}{2}$. To find the value of t, we solve the equation $\dfrac{A}{2} = \dfrac{A}{1 + Be^{-kt}}$, obtaining $1 + Be^{-kt} = 2$, $Be^{-kt} = 1$, $e^{-kt} = \dfrac{1}{B}$, $-kt = \ln\dfrac{1}{B} = -\ln B$, and so $t = \dfrac{\ln B}{k}$.

b. The quantity Q increases most rapidly at the instant of time when it reaches one-half of the maximum quantity. This occurs at $t = \dfrac{\ln B}{k}$.

37. We use the result of Exercise 36 with $t_1 = 14$, $t_2 = 21$, $A = 600$, $Q_1 = 76$, and $Q_2 = 167$ to obtain $k = \dfrac{1}{21 - 14}\ln\left[\dfrac{167(600 - 76)}{76(600 - 167)}\right] \approx 0.14$.

Using Technology page 403

1. a.

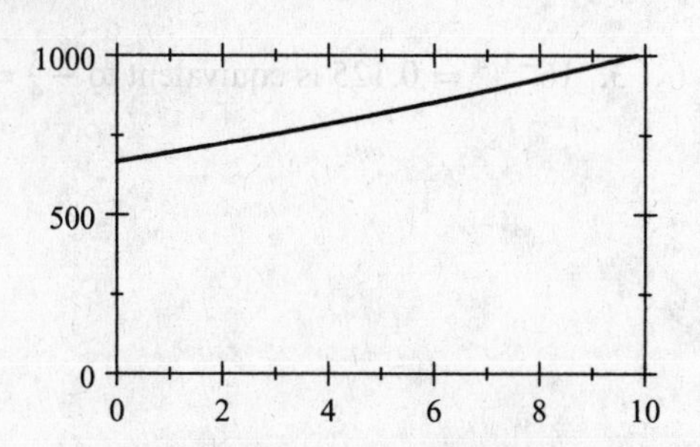

b. $T(0) = 666$ million; $T(8) \approx 926.8$ million.

c. $T'(8) \approx 38.3$ million/yr/yr.

3. a.

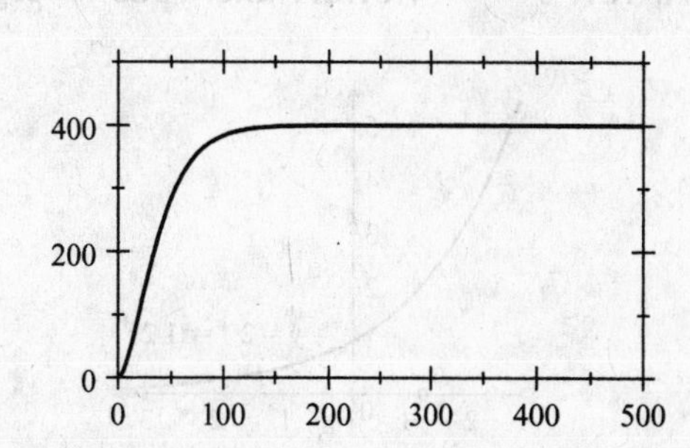

b. $\lim_{t\to\infty} N(t) = \lim_{t\to\infty} \left[-20(t+20)e^{-0.05t} + 400\right] = 400$, so Starr will eventually sell 400,000 copies of Laser Beams.

5. a.

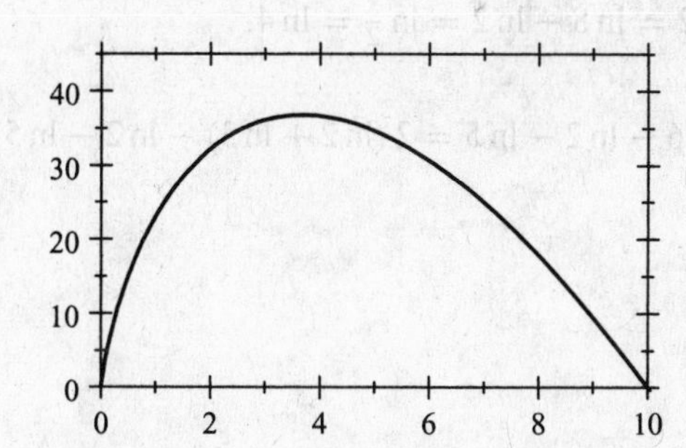

b. $R'(x) = 0$ when $x \approx 3.68$.

7. a.

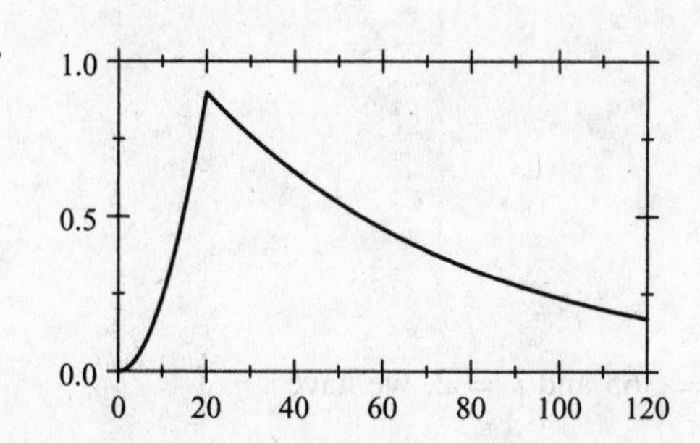

b. The initial concentration is 0.

c. $C(10) \approx 0.237$ g/cm^3.

d. $C(30) \approx 0.760$ g/cm^3.

e. $\lim_{t\to\infty} C(t) = 0$.

CHAPTER 5 Concept Review Questions page 405

1. power, 0, 1, exponential

3. a. $(0, \infty)$, $(-\infty, \infty)$, $(1, 0)$ **b.** < 1, > 1

5. accumulated amount, principal, nominal interest rate, number of conversion periods, term

7. Pe^{rt}

9. a. initially, growth **b.** decay **c.** time, one-half

CHAPTER 5 Review Exercises page 406

1.

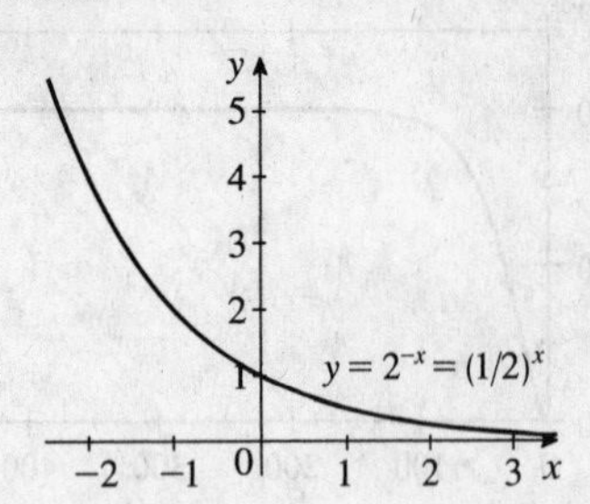

$\left(\frac{1}{2}\right)^x = \frac{1}{2^x} = 2^{-x}$, so the two graphs are the same.

3. $16^{-3/4} = 0.125$ is equivalent to $-\frac{3}{4} = \log_{16} 0.125$.

5. $\ln(x-1) + \ln 4 = \ln(2x+4) - \ln 2$, so $\ln(x-1) - \ln(2x+4) = -\ln 2 - \ln 4 = -(\ln 2 + \ln 4)$, $\ln\left(\frac{x-1}{2x+4}\right) = -\ln 8 = \ln\frac{1}{8}$, $\left(\frac{x-1}{2x+4}\right) = \frac{1}{8}$, $8x - 8 = 2x + 4$, $6x = 12$, and so $x = 2$. Check: LHS $= \ln(2-1) + \ln 4 = \ln 4$; RHS $= \ln(4+4) - \ln 2 = \ln 8 - \ln 2 = \ln\frac{8}{2} = \ln 4$.

7. $\ln 3.6 = \ln\frac{36}{10} = \ln 36 - \ln 10 = \ln 6^2 - \ln(2\cdot 5) = 2\ln 6 - \ln 2 - \ln 5 = 2(\ln 2 + \ln 3) - \ln 2 - \ln 5$

$= 2(x+y) - x - z = x + 2y - z$.

9. We first sketch the graph of $y = 2^{x+3}$, then reflect this graph with respect to the line $y = x + 3$.

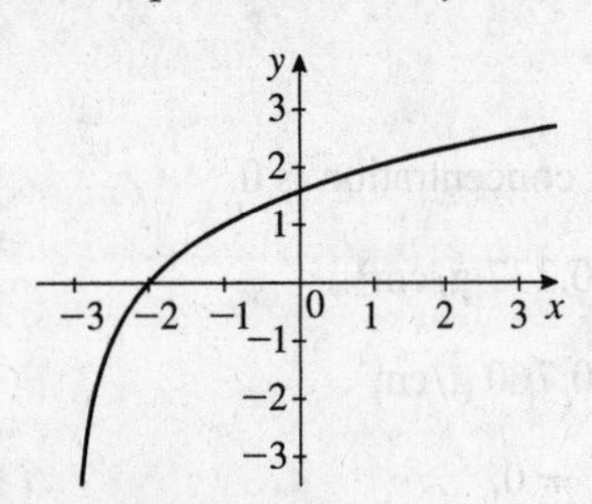

11. a. Using Formula (6) with $P = 10{,}000$, $r = 0.06$, $m = 365$ and $t = 2$, we have

$$A = 10{,}000\left(1 + \frac{0.06}{365}\right)^{365(2)} = 11{,}274.86, \text{ or } \$11{,}274.86.$$

b. Using Formula (10) with $P = 10{,}000$, $r = 0.06$, and $t = 2$, we have $A = 10{,}000e^{0.06(2)} = 11{,}274.97$, or \$11,274.97.

13. Using Formula (6) with $A = 10{,}000$, $P = 15{,}000$, $r = 0.06$, and $m = 4$, we have $A = 10{,}000\left(1 + \frac{0.06}{4}\right)^{4t} = 15{,}000$, or $\left(1 + \frac{0.06}{4}\right)^{4t} = 1.5$. Solving for t, we have $4t\ln(1.015) = \ln 1.5$, so $t = \frac{\ln 1.5}{4\ln 1.015} \approx 6.808$, or approximately 6.8 years.

15. $f(x) = xe^{2x}$, so $f'(x) = e^{2x} + xe^{2x}(2) = (1+2x)e^{2x}$.

17. $g(t) = \sqrt{t}e^{-2t}$, so $g'(t) = \frac{1}{2}t^{-1/2}e^{-2t} + \sqrt{t}e^{-2t}(-2) = \frac{1-4t}{2\sqrt{t}e^{2t}}$.

19. $y = \dfrac{e^{2x}}{1+e^{-2x}}$, so $y' = \dfrac{\left(1+e^{-2x}\right) e^{2x}(2) - e^{2x} \cdot e^{-2x}(-2)}{\left(1+e^{-2x}\right)^2} = \dfrac{2\left(e^{2x}+2\right)}{\left(1+e^{-2x}\right)^2}$.

21. $f(x) = xe^{-x^2}$, so $f'(x) = e^{-x^2} + xe^{-x^2}(-2x) = \left(1-2x^2\right)e^{-x^2}$.

23. $f(x) = x^2e^x + e^x$, so $f'(x) = 2xe^x + x^2e^x + e^x = \left(x^2+2x+1\right)e^x = (x+1)^2 e^x$.

25. $f(x) = \ln\left(e^{x^2}+1\right)$, so $f'(x) = \dfrac{e^{x^2}(2x)}{e^{x^2}+1} = \dfrac{2xe^{x^2}}{e^{x^2}+1}$.

27. $f(x) = \dfrac{\ln x}{x+1}$, so $f'(x) = \dfrac{(x+1)\left(\frac{1}{x}\right) - \ln x}{(x+1)^2} = \dfrac{1+\frac{1}{x} - \ln x}{(x+1)^2} = \dfrac{x - x\ln x + 1}{x(x+1)^2}$.

29. $y = \ln\left(e^{4x}+3\right)$, so $y' = \dfrac{e^{4x}(4)}{e^{4x}+3} = \dfrac{4e^{4x}}{e^{4x}+3}$.

31. $f(x) = \dfrac{\ln x}{1+e^x}$, so

$$f'(x) = \frac{(1+e^x)\dfrac{d}{dx}\ln x - \ln x \dfrac{d}{dx}(1+e^x)}{(1+e^x)^2} = \frac{(1+e^x)\left(\dfrac{1}{x}\right) - (\ln x)e^x}{(1+e^x)^2} = \frac{1+e^x - xe^x\ln x}{x(1+e^x)^2}$$
$$= \frac{1+e^x(1-x\ln x)}{x(1+e^x)^2}.$$

33. $y = \ln(3x+1)$, so $y' = \dfrac{3}{3x+1}$ and $y'' = 3\dfrac{d}{dx}(3x+1)^{-1} = -3(3x+1)^{-2}(3) = -\dfrac{9}{(3x+1)^2}$.

35. $h'(x) = g'(f(x))\,f'(x)$. But $g'(x) = 1 - \dfrac{1}{x^2}$ and $f'(x) = e^x$, so $f(0) = e^0 = 1$ and $f'(0) = e^0 = 1$. Therefore, $h'(0) = g'(f(0))\,f'(0) = g'(1)\,f'(0) = 0 \cdot 1 = 0$.

37. $y = \left(2x^3+1\right)\left(x^2+2\right)^3$, so $\ln y = \ln\left(2x^3+1\right) + 3\ln\left(x^2+2\right)$,

$$\frac{y'}{y} = \frac{6x^2}{2x^3+1} + \frac{3(2x)}{x^2+2} = \frac{6x^2\left(x^2+2\right) + 6x\left(2x^3+1\right)}{\left(2x^3+1\right)\left(x^2+2\right)} = \frac{6x^4+12x^2+12x^4+6x}{\left(2x^3+1\right)\left(x^2+2\right)}$$
$$= \frac{18x^4+12x^2+6x}{\left(2x^3+1\right)\left(x^2+2\right)},$$

and so $y' = 6x\left(3x^3+2x+1\right)\left(x^2+2\right)^2$.

39. $y = e^{-2x}$, so $y' = -2e^{-2x}$. This gives the slope of the tangent line to the graph of $y = e^{-2x}$ at any point (x, y). In particular, the slope of the tangent line at $\left(1, e^{-2}\right)$ is $y'(1) = -2e^{-2}$. The required equation is $y - e^{-2} = -2e^{-2}(x-1)$, or $y = \dfrac{1}{e^2}(-2x+3)$.

41. $f(x) = xe^{-2x}$. We first gather the following information on f.

1. The domain of f is $(-\infty, \infty)$.
2. Setting $x = 0$ gives 0 as the y-intercept.
3. $\lim_{x\to-\infty} xe^{-2x} = -\infty$ and $\lim_{x\to\infty} xe^{-2x} = 0$.

4. The results of part 3 show that $y = 0$ is a horizontal asymptote.

5. $f'(x) = e^{-2x} + xe^{-2x}(-2) = (1-2x)e^{-2x}$. Observe that $f'(x) = 0$ at $x = \frac{1}{2}$, a critical point of f.

The sign diagram of f' shows that f is increasing on $\left(-\infty, \frac{1}{2}\right)$ and decreasing on $\left(\frac{1}{2}, \infty\right)$.

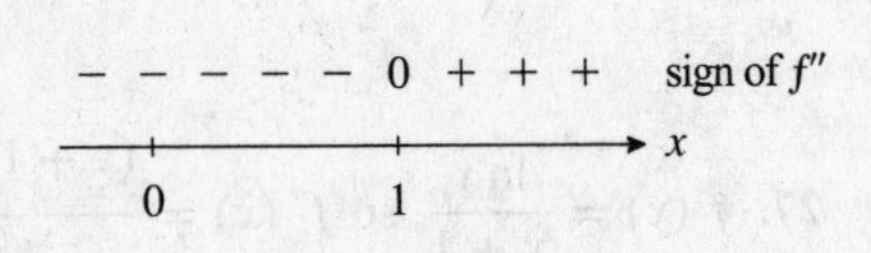

6. The results of part 5 show that $\left(\frac{1}{2}, \frac{1}{2}e^{-1}\right)$ is a relative maximum.

7. $f''(x) = -2e^{-2x} + (1-2x)e^{-2x}(-2) = 4(x-1)e^{-2x} = 0$ if $x = 1$. The sign diagram of f'' shows that the graph of f is concave downward on $(-\infty, 1)$ and concave upward on $(1, \infty)$.

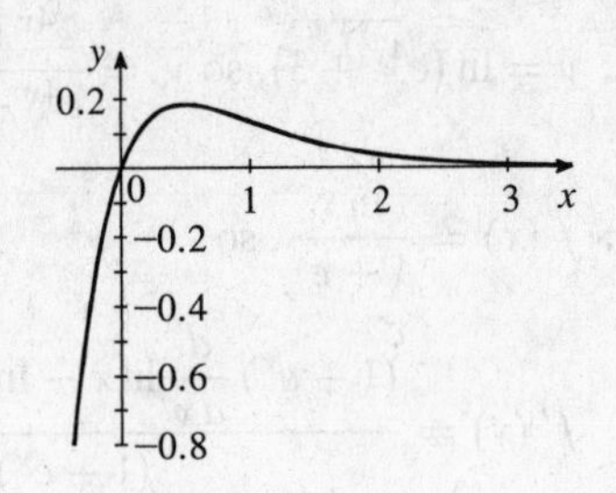

8. f has an inflection point at $(1, 1/e^2)$.

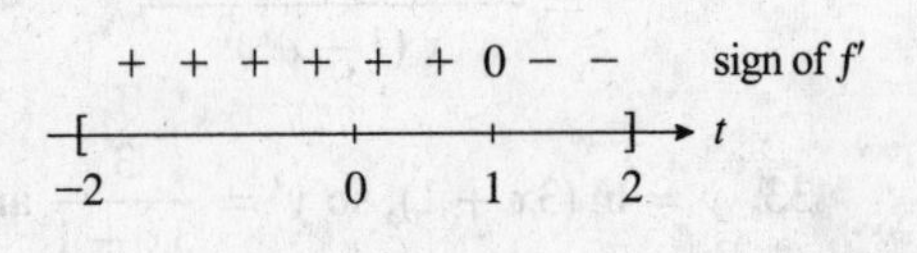

43. $f(t) = te^{-t}$, so $f'(t) = e^{-t} + t\left(-e^{-t}\right) = e^{-t}(1-t)$. Setting $f'(t) = 0$ gives $t = 1$ as the only critical point of f. From the sign diagram of f' we see that $f(1) = e^{-1} = 1/e$ is the absolute maximum value of f.

45. We want to solve the equation $8.2 = 4.5e^{r(5)}$. We have $e^{5r} = \frac{8.2}{4.5}$, so $r = \frac{1}{5}\ln\frac{8.2}{4.5} \approx 0.120$, and so the annual rate of return is 12%.

47. $P = 119{,}346e^{-0.06(4)} \approx 93{,}880.89$, or \$93,880.89.

49. a. $Q(t) = 2000e^{kt}$. Now $Q(120) = 18{,}000$ gives $2000e^{120k} = 18{,}000$, $e^{120k} = 9$, and so $120k = \ln 9$. Thus, $k = \frac{1}{120}\ln 9 \approx 0.01831$ and $Q(t) = 2000e^{0.01831t}$.

b. $Q(240) = 2000e^{0.01831(240)} \approx 161{,}992$, or approximately 162,000.

51.

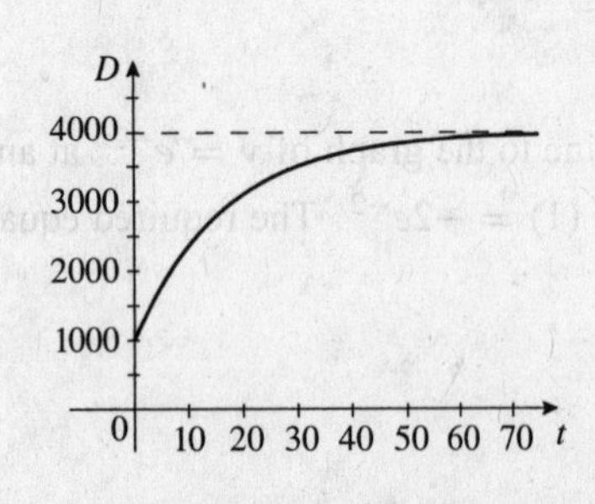

a. $D(1) = 4000 - 3000e^{-0.06} = 1175$,
$D(12) = 4000 - 3000e^{-0.72} = 2540$, and
$D(24) = 4000 - 3000e^{-1.44} = 3289$.

b. $\lim_{t\to\infty} D(t) = \lim_{t\to\infty}\left(4000 - 3000e^{-0.06t}\right) = 4000$.

53. $C(t) = 1486e^{-0.073t} + 500$.

a. The average energy consumption of the York refrigerator/freezer at the beginning of 1972 is given by $C(0) = 1486e^{-0.073(0)} + 500 = 1486 + 500 = 1986$, or 1986 kWh/yr.

b. To show that the average energy consumption of the York refrigerator is decreasing over the years in question, we compute $C'(t) = 1486e^{-0.073t}(-0.073) = -108.48e^{-0.073t}$ and note that $C'(t) < 0$ for all t. Therefore, $C(t)$ is decreasing over the interval $(0, 20)$.

c. To see if the York refrigerator/freezer satisfied the 1990 requirement, we compute $C(18) = 1486e^{-0.073(18)} + 500 = 399.35 + 500 = 899.35$, or 899.35 kWh/yr. Because this is less than 950 kWh/yr, we conclude that York satisfied the requirement.

55. a. The price at $t = 0$ is $18 - 3e^0 - 6e^0 = 9$, or \$9/unit.

b. $\frac{dp}{dt} = 6e^{-2t} + 2e^{-t/3}$, so $\left.\frac{dp}{dt}\right|_{t=0} = 6e^0 + 2e^0 = 8$. Thus, the price is increasing at the rate of \$8/unit/week.

c. The equilibrium price is given by $\lim_{t\to\infty} p = \lim_{t\to\infty}\left(18 - 3e^{-2t} - 6e^{-t/3}\right) = 18$, or \$18/unit.

57. $P(t) = \dfrac{12}{1 + 3e^{-0.2747t}}$.

a. $P(0) = \dfrac{12}{1 + 3e^0} = \dfrac{12}{4} = 3$, or 3 billion.

b. $P(8) = \dfrac{12}{1 + 3e^{-0.2747(8)}} \approx 9.00$, or approximately 9 billion.

c. $P'(t) = \dfrac{12(-0.2747)3e^{-0.2747t}}{\left(1 + 3e^{-0.2747t}\right)^2} = \dfrac{9.8892e^{-0.2747t}}{\left(1 + 3e^{-0.2747t}\right)^2}$, so $P'(6) = \dfrac{9.8892e^{-0.2747(6)}}{\left[1 + 3e^{-0.2747(6)}\right]^2} \approx 0.698567$. The population of the world is expected to be increasing by approximately 698.567 million per decade in 2030.

59. The revenue is given by $R = px = 20e^{-0.0002x}x = 20xe^{-0.0002x}$. To find the maximum of this function, we compute $R'(x) = 20e^{-0.0002x} - 20x\left(0.0002e^{-0.0002x}\right) = 20e^{-0.0002x}(1 - 0.0002x) = 0$ if $1 - 0.0002x = 0$, or $x = 5000$. From the sign diagram of R', we see that (5000, 36787.9) is a maximum, so 5000 pairs of socks should be produced to yield a maximum revenue of approximately \$36,788.

+ + + 0 − − − sign of R'

0 5000 10,000 x

CHAPTER 5 Before Moving On... page 408

1. $\dfrac{100}{1 + 2e^{0.3t}} = 40$, so $1 + 2e^{0.3t} = \dfrac{100}{40} = 2.5$, $2e^{0.3t} = 1.5$, $e^{0.3t} = \dfrac{1.5}{2} = 0.75$, $0.3t = \ln 0.75$, and so $t = \dfrac{\ln 0.75}{0.3} \approx -0.959$.

2. $A = 3000\left(1 + \frac{0.08}{52}\right)^{4(52)} = 4130.37$, or \$4130.37.

3. $f(x) = e^{\sqrt{x}}$, so $f'(x) = \dfrac{d}{dx}e^{x^{1/2}} = e^{x^{1/2}}\dfrac{d}{dx}\left(x^{1/2}\right) = e^{x^{1/2}}\left(\frac{1}{2}x^{-1/2}\right) = \dfrac{e^{\sqrt{x}}}{2\sqrt{x}}$.

4. $y = x\ln\left(x^2 + 1\right)$, so $\dfrac{dy}{dx} = x\dfrac{d}{dx}\ln\left(x^2 + 1\right) + \ln\left(x^2 + 1\right)\dfrac{d}{dx}(x) = x\cdot\dfrac{2x}{x^2 + 1} + \ln\left(x^2 + 1\right) = \dfrac{2x^2}{x^2 + 1} + \ln\left(x^2 + 1\right)$.

Thus, $\left.\dfrac{dy}{dx}\right|_{x=1} = \dfrac{2}{1 + 1} + \ln 2 = 1 + \ln 2$.

5. $y = e^{2x} \ln 3x$, so $y' = e^{2x} \dfrac{d}{dx} \ln 3x + \ln 3x \cdot \dfrac{d}{dx} e^{2x} = \dfrac{e^{2x}}{x} + 2e^{2x} \ln 3x$ and

$$y'' = \frac{d}{dx}\left(x^{-1}e^{2x}\right) + 2e^{2x}\frac{d}{dx}\ln 3x + (\ln 3x)\frac{d}{dx}\left(2e^{2x}\right) = -x^{-2}e^{2x} + 2x^{-1}e^{2x} + 2e^{2x}\left(\frac{1}{x}\right) + 4e^{2x} \cdot \ln 3x$$

$$= -\frac{1}{x^2}e^{2x} + \frac{4e^{2x}}{x} + 4(\ln 3x)e^{2x} = e^{2x}\left(\frac{4x^2 \ln 3x + 4x - 1}{x^2}\right).$$

6. $T(0) = 200$ gives $70 + ce^0 = 70 + C = 200$, so $C = 130$. Thus, $T(t) = 70 + 130e^{-kt}$. $T(3) = 180$ implies $70 + 130e^{-3k} = 180$, so $130e^{-3k} = 110$, $e^{-3k} = \dfrac{110}{130}$, $-3k = \ln \dfrac{11}{13}$, and $k = -\frac{1}{3} \ln \frac{11}{13} \approx 0.0557$. Therefore, $T(t) = 70 + 130e^{-0.0557t}$. So when $T(t) = 150$, we have $70 + 130e^{-0.0557t} = 150$, $130e^{-0.0557t} = 80$, $e^{-0.0557t} = \frac{80}{130} = \frac{8}{13}$, $-0.0557t = \ln \frac{8}{13}$, and finally $t = -\dfrac{\ln \frac{8}{13}}{0.0557} \approx 8.716$, or approximately 8.7 minutes.

6 INTEGRATION

6.1 Antiderivatives and the Rules of Integration

Problem-Solving Tips

1. Get into the habit of using the correct notation for integration. The indefinite integral of $f(x)$ with respect to x is written $\int f(x)\,dx$. It is incorrect to write $\int f(x)$ without indicating that you are integrating with respect to x. You will appreciate how important the correct notation is if you use CAS or graphic calculator with the capability to do symbolic integration. If you don't enter this information (the variable with respect to which you are performing the integration) into your calculator or computer, the integration will not be performed.

2. If you are finding an indefinite integral, be sure to include a constant of integration in your answer. Remember that $\int f(x)\,dx$ is the family of functions given by $F(x) + C$, where $F'(x) = f(x)$.

3. It's very easy to check your answer if you are finding an indefinite integral. Just take the derivative of your answer and you should get the integrand. [You are verifying that $F'(x) = f(x)$.] If not, you know immediately that you have made an error.

Concept Questions page 418

1. An antiderivative of a continuous function f on an interval I is a function F such that $F'(x) = f(x)$ for every x in I. For example, an antiderivative of $f(x) = x^2$ on $(-\infty, \infty)$ is the function $F(x) = \frac{1}{3}x^3$ on $(-\infty, \infty)$.

3. The indefinite integral of f is the family of functions $F(x) + C$, where F is an antiderivative of f and C is an arbitrary constant.

Exercises page 418

1. $F(x) = \frac{1}{3}x^3 + 2x^2 - x + 2$, so $F'(x) = x^2 + 4x - 1 = f(x)$.

3. $F(x) = (2x^2 - 1)^{1/2}$, so $F'(x) = \frac{1}{2}(2x^2 - 1)^{-1/2}(4x) = 2x(2x^2 - 1)^{-1/2} = f(x)$.

5. a. $G'(x) = \dfrac{d}{dx}(2x) = 2 = f(x)$

b. $F(x) = G(x) + C = 2x + C$

c.

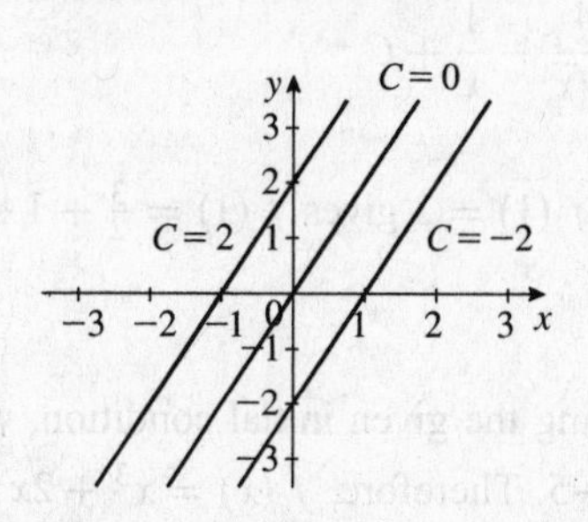

7. a. $G'(x) = \dfrac{d}{dx}\left(\frac{1}{3}x^3\right) = x^2 = f(x)$

b. $F(x) = G(x) + C = \frac{1}{3}x^3 + C$

c.

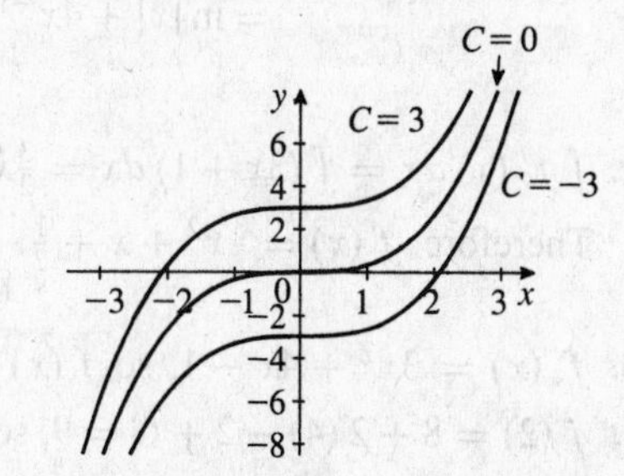

9. $\int 6\,dx = 6x + C.$

11. $\int x^3\,dx = \frac{1}{4}x^4 + C.$

13. $\int x^{-4}\,dx = -\frac{1}{3}x^{-3} + C.$

15. $\int x^{2/3}\,dx = \frac{3}{5}x^{5/3} + C.$

17. $\int x^{-5/4}\,dx = -4x^{-1/4} + C.$

19. $\displaystyle\int \frac{2}{x^3}\,dx = 2\int x^{-3}\,dx = -x^{-2} + C = -\frac{1}{x^2} + C.$

21. $\int \pi\sqrt{t}\,dt = \pi\int t^{1/2}\,dt = \pi\left(\frac{2}{3}t^{3/2}\right) + C = \frac{2\pi}{3}t^{3/2} + C.$

23. $\int (3 - 4x)\,dx = \int 3\,dx - 4\int x\,dx = 3x - 2x^2 + C.$

25. $\int \left(x^2 + x + x^{-3}\right) dx = \int x^2\,dx + \int x\,dx + \int x^{-3}\,dx = \frac{1}{3}x^3 + \frac{1}{2}x^2 - \frac{1}{2}x^{-2} + C.$

27. $\int 5e^x\,dx = 5e^x + C.$

29. $\int (1 + x + e^x)\,dx = x + \frac{1}{2}x^2 + e^x + C.$

31. $\displaystyle\int \left(4x^3 - \frac{2}{x^2} - 1\right) dx = \int \left(4x^3 - 2x^{-2} - 1\right) dx = x^4 + 2x^{-1} - x + C = x^4 + \frac{2}{x} - x + C.$

33. $\int \left(x^{5/2} + 2x^{3/2} - x\right) dx = \frac{2}{7}x^{7/2} + \frac{4}{5}x^{5/2} - \frac{1}{2}x^2 + C.$

35. $\int \left(x^{1/2} + 2x^{-1/2}\right) dx = \frac{2}{3}x^{3/2} + 4x^{1/2} + C.$

37. $\displaystyle\int \left(\frac{u^3 + 2u^2 - u}{3u}\right) du = \frac{1}{3}\int \left(u^2 + 2u - 1\right) du = \frac{1}{9}u^3 + \frac{1}{3}u^2 - \frac{1}{3}u + C.$

39. $\int (2t + 1)(t - 2)\,dt = \int \left(2t^2 - 3t - 2\right) dt = \frac{2}{3}t^3 - \frac{3}{2}t^2 - 2t + C.$

41. $\displaystyle\int \frac{1}{x^2}\left(x^4 - 2x^2 + 1\right) dx = \int \left(x^2 - 2 + x^{-2}\right) dx = \frac{1}{3}x^3 - 2x - x^{-1} + C = \frac{1}{3}x^3 - 2x - \frac{1}{x} + C.$

43. $\displaystyle\int \frac{ds}{(s+1)^{-2}} = \int (s+1)^2\,ds = \int \left(s^2 + 2s + 1\right) ds = \frac{1}{3}s^3 + s^2 + s + C.$

45. $\displaystyle\int \left(e^t + t^e\right) dt = e^t + \frac{1}{e+1}t^{e+1} + C.$

47. $\displaystyle\int \frac{x^3 + x^2 - x + 1}{x^2}\,dx = \int \left(x + 1 - \frac{1}{x} + \frac{1}{x^2}\right) dx = \frac{1}{2}x^2 + x - \ln|x| - x^{-1} + C.$

49.

$$\int \frac{\left(x^{1/2} - 1\right)^2}{x^2}\,dx = \int \frac{x - 2x^{1/2} + 1}{x^2}\,dx = \int \left(x^{-1} - 2x^{-3/2} + x^{-2}\right) dx$$
$$= \ln|x| + 4x^{-1/2} - x^{-1} + C = \ln|x| + \frac{4}{\sqrt{x}} - \frac{1}{x} + C.$$

51. $\int f'(x)\,dx = \int (3x + 1)\,dx = \frac{3}{2}x^2 + x + C$. The condition $f(1) = 3$ gives $f(1) = \frac{3}{2} + 1 + C = 3$, so $C = \frac{1}{2}$. Therefore, $f(x) = \frac{3}{2}x^2 + x + \frac{1}{2}$.

53. $f'(x) = 3x^2 + 4x - 1$, so $f(x) = x^3 + 2x^2 - x + C$. Using the given initial condition, we have $f(2) = 8 + 2(4) - 2 + C = 9$, so $16 - 2 + C = 9$, or $C = -5$. Therefore, $f(x) = x^3 + 2x^2 - x - 5$.

55. $f(x) = \int f'(x)\,dx = \int \left(1 + \frac{1}{x^2}\right) dx = \int \left(1 + x^{-2}\right) dx = x - \frac{1}{x} + C$. Using the given initial condition, we have $f(1) = 1 - 1 + C = 3$, so $C = 3$. Therefore, $f(x) = x - \frac{1}{x} + 3$.

57. $f(x) = \int \frac{x+1}{x}\,dx = \int \left(1 + \frac{1}{x}\right) dx = x + \ln|x| + C$. Using the initial condition, we have $f(1) = 1 + \ln 1 + C = 1 + C = 1$, so $C = 0$. Thus, $f(x) = x + \ln|x|$.

59. $f(x) = \int f'(x)\,dx = \int \frac{1}{2}x^{-1/2}\,dx = \frac{1}{2}\left(2x^{1/2}\right) + C = x^{1/2} + C$, and $f(2) = \sqrt{2} + C = \sqrt{2}$ implies $C = 0$. Thus, $f(x) = \sqrt{x}$.

61. $f'(x) = e^x + x$, so $f(x) = e^x + \frac{1}{2}x^2 + C$ and $f(0) = e^0 + \frac{1}{2}(0) + C = 1 + C$. Thus, $3 = 1 + C$, and so $2 = C$. Therefore, $f(x) = e^x + \frac{1}{2}x^2 + 2$.

63. The net amount on deposit in branch A is given by the area under the graph of f from $t = 0$ to $t = 180$. On the other hand, the net amount on deposit in branch B is given by the area under the graph of g over the same interval. Branch A has a larger amount on deposit because the rate at which money was deposited into branch A was always greater than the rate at which money was deposited into branch B over the period in question.

65. The number in year t is $N(t) = \int R(t)\,dt = \int 14.3\,dt = 14.3t + C$. To determine C, we use the condition $N(0) = 90.1$, giving $C = 90.1$. Therefore, $N(t) = 14.3t + 90.1$. The estimated number of users in 2015 is $N(4) = 14.3(4) + 90.1 = 147.3$, or 147.3 million.

67. Let f be the position function of the maglev. Then $f'(t) = v(t)$. Therefore, $f(t) = \int f'(t)\,dt = \int v(t)\,dt = \int (0.2t + 3)\,dt = 0.1t^2 + 3t + C$. If we measure the position of the maglev from the station, then the required function is $f(t) = 0.1t^2 + 3t$.

69. $P'(x) = -0.004x + 20$, so $P(x) = -0.002x^2 + 20x + C$. Because $C = -16{,}000$, we find that $P(x) = -0.002x^2 + 20x - 16{,}000$. The company realizes a maximum profit when $P'(x) = 0$, that is, when $x = 5000$ units. Next, $P(5000) = -0.002(5000)^2 + 20(5000) - 16{,}000 = 34{,}000$. Thus, the maximum profit of \$34,000 is realized at a production level of 5000 units.

71. a. The amount of wind energy generated in year t is $A(t) = \int r(t)\,dt = \int (5.018t - 3.204)\,dt = 2.509t^2 - 3.204t + C$. To determine C, we use the condition $A(0) = 1.8$, giving $C = 1.8$. Therefore, $A(t) = 2.509t^2 - 3.204t + 1.8$.

b. The amount of wind energy generated in 2012 was $A(7) = 2.509(7)^2 - 3.204(7) + 1.8 = 102.313$, or approximately 102.3 terawatt-hours.

c. The amount generated in 2013 was $A(8) = 2.509(8)^2 - 3.204(8) + 1.8 = 136.744$, or approximately 136.7 terawatt-hours.

73. The total number of acres grown in year t is $N(t) = \int R(t)\,dt = \int (150t + 14.82)\,dt = 75t^2 + 14.82t + C$. Using the condition $N(0) = 27.2$, we find $N(0) = C = 27.2$. Therefore, $N(t) = 75t^2 + 14.82t + 27.2$. The number of acres grown in 2012 is given by $N(6) = 75(6)^2 + 14.82(6) + 27.2 \approx 2816.12$, or approximately 2816.1 acres.

75. a. $h(t) = \int h'(t)\,dt = \int (-32t + 4)\,dt = -16t^2 + 4t + C$. But $h(0) = C = 400$, so $h(t) = -16t^2 + 4t + 400$.

b. It strikes the ground when $h(t) = 0$; that is, when $-16t^2 + 4t + 400 = 0$. Using the quadratic formula, we find that $t = \frac{-4 \pm \sqrt{16 - 4(-16)(400)}}{2(-16)} \approx 5.13$ or -4.88. We disregard the negative root since t must be nonnegative, and conclude that $t \approx 5.13$.

c. Its velocity is $-32(5.13) + 4 = 160.16$, or approximately 160.16 ft/sec downward.

77. The number of new subscribers at any time is $N(t) = \int (100 + 210t^{3/4})\, dt = 100t + 120t^{7/4} + C$. The given condition implies that $N(0) = 5000$. Using this condition, we find $C = 5000$. Therefore, $N(t) = 100t + 120t^{7/4} + 5000$. The number of subscribers 16 months from now is $N(16) = 100(16) + 120(16)^{7/4} + 5000$, or 21,960.

79. $h(t) = \int h'(t)\, dt = \int (-3t^2 + 192t)\, dt = -t^3 + 96t^2 + C = -t^3 + 96t^2 + C$. $h(0) = C = 0$ implies $h(t) = -t^3 + 96t^2$. The altitude 30 seconds after liftoff is $h(30) = -30^3 + 96(30)^2 = 59{,}400$ ft.

81. $C(x) = \int C'(x)\, dx = \int (0.000009x^2 - 0.009x + 8)\, dx = 0.000003x^3 - 0.0045x^2 + 8x + k$. $C(0) = k = 120$, and so $C(x) = 0.000003x^3 - 0.0045x^2 + 8x + 120$. Thus, $C(500) = 0.000003(500)^3 - 0.0045(500)^2 + 8(500) + 120 = \$3370$.

83. a. We have the initial-value problem $R'(t) = 8\sqrt{2}t^{1/2} - 32t^3$ with $R(0) = 0$. Integrating, we find $R(t) = \int \left(8\sqrt{2}t^{1/2} - 32t^3\right) dt = \frac{16\sqrt{2}}{3}t^{3/2} - 8t^4 + C$. $R(0) = 0$ implies that $C = 0$, so $R(t) = \frac{16\sqrt{2}}{3}t^{3/2} - 8t^4$.

b. $R\left(\frac{1}{2}\right) = \frac{16\sqrt{2}}{3}\left(\frac{1}{2}\right)^{3/2} - 8\left(\frac{1}{2}\right)^4 \approx 2.166$, so after $\frac{1}{2}$ hr, approximately 2.2 inches of rain had fallen.

85. a. The percentage of people 12 and older using social networking sites or services in year t is $P(t) = \int R(t)\, dt = \int 5.92t^{-0.158}\, dt \approx 7.031t^{0.842} + C$. To find C, we use the condition $P(1) = 7$, obtaining $7.031 + C = 7$, so $C = -0.031$. Therefore, $P(t) = 7.031t^{0.842} - 0.031$.

b. The percentage in 2013 was $P(5) = 7.031(5)^{0.842} - 0.031 \approx 27.23$, or approximately 27.2%.

87. $S'(W) = 0.131773W^{-0.575}$, so $S = \int 0.131773W^{-0.575}\, dW = 0.310054W^{0.425} + C$. Now $S(70) = 0.310054(70)^{0.425} + C = 1.886277 + C = 1.886277$, so $C = -0.000007 \approx 0$. Thus, $S(75) = 0.310054(75)^{0.425} \approx 1.9424$.

89. a. Let y denote the height of a typical preschool child. Then $R(t) = 25.8931e^{-0.993t} + 6.39$ and $y = \int R(t)\, dt = -\frac{25.8931}{0.993}e^{-0.993t} + 6.39t + C = -26.0756e^{-0.993t} + 6.39t + C$. $y\left(\frac{1}{4}\right) = -26.0756e^{-(0.993)(1/4)} + 6.39\left(\frac{1}{4}\right) + C = 60.30$. Therefore, $C = 79.045$, and so $y(t) = -26.0756e^{-0.993t} + 6.39t + 79.045$.

b. $y(1) = -26.0756e^{-0.993} + 6.39 + 79.045 \approx 75.7749$, or approximately 75.77 cm.

91. Denote the constant deceleration by k. Then $f''(t) = -k$, so $f'(t) = v(t) = -kt + C_1$. Next, the given condition implies that $v(0) = 88$. This gives $C_1 = 88$, so $f'(t) = -kt + 88$. Now $s = f(t) = \int f'(t)\, dt = \int (-kt + 88)\, dt = -\frac{1}{2}kt^2 + 88t + C_2$, and $f(0) = 0$ gives $s = f(t) = -\frac{1}{2}kt^2 + 88t$. Because the car is brought to rest in 9 seconds, we have $v(9) = -9k + 88 = 0$, or $k = \frac{88}{9} \approx 9.78$, so the deceleration is 9.78 ft/sec^2. The distance covered is $s = f(9) = -\frac{1}{2}\left(\frac{88}{9}\right)(81) + 88(9) = 396$, so the stopping distance is 396 ft.

93. The time taken by runner A to cross the finish line is $t = \frac{200}{22} = \frac{100}{11}$ sec. Let a be the constant acceleration of runner B as he begins to spurt. Then $\frac{dv}{dt} = a$, so the velocity of runner B as he runs towards the finish line is $v = \int a\,dt = at + c$. At $t = 0$, $v = 20$ and so $v = at = 20$. Now $\frac{ds}{dt} = v = at + 20$, so $s = \int (at + 20)\,dt = \frac{1}{2}at^2 + 20t + k$, where k is the constant of integration. Next, $s(0) = 0$ gives $s = \frac{1}{2}at^2 + 20t = \left(\frac{1}{2}at + 20\right)t$. In order for runner B to cover 220 ft in $\frac{100}{11}$ sec, we must have $\left[\frac{1}{2}a\left(\frac{100}{11}\right) + 20\right]\frac{100}{11} = 220$, so $\frac{50}{11}a + 20 = \frac{220\cdot 11}{100} = \frac{121}{5}$, $\frac{50}{11}a = \frac{121}{5} - 20 = \frac{21}{5}$, and $a = \frac{21}{5}\cdot\frac{11}{50} = 0.924\text{ ft/sec}^2$. Therefore, runner B must have an acceleration of at least 0.924 ft/sec^2.

95. Suppose the acceleration is k. The distance covered is $s = f(t)$ and satisfies $f''(t) = k$. Thus, $f'(t) = v(t) = \int k\,dt = kt + C_1$. Next, $v(0) = 0$ gives $v(t) = kt$, and so $s = f(t) = \int kt\,dt = \frac{1}{2}kt^2 + C_2$. Now $f(0) = 0$ gives $s = \frac{1}{2}kt^2$. If it traveled 800 ft, we have $800 = \frac{1}{2}kt^2$, so $t = \frac{40}{\sqrt{k}}$. Its speed at this time is $v(t) = kt = k\left(\frac{40}{\sqrt{k}}\right) = 40\sqrt{k}$. We want the speed to be at least 240 ft/sec, so we require $40\sqrt{k} > 240$, implying that $k > 36$. In other words, the acceleration must be at least 36 ft/sec^2.

97. False. $\int f(x)\,dx = F(x) + C$, where C is an arbitrary constant.

99. False. $\int \frac{d}{dx}[f(x)]\,dx = \int f'(x)\,dx = f(x) + C$, where C is a constant of integration.

6.2 Integration by Substitution

Problem-Solving Tips

1. Here are some tips for using the method of substitution.

   a. The idea is to replace the given integral by a simpler integral, so look for a substitution $u = g(x)$ that simplifies the integral.

   b. Check to see that $du = g'(x)\,dx$ appears in the integral.

2. Look through Problems 1–50 to familiarize yourself with the types of functions that can be integrated using the method of substitution. Even if you don't complete every problem, check to see if you can set up the given integral so that the method of substitution can be used to complete the integration.

Concept Questions page 430

1. To find $I = \int f(g(x))g'(x)\,dx$ by the Method of Substitution, let $u = g(x)$, so that $du = g'(x)\,dx$. Making the substitution, we obtain $I = \int f(u)\,du$, which can be integrated with respect to u. Finally, replace u by $u = g(x)$ to evaluate the integral.

Exercises page 430

1. Put $u = 4x + 3$, so $du = 4\,dx$ and $dx = \frac{1}{4}\,du$. Then $\int 4(4x+3)^4\,dx = \int u^4\,du = \frac{1}{5}u^5 + C = \frac{1}{5}(4x+3)^5 + C$.

3. Let $u = x^3 - 2x$, so $du = (3x^2 - 2)\,dx$. Then $\int (x^3 - 2x)^2(3x^2 - 2)\,dx = \int u^2\,du = \frac{1}{3}u^3 + C = \frac{1}{3}(x^3 - 2x)^3 + C$.

5. Let $u = 2x^2 + 3$, so $du = 4x\,dx$. Then
$$\int \frac{4x}{(2x^2+3)^3}\,dx = \int \frac{1}{u^3}\,du = \int u^{-3}\,du = -\tfrac{1}{2}u^{-2} + C = -\frac{1}{2(2x^2+3)^2} + C.$$

7. Put $u = t^3 + 2$, so $du = 3t^2\,dt$ and $t^2\,dt = \frac{1}{3}\,du$. Then
$\int 3t^2\sqrt{t^3+2}\,dt = \int u^{1/2}\,du = \frac{2}{3}u^{3/2} + C = \frac{2}{3}(t^3+2)^{3/2} + C.$

9. Let $u = x^2 - 1$, so $du = 2x\,dx$ and $x\,dx = \frac{1}{2}\,du$. Then
$\int 2(x^2-1)^9\,x\,dx = 2\int \frac{1}{2}u^9\,du = \frac{1}{10}u^{10} + C = \frac{1}{10}(x^2-1)^{10} + C.$

11. Let $u = 1 - x^5$, so $du = -5x^4\,dx$ and $x^4\,dx = -\frac{1}{5}\,du$. Then
$$\int \frac{x^4}{1-x^5}\,dx = -\frac{1}{5}\int \frac{du}{u} = -\tfrac{1}{5}\ln|u| + C = -\tfrac{1}{5}\ln\left|1-x^5\right| + C.$$

13. Let $u = x - 2$, so $du = dx$. Then $\displaystyle\int \frac{2}{x-2}\,dx = 2\int \frac{du}{u} = 2\ln|u| + C = \ln u^2 + C = \ln(x-2)^2 + C.$

15. Let $u = 0.3x^2 - 0.4x + 2$. Then $du = (0.6x - 0.4)\,dx = 2(0.3x - 0.2)\,dx$. Thus,
$$\int \frac{0.3x - 0.2}{0.3x^2 - 0.4x + 2}\,dx = \int \frac{1}{2u}\,du = \tfrac{1}{2}\ln|u| + C = \tfrac{1}{2}\ln\left(0.3x^2 - 0.4x + 2\right) + C.$$

17. Let $u = 3x^2 - 1$, so $du = 6x\,dx$ and $x\,dx = \frac{1}{6}\,du$. Then
$$\int \frac{2x}{3x^2-1}\,dx = 2\int \frac{x}{3x^2-1}\,dx = \frac{1}{3}\int \frac{du}{u} = \tfrac{1}{3}\ln|u| + C = \tfrac{1}{3}\ln\left|3x^2-1\right| + C.$$

19. Let $u = -2x$, so $du = -2\,dx$ and $dx = -\frac{1}{2}\,du$. Then $\int e^{-2x}\,dx = -\frac{1}{2}\int e^u\,du = -\frac{1}{2}e^u + C = -\frac{1}{2}e^{-2x} + C.$

21. Let $u = 2 - x$, so $du = -dx$ and $dx = -du$. Then $\int e^{2-x}\,dx = -\int e^u\,du = -e^u + C = -e^{2-x} + C.$

23. Let $u = -x^2$, so $du = -2x\,dx$ and $x\,dx = -\frac{1}{2}\,du$. Then $\int xe^{-x^2}\,dx = \int -\frac{1}{2}e^u\,du = -\frac{1}{2}e^u + C = -\frac{1}{2}e^{-x^2} + C.$

25. $\int (e^x - e^{-x})\,dx = \int e^x\,dx - \int e^{-x}\,dx = e^x - \int e^{-x}\,dx$. To evaluate the second integral on the right, let $u = -x$ so $du = -dx$ and $dx = -du$. Then $\int (e^x - e^{-x})\,dx = e^x + \int e^u\,du = e^x + e^u + C = e^x + e^{-x} + C.$

27. Let $u = 1 + e^x$, so $du = e^x\,dx$. Then $\displaystyle\int \frac{2e^x}{1+e^x}\,dx = 2\int \frac{e^x}{1+e^x}\,dx = 2\int \frac{du}{u} = 2\ln|u| + C = 2\ln(1+e^x) + C.$

29. Let $u = \sqrt{x} = x^{1/2}$. Then $du = \frac{1}{2}x^{-1/2}\,dx$ and $2\,du = x^{-1/2}\,dx$, so
$$\int \frac{e^{\sqrt{x}}}{\sqrt{x}}\,dx = \int 2e^u\,du = 2e^u + C = 2e^{\sqrt{x}} + C.$$

31. Let $u = e^{3x} + x^3$, so $du = (3e^{3x} + 3x^2)\,dx = 3(e^{3x} + x^2)\,dx$ and $(e^{3x} + x^2)\,dx = \frac{1}{3}du$. Then
$$\int \frac{e^{3x}+x^2}{(e^{3x}+x^3)^3}\,dx = \frac{1}{3}\int \frac{du}{u^3} = \frac{1}{3}\int u^{-3}\,du = -\frac{u^{-2}}{6} + C = -\frac{1}{6(e^{3x}+x^3)^2} + C.$$

33. Let $u = e^{2x} + 1$, so $du = 2e^{2x}\,dx$ and $\frac{1}{2}\,du = e^{2x}\,dx$. Then
$\int e^{2x}(e^{2x}+1)^3\,dx = \int \frac{1}{2}u^3\,du = \frac{1}{8}u^4 + C = \frac{1}{8}(e^{2x}+1)^4 + C.$

35. Let $u = \ln 5x$, so $du = \frac{1}{x}\,dx$. Then $\int \frac{\ln 5x}{x}\,dx = \int u\,du = \frac{1}{2}u^2 + C = \frac{1}{2}(\ln 5x)^2 + C$.

37. Let $u = 3\ln x$, so $du = \frac{3}{x}\,dx$. Then $3\int \frac{1}{x\ln x}\,dx = 3\int \frac{du}{u} = 3\ln|u| + C = 3\ln|\ln x| + C$.

39. Let $u = \ln x$, so $du = \frac{1}{x}\,dx$. Then $\int \frac{\sqrt{\ln x}}{x}\,dx = \int \sqrt{u}\,du = \frac{2}{3}u^{3/2} + C = \frac{2}{3}(\ln x)^{3/2} + C$.

41. $\int \left(xe^{x^2} - \frac{x}{x^2+2}\right)dx = \int xe^{x^2}\,dx - \int \frac{x}{x^2+2}\,dx$. To evaluate the first integral, let $u = x^2$, so $du = 2x\,dx$ and $x\,dx = \frac{1}{2}\,du$. Then $\int xe^{x^2}\,dx = \frac{1}{2}\int e^u\,du + C_1 = \frac{1}{2}e^u + C_1 = \frac{1}{2}e^{x^2} + C_1$. To evaluate the second integral, let $u = x^2 + 2$, so $du = 2x\,dx$ and $x\,dx = \frac{1}{2}\,du$. Then $\int \frac{x}{x^2+2}\,dx = \frac{1}{2}\int \frac{du}{u} = \frac{1}{2}\ln|u| + C_2 = \frac{1}{2}\ln(x^2+2) + C_2$.

Therefore, $\int \left(xe^{x^2} - \frac{x}{x^2+2}\right)dx = \frac{1}{2}e^{x^2} - \frac{1}{2}\ln(x^2+2) + C$.

43. Let $u = \sqrt{x} - 1$, so $du = \frac{1}{2}x^{-1/2}\,dx = \frac{1}{2\sqrt{x}}\,dx$ and $dx = 2\sqrt{x}\,du$. Also, we have $\sqrt{x} = u + 1$, so $x = (u+1)^2 = u^2 + 2u + 1$ and $dx = 2(u+1)\,du$. Thus,

$$\begin{aligned}\int \frac{x+1}{\sqrt{x}-1}\,dx &= \int \frac{u^2+2u+2}{u}\cdot 2(u+1)\,du = 2\int \frac{(u^3+3u^2+4u+2)}{u}\,du\\ &= 2\int \left(u^2+3u+4+\frac{2}{u}\right)du = 2\left(\frac{1}{3}u^3 + \frac{3}{2}u^2 + 4u + 2\ln|u|\right) + C\\ &= 2\left[\frac{1}{3}\left(\sqrt{x}-1\right)^3 + \frac{3}{2}\left(\sqrt{x}-1\right)^2 + 4\left(\sqrt{x}-1\right) + 2\ln\left|\sqrt{x}-1\right|\right] + C.\end{aligned}$$

45. Let $u = x - 1$, so $du = dx$. Also, $x = u + 1$, and so

$$\begin{aligned}\int x(x-1)^5\,dx &= \int (u+1)u^5\,du = \int (u^6+u^5)\,du = \frac{1}{7}u^7 + \frac{1}{6}u^6 + C = \frac{1}{7}(x-1)^7 + \frac{1}{6}(x-1)^6 + C\\ &= \frac{(6x+1)(x-1)^6}{42} + C.\end{aligned}$$

47. Let $u = 1 + \sqrt{x}$, so $du = \frac{1}{2}x^{-1/2}\,dx$ and $dx = 2\sqrt{x} = 2(u-1)\,du$. Then

$$\begin{aligned}\int \frac{1-\sqrt{x}}{1+\sqrt{x}}dx &= \int \left(\frac{1-(u-1)}{u}\right)\cdot 2(u-1)\,du = 2\int \frac{(2-u)(u-1)}{u}\,du = 2\int \frac{-u^2+3u-2}{u}\,du\\ &= 2\int \left(-u+3-\frac{2}{u}\right)du = -u^2 + 6u - 4\ln|u| + C\\ &= -\left(1+\sqrt{x}\right)^2 + 6\left(1+\sqrt{x}\right) - 4\ln\left(1+\sqrt{x}\right) + C\\ &= -1 - 2\sqrt{x} - x + 6 + 6\sqrt{x} - 4\ln\left(1+\sqrt{x}\right) + C = -x + 4\sqrt{x} + 5 - 4\ln\left(1+\sqrt{x}\right) + C.\end{aligned}$$

49. $I = \int v^2(1-v)^6\,dv$. Let $u = 1 - v$, so $du = -dv$. Also, $1 - u = v$, and so $(1-u)^2 = v^2$. Therefore,

$$\begin{aligned}I &= \int -\left(1-2u+u^2\right)u^6 du = \int -\left(u^6 - 2u^7 + u^8\right)du = -\left(\frac{1}{7}u^7 - \frac{1}{4}u^8 + \frac{1}{9}u^9\right) + C\\ &= -u^7\left(\frac{1}{7} - \frac{1}{4}u + \frac{1}{9}u^2\right) + C = -\frac{1}{252}(1-v)^7\left[36 - 63(1-v) + 28\left(1 - 2v + v^2\right)\right]\\ &= -\frac{1}{252}(1-v)^7\left[36 - 63 + 63v + 28 - 56v + 28v^2\right] = -\frac{1}{252}(1-v)^7\left(28v^2 + 7v + 1\right) + C.\end{aligned}$$

51. $f(x) = \int f'(x)\,dx = 5\int (2x-1)^4\,dx$. Let $u = 2x-1$, so $du = 2\,dx$ and $dx = \frac{1}{2}\,du$. Then $f(x) = \frac{5}{2}\int u^4\,du = \frac{1}{2}u^5 + C = \frac{1}{2}(2x-1)^5 + C$. Next, $f(1) = 3$ implies $\frac{1}{2} + C = 3$, so $C = \frac{5}{2}$. Therefore, $f(x) = \frac{1}{2}(2x-1)^5 + \frac{5}{2}$.

53. $f(x) = \int -2xe^{-x^2+1}\,dx$. Let $u = -x^2+1$, so $du = -2x\,dx$. Then $f(x) = \int e^u\,du = e^u + C = e^{-x^2+1} + C$. The condition $f(1) = 0$ implies $f(1) = 1 + C = 0$, so $C = -1$. Therefore, $f(x) = e^{-x^2+1} - 1$.

55. $N'(t) = 2000(1+0.2t)^{-3/2}$. Let $u = 1+0.2t$, so $du = 0.2\,dt$ and $5\,du = dt$. Then $N(t) = (5)(2000)\int u^{-3/2}\,du = -20{,}000u^{-1/2} + C = -20{,}000(1+0.2t)^{-1/2} + C$. Next, $N(0) = -20{,}000(1)^{-1/2} + C = 1000$. Therefore, $C = 21{,}000$ and $N(t) = -\dfrac{20{,}000}{\sqrt{1+0.2t}} + 21{,}000$. In particular, $N(5) = -\frac{20{,}000}{\sqrt{2}} + 21{,}000 \approx 6858$.

57. The amount of CNP fraud in year t is $A(t) = \displaystyle\int [-R(t)]\,dt = -\int \frac{92.07}{t+1}\,dt = -92.07\ln(t+1) + C$. To determine C, we use the condition $A(0) = 328.9$ to obtain $C = 328.9$. Thus, $A(t) = 328.9 - 92.07\ln(t+1)$. The amount of CNP fraud in 2012 was $A(4) = 328.9 - 92.07\ln 5 \approx 180.72$, or approximately 180.7 million GBP.

59. The population t years from now will be

$P(t) = \displaystyle\int r(t)\,dt = \int 400\left(1 + \frac{2t}{24+t^2}\right)dt = \int 400\,dt + 800\int \frac{t}{24+t^2}\,dt$. In order to evaluate the second integral on the right, let $u = 24+t^2$, so $du = 2t\,dt$. We obtain

$P(t) = 400t + 800\displaystyle\int \frac{\frac{1}{2}\,du}{u} = 400t + 400\ln|u| + C = 400\left[t + \ln\left(24+t^2\right)\right] + C$. To find C, use the condition $P(0) = 60{,}000$, giving $400(0 + \ln 24) + C = 60{,}000$ and $C = 60{,}000 - 400\ln 24 = 58{,}728.78$. Thus, $P(t) = 400\left[t + \ln\left(24+t^2\right)\right] + 58{,}728.78$, and so the population 5 years from now will be $400\left[5 + \ln\left(24+5^2\right)\right] + 58{,}728.78 \approx 62{,}285.51$, or approximately 62,286.

61. Let $u = 1 + 1.09t$. Then $du = 1.09\,dt$, so

$$\int \frac{5.45218}{(1+1.09t)^{0.9}}\,dt = 5.45218\int (1+1.09t)^{-0.9}\,dt = \frac{5.45218}{1.09}\int u^{-0.9}\,du = 50.02u^{0.1} + C$$
$$= 50.02(1+1.09t)^{0.1} + C.$$

Then $g(0) = 50.02 + C = 50.02$ implies that $C = 0$, so $g(t) = 50.02(1+1.09t)^{0.1}$ and $g(100) = 50.02(110)^{0.1} \approx 80.04$.

63. a. The number of online viewers in year t is $N(t) = \int r(t)\,dt = \int 9.045e^{0.067t}\,dt = 135e^{0.067t} + C$. To determine C, we use the condition $N(0) = 135$, giving $C = 0$. Therefore, $N(t) = 135e^{0.067t}$.

b. The number of viewers in 2012 was $N(4) = 135e^{0.067(4)} \approx 176.492$, or approximately 176.5 million. The number in 2013 was $N(5) = 135e^{0.067(5)} \approx 188.722$, or approximately 188.7 million.

65. a. $R(t) = \int r(t)\,dt = \int 3.1182e^{0.163(t+1)}\,dt \approx 19.1301e^{0.163(t+1)} + C$. To find C, we use the condition $R(0) = 22.7$, obtaining $19.1301 + C = 22.7$, so $C = 3.5699$. Thus, $R(t) = 19.1301e^{0.163(t+1)} + 3.5699$.

b. Online ad revenue in 2012 was $R(3) \approx 40.2877$, or approximately \$40.3 billion.

67. $A(t) = \int A'(t)\,dt = r\int e^{-at}\,dt$. Let $u = -at$, so $du = -a\,dt$ and $dt = -\frac{1}{a}\,du$. Then $A(t) = r\left(-\frac{1}{a}\right)\int e^{u}\,du = -\frac{r}{a}e^{u} + C = -\frac{r}{a}e^{-at} + C$. $A(0) = 0$ implies $-\frac{r}{a} + C = 0$, so $C = \frac{r}{a}$. Therefore, $A(t) = -\frac{r}{a}e^{-at} + \frac{r}{a} = \frac{r}{a}\left(1 - e^{-at}\right)$.

69. True. Let $I = \int x f\left(x^2\right)dx$ and put $u = x^2$. Then $du = 2x\,dx$ and $x\,dx = \frac{1}{2}\,du$, so $I = \frac{1}{2}\int f(u)\,du = \frac{1}{2}\int f(x)\,dx$.

71. True. Put $u = kx$. Then $du = k\,dx$ and $dx = (1/k)\,du$. Thus, $I = \int e^{kx} f\left(e^{kx}\right)dx = \int e^{u} f\left(e^{u}\right)(1/k)\,du = \frac{1}{k}\int e^{u} f\left(e^{u}\right)du$. Next, we put $w = e^{u}$, so $dw = e^{u}du$. Then $I = \frac{1}{k}\int f(w)\,dw = \frac{1}{k}\int f(x)\,dx$.

6.3 Area and the Definite Integral

Problem-Solving Tips

1. In Sections 6.1 and 6.2, we found the indefinite integral of a function, that is, $\int f(x)\,dx = F(x) + C$. Note that our answer is a *family of functions* $F(x) + C$ for which $F'(x) = f(x)$. In this section, we found the definite integral of a function, $\int_a^b f(x)\,dx = \lim_{n\to\infty}\left[f(x_1)\,\Delta x + f(x_2)\,\Delta x + \cdots + f(x_n)\,\Delta x\right]$. Note that the answer here is a *number*.

2. The geometric interpretation of a definite integral is as follows: If f is continuous on $[a, b]$, then $\int_a^b f(x)\,dx$ is equal to the area of the region above the x-axis between the x-axis and the graph of f over $[a, b]$ minus the area of the corresponding region below the x-axis.

Concept Questions page 442

1. See page 438 in the text.

Exercises page 442

1. $\frac{1}{3}(1.9 + 1.5 + 1.8 + 2.4 + 2.7 + 2.5) = \frac{12.8}{3} \approx 4.27$.

3. a.

$A = \frac{1}{2}(2)(6) = 6$.

b. $\Delta x = \frac{2}{4} = \frac{1}{2}$, so $x_1 = 0$, $x_2 = \frac{1}{2}$, $x_3 = 1$, $x_4 = \frac{3}{2}$. Thus,

$$A \approx \frac{1}{2}\left[3(0) + 3\left(\frac{1}{2}\right) + 3(1) + 3\left(\frac{3}{2}\right)\right] = \frac{9}{2} = 4.5.$$

c. $\Delta x = \frac{2}{8} = \frac{1}{4}$, so $x_1 = 0, \ldots, x_8 = \frac{7}{4}$. Thus,

$$A \approx \frac{1}{4}\left[3(0) + 3\left(\frac{1}{4}\right) + 3\left(\frac{1}{2}\right) + 3\left(\frac{3}{4}\right) + 3(1) + 3\left(\frac{5}{4}\right) + 3\left(\frac{3}{2}\right) + 3\left(\frac{7}{4}\right)\right] = \frac{21}{4} = 5.25.$$

d. Yes.

5. a.

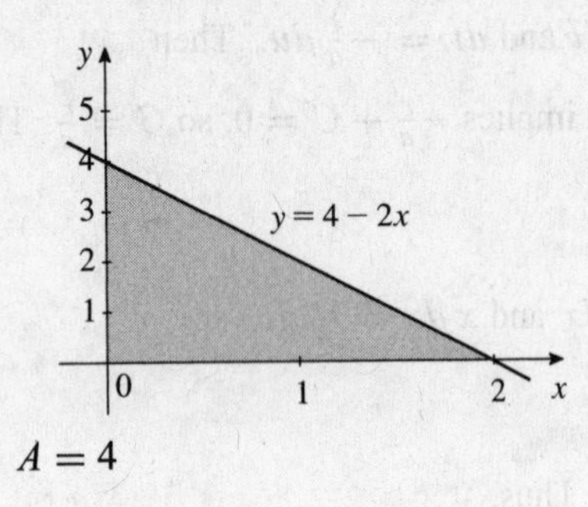

$A = 4$

b. $\Delta x = \frac{2}{5} = 0.4$, so $x_1 = 0, x_2 = 0.4, x_3 = 0.8, x_4 = 1.2,$ $x_5 = 1.6$. Thus,

$$A \approx 0.4\{[4-2(0)]+[4-2(0.4)]+[4-2(0.8)] + [4-2(1.2)]+[4-2(1.6)]\} = 4.8.$$

c. $\Delta x = \frac{2}{10} = 0.2$, so $x_1 = 0, x_2 = 0.2, x_3 = 0.4, \ldots, x_{10} = 1.8$. Thus,

$$A \approx 0.2\{[4-2(0)]+[4-2(0.2)]+[4-2(0.4)] + [4-2(0.6)]+[4-2(0.8)]+[4-2(1.0)]+[4-2(1.2)] + [4-2(1.4)]+[4-2(1.6)]+[4-2(1.8)]\} = 4.4.$$

d. Yes.

7. a. $\Delta x = \frac{4-2}{2} = 1$, so $x_1 = 2.5, x_2 = 3.5$. The Riemann sum is $1(2.5^2+3.5^2) = 18.5$.

b. $\Delta x = \frac{4-2}{5} = 0.4$, so $x_1 = 2.2, x_2 = 2.6, x_3 = 3.0, x_4 = 3.4, x_5 = 3.8$. The Riemann sum is $0.4(2.2^2+2.6^2+3.0^2+3.4^2+3.8^2) = 18.64$.

c. $\Delta x = \frac{4-2}{10} = 0.2$, so $x_1 = 2.1, x_2 = 2.3, x_2 = 2.5, \ldots, x_{10} = 3.9$. The Riemann sum is $0.2(2.1^2+2.3^2+2.5^2+2.7^2+2.9^2+3.1^2+3.3^2+3.5^2+3.7^2+3.9^2) = 18.66$.

d. The area appears to be $18\frac{2}{3}$.

9. a. $\Delta x = \frac{4-2}{2} = 1$, so $x_1 = 3, x_2 = 4$. The Riemann sum is $(1)(3^2+4^2) = 25$.

b. $\Delta x = \frac{4-2}{5} = 0.4$, so $x_1 = 2.4, x_2 = 2.8, x_3 = 3.2, x_4 = 3.6, x_5 = 4$. The Riemann sum is $0.4(2.4^2+2.8^2+\cdots+4^2) = 21.12$.

c. $\Delta x = \frac{4-2}{10} = 0.2$, so $x_1 = 2.2, x_2 = 2.4, x_3 = 2.6, \ldots, x_{10} = 4$. The Riemann sum is $0.2(2.2^2+2.4^2+2.6^2+2.8^2+3.0^2+3.2^2+3.4^2+3.6^2+3.8^2+4^2) = 19.88$.

d. The area appears to be 19.9.

11. a. $\Delta x = \frac{1}{2}$, so $x_1 = 0, x_2 = \frac{1}{2}$. The Riemann sum is $f(x_1)\Delta x + f(x_2)\Delta x = \left[(0)^3 + \left(\frac{1}{2}\right)^3\right]\frac{1}{2} = \frac{1}{16} = 0.0625$.

b. $\Delta x = \frac{1}{5}$, so $x_1 = 0, x_2 = \frac{1}{5}, x_3 = \frac{2}{5}, x_4 = \frac{3}{5}, x_5 = \frac{4}{5}$. The Riemann sum is $f(x_1)\Delta x + f(x_2)\Delta x + \cdots + f(x_5)\Delta x = \left[\left(\frac{1}{5}\right)^3 + \left(\frac{2}{5}\right)^3 + \cdots + \left(\frac{4}{5}\right)^3\right]\frac{1}{5} = \frac{100}{625} = 0.16$.

c. $\Delta x = \frac{1}{10}$, so $x_1 = 0, x_2 = \frac{1}{10}, x_3 = \frac{2}{10}, \ldots, x_{10} = \frac{9}{10}$. The Riemann sum is

$$\begin{aligned} &f(x_1)\Delta x + f(x_2)\Delta x + \cdots + f(x_{10})\Delta x \\ &= \left[0^3 + \left(\tfrac{1}{10}\right)^3 + \left(\tfrac{2}{10}\right)^3 + \left(\tfrac{3}{10}\right)^3 + \left(\tfrac{4}{10}\right)^3 + \left(\tfrac{5}{10}\right)^3 + \left(\tfrac{6}{10}\right)^3 + \left(\tfrac{7}{10}\right)^3 + \left(\tfrac{8}{10}\right)^3 + \left(\tfrac{9}{10}\right)^3\right]\tfrac{1}{10} \\ &= \tfrac{2025}{10{,}000} = 0.2025 \approx 0.2. \end{aligned}$$

d. The Riemann sums seem to approach 0.2.

13. $\Delta x = \frac{2-0}{5} = \frac{2}{5}$, so $x_1 = \frac{1}{5}, x_2 = \frac{3}{5}, x_3 = \frac{5}{5}, x_4 = \frac{7}{5}, x_5 = \frac{9}{5}$. Thus,

$$A \approx \left\{\left[\left(\tfrac{1}{5}\right)^2+1\right]+\left[\left(\tfrac{3}{5}\right)^2+1\right]+\left[\left(\tfrac{5}{5}\right)^2+1\right]+\left[\left(\tfrac{7}{5}\right)^2+1\right]+\left[\left(\tfrac{9}{5}\right)^2+1\right]\right\}\left(\tfrac{2}{5}\right) = \tfrac{580}{125} = 4.64.$$

15. $\Delta x = \frac{3-1}{4} = \frac{1}{2}$, so $x_1 = \frac{3}{2}, x_2 = \frac{4}{2} = 2, x_3 = \frac{5}{2}, x_4 = 3$. Thus, $A \approx \left(\frac{1}{3/2} + \frac{1}{2} + \frac{1}{5/2} + \frac{1}{3}\right)\frac{1}{2} \approx 0.95$.

17. $A \approx 20\left[f(10)+f(30)+f(50)+f(70)+f(90)\right] = 20(80+100+110+100+80) = 9400\text{ ft}^2$.

19. False. Take $f(x)=x$, $a=-1$, and $b=2$. Then $\int_{-1}^{2} f(x)\,dx = \int_{-1}^{2} x\,dx = \frac{1}{2}x^2\Big|_{-1}^{2} = \frac{1}{2}(4-1) = \frac{3}{2} > 0$, but $f(-1) = -1 < 0$.

6.4 The Fundamental Theorem of Calculus

Concept Questions page 452

1. See the Fundamental Theorem of Calculus on page 444 of the text.

Exercises page 453

1. $A = \int_1^4 2\,dx = 2x|_1^4 = 2(4-1) = 6$. The region is a rectangle with area $3 \cdot 2 = 6$.

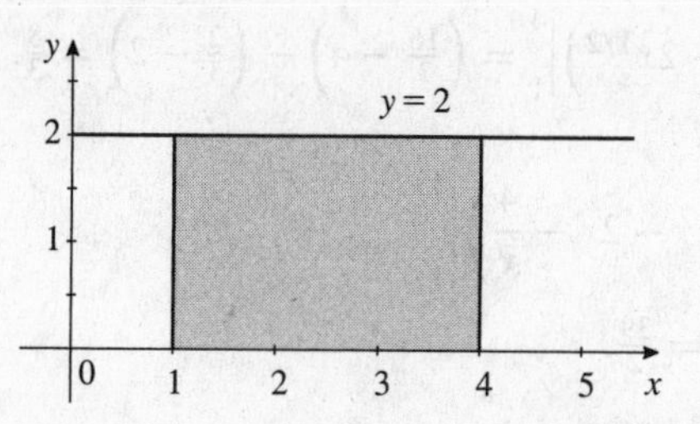

3. $A = \int_1^3 2x\,dx = x^2\Big|_1^3 = 9-1 = 8$. The region is a parallelogram with area $\frac{1}{2}(3-1)(2+6) = 8$.

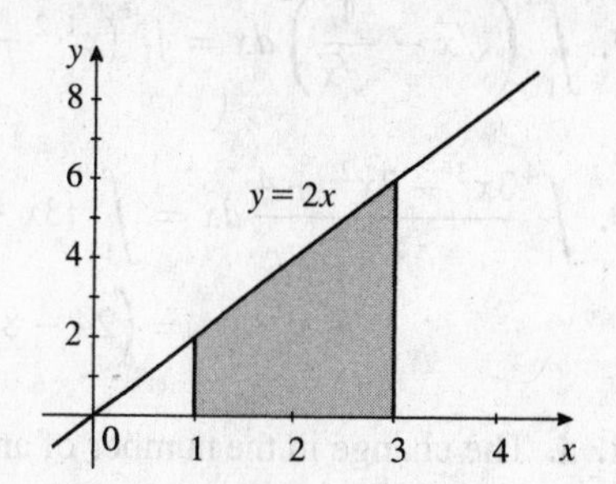

5. $A = \int_{-1}^{2}(2x+3)\,dx = (x^2+3x)\Big|_{-1}^{2} = (4+6)-(1-3) = 12$.

7. $A = \int_{-1}^{2}(-x^2+4)\,dx = \left(-\frac{1}{3}x^3+4x\right)\Big|_{-1}^{2} = \left(-\frac{8}{3}+8\right)-\left(\frac{1}{3}-4\right) = 9$.

9. $A = \displaystyle\int_1^2 \frac{1}{x}\,dx = \ln|x|\,|_1^2 = \ln 2 - \ln 1 = \ln 2$.

11. $A = \int_1^9 \sqrt{x}\,dx = \frac{2}{3}x^{3/2}\Big|_1^9 = \frac{2}{3}(27-1) = \frac{52}{3}$.

13. $A = \int_{-8}^{-1}\left(1-x^{1/3}\right)dx = \left(x-\frac{3}{4}x^{4/3}\right)\Big|_{-8}^{-1} = \left(-1-\frac{3}{4}\right)-(-8-12) = \frac{73}{4}$.

15. $A = \int_0^2 e^x\,dx = e^x|_0^2 = e^2 - 1 \approx 6.39$.

17. $\int_2^4 3\,dx = 3x|_2^4 = 3(4-2) = 6$.

19. $\int_1^4(2x+3)\,dx = (x^2+3x)\Big|_1^4 = (16+12)-(1+3) = 24$.

21. $\int_{-1}^{3} 2x^2\,dx = \frac{2}{3}x^3\Big|_{-1}^{3} = \frac{2}{3}(27) - \frac{2}{3}(-1) = \frac{56}{3}$.

23. $\int_{-2}^{2}(x^2-1)dx = \left(\frac{1}{3}x^3-x\right)\Big|_{-2}^{2} = \left(\frac{8}{3}-2\right)-\left(-\frac{8}{3}+2\right) = \frac{4}{3}$.

25. $\int_1^8 2x^{1/3}\,dx = 2 \cdot \frac{3}{4}x^{4/3}\Big|_1^8 = \frac{3}{2}(16-1) = \frac{45}{2}$.

27. $\int_0^1 (x^3 - 2x^2 + 1)\,dx = \left(\frac{1}{4}x^4 - \frac{2}{3}x^3 + x\right)\Big|_0^1 = \frac{1}{4} - \frac{2}{3} + 1 = \frac{7}{12}$.

29. $\int_1^4 \frac{1}{x}\,dx = \ln|x|\,|_1^4 = \ln 4 - \ln 1 = \ln\frac{4}{1} = \ln 4$.

31. $\int_0^4 x(x^2-1)\,dx = \int_0^4 (x^3 - x)\,dx = \left(\frac{1}{4}x^4 - \frac{1}{2}x^2\right)\Big|_0^4 = 64 - 8 = 56$.

33. $\int_1^3 (t^2 - t)^2\,dt = \int_1^3 (t^4 - 2t^3 + t^2)\,dt = \left(\frac{1}{5}t^5 - \frac{1}{2}t^4 + \frac{1}{3}t^3\right)\Big|_1^3 = \left(\frac{243}{5} - \frac{81}{2} + \frac{27}{3}\right) - \left(\frac{1}{5} - \frac{1}{2} + \frac{1}{3}\right) = \frac{512}{30} = \frac{256}{15}$.

35. $\int_{-3}^{-1} x^{-2}dx = -\frac{1}{x}\Big|_{-3}^{-1} = 1 - \frac{1}{3} = \frac{2}{3}$.

37. $\int_1^4 \left(\sqrt{x} - \frac{1}{\sqrt{x}}\right)dx = \int_1^4 (x^{1/2} - x^{-1/2})\,dx = \left(\frac{2}{3}x^{3/2} - 2x^{1/2}\right)\Big|_1^4 = \left(\frac{16}{3} - 4\right) - \left(\frac{2}{3} - 2\right) = \frac{8}{3}$.

39. $\int_1^4 \frac{3x^3 - 2x^2 + 4}{x^2}\,dx = \int_1^4 (3x - 2 + 4x^{-2})\,dx = \left(\frac{3}{2}x^2 - 2x - \frac{4}{x}\right)\Big|_1^4$
$= (24 - 8 - 1) - \left(\frac{3}{2} - 2 - 40\right) = \frac{39}{2}$.

41. a. The change in the number of annual personal bankruptcy filings between September 30, 2010 and September 2012 was $N(t) = \int_0^2 [-R(t)]\,dt = -\int_0^2 (0.077t + 0.0825)\,dt = \left[-0.0385t^2 + 0.0825t\right]_0^2 \approx -0.319$, a decline of approximately 319,000.

b. The approximate number of personal bankruptcy filings in 2012 was $N(2) = N(0) + \int_0^2 N'(t)\,dt = 1.538 - 0.319 = 1.219$, or 1,219,000.

43. a. $C(300) - C(0) = \int_0^{300} (0.0003x^2 - 0.12x + 20)\,dx = (0.0001x^3 - 0.06x^2 + 20x)\big|_0^{300}$
$= 0.0001(300)^3 - 0.06(300)^2 + 20(300) = 3300$.

Therefore, $C(300) = 3300 + C(0) = 3300 + 800 = \$4100$.

b. $\int_{200}^{300} C'(x)\,dx = (0.0001x^3 - 0.06x^2 + 20x))\big|_{200}^{300}$
$= \left[0.0001(300)^3 - 0.06(300)^2 + 20(300)\right] - \left[0.0001(200)^3 - 0.06(200)^2 + 20(200)\right] = \$900$.

45. a. The profit is
$\int_0^{200} (-0.0003x^2 + 0.02x + 20)\,dx + P(0) = (-0.0001x^3 + 0.01x^2 + 20x)\big|_0^{200} + P(0)$
$= 3600 + P(0) = 3600 - 800$, or \$2800.

b. $\int_{200}^{220} P'(x)\,dx = P(220) - P(200) = (-0.0001x^3 + 0.01x^2 + 20x)\big|_{200}^{220} = 219.20$, or \$219.20.

47. The distance is $\int_0^{20} v(t)\,dt = \int_0^{20} (-t^2 + 20t + 440)\,dt = \left(-\frac{1}{3}t^3 + 10t^2 + 440t\right)\Big|_0^{20} \approx 10{,}133.3$ ft.

49. The average U.S. household credit card debt at the beginning of 2012 was $D(4) = D(0) + \int_0^4 (258t^2 - 680t - 316)\,dt = 8382 + \left[86t^3 - 340t^2 - 316t\right]_0^4 = 7182$, or \$7182.

51. The amount of smoke left after 5 minutes is
$100 - \int_0^5 R(t)\,dt = 100 - \int_0^5 (0.00032t^4 - 0.01872t^3 + 0.3948t^2 - 3.83t + 17.63)\,dt$
$= 100 - (0.000064t^5 - 0.00468t^4 + 0.1316t^3 - 1.915t^2 + 17.63t)\big|_0^5 = 46$, or 46%.
The amount of smoke left after 10 minutes is
$100 - \int_0^{10} R(t)\,dt = 100 - (0.000064t^5 - 0.00468t^4 + 0.1316t^3 - 1.915t^2 + 17.63t)\big|_0^{10} = 24$, or 24%.

53. a. $f(t) = \int R(t)\,dt = \int 0.8256t^{-0.04}\,dt = \frac{0.8256}{0.96}t^{0.96} + C = 0.86t^{0.96} + C$. $f(1) = 0.9$, and so $0.86 + C = 0.9$ and $C = 0.04$. Thus, $f(t) = 0.86t^{0.96} + 0.04$.
b. In 2014, mobile phone ad spending is projected to be $f(8) = 0.86(8)^{0.96} + 0.04 \approx 6.37$, or approximately \$6.37 billion.

55. a. The revenue of the company increased by $\int_0^6 R'(t)\,dt = \int_0^6 0.545043e^{0.291t}\,dt = 1.873e^{0.291t}\big|_0^6 \approx 8.862$, or approximately \$8.86 billion.
b. The revenue of the company in 2012 was $R(0) + \int_0^6 R'(t)\,dt \approx 10.71 + 8.86 \approx 19.57$, or approximately \$19.57 billion.

57. The increase in the senior population over the period in question is $\displaystyle\int_0^3 f(t)\,dt = \int_0^3 \frac{85}{1 + 1.859e^{-0.66t}}\,dt$.
Multiplying the numerator and denominator of the integrand by $e^{0.66t}$ gives $\displaystyle\int_0^3 f(t)\,dt = 85\int_0^3 \frac{e^{0.66t}}{e^{0.66t} + 1.859}\,dt$.
Now let $u = 1.859 + e^{0.66t}$, so $du = 0.66e^{0.66t}\,dt$ and $e^{0.66t}\,dt = \dfrac{du}{0.66}$. If $t = 0$, then $u = 2.859$, and if $t = 3$, then $u = 9.1017$. Substituting, we have
$\displaystyle\int_0^3 f(t)\,dt = 85\int_{2.859}^{9.1017} \frac{du}{0.66u} = \frac{85}{0.66}\ln u\Big|_{2.859}^{9.1017} = \frac{85}{0.66}(\ln 9.1017 - \ln 2.859) = \frac{85}{0.66}\ln\frac{9.1017}{2.859} \approx 149.135$, or approximately 149.14 million people.

59. $f(x) = x^4 - 2x^2 + 2$, so $f'(x) = 4x^3 - 4x = 4x(x^2 - 1) = 4x(x+1)(x-1)$. Setting $f'(x) = 0$ gives $x = -1$, 0, and 1 as critical numbers. Now calculate $f''(x) = 12x^2 - 4 = 4(3x^2 - 1)$ and use the second derivative test: $f''(-1) = 8 > 0$, so $(-1, 1)$ is a relative minimum; $f''(0) = -4 < 0$, so $(0, 2)$ is a relative maximum; and $f''(1) = 8 > 0$, so $(1, 1)$ is a relative minimum. The graph of f is symmetric with respect to the y-axis because $f(-x) = (-x)^4 - 2(-x)^2 + 2 = x^4 - 2x^2 + 2 = f(x)$. Thus, the required area is the area under the graph of f between $x = 0$ and $x = 1$, that is, $A = \int_0^1 (x^4 - 2x^2 + 2)\,dx = \left(\frac{1}{5}x^5 - \frac{2}{3}x^3 + 2x\right)\Big|_0^1 = \frac{1}{5} - \frac{2}{3} + 2 = \frac{23}{15}$.

61. False. The integrand $f(x) = 1/x^3$ is discontinuous at $x = 0$.

63. False. $f(x)$ is not nonnegative on $[0, 2]$.

Using Technology page 456

1. 6.1787 **3.** 0.7873 **5.** −0.5888 **7.** 2.7044

9. 3.9973 **11.** 46%, 24% **13.** 60,156

6.5 Evaluating Definite Integrals

Concept Questions page 463

1. *Approach I:* We first find the indefinite integral. Let $u = x^3 + 1$, so that $du = 3x^2 dx$ and or $x^2 dx = \frac{1}{3} du$. Then $\int x^2 (x^3 + 1)^2 dx = \frac{1}{3} \int u^2 du = \frac{1}{9} u^3 + C = \frac{1}{9} (x^3 + 1)^3 + C$. Therefore, $\int_0^1 x^2 (x^3 + 1)^2 dx = \frac{1}{9} \left(x^3 + 1^3\right)\Big|_0^1 = \frac{1}{9} (8 - 1) = \frac{7}{9}$.

Approach II: Transform the definite integral in x into an integral in u: Let $u = x^3 + 1$, so that $du = 3x^2 dx$ and $x^2 dx = \frac{1}{3} du$. Next, find the limits of integration with respect to u. If $x = 0$, then $u = 0^3 + 1 = 1$ and if $x = 1$, then $u = 1^3 + 1 = 2$. Therefore, $\int_0^1 x^2 (x^3 + 1)^2 dx = \frac{1}{3} \int_1^2 u^2 du = \frac{1}{9} u^3 \Big|_1^2 = \frac{1}{9} (8 - 1) = \frac{7}{9}$.

Exercises page 463

1. Let $u = x^2 - 1$, so $du = 2x\,dx$ and $x\,dx = \frac{1}{2} du$. If $x = 0$, then $u = -1$ and if $x = 2$, then $u = 3$, so $\int_0^2 x (x^2 - 1)^3 dx = \frac{1}{2} \int_{-1}^3 u^3 du = \frac{1}{8} u^4 \Big|_{-1}^3 = \frac{1}{8} (81) - \frac{1}{8} (1) = 10$.

3. Let $u = 5x^2 + 4$, so $du = 10x\,dx$ and $x\,dx = \frac{1}{10} du$. If $x = 0$, then $u = 4$ and if $x = 1$, then $u = 9$, so $\int_0^1 x\sqrt{5x^2 + 4}\,dx = \frac{1}{10} \int_4^9 u^{1/2} du = \frac{1}{15} u^{3/2} \Big|_4^9 = \frac{1}{15} (27) - \frac{1}{15} (8) = \frac{19}{15}$.

5. Let $u = x^3 + 1$, so $du = 3x^2 dx$ and $x^2 dx = \frac{1}{3} du$. If $x = 0$, then $u = 1$ and if $x = 2$, then $u = 9$, so $\int_0^2 x^2 (x^3 + 1)^{3/2} dx = \frac{1}{3} \int_1^9 u^{3/2} du = \frac{2}{15} u^{5/2} \Big|_1^9 = \frac{2}{15} (243) - \frac{2}{15} (1) = \frac{484}{15}$.

7. Let $u = 2x + 1$, so $du = 2\,dx$ and $dx = \frac{1}{2} du$. If $x = 0$, then $u = 1$ and if $x = 1$ then $u = 3$, so

$$\int_0^1 \frac{1}{\sqrt{2x + 1}} dx = \frac{1}{2} \int_1^3 \frac{1}{\sqrt{u}} du = \frac{1}{2} \int_1^3 u^{-1/2} du = u^{1/2} \big|_1^3 = \sqrt{3} - 1.$$

9. Let $u = 2x - 1$, so $du = 2\,dx$ and $dx = \frac{1}{2} du$. If $x = 1$, then $u = 1$ and if $x = 3$, then $u = 5$, so $\int_1^3 (2x - 1)^4 dx = \frac{1}{2} \int_1^5 u^4 du = \frac{1}{10} u^5 \Big|_1^5 = \frac{1}{10} (3125 - 1) = \frac{1562}{5}$.

11. Let $u = x^3 + 1$, so $du = 3x^2 dx$ and $x^2 dx = \frac{1}{3} du$. If $x = -1$, then $u = 0$ and if $x = 1$, then $u = 2$, so $\int_{-1}^1 x^2 (x^3 + 1)^4 dx = \frac{1}{3} \int_0^2 u^4 du = \frac{1}{15} u^5 \Big|_0^2 = \frac{32}{15}$.

13. Let $u = x - 1$, so $du = dx$. If $x = 1$, then $u = 0$ and if $x = 5$, then $u = 4$, so $\int_1^5 x\sqrt{x - 1}\,dx = \int_0^4 (u + 1) u^{1/2} du = \int_0^4 (u^{3/2} + u^{1/2}) du = \left(\frac{2}{5} u^{5/2} + \frac{2}{3} u^{3/2}\right)\Big|_0^4 = \frac{2}{5} (32) + \frac{2}{3} (8) = \frac{272}{15}$.

15. Let $u = x^2$, so $du = 2x\,dx$ and $x\,dx = \frac{1}{2} du$. If $x = 0$, then $u = 0$ and if $x = 2$, then $u = 4$, so $\int_0^2 2xe^{x^2} dx = \int_0^4 e^u du = e^u \big|_0^4 = e^4 - 1$.

17. $\int_0^1 (e^{2x} + x^2 + 1) dx = \left(\frac{1}{2} e^{2x} + \frac{1}{3} x^3 + x\right)\Big|_0^1 = \left(\frac{1}{2} e^2 + \frac{1}{3} + 1\right) - \frac{1}{2} = \frac{1}{2} e^2 + \frac{5}{6}$.

19. Put $u = x^2 + 1$, so $du = 2x\,dx$ and $x\,dx = \frac{1}{2}\,du$. Then $\int_{-1}^{1} xe^{x^2+1}\,dx = \frac{1}{2}\int_2^2 e^u\,du = \frac{1}{2}e^u\Big|_2^2 = 0$ because the upper and lower limits are equal.

21. Let $u = x - 2$, so $du = dx$. If $x = 3$, then $u = 1$ and if $x = 6$, then $u = 4$, so
$$\int_3^6 \frac{1}{x-2}\,dx = \int_1^4 \frac{du}{u} = \ln|u|\big|_1^4 = \ln 4.$$

23. Let $u = x^3 + 3x^2 - 1$, so $du = (3x^2 + 6x)\,dx = 3\left(x^2 + 2x\right)dx$. If $x = 1$, then $u = 3$, and if $x = 2$, then $u = 19$, so $\displaystyle\int_1^2 \frac{x^2 + 2x}{x^3 + 3x^2 - 1}\,dx = \frac{1}{3}\int_3^{19} \frac{du}{u} = \frac{1}{3}\ln u\Big|_3^{19} = \frac{1}{3}\left(\ln 19 - \ln 3\right).$

25. $\displaystyle\int_1^2 \left(4e^{2u} - \frac{1}{u}\right)du = 2e^{2u} - \ln u\big|_1^2 = \left(2e^4 - \ln 2\right) - \left(2e^2 - 0\right) = 2e^4 - 2e^2 - \ln 2.$

27. $\int_1^2 \left(2e^{-4x} - x^{-2}\right)dx = \left(-\frac{1}{2}e^{-4x} + \frac{1}{x}\right)\Big|_1^2 = \left(-\frac{1}{2}e^{-8} + \frac{1}{2}\right) - \left(-\frac{1}{2}e^{-4} + 1\right) = -\frac{1}{2}e^{-8} + \frac{1}{2}e^{-4} - \frac{1}{2}$
$= \frac{1}{2}\left(e^{-4} - e^{-8} - 1\right).$

29. $A = \int_{-1}^{2}\left(x^2 - 2x + 2\right)dx = \left(\frac{1}{3}x^3 - x^2 + 2x\right)\Big|_{-1}^{2} = \left(\frac{8}{3} - 4 + 4\right) - \left(-\frac{1}{3} - 1 - 2\right) = 6.$

31. $\displaystyle A = \int_1^2 \frac{dx}{x^2} = \int_1^2 x^{-2}\,dx = -\frac{1}{x}\Big|_1^2 = \frac{1}{4} - (-1) = \frac{1}{2}.$

33. $A = \int_{-1}^{2} e^{-x/2}\,dx = -2e^{-x/2}\big|_{-1}^{2} = -2\left(e^{-1} - e^{1/2}\right) = 2\left(\sqrt{e} - 1/e\right).$

35. The average value is $\frac{1}{2}\int_0^2 (2x + 3)\,dx = \frac{1}{2}\left(x^2 + 3x\right)\Big|_0^2 = \frac{1}{2}(10) = 5.$

37. The average value is $\frac{1}{2}\int_1^3 \left(2x^2 - 3\right)dx = \frac{1}{2}\left(\frac{2}{3}x^3 - 3x\right)\Big|_1^3 = \frac{1}{2}\left(9 + \frac{7}{3}\right) = \frac{17}{3}.$

39. The average value is
$\frac{1}{3}\int_{-1}^{2}\left(x^2 + 2x - 3\right)dx = \frac{1}{3}\left(\frac{1}{3}x^3 + x^2 - 3x\right)\Big|_{-1}^{2} = \frac{1}{3}\left[\left(\frac{8}{3} + 4 - 6\right) - \left(-\frac{1}{3} + 1 + 3\right)\right]$
$= \frac{1}{3}\left(\frac{8}{3} - 2 + \frac{1}{3} - 4\right) = -1.$

41. The average value is $\frac{1}{4}\int_0^4 (2x + 1)^{1/2}\,dx = \left(\frac{1}{4}\right)\left(\frac{1}{2}\right)\left(\frac{2}{3}\right)(2x + 1)^{3/2}\Big|_0^4 = \frac{1}{12}(27 - 1) = \frac{13}{6}.$

43. The average value is $\frac{1}{2}\int_0^2 xe^{x^2}\,dx = \frac{1}{4}e^{x^2}\Big|_0^2 = \frac{1}{4}\left(e^4 - 1\right).$

45. The distance traveled is $\int_0^4 3t\sqrt{16 - t^2}\,dt = 3\left(-\frac{1}{2}\right)\left(\frac{2}{3}\right)\left(16 - t^2\right)^{3/2}\Big|_0^4 = 64$ ft.

47. The amount is $\int_1^2 t\left(\frac{1}{2}t^2 + 1\right)^{1/2} dt$. Let $u = \frac{1}{2}t^2 + 1$, so $du = t\,dt$. Then
$\int_1^2 t\left(\frac{1}{2}t^2 + 1\right)^{1/2} dt = \int_{3/2}^{3} u^{1/2}\,du = \frac{2}{3}u^{3/2}\Big|_{3/2}^{3} = \frac{2}{3}\left[(3)^{3/2} - \left(\frac{3}{2}\right)^{3/2}\right] \approx \$2.24$ million.

49. Using the substitution $u = 0.05t$, we find that the amount produced was $\int_0^{20} 3.5e^{0.05t}\,dt = \frac{3.5}{0.05}e^u\Big|_0^{20} = 70\,(e-1) \approx 120.3$, or 120.3 billion metric tons.

51. The average spending per year between 2005 and 2011 is

$$A = \frac{1}{7-1}\int_1^7 0.86t^{0.96}\,dt = \frac{0.86}{6}\cdot\frac{1}{1.96}t^{1.96}\Big|_1^7 = \frac{0.86}{6\,(1.96)}\left(7^{1.96}-1\right) \approx 3.24, \text{ or \$3.24 billion per year.}$$

53. a. The gasoline consumption in 2017 is given by $A\,(10) = 0.014\,(10^2) + 1.93\,(10) + 140 = 160.7$, or 160.7 billion gallons per year.

b. The average consumption per year between 2007 and 2017 is given by

$$A = \frac{1}{10-0}\int_0^{10}\left(0.014t^2 + 1.93t + 140\right)dt = \frac{1}{10}\left(\frac{0.014}{3}t^3 + \frac{1.93}{2}t^2 + 140t\right)\Big|_0^{10} \approx 150.12, \text{ or 150.1 billion gallons}$$

per year per year.

55. $I = \int_0^{0.02}\left[-9{,}400{,}000\,(t^2 - 0.02t)\right]dt = -9{,}400{,}000\left[\frac{1}{3}t^3 - 0.01t^2\right]_0^{0.02} \approx 12.5333$, or approximately 12.53 newton-seconds. The average force acting on the baseball is $F = \frac{1}{0.02-0}\int_0^{0.02} F\,(t)\,dt \approx \frac{12.53}{0.02} = 626.5$, or 626.5 N.

57. The average velocity of the blood is

$$\frac{1}{R}\int_0^R k\left(R^2 - r^2\right)dr = \frac{k}{R}\int_0^R\left(R^2 - r^2\right)dr = \frac{k}{r}\left(R^2r - \frac{1}{3}r^3\right)\Big|_0^4 = \frac{k}{R}\left(R^3 - \frac{1}{3}R^3\right) = \frac{k}{R}\cdot\frac{2}{3}R^3 = \frac{2}{3}kR^2 \text{ cm/sec.}$$

59. The average concentration of the drug is

$$\frac{1}{4}\int_0^4 \frac{0.2t}{t^2+1}\,dt = \frac{0.2}{4}\int_0^4 \frac{t}{t^2+1}\,dt = \frac{0.2}{(4)\,(2)}\ln\left(t^2+1\right)\Big|_0^4 = 0.025\ln 17 \approx 0.071,$$

or 0.071 milligrams per cm^3.

61. The average yearly sales of the company over its first 5 years of operation is given by

$$\frac{1}{5-0}\int_0^5 t\left(0.2t^2+4\right)^{1/2}dt = \frac{1}{5}\left[\left(\frac{5}{2}\right)\left(\frac{2}{3}\right)\left(0.2t^2+4\right)^{3/2}\right]_0^5 \quad (\text{let } u = -0.2t^2+4)$$

$$= \frac{1}{5}\left[\frac{5}{3}\,(5+4)^{3/2} - \frac{5}{3}\,(4)^{3/2}\right] = \frac{1}{3}\,(27-8) = \frac{19}{3}, \text{ or about \$6.33 million.}$$

63. The average increase in the number of credit cards issued in China between 2009 and 2012 was

$$\frac{1}{4-1}\int_1^4 153e^{0.21t}\,dt = \frac{1}{3}\left[\frac{153}{0.21}e^{0.21t}\right]_1^4 \approx 262.9, \text{ or approximately 262.9 million cards per year.}$$

65. a. The projected number of passengers in year t is $N\,(t) = \int R\,(t)\,dt = \int 6.69e^{0.0456t}\,dt \approx 146.711e^{0.0456t} + C$. To find C, we use the condition $N\,(2) = 160$; giving $C \approx -0.720$. Therefore, $N\,(t) \approx 146.711e^{0.0456t} - 0.72$.

b. The number of passengers in 2030 is projected to be $N\,(20) \approx 146.711e^{0.0456(20)} - 0.72 \approx 364.49$, or approximately 364.5 million.

c. The average growth rate from 2012 to 2030 is approximately $\dfrac{1}{20-2}\displaystyle\int_2^{20} R\,(t)\,dt \approx \dfrac{364.49-160}{18} \approx 11.36$, or approximately 11.4 million per year.

67. $\frac{1}{h}\int_0^h (2gx)^{1/2}\,dx = \frac{1}{3h}\,(2gx)^{3/2}\Big|_0^h = \frac{2}{3}\sqrt{2gh}$; that is, $\frac{2}{3}\sqrt{2gh}$ ft/sec.

69. $\int_a^a f(x)\,dx = F(x)|_a^a = F(a) - F(a) = 0$, where $F'(x) = f(x)$.

71. $\int_1^3 x^2\,dx = \frac{1}{3}x^3\Big|_1^3 = 9 - \frac{1}{3} = \frac{26}{3} = -\int_3^1 x^2\,dx = -\frac{1}{3}x^3\Big|_3^1 = -\frac{1}{3} + 9 = \frac{26}{3}$.

73. $\int_1^9 2\sqrt{x}\,dx = \frac{4}{3}x^{3/2}\Big|_1^9 = \frac{4}{3}(27-1) = \frac{104}{3}$ and $2\int_1^9 \sqrt{x}\,dx = 2\left(\frac{2}{3}x^{3/2}\right)\Big|_1^9 = \frac{104}{3}$.

75. $\int_0^3 (1+x^3)\,dx = x + \frac{1}{4}x^4\Big|_0^3 = 3 + \frac{81}{4} = \frac{93}{4}$ and

$\int_0^1 (1+x^3)\,dx + \int_1^3 (1+x^3)\,dx = \left(x + \frac{1}{4}x^4\right)\Big|_0^1 + \left(x + \frac{1}{4}x^4\right)\Big|_1^3 = \left(1 + \frac{1}{4}\right) + \left(3 + \frac{81}{4}\right) - \left(1 + \frac{1}{4}\right) = \frac{93}{4}$,

demonstrating Property 5.

77. $\int_3^3 (1+\sqrt{x})\,e^{-x}\,dx = 0$ by Property 1 of the definite integral.

79. a. $\int_{-1}^2 [2f(x) + g(x)]\,dx = 2\int_{-1}^2 f(x)\,dx + \int_{-1}^2 g(x)\,dx = 2(-2) + 3 = -1$.

b. $\int_{-1}^2 [g(x) - f(x)]\,dx = \int_{-1}^2 g(x)\,dx - \int_{-1}^2 f(x)\,dx = 3 - (-2) = 5$.

c. $\int_{-1}^2 [2f(x) - 3g(x)]\,dx = 2\int_{-1}^2 f(x)\,dx - 3\int_{-1}^2 g(x)\,dx = 2(-2) - 3(3) = -13$.

81. True. This follows from Property 1 of the definite integral.

83. False. Only a constant can be "moved out" of the integral.

85. True. This follows from Properties 3 and 4 of the definite integral.

Using Technology page 468

1. 7.716667 **3.** 17.564865 **5.** 159/bean stem **7.** 0.48 g/cm^3/day

6.6 Area between Two Curves

Problem-Solving Tips

Note that the formula for the area between the graphs of two continuous functions f and g, where $f(x) \geq g(x)$ on $[a, b]$, is given by $\int_a^b [f(x) - g(x)]\,dx$. The condition $f(x) \geq g(x)$ on $[a, b]$ tells us that we cannot interchange $f(x)$ and $g(x)$ in this formula, as that would yield a negative answer, and area cannot be negative.

Concept Questions page 475

1. $\int_a^b [f(x) - g(x)]\,dx$

Exercises page 475

1. $-\int_0^6 (x^3 - 6x^2)\,dx = \left(-\frac{1}{4}x^4 + 2x^3\right)\Big|_0^6 = -\frac{1}{4}(6)^4 + 2(6)^3 = 108$.

3. $A = -\int_{-1}^0 x\sqrt{1-x^2}\,dx + \int_0^1 x\sqrt{1-x^2}dx = 2\int_0^1 x\left(1-x^2\right)^{1/2} dx$ by symmetry. Let $u = 1 - x^2$, so $du = -2x\,dx$ and $x\,dx = -\frac{1}{2}\,du$. If $x = 0$, then $u = 1$ and if $x = 1$, then $u = 0$, so

$A = (2)\left(-\frac{1}{2}\right)\int_0^1 u^{1/2}\,du = -\frac{2}{3}u^{3/2}\Big|_1^0 = \frac{2}{3}$.

5. $A = -\int_0^4 \left(x - 2\sqrt{x}\right) dx = \int_0^4 \left(-x + 2x^{1/2}\right) dx = \left(-\frac{1}{2}x^2 + \frac{4}{3}x^{3/2}\right)\Big|_0^4 = 8 + \frac{32}{3} = \frac{8}{3}.$

7. The required area is given by

$\int_{-1}^0 \left(x^2 - x^{1/3}\right) dx + \int_0^1 \left(x^{1/3} - x^2\right) dx = \left(\frac{1}{3}x^3 - \frac{3}{4}x^{4/3}\right)\Big|_{-1}^0 + \left(\frac{3}{4}x^{4/3} - \frac{1}{3}x^3\right)\Big|_0^1 = -\left(-\frac{1}{3} - \frac{3}{4}\right) + \left(\frac{3}{4} - \frac{1}{3}\right) = \frac{3}{2}.$

9. The required area is given by $-\int_{-1}^2 -x^2 dx = \frac{1}{3}x^3\Big|_{-1}^2 = \frac{8}{3} + \frac{1}{3} = 3.$

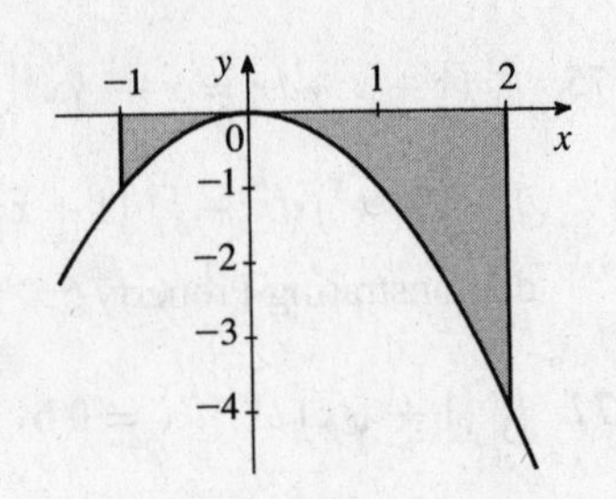

11. $y = x^2 - 5x + 4 = (x - 4)(x - 1) = 0$ if $x = 1$ or 4, the x-intercepts of the graph of f. Thus,

$$A = -\int_1^3 \left(x^2 - 5x + 4\right) dx = \left(-\frac{1}{3}x^3 + \frac{5}{2}x^2 - 4x\right)\Big|_1^3$$
$$= \left(-9 + \frac{45}{2} - 12\right) - \left(-\frac{1}{3} + \frac{5}{2} - 4\right) = \frac{10}{3}.$$

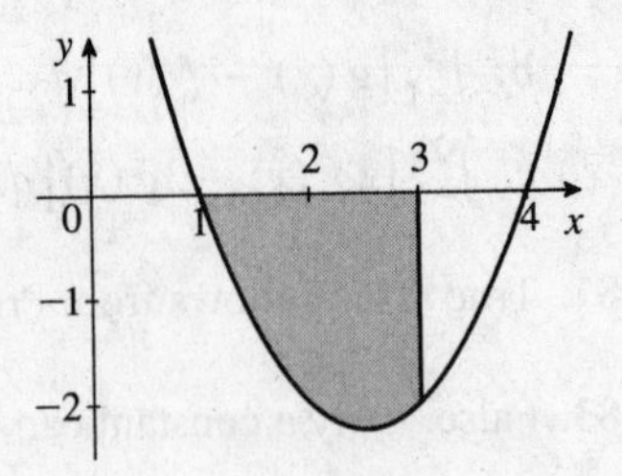

13. The required area is given by

$$-\int_0^9 -\left(1 + \sqrt{x}\right) dx = \left(x + \frac{2}{3}x^{3/2}\right)\Big|_0^9 = 9 + 18 = 27.$$

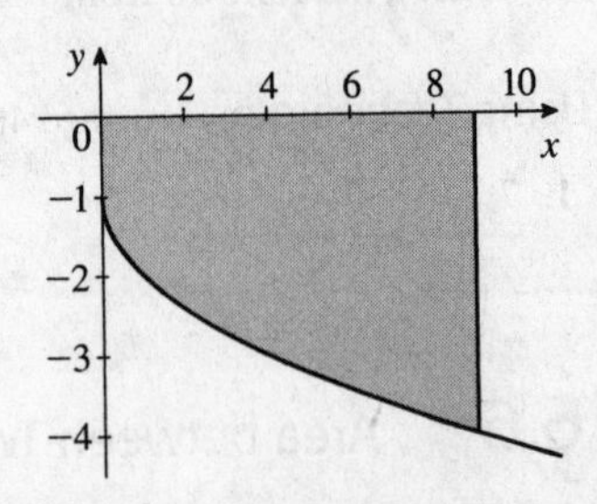

15. $-\int_{-2}^4 \left(-e^{x/2}\right) dx = 2e^{x/2}\Big|_{-2}^4 = 2\left(e^2 - e^{-1}\right).$

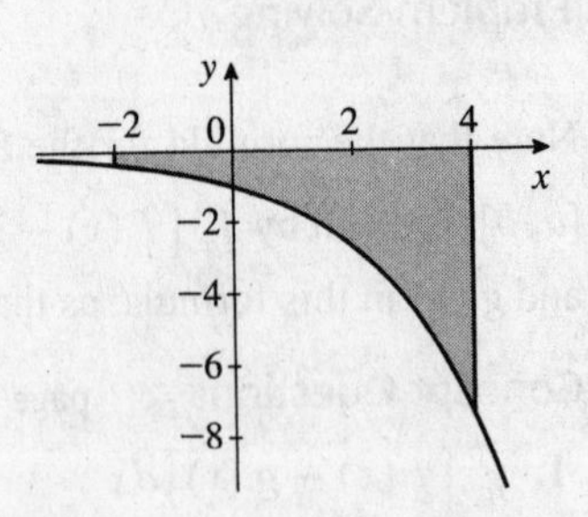

17. $A = \int_1^3 \left[\left(x^2 + 3\right) - 1\right] dx = \int_1^3 \left(x^2 + 2\right) dx = \left(\frac{1}{3}x^3 + 2x\right)\Big|_1^3$

$= (9 + 6) - \left(\frac{1}{3} + 2\right) = \frac{38}{3}.$

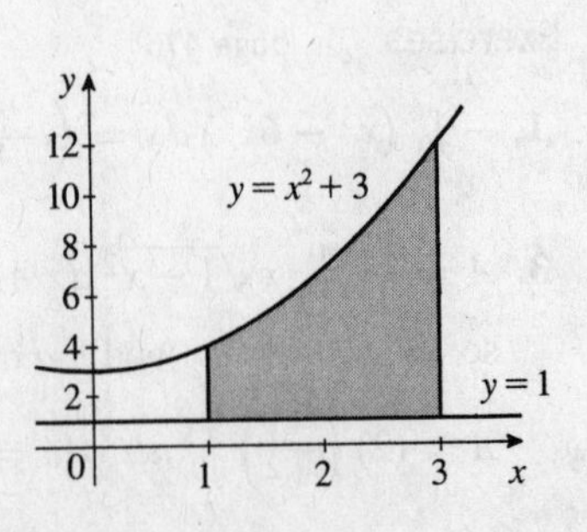

19. $A = \int_0^2 \left(-x^2 + 2x + 3 + x - 3\right) dx = \int_0^2 \left(-x^2 + 3x\right) dx$
$= \left(-\frac{1}{3}x^3 + \frac{3}{2}x^2\right)\Big|_0^2 = -\frac{1}{3}(8) + \frac{3}{2}(4) = 6 - \frac{8}{3} = \frac{10}{3}.$

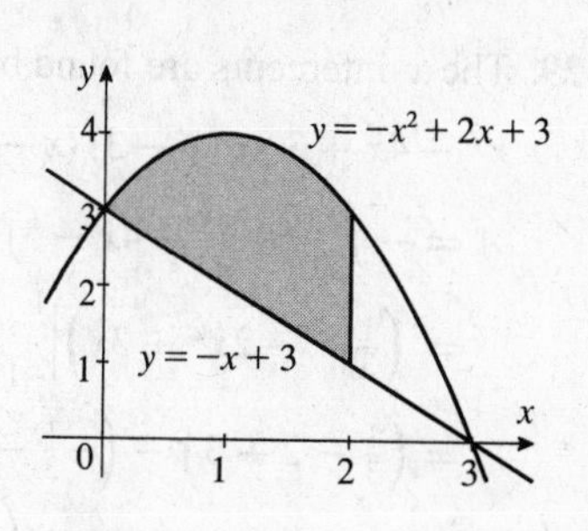

21. $A = \int_{-1}^2 \left[\left(x^2 + 1\right) - \frac{1}{3}x^3\right] dx = \int_{-1}^2 \left(-\frac{1}{3}x^3 + x^2 + 1\right) dx$
$= \left(-\frac{1}{12}x^4 + \frac{1}{3}x^3 + x\right)\Big|_{-1}^2$
$= \left(-\frac{4}{3} + \frac{8}{3} + 2\right) - \left(-\frac{1}{12} - \frac{1}{3} - 1\right) = \frac{19}{4}.$

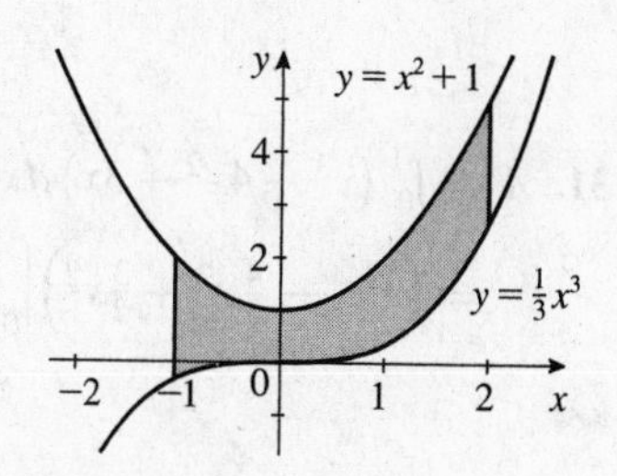

23. $A = \int_1^4 \left[(2x - 1) - \frac{1}{x}\right] dx = \int_1^4 \left(2x - 1 - \frac{1}{x}\right) dx$
$= \left(x^2 - x - \ln x\right)\Big|_1^4 = (16 - 4 - \ln 4) - (1 - 1 - \ln 1)$
$= 12 - \ln 4 \approx 10.6.$

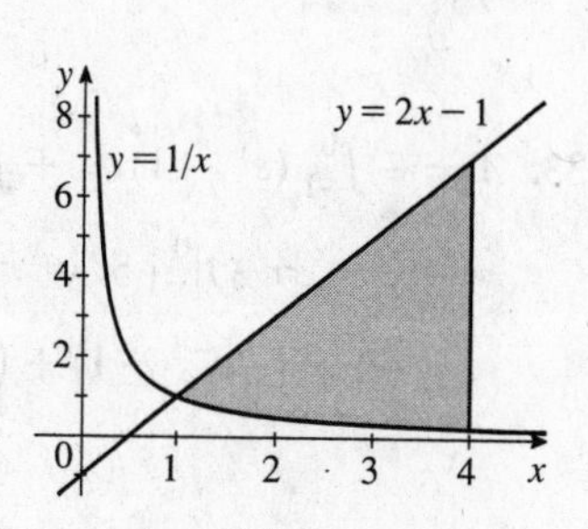

25. $A = \int_1^2 \left(e^x - \frac{1}{x}\right) dx = \left(e^x - \ln x\right)\Big|_1^2 = \left(e^2 - \ln 2\right) - e = e^2 - e - \ln 2.$

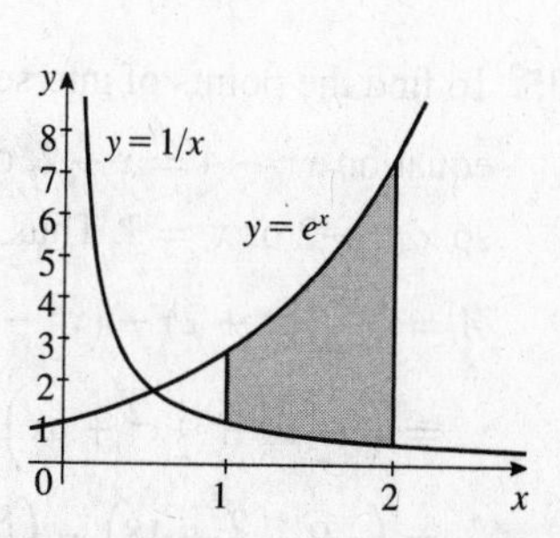

27. $A = -\int_{-1}^0 x\,dx + \int_0^2 x\,dx = -\frac{1}{2}x^2\Big|_{-1}^0 + \frac{1}{2}x^2\Big|_0^2 = \frac{1}{2} + 2 = \frac{5}{2}.$

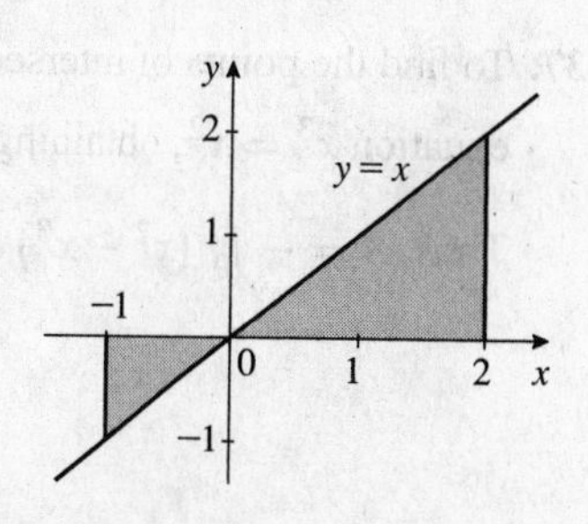

29. The x-intercepts are found by solving $x^2 - 4x + 3 = (x - 3)(x - 1) = 0$, giving $x = 1$ or 3. Thus,

$$A = -\int_{-1}^{1} (-x^2 + 4x - 3)\,dx + \int_{1}^{2} (-x^2 + 4x - 3)\,dx$$
$$= \left(\tfrac{1}{3}x^3 - 2x^2 + 3x\right)\Big|_{-1}^{1} + \left(-\tfrac{1}{3}x^3 + 2x^2 - 3x\right)\Big|_{1}^{2}$$
$$= \left(\tfrac{1}{3} - 2 + 3\right) - \left(-\tfrac{1}{3} - 2 - 3\right) + \left(-\tfrac{8}{3} + 8 - 6\right) - \left(-\tfrac{1}{3} + 2 - 3\right)$$
$$= \tfrac{22}{3}.$$

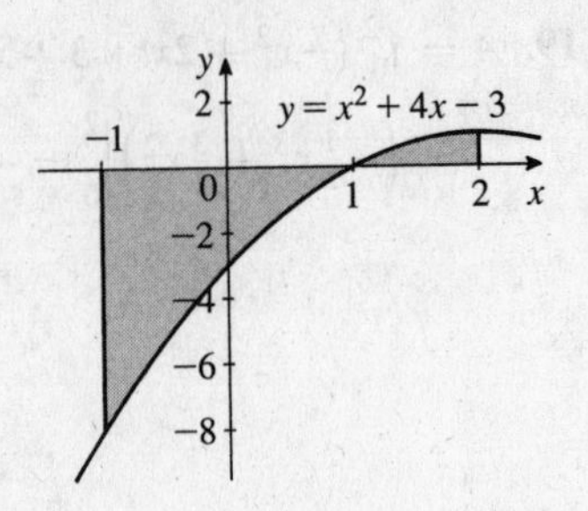

31. $A = \int_{0}^{1} (x^3 - 4x^2 + 3x)\,dx - \int_{1}^{2} (x^3 - 4x^2 + 3x)\,dx$
$$= \left(\tfrac{1}{4}x^4 - \tfrac{4}{3}x^3 + \tfrac{3}{2}x^2\right)\Big|_{0}^{1} - \left(\tfrac{1}{4}x^4 - \tfrac{4}{3}x^3 + \tfrac{3}{2}x^2\right)\Big|_{1}^{2} = \tfrac{3}{2}.$$

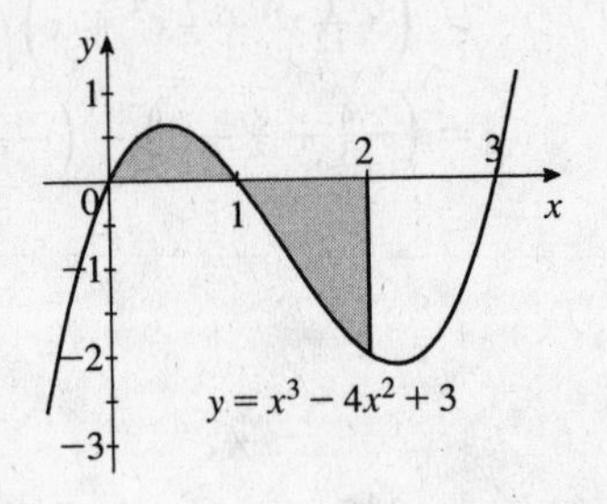

33. $A = -\int_{-1}^{0} (e^x - 1)\,dx + \int_{0}^{3} (e^x - 1)dx$
$$= (-e^x + x)\big|_{-1}^{0} + (e^x - x)\big|_{0}^{3}$$
$$= -1 - (-e^{-1} - 1) + (e^3 - 3) - 1 = e^3 - 4 + \tfrac{1}{e} \approx 16.5.$$

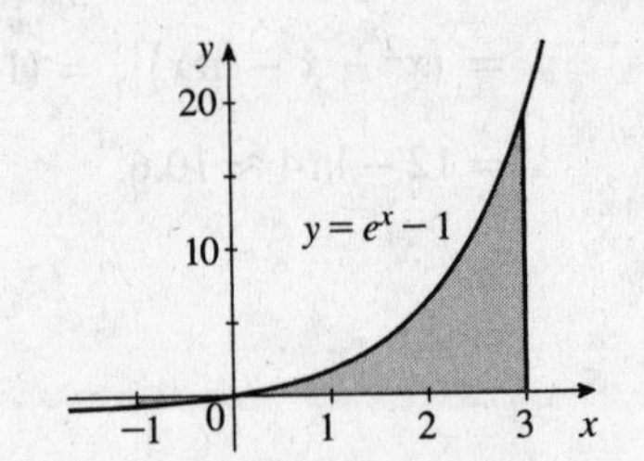

35. To find the points of intersection of the two curves, we solve the equation $x^2 - 4 = x + 2$, obtaining $x^2 - x - 6 = (x - 3)(x + 2) = 0$, so $x = -2$ or $x = 3$. Thus,

$$A = \int_{-2}^{3} [(x + 2) - (x^2 - 4)]\,dx = \int_{-2}^{3} (-x^2 + x + 6)\,dx$$
$$= \left(-\tfrac{1}{3}x^3 + \tfrac{1}{2}x^2 + 6x\right)\Big|_{-2}^{3}$$
$$= \left(-9 + \tfrac{9}{2} + 18\right) - \left(\tfrac{8}{3} + 2 - 12\right) = \tfrac{125}{6}.$$

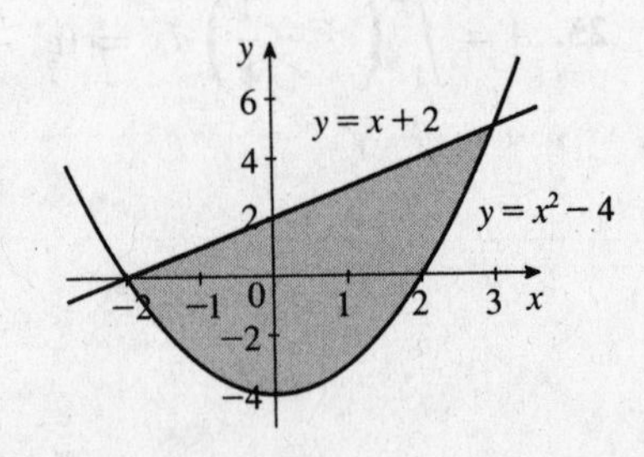

37. To find the points of intersection of the two curves, we solve the equation $x^3 = x^2$, obtaining $x^3 - x^2 = x^2(x - 1) = 0$, so $x = 0$ or 1.

Thus, $A = -\int_{0}^{1} (x^2 - x^3)\,dx = \left(\tfrac{1}{3}x^3 - \tfrac{1}{4}x^4\right)\Big|_{0}^{1} = \tfrac{1}{3} - \tfrac{1}{4} = \tfrac{1}{12}$.

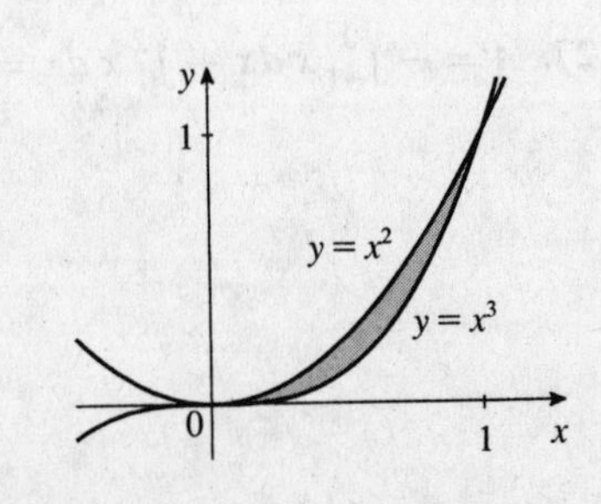

39. To find the points of intersection of the two curves, we solve the equation $x^3 - 6x^2 + 9x = x^2 - 3x$, obtaining $x^3 - 7x^2 + 12x = x(x-4)(x-3) = 0$, so $x = 0, 3$, or 4. Thus,

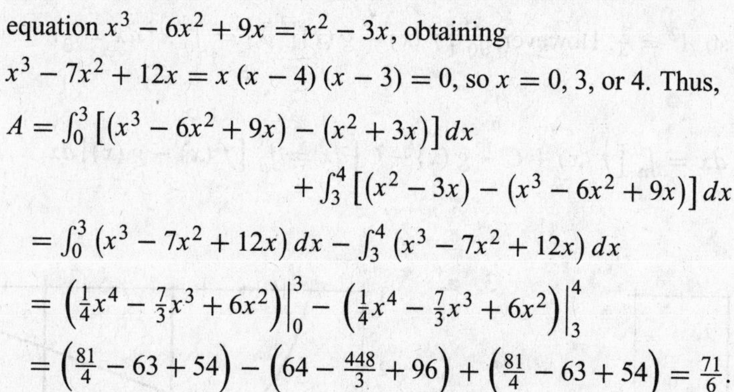

$$A = \int_0^3 \left[\left(x^3 - 6x^2 + 9x\right) - \left(x^2 + 3x\right)\right] dx + \int_3^4 \left[\left(x^2 - 3x\right) - \left(x^3 - 6x^2 + 9x\right)\right] dx$$
$$= \int_0^3 \left(x^3 - 7x^2 + 12x\right) dx - \int_3^4 \left(x^3 - 7x^2 + 12x\right) dx$$
$$= \left(\tfrac{1}{4}x^4 - \tfrac{7}{3}x^3 + 6x^2\right)\Big|_0^3 - \left(\tfrac{1}{4}x^4 - \tfrac{7}{3}x^3 + 6x^2\right)\Big|_3^4$$
$$= \left(\tfrac{81}{4} - 63 + 54\right) - \left(64 - \tfrac{448}{3} + 96\right) + \left(\tfrac{81}{4} - 63 + 54\right) = \tfrac{71}{6}.$$

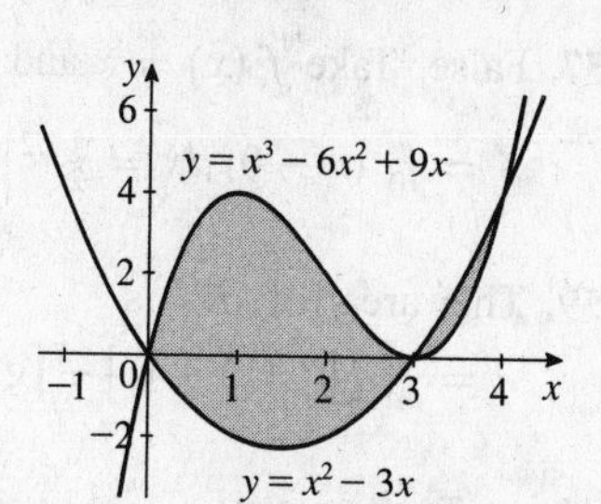

41. By symmetry, $A = 2\int_0^3 x\left(9 - x^2\right)^{1/2} dx$. We integrate using the substitution $u = 9 - x^2$, so $du = -2x\,dx$. If $x = 0$, then $u = 9$ and if $x = 3$, then $u = 0$, so

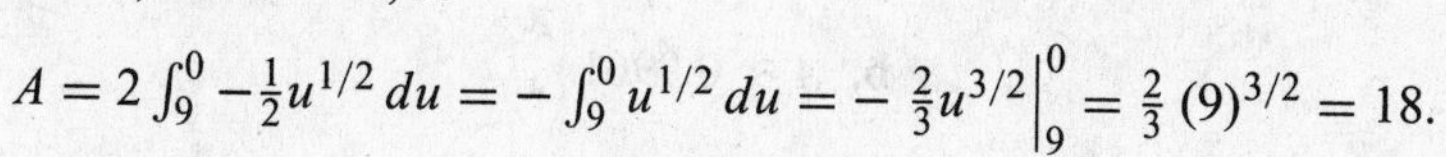

$$A = 2\int_9^0 -\tfrac{1}{2}u^{1/2}\,du = -\int_9^0 u^{1/2}\,du = -\tfrac{2}{3}u^{3/2}\Big|_9^0 = \tfrac{2}{3}(9)^{3/2} = 18.$$

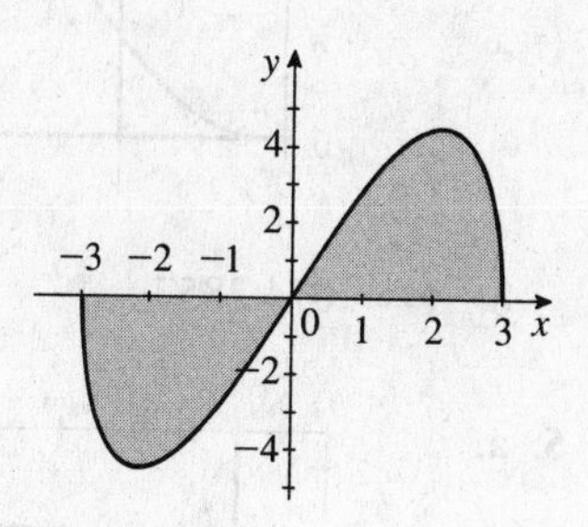

43. $S = \int_0^b \left[g(x) - f(x)\right] dx$ gives the additional revenue that the company would realize if it used a different advertising agency.

45. The shortfall is $\int_{2010}^{2050} \left[f(t) - g(t)\right] dt$.

47. a. $\int_{T_1}^{T} \left[g(t) - f(t)\right] dt - \int_0^{T_1} \left[f(t) - g(t)\right] dt = A_2 - A_1$.

b. The number $A_2 - A_1$ gives the distance car 2 is ahead of car 1 after T seconds.

49. Mexican oil profits from hedging in 2009 are given by
$P = 70 \cdot 8 - \int_0^8 f(t)\,dt - \left[\int_8^{12} f(t)\,dt - 70 \cdot 4\right] = 840 - \int_0^{12} f(t)\,dt$ (dollars).

51. The additional amount of coal that will be produced is

$$\int_0^{20} \left(3.5e^{0.05t} - 3.5e^{0.01t}\right) dt = 3.5\int_0^{20} \left(e^{0.05t} - e^{0.01t}\right) dt = 3.5\left(20e^{0.05t} - 100e^{0.01t}\right)\Big|_0^{20}$$
$$= 3.5\left[\left(20e - 100e^{0.2}\right) - (20 - 100)\right] = 42.8 \text{ billion metric tons.}$$

53. If the campaign is mounted, there will be
$\int_0^5 \left(60e^{0.02t} + t^2 - 60\right) dt = \left(3000e^{0.02t} + \tfrac{1}{3}t^3 - 60t\right)\Big|_0^5 = 3315.5 + \tfrac{125}{3} - 300 - 3000 \approx 57.179$, or 57,179 fewer people.

55. True. If $f(x) \geq g(x)$ on $[a, b]$, then the area of the region is $\int_a^b \left[f(x) - g(x)\right] dx = \int_a^b \left|f(x) - g(x)\right| dx$. If $f(x) \leq g(x)$ on $[a, b]$, then the area of the region is $\int_a^b \left[g(x) - f(x)\right] dx = \int_a^b \left\{-\left[f(x) - g(x)\right]\right\} dx = \int_a^b \left|f(x) - g(x)\right| dx$.

57. False. Take $f(x) = x$ and $g(x) = 0$ on $[0, 1]$. Then the area bounded by the graphs of f and g on $[0, 1]$ is $A = \int_0^1 (x - 0)\,dx = \frac{1}{2}x^2\Big|_0^1 = \frac{1}{2}$ and so $A^2 = \frac{1}{4}$. However, $\int_0^1 [f(x) - g(x)]^2\,dx = \int_0^1 x^2\,dx = \frac{1}{3}$.

59. The area of R' is
$A = \int_a^b \{[f(x) + C] - [g(x) + C]\}\,dx = \int_a^b [f(x) + C - g(x) - C]\,dx = \int_a^b [f(x) - g(x)]\,dx.$

Using Technology page 480

1. a.

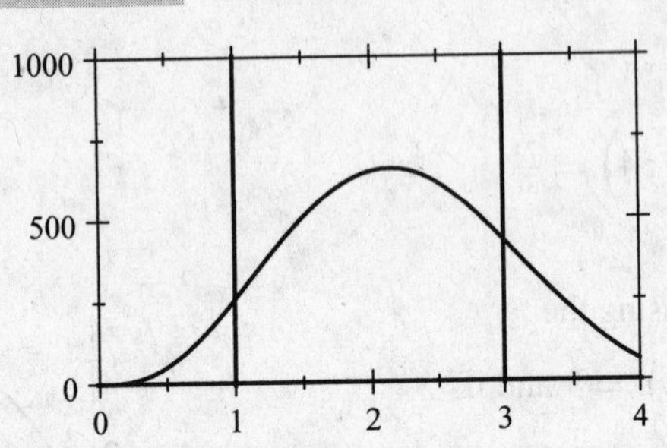

b. $A \approx 1074.2857$.

3. a.

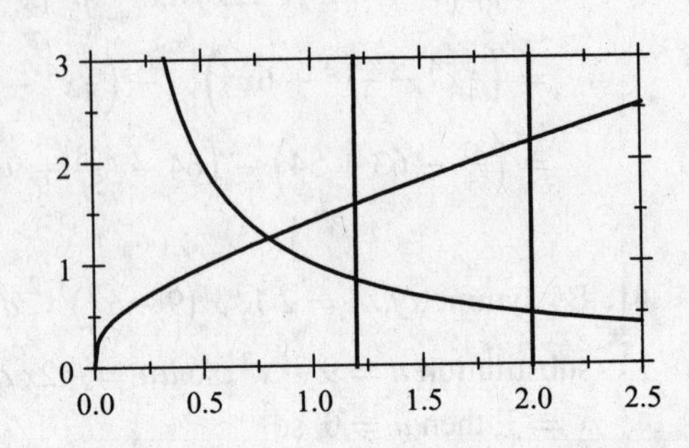

b. $A \approx 0.9961$.

5. a.

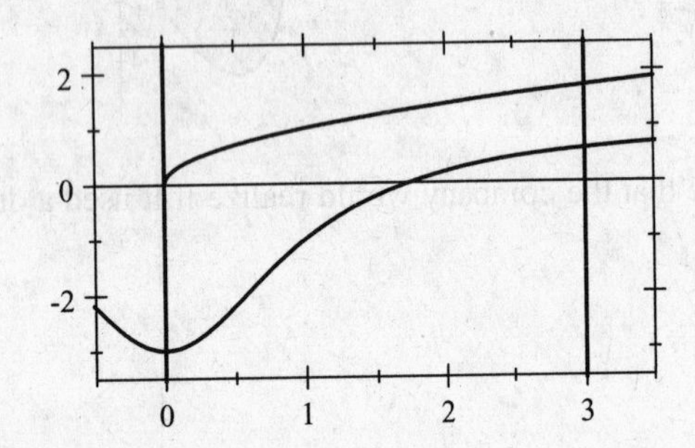

b. $A \approx 5.4603$.

7. a.

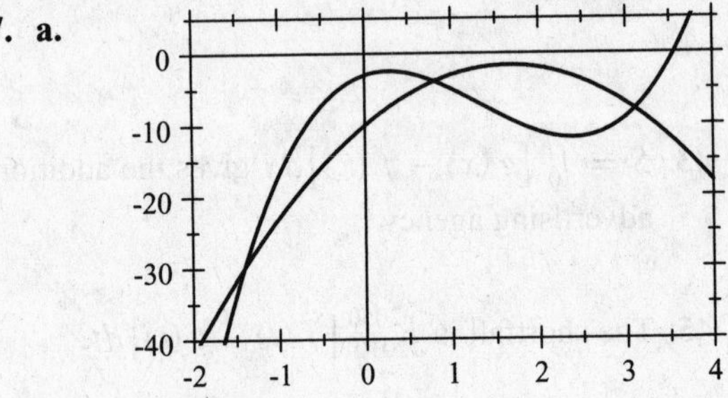

b. $A \approx 25.8549$.

9. a.

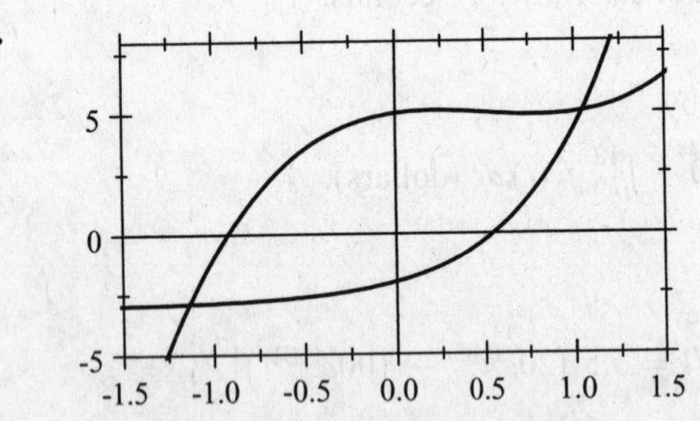

b. $A \approx 10.5144$.

11. a.

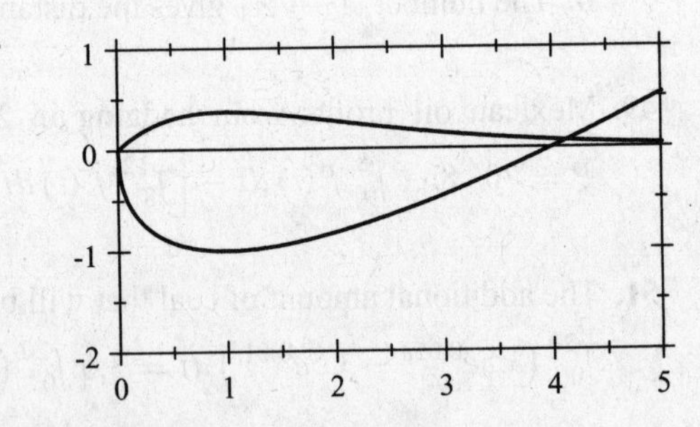

b. $A \approx 3.5799$.

13. The area of the larger region is 207.43.

6.7 Applications of the Definite Integral to Business and Economics

Concept Questions page 490

1. a. See the definition on page 482 of the text. **b.** See the definition on page 482 of the text.

3. See the definition on page 487 of the text.

Exercises page 490

1. When $p = 4$, $-0.01x^2 - 0.1x + 6 = 4$, so $x^2 + 10x - 200 = 0$, and therefore $(x - 10)(x + 20) = 0$, giving $x = 10$ or -20. We reject the root $x = -20$ and find that the equilibrium price occurs at $x = 10$. The consumers' surplus is thus $CS = \int_0^{10} (-0.01x^2 - 0.1x + 6)\, dx - (4)(10) = \left(-\frac{0.01}{3}x^3 - 0.05x^2 + 6x\right)\Big|_0^{10} - 40 \approx 11.667$, or \$11,667.

3. Setting $p = 10$, we have $\sqrt{225 - 5x} = 10$, $225 - 5x = 100$, and so $x = 25$. Then $CS = \int_0^{25} \sqrt{225 - 5x}\, dx - (10)(25) = \int_0^{25} (225 - 5x)^{1/2}\, dx - 250$. To evaluate the integral, let $u = 225 - 5x$, so $du = -5\, dx$ and $dx = -\frac{1}{5}\, du$. If $x = 0$, then $u = 225$ and if $x = 25$, then $u = 100$, so $CS = -\frac{1}{5}\int_{225}^{100} u^{1/2}\, du - 250 = -\frac{2}{15}u^{3/2}\Big|_{225}^{100} - 250 = -\frac{2}{15}(1000 - 3375) - 250 = 66.667$, or \$6,667.

5. To find the equilibrium point, we solve $0.01x^2 + 0.1x + 3 = -0.01x^2 - 0.2x + 8$, finding $0.02x^2 + 0.3x - 5 = 0$, $2x^2 + 30x - 500 = (2x - 20)(x + 25) = 0$, and so $x = -25$ or 10. Thus, the equilibrium point is $(10, 5)$. Then $PS = (5)(10) - \int_0^{10} (0.01x^2 + 0.1x + 3)\, dx = 50 - \left(\frac{0.01}{3}x^3 + 0.05x^2 + 3x\right)\Big|_0^{10} = 50 - \frac{10}{3} - 5 - 30 = \frac{35}{3}$, or approximately \$11,667.

7. a. Setting $p = 250$, we have $100 + 80e^{0.05x} = 250$, $e^{0.05x} = \frac{150}{80} = \frac{15}{8}$, $\ln e^{0.05x} = \ln \frac{15}{8}$, $0.05x = \ln \frac{15}{8}$, and $x \approx 12.572$. The number of matresses the supplier will make available in the market per month is approximately 1257.

b. Taking $\overline{p} = 250$ and $\overline{x} = 12.572$ and using Formula (17), we find $PS \approx 12.572 \cdot 250 - \int_0^{12.572} (100 + 80e^{0.05x})\, dx = 3143 - \left(100x + \frac{80}{0.05}e^{0.05x}\right)\Big|_0^{12.572} \approx 485.826$, and so the producers' surplus is approximately \$48,583.

9. If $p = 160$, then we have $100\left(0.5x + \dfrac{0.4}{1 + x}\right) = 160$, so $50x + \dfrac{40}{1 + x} = 160$, $50x^2 + 50x + 40 = 160 + 160x$, $50x^2 - 110x - 120 = 0$, $5x^2 - 11x - 12 = 0$, and $(5x + 4)(x - 3) = 0$. Thus, $x = -\frac{4}{5}$ or $x = 3$. We reject the negative root, and using Formula (17) with $\overline{p} = 160$ and $\overline{x} = 3$, we have $PS = 3 \cdot 160 - \displaystyle\int_0^3 100\left(0.5x + \frac{0.4}{1 + x}\right) dx = 480 - 100\left[0.25x^2 + 0.4\ln(1 + x)\right]_0^3 \approx 199.548$. Therefore, the producers' surplus is approximately \$199,548.

11. To find the market equilibrium, we solve $-0.2x^2 + 80 = 0.1x^2 + x + 40$, obtaining $0.3x^2 + x - 40 = 0$, $3x^2 + 10x - 400 = 0$, $(3x + 40)(x - 10) = 0$, and so $x = -\frac{40}{3}$ or $x = 10$. We reject the negative root. The corresponding equilibrium price is \$60, the consumers' surplus is $CS = \int_0^{10} (-0.2x^2 + 80)\,dx - (60)(10) = \left(-\frac{0.2}{3}x^3 + 80x\right)\Big|_0^{10} - 600 \approx 133.33$, or \$13,333, and the producers' surplus is $PS = 600 - \int_0^{10} (0.1x^2 + x + 40)\,dx = 600 - \left(\frac{0.1}{3}x^3 + \frac{1}{2}x^2 + 40x\right)\Big|_0^{10} \approx 116.67$, or \$11,667.

13. Here $R(t) = 120{,}000$, $r = 0.035$, and $T = 4$, so the required future value is $A = e^{0.035(4)} \int_0^4 (120{,}000)\,e^{-0.035t}\,dt = 120{,}000e^{0.14}\left[-\frac{1}{0.035}e^{-0.035t}\right]_0^4 \approx 515{,}224.453$, or approximately \$515,224.45.

15. Here $P = 200{,}000$, $r = 0.08$, and $T = 5$, so $PV = \int_0^5 200{,}000e^{-0.08t}\,dt = -\frac{200{,}000}{0.08}e^{-0.08t}\Big|_0^5 = -2{,}500{,}000\left(e^{-0.4} - 1\right) \approx 824{,}199.85$, or \$824,200.

17. Here $P = 250$, $m = 12$, $T = 20$, and $r = 0.04$, so $A = \dfrac{mP}{r}\left(e^{rT} - 1\right) = \dfrac{12(250)}{0.04}\left(e^{0.8} - 1\right) \approx 91{,}915.57$, or approximately \$91,916.

19. Here $R(t) = 20e^{0.08t}$, $r = 0.03$, and $T = 5$.

a. The future value is $A = e^{0.03(5)} \int_0^5 20e^{0.08t}e^{-0.03t}\,dt = 20e^{0.15} \int_0^5 e^{0.05t}\,dt = 20e^{0.15}\left[\frac{1}{0.05}e^{0.05t}\right]_0^5 \approx 131.9962$, or approximately \$131,996.

b. The present value is $PV = \int_0^5 20e^{0.08t}e^{-0.03t}\,dt = 20\int_0^5 e^{0.05t}\,dt = \left[\frac{20}{0.05}e^{0.05t}\right]_0^5 \approx 113.61017$, or approximately \$113,610.

21. Here $P = 150$, $m = 12$, $T = 15$, and $r = 0.05$, so $A = \dfrac{12(150)}{0.05}\left(e^{0.75} - 1\right) \approx 40{,}212.00$, or approximately \$40,212.

23. Here $P = 2000$, $m = 1$, $T = 15.75$, and $r = 0.05$, so $A = \dfrac{1(2000)}{0.05}\left(e^{0.7875} - 1\right) \approx 47{,}915.79$, or approximately \$47,916.

25. Here $P = 1200$, $m = 12$, $T = 15$, and $r = 0.06$, so $PV = \dfrac{12(1200)}{0.06}\left(1 - e^{-0.9}\right) \approx 142{,}423.28$, or approximately \$142,423.

27. We want the present value of an annuity with $P = 300$, $m = 12$, $T = 10$, and $r = 0.05$, so $PV = \dfrac{12(300)}{0.05}\left(1 - e^{-0.5}\right) \approx 28{,}329.79$, or approximately \$28,330.

29. $L = 2\int_0^1 \left[x - \left(0.3x^{1.5} + 0.7x^{2.5}\right)\right]dx = 2\int_0^1 \left(x - 0.3x^{1.5} - 0.7x^{2.5}\right)dx = 2\left[\frac{1}{2}x^2 - \frac{0.3}{2.5}x^{2.5} - \frac{0.7}{3.5}x^{3.5}\right]_0^1 \approx 0.36$.

31. a.

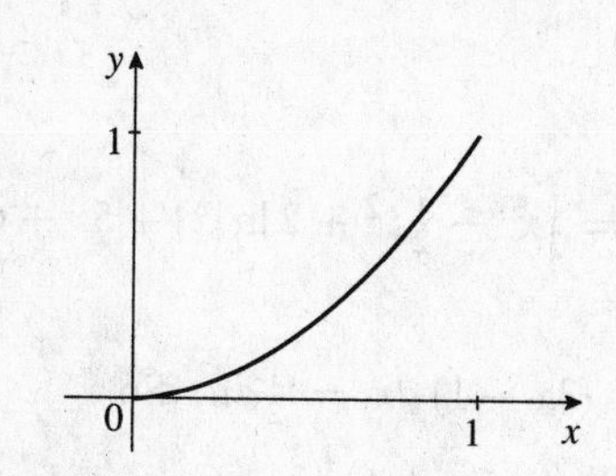

b. $f(0.4) = \frac{15}{16}(0.4)^2 + \frac{1}{16}(0.4) \approx 0.175$ and $f(0.9) = \frac{15}{16}(0.9)^2 + \frac{1}{16}(0.9) \approx 0.816$. Thus, the lowest 40% of earners receive 17.5% of the total income and the lowest 90% of earners receive 81.6%.

33. a. $L_1 = 2\int_0^1 [x - f(x)]\,dx = 2\int_0^1 \left(x - \frac{13}{14}x^2 - \frac{1}{14}x\right)dx = 2\int_0^1 \left(\frac{13}{14}x - \frac{13}{14}x^2\right)dx = \frac{13}{7}\int_0^1 (x - x^2)\,dx$

$= \frac{13}{7}\left(\frac{1}{2}x^2 - \frac{1}{3}x^3\right)\Big|_0^1 = \frac{13}{7}\left(\frac{1}{2} - \frac{1}{3}\right) = \frac{13}{7} \cdot \frac{1}{6} = \frac{13}{42} = 0.3095$ and

$L_2 = 2\int_0^1 \left(x - \frac{9}{11}x^4 - \frac{2}{11}x\right)dx = 2\int_0^1 \left(\frac{9}{11}x - \frac{9}{11}x^4\right)dx = 2\left(\frac{9}{11}\right)\int_0^1 (x - x^4)\,dx$

$= \frac{18}{11}\left(\frac{1}{2}x^2 - \frac{1}{5}x^5\right)\Big|_0^1 = \frac{18}{11}\left(\frac{1}{2} - \frac{1}{5}\right) = 0.4909.$

b. College teachers have a more equitable income distribution.

Using Technology page 493

1. The consumer's surplus is \$18,000,000 and the producer's surplus is \$11,700,000.

3. The consumer's surplus is \$33,120 and the producer's surplus is \$2880.

5. Investment A will generate a higher net income.

CHAPTER 6 Concept Review Questions page 496

1. a. $F'(x) = f(x)$ **b.** $F(x) + C$

3. a. unknown **b.** function

5. a. $\int_a^b f(x)\,dx$ **b.** minus

7. a. $\dfrac{1}{b-a}\int_a^b f(x)\,dx$ **b.** area, area

9. a. $\int_0^{\bar{x}} D(x)\,dx - \bar{p}\bar{x}$ **b.** $\bar{p}\bar{x} - \int_0^{\bar{x}} S(x)\,dx$

11. $\dfrac{mP}{r}\left(e^{rT} - 1\right)$

CHAPTER 6 Review Exercises page 496

1. $\int (x^3 + 2x^2 - x)\,dx = \frac{1}{4}x^4 + \frac{2}{3}x^3 - \frac{1}{2}x^2 + C.$

3. $\displaystyle\int \left(x^4 - 2x^3 + \frac{1}{x^2}\right)dx = \frac{x^5}{5} - \frac{x^4}{2} - \frac{1}{x} + C.$

5. $\int x\left(2x^2+x^{1/2}\right)dx=\int\left(2x^3+x^{3/2}\right)dx=\frac{1}{2}x^4+\frac{2}{5}x^{5/2}+C.$

7. $\int\left(x^2-x+\frac{2}{x}+5\right)dx=\int x^2\,dx-\int x\,dx+2\int\frac{dx}{x}+5\int dx=\frac{1}{3}x^3-\frac{1}{2}x^2+2\ln|x|+5x+C.$

9. Let $u=3x^2-2x+1$, so $du=(6x-2)\,dx=2(3x-1)\,dx$ or $(3x-1)\,dx=\frac{1}{2}du$. So $\int(3x-1)\left(3x^2-2x+1\right)^{1/3}dx=\frac{1}{2}\int u^{1/3}du=\frac{3}{8}u^{4/3}+C=\frac{3}{8}\left(3x^2-2x+1\right)^{4/3}+C.$

11. Let $u=x^2-2x+5$, so $du=2(x-1)\,dx$ and $(x-1)\,dx=\frac{1}{2}\,du$. Then $\int\frac{x-1}{x^2-2x+5}\,dx=\frac{1}{2}\int\frac{du}{u}=\frac{1}{2}\ln|u|+C=\frac{1}{2}\ln\left(x^2-2x+5\right)+C.$

13. Put $u=x^2+x+1$, so $du=(2x+1)\,dx=2\left(x+\frac{1}{2}\right)dx$ and $\left(x+\frac{1}{2}\right)dx=\frac{1}{2}du$. Then $\int\left(x+\frac{1}{2}\right)e^{x^2+x+1}\,dx=\frac{1}{2}\int e^u\,du=\frac{1}{2}e^u+C=\frac{1}{2}e^{x^2+x+1}+C.$

15. Let $u=\ln x$, so $du=\frac{1}{x}\,dx$. Then $\int\frac{(\ln x)^5}{x}\,dx=\int u^5\,du=\frac{1}{6}u^6+C=\frac{1}{6}(\ln x)^6+C.$

17. Let $u=x^2+1$, so $x^2=u-1$, $du=2x\,dx$, $x\,dx=\frac{1}{2}\,du$. Then

$$\int x^3\left(x^2+1\right)^{10}dx=\frac{1}{2}\int(u-1)\,u^{10}\,du=\frac{1}{2}\int\left(u^{11}-u^{10}\right)du=\frac{1}{2}\left(\frac{1}{12}u^{12}-\frac{1}{11}u^{11}\right)+C$$
$$=\frac{1}{264}u^{11}(11u-12)+C=\frac{1}{264}\left(x^2+1\right)^{11}\left(11x^2-1\right)+C.$$

19. Put $u=x-2$, so $du=dx$. Then $x=u+2$ and

$$\int\frac{x}{\sqrt{x-2}}\,dx=\int\frac{u+2}{\sqrt{u}}\,du=\int\left(u^{1/2}+2u^{-1/2}\right)du=\int u^{1/2}\,du+2\int u^{-1/2}\,du=\frac{2}{3}u^{3/2}+4u^{1/2}+C$$
$$=\frac{2}{3}u^{1/2}(u+6)+C=\frac{2}{3}\sqrt{x-2}\,(x-2+6)+C=\frac{2}{3}(x+4)\sqrt{x-2}+C.$$

21. $\int_0^1\left(2x^3-3x^2+1\right)dx=\left.\left(\frac{1}{2}x^4-x^3+x\right)\right|_0^1=\frac{1}{2}-1+1=\frac{1}{2}.$

23. $\int_1^4\left(x^{1/2}+x^{-3/2}\right)dx=\left.\left(\frac{2}{3}x^{3/2}-2x^{-1/2}\right)\right|_1^4=\left.\left(\frac{2}{3}x^{3/2}-\frac{2}{\sqrt{x}}\right)\right|_1^4=\left(\frac{16}{3}-1\right)-\left(\frac{2}{3}-2\right)=\frac{17}{3}.$

25. Put $u=x^3-3x^2+1$, so $du=\left(3x^2-6x\right)dx=3\left(x^2-2x\right)dx$ and $\left(x^2-2x\right)dx=\frac{1}{3}\,du$. If $x=-1$, $u=-3$ and if $x=0$, $u=1$, so $\int_{-1}^0 12\left(x^2-2x\right)\left(x^3-3x^2+1\right)^3dx=(12)\left(\frac{1}{3}\right)\int_{-3}^1 u^3\,du=4\left(\frac{1}{4}\right)\left.u^4\right|_{-3}^1=1-81=-80.$

27. Let $u=x^2+1$, so $du=2x\,dx$ and $x\,dx=\frac{1}{2}du$. If $x=0$, then $u=1$, and if $x=2$, then $u=5$, so $\int_0^2\frac{x}{x^2+1}\,dx=\frac{1}{2}\int_1^5\frac{du}{u}=\left.\frac{1}{2}\ln u\right|_1^5=\frac{1}{2}\ln 5.$

29. Let $u=1+2x^2$, so $du=4x\,dx$ and $x\,dx=\frac{1}{4}\,du$. If $x=0$, then $u=1$ and if $x=2$, then $u=9$, so $\int_0^2\frac{4x}{\sqrt{1+2x^2}}\,dx=\int_1^9\frac{du}{u^{1/2}}=\left.2u^{1/2}\right|_1^9=2(3-1)=4.$

31. Let $u = 1 + e^{-x}$, so $du = -e^{-x}\,dx$ and $e^{-x}\,dx = -du$. Then
$$\int_{-1}^{0} \frac{e^{-x}}{\left(1+e^{-x}\right)^2}\,dx = -\int_{1+e}^{2} \frac{du}{u^2} = \frac{1}{u}\bigg|_{1+e}^{2} = \frac{1}{2} - \frac{1}{1+e} = \frac{e-1}{2\,(1+e)}.$$

33. $f(x) = \int f'(x)\,dx = \int \left(3x^2 - 4x + 1\right) dx = 3\int x^2\,dx - 4\int x\,dx + \int dx = x^3 - 2x^2 + x + C$. The given condition implies that $f(1) = 1$, so $1 - 2 + 1 + C = 1$, and thus $C = 1$. Therefore, the required function is $f(x) = x^3 - 2x^2 + x + 1$.

35. $f(x) = \int f'(x)\,dx = \int \left(1 - e^{-x}\right) dx = x + e^{-x} + C$. Now $f(0) = 2$ implies $0 + 1 + C = 2$, so $C = 1$ and the required function is $f(x) = x + e^{-x} + 1$.

37. a. The integral $\int_0^T \left[f(t) - g(t)\right] dt$ represents the distance in feet between Car A and Car B at time T. If Car A is ahead, the integral is positive and if Car B is ahead, it is negative.

b. The distance is greatest at $t = 10$, at which point Car B's velocity exceeds that of Car A and Car B starts catching up. At that instant, the distance between the cars is $\int_0^{10} \left[f(t) - g(t)\right] dt$.

39. $\Delta x = \frac{2-1}{5} = \frac{1}{5}$, so $x_1 = \frac{6}{5}$, $x_2 = \frac{7}{5}$, $x_3 = \frac{8}{5}$, $x_4 = \frac{9}{5}$, $x_5 = \frac{10}{5}$. The Riemann sum is
$$f(x_1)\,\Delta x + \cdots + f(x_5)\,\Delta x = \left\{\left[-2\left(\tfrac{6}{5}\right)^2 + 1\right] + \left[-2\left(\tfrac{7}{5}\right)^2 + 1\right] + \cdots + \left[-2\left(\tfrac{10}{5}\right)^2 + 1\right]\right\}\left(\tfrac{1}{5}\right)$$
$$= \tfrac{1}{5}\left(-1.88 - 2.92 - 4.12 - 5.48 - 7\right) = -4.28.$$

41. $V(t) = \int V'(t)\,dt = 3800\int (t-10)\,dt = 1900\,(t-10)^2 + C$. The initial condition implies that $V(0) = 200{,}000$, that is, $190{,}000 + C = 200{,}000$, and so $C = 10{,}000$. Therefore, $V(t) = 1900\,(t-10)^2 + 10{,}000$. The resale value of the computer after 6 years is given by $V(6) = 1900\,(-4)^2 + 10{,}000 = 40{,}400$, or \$40,400.

43. a. $R(x) = \int R'(x)\,dx = \int (-0.03x + 60)\,dx = -0.015x^2 + 60x + C$. $R(0) = 0$ implies that $C = 0$, so $R(x) = -0.015x^2 + 60x$.

b. From $R(x) = px$, we have $-0.015x^2 + 60x = px$, and so $p = -0.015x + 60$.

45. a. The total number of DVDs sold as of year t is
$T(t) = \int R(t)\,dt = \int \left(-0.03t^2 + 0.218t - 0.032\right) dt = -0.01t^3 + 0.109t^2 - 0.032t + C$. Using the condition $T(0) = 0.1$, we find $T(0) = C = 0.1$. Therefore, $T(t) = -0.01t^3 + 0.109t^2 - 0.032t + 0.1$.

b. The total number of DVDs sold in 2003 is $T(4) = -0.01\,(4)^3 + 0.109\,(4)^2 - 0.032\,(4) + 0.1 = 1.076$, or 1.076 billion.

47. $C(x) = \int C'(x)\,dx = \int \left(0.00003x^2 - 0.03x + 10\right) dx = 0.00001x^3 - 0.015x^2 + 10x + k$. Now $C(0) = 600$ implies that $k = 600$, so $C(x) = 0.00001x^3 - 0.015x^2 + 10x + 600$. The total cost incurred in producing the first 500 corn poppers is $C(500) = 0.00001\,(500)^3 - 0.015\,(500)^2 + 10\,(500) + 600 = 3100$, or \$3100.

49. Using the substitution $u = 1 + 0.4t$, we find that
$N(t) = \int 3000\,(1 + 0.4t)^{-1/2}\,dt = \frac{3000}{0.4} \cdot 2\,(1 + 0.4t)^{1/2} + C = 15{,}000\sqrt{1 + 0.4t} + C$. $N(0) = 100{,}000$ implies $15{,}000 + C = 100{,}000$, so $C = 85{,}000$. Therefore, $N(t) = 15{,}000\sqrt{1 + 0.4t} + 85{,}000$. The number using the subway six months from now will be $N(6) = 15{,}000\sqrt{1 + 2.4} + 85{,}000 \approx 112{,}659$.

51. a. The online retail sales will be

$S(t) = \int R(t)\,dt = 15.82 \int e^{-0.176t}\,dt = -\frac{15.82}{0.176}e^{-0.176t} + C = -89.89e^{-0.176t} + C.$

$S(0) = 116$ implies that $-89.89 + C = 116$, so $C = 205.89$. Therefore, $S(t) = 205.89 - 89.89e^{-0.176t}$.

b. The sales will be $S(4) = 205.89 - 89.89e^{-0.176(4)} \approx 161.43$, or \$161.43 billion.

53. The number of speakers sold at the end of 5 years is

$f(t) = \int f'(t)\,dt = \int_0^5 2000\left(3 - 2e^{-t}\right)dt = 2000\left[3(5) - 2e^{-5}\right] - 2000\left[3 - 2(1)\right] = 26{,}027.$

55. $A = \int_0^2 e^{2x}\,dx = \frac{1}{2}e^{2x}\Big|_0^2 = \frac{1}{2}\left(e^4 - 1\right).$

57. The graph of y intersects the x-axis at $x = -2$ and $x = 1$, so

$A = \int_{-2}^{1} \left(-x^2 - x + 2\right)dx = \left(-\frac{1}{3}x^3 - \frac{1}{2}x^2 + 2x\right)\Big|_{-2}^{1}$

$= \left(-\frac{1}{3} - \frac{1}{2} + 2\right) - \left(\frac{8}{3} - \frac{4}{2} - 4\right) = \frac{7}{6} + \frac{10}{3} = \frac{9}{2}.$

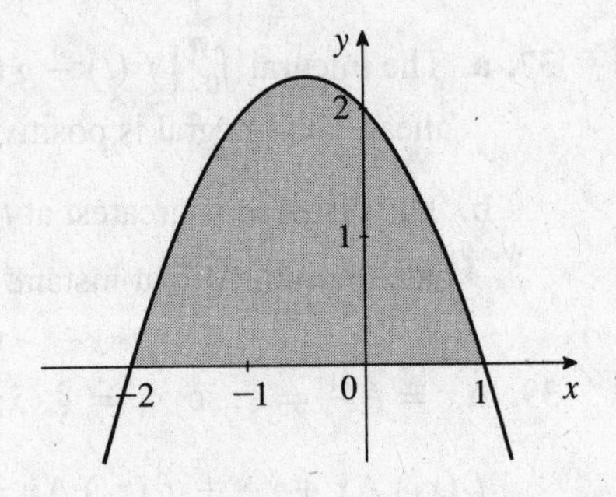

59. To find the points of intersection of the two curves, we solve $x^4 = x$, obtaining $x\left(x^3 - 1\right) = 0$, and so $x = 0$ or 1. Thus,

$A = \int_0^1 \left(x - x^4\right)dx = \left(\frac{1}{2}x^2 - \frac{1}{5}x^5\right)\Big|_0^1 = \frac{1}{2} - \frac{1}{5} = \frac{3}{10}.$

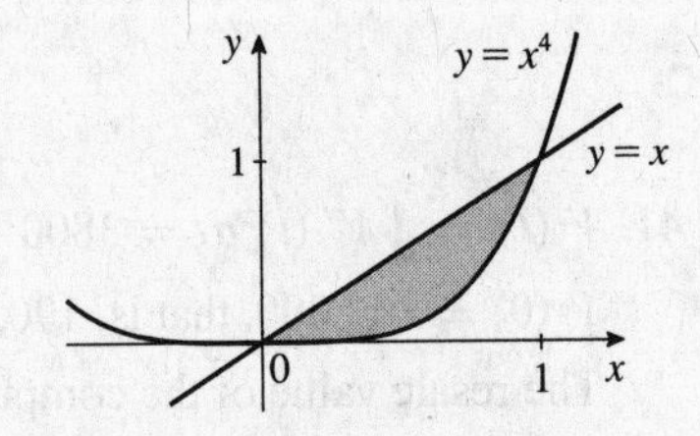

61. The amount of additional oil that will be produced over the next ten years is given by

$\int_0^{10} \left[R_2(t) - R_1(t)\right]dt = \int_0^{10} \left(100e^{0.08t} - 100e^{0.05t}\right)dt = 100\int_0^{10} \left(e^{0.08t} - e^{0.05t}\right)dt$

$= \left(\frac{100}{0.08}e^{0.08t} - \frac{100}{0.05}e^{0.05t}\right)\Big|_0^{10} = 1250e^{0.8} - 2000e^{0.5} - 1250 + 2000$

$= 2781.9 - 3297.4 - 1250 + 2000 = 234.5$, or 234,500 barrels.

63. The average temperature is

$\frac{1}{12}\int_0^{12} \left(-0.05t^3 + 0.4t^2 + 3.8t + 5.6\right)dt = \frac{1}{12}\left(-\frac{0.05}{4}t^4 + \frac{0.4}{3}t^3 + 1.9t^2 + 5.6t\right)\Big|_0^{12} = 26^\circ\text{F}.$

65. The average rate of growth between $t = 0$ and $t = 9$ is

$\frac{1}{9-0}\int_0^9 R(t)\,dt = \frac{1}{9}\int_0^9 \left(-0.0039t^2 + 0.0374t + 0.0046\right)dt = \frac{1}{9}\left(-0.0013t^3 + 0.0187t^2 + 0.0046t\right)\Big|_0^9$

$= \frac{1}{9}\left[-0.0013\left(9^3\right) + 0.0187\left(9^2\right) + 0.0046(9)\right] = 0.0676$, or 67,600/yr.

67. To find the equilibrium point, we solve $0.1x^2 + 2x + 20 = -0.1x^2 - x + 40$, obtaining $0.2x^2 + 3x - 20 = 0$, $x^2 + 15x - 100 = 0$, $(x + 20)(x - 5) = 0$, and so $x = 5$. Therefore, $p = -0.1(25) - 5 + 40 = 32.5$, and

$CS = \int_0^5 \left(-0.1x^2 - x + 40\right)dx - (5)(32.5) = \left(-\frac{0.1}{3}x^3 - \frac{1}{2}x^2 + 40x\right)\Big|_0^5 - 162.5 = 20.833$, or \$2083. Also,

$PS = 5(32.5) - \int_0^5 \left(0.1x^2 + 2x + 20\right)dx = 162.5 - \left(\frac{0.1}{3}x^3 + x^2 + 20x\right)\Big|_0^5 \approx 33.33$, or \$3333.

69. Use Equation (18) with $P = 925$, $m = 12$, $T = 30$, and $r = 0.06$ to get
$PV = \frac{mP}{r}\left(1 - e^{-rT}\right) = \frac{12 \cdot 925}{0.06}\left(1 - e^{-0.06 \cdot 30}\right) = 154{,}419.71$. We conclude that the present value of the purchase price of the house is $154{,}419.71 + 20{,}000$, or approximately \$174,420.

71. a.

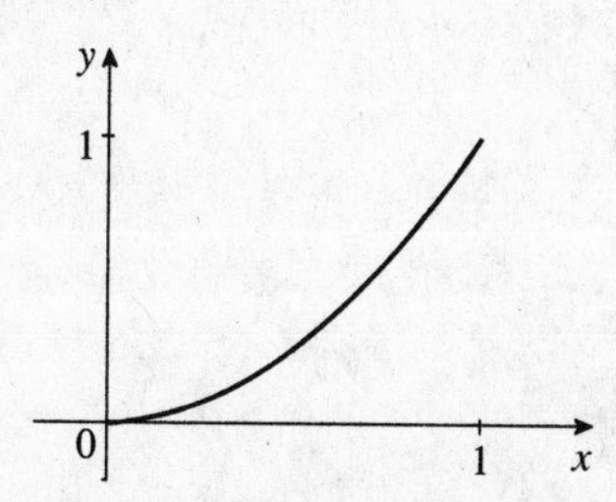

b. $f(0.3) = \frac{17}{18}(0.3)^2 + \frac{1}{18}(0.3) \approx 0.1$. Thus, 30% of the people receive 10% of the total income.
$f(0.6) = \frac{17}{18}(0.6)^2 + \frac{1}{18}(0.6) \approx 0.37$, so 60% of the people receive 37% of the total income.

c. The coefficient of inequality for this curve is
$L = 2\int_0^1 \left(x - \frac{17}{18}x^2 - \frac{1}{18}x\right) dx = \frac{17}{9}\int_0^1 \left(x - x^2\right) dx = \frac{17}{9}\left(\frac{1}{2}x^2 - \frac{1}{3}x^3\right)\Big|_0^1 = \frac{17}{54} \approx 0.315$.

CHAPTER 6 Before Moving On... page 500

1. $\int \left(2x^3 + \sqrt{x} + \frac{2}{x} - \frac{2}{\sqrt{x}}\right) dx = 2\int x^3\,dx + \int x^{1/2}\,dx + 2\int \frac{1}{x}\,dx - 2\int x^{-1/2}\,dx$
$= \frac{1}{2}x^4 + \frac{2}{3}x^{3/2} + 2\ln|x| - 4x^{1/2} + C$.

2. $f(x) = \int f'(x)\,dx = \int (e^x + x)\,dx = e^x + \frac{1}{2}x^2 + C$. $f(0) = 2$ implies $f(0) = e^0 + 0 + C = 2$, so $C = 1$. Therefore, $f(x) = e^x + \frac{1}{2}x^2 + 1$.

3. Let $u = x^2 + 1$, so $du = 2x\,dx$ or $x\,dx = \frac{1}{2}du$. Then
$\int \frac{x}{\sqrt{x^2+1}}\,dx = \frac{1}{2}\int \frac{du}{\sqrt{u}} = \frac{1}{2}\int u^{-1/2}\,du = \frac{1}{2}\left(2u^{1/2}\right) + C = \sqrt{u} + C = \sqrt{x^2+1} + C$.

4. Let $u = 2 - x^2$, so $du = -2x\,dx$ and $x\,dx = -\frac{1}{2}\,du$. If $x = 0$, then $u = 2$ and if $x = 1$, then $u = 1$. Therefore,
$\int_0^1 x\sqrt{2 - x^2}\,dx = -\frac{1}{2}\int_2^1 u^{1/2}\,du = -\frac{1}{2}\left(\frac{2}{3}u^{3/2}\right)\Big|_2^1 = -\frac{1}{3}u^{3/2}\Big|_2^1 = -\frac{1}{3}\left(1 - 2^{3/2}\right) = \frac{1}{3}\left(2\sqrt{2} - 1\right)$.

5. To find the points of intersection, we solve $x^2 - 1 = 1 - x$, obtaining $x^2 + x - 2 = 0$, $(x+2)(x-1) = 0$, and so $x = -2$ or $x = 1$. The points of intersection are $(-2, 3)$ and $(1, 0)$. Thus, the required area is
$A = \int_{-2}^1 \left[(1 - x) - \left(x^2 - 1\right)\right] dx = \int_{-2}^1 \left(2 - x - x^2\right) dx = \left(2x - \frac{1}{2}x^2 - \frac{1}{3}x^3\right)\Big|_{-2}^1$
$= \left(2 - \frac{1}{2} - \frac{1}{3}\right) - \left(-4 - 2 + \frac{8}{3}\right) = \frac{9}{2}$.

7 ADDITIONAL TOPICS IN INTEGRATION

7.1 Integration by Parts

Problem-Solving Tips

1. When you integrate by parts, remember to choose u and dv so that du is simpler than u and dv is easy to integrate.

2. It is helpful to follow the pattern used in the examples in this section: $\int u\,dv = uv - \int v\,du$, where

$$u = \_\_\_\_ \qquad dv = \_\_\_\_$$
$$du = \_\_\_\_ \qquad v = \_\_\_\_$$

Concept Questions page 507

1. $\int u\,dv = uv - \int v\,du$

Exercises page 507

1. $I = \int xe^{2x}\,dx$. Let $u = x$ and $dv = e^{2x}\,dx$, so $du = dx$ and $v = \frac{1}{2}e^{2x}$. Then
$I = uv - \int v\,du = \frac{1}{2}xe^{2x} - \int \frac{1}{2}e^{2x}\,dx = \frac{1}{2}xe^{2x} - \frac{1}{4}e^{2x} = \frac{1}{4}e^{2x}(2x-1) + C$.

3. $I = \int \frac{1}{2}xe^{x/4}\,dx$. Let $u = \frac{1}{2}x$ and $dv = e^{x/4}\,dx$, so $du = \frac{1}{2}dx$ and $v = 4e^{x/4}$. Then
$\int \frac{1}{2}xe^{x/4}\,dx = uv - \int v\,du = 2xe^{x/4} - 2\int e^{x/4}\,dx = 2xe^{x/4} - 8e^{x/4} + C = 2(x-4)e^{x/4} + C$.

5. $\int (e^x - x)^2\,dx = \int (e^{2x} - 2xe^x + x^2)\,dx = \int e^{2x}\,dx - 2\int xe^x\,dx + \int x^2\,dx$. Using the result $\int xe^x\,dx = (x-1)e^x + k$, from Example 1, we see that $\int (e^x - x)^2\,dx = \frac{1}{2}e^{2x} - 2(x-1)e^x + \frac{1}{3}x^3 + C$.

7. $I = \int (x+1)e^x\,dx$. Let $u = x+1$ and $dv = e^x\,dx$, so $du = dx$ and $v = e^x$. Then
$I = (x+1)e^x - \int e^x\,dx = (x+1)e^x - e^x + C = xe^x + C$.

9. Let $u = x$ and $dv = (x+1)^{-3/2}\,dx$, so $du = dx$ and $v = -2(x+1)^{-1/2}$. Then

$$\int x(x+1)^{-3/2}\,dx = uv - \int v\,du = -2x(x+1)^{-1/2} + 2\int (x+1)^{-1/2}\,dx$$
$$= -2x(x+1)^{-1/2} + 4(x+1)^{1/2} + C = 2(x+1)^{-1/2}[-x + 2(x+1)] + C = \frac{2(x+2)}{\sqrt{x+1}} + C.$$

11. $I = \int x(x-5)^{1/2}\,dx$. Let $u = x$ and $dv = (x-5)^{1/2}\,dx$, so $du = dx$ and $v = \frac{2}{3}(x-5)^{3/2}$. Then

$$I = \frac{2}{3}x(x-5)^{3/2} - \int \frac{2}{3}(x-5)^{3/2}\,dx = \frac{2}{3}x(x-5)^{3/2} - \frac{2}{3}\cdot\frac{2}{5}(x-5)^{5/2} + C$$
$$= \frac{2}{3}(x-5)^{3/2}\left[x - \frac{2}{5}(x-5)\right] + C = \frac{2}{15}(x-5)^{3/2}(5x - 2x + 10) + C = \frac{2}{15}(x-5)^{3/2}(3x+10) + C.$$

13. $I = \int x\ln 2x\,dx$. Let $u = \ln 2x$ and $dv = x\,dx$, so $du = \frac{1}{x}\,dx$ and $v = \frac{1}{2}x^2$. Then
$I = \frac{1}{2}x^2\ln 2x - \int \frac{1}{2}x\,dx = \frac{1}{2}x^2\ln 2x - \frac{1}{4}x^2 + C = \frac{1}{4}x^2(2\ln 2x - 1) + C$.

15. Let $u = \ln x$ and $dv = x^3\,dx$, so $du = \frac{1}{x}\,dx$, and $v = \frac{1}{4}x^4$. Then
$\int x^3 \ln x\,dx = \frac{1}{4}x^4 \ln x - \frac{1}{4}\int x^3\,dx = \frac{1}{4}x^4 \ln x - \frac{1}{16}x^4 + C = \frac{1}{16}x^4\,(4\ln x - 1) + C.$

17. Let $u = \ln x^{1/2}$ and $dv = x^{1/2}\,dx$, so $du = \frac{1}{2x}\,dx$ and $v = \frac{2}{3}x^{3/2}$. Then
$\int \sqrt{x} \ln \sqrt{x}\,dx = uv - \int v\,du = \frac{2}{3}x^{3/2} \ln x^{1/2} - \frac{1}{3}\int x^{1/2}\,dx = \frac{2}{3}x^{3/2} \ln x^{1/2} - \frac{2}{9}x^{3/2} + C$
$= \frac{2}{9}x\sqrt{x}\,(3\ln\sqrt{x} - 1) + C.$

19. Let $u = \ln x$ and $dv = x^{-2}\,dx$, so $du = \frac{1}{x}\,dx$ and $v = -x^{-1}$. Then
$\displaystyle\int \frac{\ln x}{x^2}\,dx = uv - \int v\,du = -\frac{\ln x}{x} + \int x^{-2}\,dx = -\frac{\ln x}{x} - \frac{1}{x} + C = -\frac{1}{x}\,(\ln x + 1) + C.$

21. $\int xe^{(x+1)^2}\,dx = \int xe^{x^2+2x+1}\,dx = \int \left(xe^{x^2} + xe^{2x} + ex\right)dx = \int xe^{x^2}\,dx + \int xe^{2x}\,dx + e\int x\,dx$
$= \frac{1}{2}e^{x^2} + I + \frac{1}{2}ex^2 + C_1.$

To find $I = \int xe^{2x}\,dx$, we integrate by parts with $u = x$ and $dv = e^{2x}\,dx$, so $du = dx$ and $v = \frac{1}{2}e^{2x}$. Then $I = \frac{1}{2}xe^{2x} - \int \frac{1}{2}e^{2x}\,dx = \frac{1}{2}xe^{2x} - \frac{1}{4}e^{2x} + C_2$. Finally,
$\int xe^{(x+1)^2}\,dx = \frac{1}{2}e^{x^2} + \left(\frac{1}{2}xe^{2x} - \frac{1}{4}e^{2x} + C_2\right) + \frac{1}{2}ex^2 + C_1 = \frac{1}{2}e^{x^2} + \frac{1}{2}xe^{2x} - \frac{1}{4}e^{2x} + \frac{1}{2}ex^2 + C$, where $C = C_1 + C_2$.

23. Let $u = \ln x$ and $dv = dx$, so $du = \frac{1}{x}\,dx$ and $v = x$. Then
$\int \ln x\,dx = uv - \int v\,du = x\ln x - \int dx = x\ln x - x + C = x\,(\ln x - 1) + C.$

25. Let $u = x^2$ and $dv = e^{-x}\,dx$, so $du = 2x\,dx$ and $v = -e^{-x}$. Then
$\int x^2 e^{-x}\,dx = uv - \int v\,du = -x^2 e^{-x} + 2\int xe^{-x}\,dx$. We can integrate by parts again or, using the result of Exercise 2, we can write
$\int x^2 e^{-x}\,dx = -x^2 e^{-x} + 2\left[-(x+1)\,e^{-x}\right] + C = -x^2 e^{-x} - 2\,(x+1)\,e^{-x} + C$
$= -\,(x^2 + 2x + 2)\,e^{-x} + C.$

27. $I = \int x\,(\ln x)^2\,dx$. Let $u = (\ln x)^2$ and $dv = x\,dx$, so $du = 2\,(\ln x)\left(\dfrac{1}{x}\right) = \dfrac{2\ln x}{x}$ and $v = \frac{1}{2}x^2$. Then
$I = \frac{1}{2}x^2\,(\ln x)^2 - \int x\ln x\,dx$. Next, we evaluate $\int x\ln x\,dx$ by letting $u = \ln x$ and $dv = x\,dx$, so $du = \frac{1}{x}\,dx$ and $v = \frac{1}{2}x^2$. Then $\int x\ln x\,dx = \frac{1}{2}x^2\,(\ln x) - \frac{1}{2}\int x\,dx = \frac{1}{2}x^2 \ln x - \frac{1}{4}x^2 + C$. Therefore,
$\int x\,(\ln x)^2\,dx = \frac{1}{2}x^2\,(\ln x)^2 - \frac{1}{2}x^2 \ln x + \frac{1}{4}x^2 + C = \frac{1}{4}x^2\left[2\,(\ln x)^2 - 2\ln x + 1\right] + C.$

29. $\int_0^{\ln 3} xe^x\,dx = (x - 1)\,e^x\big|_0^{\ln 3}$ (using the results of Example 1)
$= (\ln 3 - 1)\,e^{\ln 3} - (-e^0) = 3\,(\ln 3 - 1) + 1 = 3\ln 3 - 2$ (because $e^{\ln 3} = 3$).

31. We first integrate $I = \int \ln x\,dx$. Using parts with $u = \ln x$ and $dv = dx$, so $du = \frac{1}{x}\,dx$ and $v = x$, we have $I = x\ln x - \int dx = x\ln x - x + C = x\,(\ln x - 1) + C$. Therefore,
$3\int_1^4 \ln x\,dx = 3x\,(\ln x - 1)\big|_1^4 = 3\left[4\,(\ln 4 - 1) - 1\,(\ln 1 - 1)\right] = 12\ln 4 - 9 = 3\,(4\ln 4 - 3).$

33. Let $u = x$ and $dv = e^{2x}\,dx$. Then $du = dx$ and $v = \frac{1}{2}e^{2x}$, so
$\int_0^2 xe^{2x}\,dx = \frac{1}{2}xe^{2x}\Big|_0^2 - \frac{1}{2}\int_0^2 e^{2x}\,dx = e^4 - \left(\frac{1}{4}e^{2x}\right)\Big|_0^2 = e^4 - \frac{1}{4}e^4 + \frac{1}{4} = \frac{1}{4}\,(3e^4 + 1).$

35. Let $u = x$ and $dv = e^{-2x}\,dx$, so $du = dx$ and $v = -\frac{1}{2}e^{-2x}$. Then $f(x) = \int xe^{-2x}\,dx = -\frac{1}{2}xe^{-2x} + \frac{1}{2}\int e^{-2x}\,dx = -\frac{1}{2}xe^{-2x} - \frac{1}{4}e^{-2x} + C$. Solving $f(0) = -\frac{1}{4} + C = 3$, we find that $C = \frac{13}{4}$, so $y = -\frac{1}{2}xe^{-2x} - \frac{1}{4}e^{-2x} + \frac{13}{4}$.

37. The required area is given by $\int_1^5 \ln x\,dx$. We first find $\int \ln x\,dx$. Using parts with $u = \ln x$ and $dv = dx$, so $du = \frac{1}{x}\,dx$ and $v = x$, we have $\int \ln x\,dx = x\ln x - \int dx = x\ln x - x = x(\ln x - 1) + C$. Therefore, $\int_1^5 \ln x\,dx = x(\ln x - 1)|_1^5 = 5(\ln 5 - 1) - 1(\ln 1 - 1) = 5\ln 5 - 4$, and the required area is $5\ln 5 - 4$.

39. The distance covered is given by $\int_0^{10} 100te^{-0.2t}\,dt = 100\int_0^{10} te^{-0.2t}\,dt$. We integrate by parts with $u = t$ and $dv = e^{-0.2t}\,dt$, so $du = dt$ and $v = -\frac{1}{0.2}e^{-0.2t} = -5e^{-0.2t}$. Therefore,

$$100\int_0^{10} te^{-0.2t}dt = 100\left(-5te^{-0.2t}\right)\Big|_0^{10} + 500\int_0^{10} e^{-0.2t}\,dt = -5000e^{-2} - \left(2500e^{-0.2t}\right)\Big|_0^{10}$$
$$= -5000e^{-2} - \left(2500e^{-2} - 2500\right) = 2500 - 7500e^{-2} \approx 1485, \text{ or } 1485 \text{ feet.}$$

41. $N = 2\int te^{-0.1t}dt$. Let $u = t$ and $dv = e^{-0.1t}$, so $du = dt$ and $v = -10e^{-0.1t}$. Then $N(t) = 2\left(-10te^{-0.1t} + 10\int e^{-0.1t}\,dt\right) = 2\left(-10te^{-0.1t} - 100e^{-0.1t}\right) + C = -20e^{-0.1t}(t + 10) + C$. Because $N(0) = -20(10) + C = 0$, $C = 200$. Therefore, $N(t) = -20e^{-0.1t}(t + 10) + 200$.

43. The number of accidents is expected to be

$$E = 982 + \int_0^{12}\left(-10 - te^{0.1t}\right)dt = 982 - \int_0^{12} 10\,dt - \int_0^{12} te^{0.1t}\,dt = 982 - 120 - \int_0^{12} te^{0.1t}\,dt$$
$$= 862 - \int_0^{12} te^{0.1t}\,dt.$$

To evaluate the last integral, let $u = t$ and $dv = e^{0.1t}\,dt$, so $du = dt$ and $v = 10e^{0.1t}$. Then $\int_0^{12} te^{0.1t}dt = 10te^{0.1t}\Big|_0^{12} - 10\int_0^{12} e^{0.1t}dt = 120e^{1.2} - \left(100e^{0.1t}\right)\Big|_0^{12} = 120e^{1.2} - 100e^{1.2} + 100 \approx 166$. Therefore, $E \approx 862 - 166 \approx 696$, or approximately 696 accidents.

45. The membership will be $N(5) = N(0) + \int_0^5 9\sqrt{t+1}\ln\sqrt{t+1}\,dt = 50 + 9\int_0^5 \sqrt{t+1}\ln\sqrt{t+1}\,dt$. To evaluate the integral, let $u = t + 1$, so $du = dt$. If $t = 0$, then $u = 1$ and if $t = 5$, then $u = 6$, so $9\int_0^5 \sqrt{t+1}\ln\sqrt{t+1}\,dt = 9\int_1^6 \sqrt{u}\ln\sqrt{u}\,du$. Using the results of Exercise 17, we have $\int \sqrt{x}\ln\sqrt{x}\,dx = \frac{2}{9}x\sqrt{x}\left(3\ln\sqrt{x} - 1\right) + C$. Thus, $9\int_1^6 \sqrt{u}\ln\sqrt{u}\,du = \left[2u\sqrt{u}\left(3\ln\sqrt{u} - 1\right)\right]_1^6 = 2(6)\sqrt{6}\left(3\ln\sqrt{6} - 1\right) - 2(-1) \approx 51.606$, and so $N = 50 + 51.606 \approx 101.606$, or 101,606 people.

47. The average annual fraud over the period from 2008 through 2012 is $A = \frac{1}{4-0}\int_0^4 [328.9 - 92.07\ln(t+1)]\,dt$. To find $I = \int \ln(t+1)\,dt$, let $u = t + 1$, so $du = dt$ and $t = u - 1$. Then $I = \int \ln u\,du = u\ln u - u + C$ (See Exercise 23). Thus, $A = \frac{1}{4}\left[328.9t - 92.07\left[(t+1)\ln(t+1) - (t+1)\right]\right]_0^4 \approx 235.74$ (million GBP).

49. The value of the income stream at the end of 5 years is

$F = e^{0.05(5)} \int_0^5 (100 + 20t) e^{-0.05t} \, dt = e^{0.25} \left[\int_0^5 100e^{-0.05t} \, dt + 20 \int_0^5 te^{-0.05t} \right] dt$

$= e^{0.25} \left\{ \left[-2000e^{-0.05t} \right]_0^5 + 20 \int_0^5 te^{-0.05t} \, dt \right\}$.

To find $I = \int te^{-0.05t} \, dt$, use parts with $u = t$ and $dv = e^{-0.05t} \, dt$, so $du = dt$ and $v = -\frac{1}{0.05} e^{-0.05t} = -20e^{-0.05t}$.

Thus, $I = -20te^{-0.05t} + \int 20e^{-0.05t} \, dt = -20te^{-0.05t} + 20 \left(\frac{1}{-0.05} \right) e^{-0.05t} + C = -20te^{-0.05t} - 400e^{-0.05t} + C$.

Finally, $F = e^{0.25} \left[-2000e^{-0.05t} + 20 \left(-20te^{-0.05t} - 400e^{-0.05t} \right) \right]_0^5 \approx 840.254$, or approximately \$840,254.

51. $PV = \int_0^5 (30{,}000 + 800t) e^{-0.05t} \, dt = 30{,}000 \int_0^5 e^{-0.05t} \, dt + 800 \int_0^5 te^{-0.05t} \, dt$. Let $I = \int te^{-0.05t} \, dt$. To evaluate I by parts, let $u = t$ and $dv = e^{-0.05t} \, dt$, so $du = dt$ and $v = -\frac{1}{0.05} e^{-0.05t} = -20e^{-0.05t}$. Then $I = -20te^{-0.05t} + 20 \int e^{-0.05t} dt = -20te^{-0.05t} - 400e^{-0.05t} + C$. Thus,

$PV = \left[-\frac{30{,}000}{0.05} e^{-0.05t} - 800\,(20)\, te^{-0.05t} - 800\,(400)\, e^{-0.05t} \right]_0^5$

$= -600{,}000e^{-0.25} + 600{,}000 - 80{,}000e^{-0.25} - 320{,}000e^{-0.25} + 320{,}000 = 920{,}000 - 1{,}000{,}000e^{-0.25}$

$= 141{,}199.22$, or approximately \$141,199.

53. $p = 300 - 2(x + 1) \ln(x + 1)$. If $x = 20$, then we have $p = 300 - 2(21) \ln 21 \approx 172.130$. Therefore, with $\overline{x} = 20$ and $\overline{p} = 172.130$ we find, using Formula (16) from Section 6.7, $CS \approx \int_1^{20} (300 - 2(x + 1) \ln(x + 1)) \, dx - 172.130 \cdot 20$. Using the result of Example 2, $\int (x + 1) \ln(x + 1) \, dx = \frac{1}{4} (x + 1)^2 (2 \ln(x + 1) - 1) + C$, so we have

$CS \approx \left[300x - 2 \left(\frac{1}{4} \right) (x + 1)^2 (2 \ln(x + 1) - 1) \right]_1^{20} - 3442.60 \approx 4877.87 - 299.23 - 3442.60 = 1136.04$, or approximately \$113,604.

55. $L = 2 \int_0^1 \left[x - xe^{2(x-1)} \right] dx = 2 \int_0^1 \left(x - xe^{2x}e^{-2} \right) dx = 2 \left[\frac{1}{2}x^2 - e^{-2} \int_0^1 xe^{2x} \, dx \right]$. Let $I = \int xe^{2x} \, dx$. To find I, use parts with $u = x$ and $dv = e^{2x} \, dx$. Then $du = dx$ and $v = \frac{1}{2}e^{2x}$, so $I = \frac{1}{2}xe^{2x} - \frac{1}{2} \int e^{2x} \, dx = \frac{1}{2}xe^{2x} - \frac{1}{4}e^{2x} + C$, so $L = 2 \left[\frac{1}{2}x^2 - e^{-2} \left(\frac{1}{2}xe^{2x} - \frac{1}{4}e^{2x} \right) \right]_0^1 \approx 0.432$.

57. **a.** $A(0) = 180 (1 - e^0) - 6(0) e^0 = 0$, or 0 lb.

b. $A(180) = 180 \left(1 - e^{-180/30} \right) - 6(180) e^{-180/30} \approx 176.877$, or 176.9 lb.

c. The average amount of salt over the first 3 hours is

$\overline{A} = \frac{1}{180} \int_0^{180} \left[180 \left(1 - e^{-t/30} \right) - 6te^{-t/30} \right] dt = \int_0^{180} \left(1 - e^{-t/30} \right) dt - \frac{1}{30} \int_0^{180} te^{-t/30} \, dt$. Call the first integral I_1 and the second integral I_2. Then

$I_1 = \int_0^{180} \left(1 - e^{-t/30} \right) dt = \left(t + 30e^{-t/30} \right) \Big|_0^{180}$ [substitute $u = -t/30$] $= \left(180 + 30e^{-6} \right) - \left(0 + 30e^0 \right)$

≈ 150.074.

To evaluate I_2, consider the corresponding indefinite integral $J = \int te^{-t/30} \, dt$. Integrating by parts with $u = t$ and $dv = e^{-t/30} dt$, so $du = dt$ and $v = -30e^{-t/30}$, we find

$J = -30te^{-t/30} - \int \left(-30e^{-t/30} \right) dt = -30te^{-t/30} - 900e^{-t/30} + C = -30(t + 30) e^{-t/30} + C$.

Therefore, $I_2 = \frac{1}{30} J \Big|_0^{180} = -(t + 30) e^{-t/30} \Big|_0^{180} = -(180 + 30) e^{-6} + 30e^0 \approx 29.479$. Thus,

$\overline{A} = I_1 - I_2 \approx 150.074 - 29.479 = 120.595$, or approximately 120.6 lb.

59. True. This is the integration by parts formula.

61. True. Let $U = uv$ and $dV = dw$, so $dU = u\,dv + v\,du$ and $V = w$. Then $\int uv\,dw = \int U\,dV = UV - \int V\,dU = uvw - \int w\,(u\,dv + v\,du) = uvw - \int uw\,dv - \int vw\,du$.

7.2 Integration Using Tables of Integrals

Problem-Solving Tips

The integrals in the exercise set may not have exactly the same form as those in the table of integrals. Sometimes you may need to rewrite the integral, as in Example 2 on page 512 of the text, or you may need to apply a rule more than once, as in Example 5 on page 514 of the text.

Concept Questions page 515

1. a. Formula (19) seems appropriate.

b. Put $a = \sqrt{2}$ and $x = u$. Then using Formula (19), we have

$$\int \frac{\sqrt{2-x^2}}{x}\,dx = \int \frac{\sqrt{\left(\sqrt{2}\right)^2 - x^2}}{x}\,dx = \sqrt{2-x^2} - \sqrt{2}\ln\left|\frac{\sqrt{a}+\sqrt{2-x^2}}{x}\right| + C.$$

Exercises page 515

1. First we note that $\int \frac{2x}{2+3x}\,dx = 2\int \frac{x}{2+3x}\,dx$. Next, we use Formula (1) with $a = 2$, $b = 3$, and $u = x$. Then $\int \frac{2x}{2+3x}\,dx = \frac{2}{9}\left(2+3x-2\ln|2+3x|\right) + C$.

3. $\int \frac{3x^2}{2+4x}\,dx = \frac{3}{2}\int \frac{x^2}{1+2x}\,dx$. Use Formula (2) with $a = 1$, $b = 2$, and $u = x$, obtaining $\int \frac{3x^2}{2+4x}\,dx = \frac{3}{32}\left[(1+2x)^2 - 4(1+2x) + 2\ln|1+2x|\right] + C$.

5. $\int x^2\sqrt{9+4x^2}\,dx = \int x^2\sqrt{4\left(\frac{9}{4}+x^2\right)}\,dx = 2\int x^2\sqrt{\left(\frac{3}{2}\right)^2 + x^2}\,dx$. Using Formula (8) with $a = \frac{3}{2}$ and $u = x$, we find that $\int x^2\sqrt{9+4x^2}\,dx = 2\left[\left(\frac{x}{8}\right)\left(\frac{9}{4}+2x^2\right)\sqrt{\frac{9}{4}+x^2} - \frac{81}{128}\ln\left|x+\sqrt{\frac{9}{4}+x^2}\right|\right] + C$.

7. Use Formula (6) with $a = 1$, $b = 4$, and $u = x$. Then $\int \frac{dx}{x\sqrt{1+4x}} = \ln\left|\frac{\sqrt{1+4x}-1}{\sqrt{1+4x}+1}\right| + C$.

9. Use Formula (9) with $a = 3$ and $u = 2x$, so $du = 2\,dx$. Then $\int_0^2 \frac{dx}{\sqrt{9+4x^2}} = \frac{1}{2}\int_0^4 \frac{du}{\sqrt{3^2+u^2}} = \frac{1}{2}\ln\left|u+\sqrt{9+u^2}\right|\Big|_0^4 = \frac{1}{2}(\ln 9 - \ln 3) = \frac{1}{2}\ln 3$. Note that the limits of integration change from $x = 0$ and $x = 2$ to $u = 0$ and $u = 4$ respectively.

11. Using Formula (22) with $a = 3$ and $u = x$, we see that $\int \frac{dx}{\left(9-x^2\right)^{3/2}} = \frac{x}{9\sqrt{9-x^2}} + C$.

13. $I = \int x^2\sqrt{x^2-4}\,dx$. Use Formula (14) with $a = 2$ and $u = x$ to obtain
$I = \frac{x}{8}\left(2x^2-4\right)\sqrt{x^2-4} - 2\ln\left|x+\sqrt{x^2-4}\right| + C$.

15. Using Formula (19) with $a = 2$ and $u = x$, we have $\displaystyle\int \frac{\sqrt{4-x^2}}{x}\,dx = \sqrt{4-x^2} - 2\ln\left|\frac{2+\sqrt{4-x^2}}{x}\right| + C$.

17. $I = \int xe^{2x}\,dx$. Use Formula (23) with $a = 2$ and $u = x$ to obtain $I = \frac{1}{4}(2x-1)e^{2x} + C$.

19. $\displaystyle I = \int \frac{dx}{(x+1)\ln(x+1)}$. Let $u = x+1$, so $du = dx$. Then $\displaystyle\int \frac{dx}{(x+1)\ln(x+1)} = \int \frac{du}{u\ln u}$. Now use Formula (28) with $u = x$ to obtain $\displaystyle\int \frac{du}{u\ln u} = \ln|\ln u| + C$. Therefore, $\displaystyle\int \frac{dx}{(x+1)\ln(x+1)} = \ln|\ln(x+1)| + C$.

21. $\displaystyle I = 3\int \frac{e^{2x}}{(1+3e^x)^2}\,dx$. Put $u = e^x$, so $du = e^x\,dx$. Then use Formula (3) with $a = 1$ and $b = 3$. Thus,
$\displaystyle I = 3\int \frac{u}{(1+3u)^2}\,du = \frac{1}{3}\left(\frac{1}{1+3u} + \ln|1+3u|\right) + C = \frac{1}{3}\left[\frac{1}{1+3e^x} + \ln(1+3e^x)\right] + C$.

23. $\displaystyle\int \frac{3e^x}{1+e^{x/2}}\,dx = 3\int \frac{e^{x/2}}{e^{-x/2}+1}\,dx$. Let $v = e^{x/2}$, so $dv = \frac{1}{2}e^{x/2}\,dx$ and $e^{x/2}\,dx = 2\,dv$. Then $\displaystyle\int \frac{3e^x}{1+e^{x/2}}\,dx = 6\int \frac{dv}{(1/v)+1} = 6\int \frac{v}{v+1}\,dv$. Use Formula (1) with $a = 1$, $b = 1$, and $u = v$, obtaining $\displaystyle 6\int \frac{v}{v+1}\,dv = 6\left(1+v-\ln|1+v|\right) + C$. Thus, $\displaystyle\int \frac{3e^x}{1+e^{x/2}}\,dx = 6\left[1+e^{x/2}-\ln\left(1+e^{x/2}\right)\right] + C$. This answer may be written in the form $6\left[e^{x/2}-\ln\left(1+e^{x/2}\right)\right] + C_1$, where $C_1 = C+6$ is an arbitrary constant.

25. $\displaystyle I = \int \frac{4\ln x}{x(2+3\ln x)}\,dx$. Let $v = \ln x$ so that $dv = \frac{1}{x}\,dx$. Then $\displaystyle I = 4\int \frac{v}{2+3v}\,dv$. Now use Formula (1) with $a = 2$, $b = 3$, and $u = v$ to obtain $\displaystyle 4\int \frac{v}{2+3v}\,dv = \frac{4}{9}\left(2+3\ln v - 2\ln|2+3\ln v|\right) + C$. Thus,
$\displaystyle\int \frac{4\ln x}{x(2+3\ln x)}\,dx = \frac{4}{9}\left(2+3\ln x - 2\ln|2+3\ln x|\right) + C$.

27. Using Formula (24) with $a = 1$, $n = 2$, and $u = x$, we have
$\int_0^1 x^2e^x\,dx = \left(x^2e^x\right)\Big|_0^1 - 2\int_0^1 xe^x\,dx = \left[x^2e^x - 2\left(xe^x - e^x\right)\right]_0^1 = \left(x^2e^x - 2xe^x + 2e^x\right)\Big|_0^1$
$= e - 2e + 2e - 2 = e - 2$.

29. $I = \int x^2\ln x\,dx$. Use Formula (27) with $n = 2$ and $u = x$ to obtain $\displaystyle I = \int x^2\ln x\,dx = \frac{x^3}{9}(3\ln x - 1) + C$.

31. $I = \int (\ln x)^3\,dx$. Use Formula (29) with $n = 3$ to write $I = x(\ln x)^3 - 3\int(\ln x)^2\,dx$. Now use Formula (29) again with $n = 2$ to obtain $I = x(\ln x)^3 - 3\left[x(\ln x)^2 - 2\int \ln x\,dx\right]$. Using Formula (29) one more time with $n = 1$ gives $\int(\ln x)^3\,dx = x(\ln x)^3 - 3x(\ln x)^2 + 6(x\ln x - x) + C = x(\ln x)^3 - 3x(\ln x)^2 + 6x\ln x - 6x + C$.

33. The number of visitors admitted to the amusement park by noon is found by evaluating the integral $\int_0^3 \frac{60}{(2+t^2)^{3/2}}\,dt = 60\int_0^3 \frac{dt}{(2+t)^{3/2}}$. Using Formula (12) with $a = \sqrt{2}$ and $u = t$, we find

$$60\int_0^3 \frac{dt}{(2+t^2)^{3/2}} = 60\left(\frac{t}{2\sqrt{2+t^2}}\right)\Bigg|_0^3 = 60\left(\frac{3}{2\sqrt{11}-0}\right) = \frac{90}{\sqrt{11}} \approx 27.136,$$ or approximately 27,136.

35. To find the average number of fruit flies over the first 10 days, use Formula (25) with $a = -0.02$, $b = 24$, and $u = t$. Thus,

$$\frac{1}{10}\int \frac{1000}{1+24e^{-0.02t}}\,dt = 100\int_0^{10} \frac{1}{1+24e^{-0.02t}}\,dt = 100\left[t + \frac{1}{0.02}\ln\left(1+24e^{-0.02t}\right)\right]_0^{10}$$
$$= 100\,(10 + 50\ln 20.6495 - 50\ln 25) \approx 44.08,$$ or approximately 44 fruit flies.

Over the first 20 days, the average is

$$\frac{1}{20}\int_0^{20} \frac{1000}{1+24e^{-0.02t}}\,dt = 50\int_0^{20} \frac{1}{1+24e^{-0.02t}}\,dt = 50\left[t + \frac{1}{0.02}\ln\left(1+24e^{-0.02t}\right)\right]_0^{20}$$
$$\approx 50\,(20 + 50\ln 17.088 - 50\ln 25) \approx 48.75,$$ or approximately 49 fruit flies.

37. To find the average average life expectancy for women from 1907 through 2007, use Formula (26) with $u = t$. Then $\frac{1}{6-1}\int_1^6 (49.9 + 17.1\ln t)\,dt = \frac{1}{5}\left[49.9t + 17.1\,(t\ln t - t)\right]_1^6$

$$= \tfrac{1}{5}\left\{[49.9\,(6) + 17.1\,(6\ln 6 - 6)] - [49.9\,(1) + 17.1\,(\ln 1 - 1)]\right\} \approx 69.6,$$ or 69.6 years.

39. Letting $p = 50$ gives $50 = \frac{30x}{5-x}$, so $5 = \frac{3x}{5-x}$, $25 - 5x = 3x$, and $x = \frac{25}{8} = 3.125$. Using Formula (1) with $a = 5$, $b = -1$, and $u = x$, we have

$$PS = (50)\,(3.125) - \int_0^{3.125} \frac{30x}{5-x}\,dx = 156.25 - 30\int_0^{3.125} \frac{x}{5-x}\,dx$$

$$= 156.25 - 30\,(5 - x - 5\ln|5-x|)\big|_0^{3.125} = 156.25 - 30\,[(5 - 3.125 - 5\ln 1.875) - 5 + 5\ln 5]$$

≈ 102.875, or approximately \$10,288.

41. If $p = 74$, then we have $74 = 50 + x\sqrt{1+x}$, so $x\sqrt{1+x} = 24$, $x^2\,(1+x) = 576$, and $x^3 + x^2 = 576$. If $x = 8$, then $8^3 + 8^2 = 512 + 64 = 576$, so $x = 8$ is a possible solution, and does in fact satisfy the original equation. Since $x^3 + x^2 - 576 = (x-8)\,(x^2 + 9x + 72)$, we see that $x = 8$ is the only real root. Thus, $PS = \int_0^8 \left[74 - \left(50 + x\sqrt{1+x}\right)\right]dx = \int_0^8 \left(24 - x\sqrt{1+x}\right)dx$. Using Formula 4 with $a = 1$, $b = -1$, and $u = x$, we find $PS = \left[24x - \frac{2}{15}\,(3x-2)\,(1+x)^{3/2}\right]_0^8 \approx 112.533$, or approximately \$112,533.

43. $I = \int_0^{10} (250{,}000 + 2000t^2)\,e^{-0.1t}\,dt = -2{,}500{,}000e^{-0.1t}\big|_0^{10} + 2000\int_0^{10} t^2e^{-0.1t}\,dt$. Using Formula (24) with $a = -0.1$, $n = 2$, and $u = t$ to evaluate the second integral, we have

$I = 2{,}500{,}000\,(1 - e^{-1}) + 2000\left[\left(-10t^2e^{-0.1t}\right)\big|_0^{10} + \frac{2}{0.1}\int_0^{10} te^{-0.1t}\,dt\right]$. Next, using Formula (23) to evaluate the last integral, we obtain

$$I \approx 1{,}580{,}301.397 + 2000\left\{-1000e^{-1} + 20\left[\frac{1}{0.01}\,(-0.1t - 1)\,e^{-0.1t}\right]_0^{10}\right\}$$

$$\approx 1{,}580{,}301.397 + 2000\left[-1000e^{-1} - 2000\,(-2e^{-1} + 1)\right]$$

$$\approx 1{,}580{,}301.397 + 2000\,(160.60) = \$1{,}901{,}507,$$ or approximately \$1,901,507.

45. Using Formula (22) from Section 6.7 and Formula (4) from the table of integrals with $a = 1$, $b = 8$, and $u = x$, we obtain

$$L = 2\int_0^1 \left(x - \tfrac{1}{3}x\sqrt{1+8x}\right) dx = x^2\Big|_0^1 - \tfrac{2}{3}\int_0^1 x\sqrt{1+8x}\,dx = 1 - \left[\tfrac{2}{3}\left(\tfrac{2}{15\cdot 64}\right)(24x-2)(1+8x)^{3/2}\right]_0^1$$

$$= 1 - \tfrac{1}{720}\left[(22)(27) - (-2)\right] \approx 0.1722.$$

47. The Gini Index for country A is $I_A = 2\int_0^1 \left(x - x^2 e^{x-1}\right) dx = 2\int_0^1 x\,dx - 2e^{-1}\int_0^1 x^2 e^x\,dx$.

To evaluate the second integral, we use Formula 24 followed by Formula 23, obtaining

$$I_A = 1 - 2e^{-1}\left[x^2 e^x - 2\int x e^x\,dx\right]_0^1 = 1 - 2e^{-1}\left[x^2 e^x - 2(x-1)e^x\right]_0^1 \approx 0.472.$$

For the second country, $I_B = 2\int_0^1 \left(x - \tfrac{1}{2}x^2\sqrt{3+x^2}\right) dx = 2\int_0^1 x\,dx - \int_0^1 x^2\sqrt{3+x^2}\,dx$.

To evaluate the second integral, we use Formula 8 with $a = \sqrt{3}$ and $u = x$, giving

$$I_B = x^2\Big|_0^1 - \left[\tfrac{1}{8}x\left(3+2x^2\right)\sqrt{3+x^2} - \tfrac{9}{8}\ln\left|x + \sqrt{3+x^2}\right|\right]_0^1 = 1 - \tfrac{1}{8}(5)(2) + \tfrac{9}{8}\ln(1+2) \approx 0.368.$$

From these results, we see that country B has a more equitable income distribution.

7.3 Numerical Integration

Concept Questions page 528

1. In the trapezoidal rule, each region beneath (or above) the graph of f is approximated by the area of a trapezoid whose base consists of two consecutive points in the partition. Therefore, n can be odd or even. In Simpson's Rule, the area of each subregion is approximated by part of a parabola passing through those points. Therefore, there are two subintervals involved in the approximations, and so n must be even.

3. If we use the trapezoidal rule and f is a linear function, then $f''(x) = 0$, so $M = 0$, and consequently the maximum error is 0. If we use Simpson's Rule, then $f^{(4)}(x) = 0$, so $M = 0$, and once again the maximum error is 0.

Exercises page 528

1. $\Delta x = \frac{b-a}{n} = \frac{2-0}{6} = \frac{1}{3}$, so $x_0 = 0$, $x_1 = \frac{1}{3}$, $x_2 = \frac{2}{3}$, $x_3 = 1$, $x_4 = \frac{4}{3}$, $x_5 = \frac{5}{3}$, $x_6 = 2$.

Trapezoidal Rule:

$$\int_0^2 x^2\,dx \approx \tfrac{1}{6}\left[0 + 2\left(\tfrac{1}{3}\right)^2 + 2\left(\tfrac{2}{3}\right)^2 + 2(1)^2 + 2\left(\tfrac{4}{3}\right)^2 + 2\left(\tfrac{5}{3}\right)^2 + 2^2\right]$$

$$\approx \tfrac{1}{6}(0.22222 + 0.88889 + 2 + 3.55556 + 5.55556 + 4) \approx 2.7037.$$

Simpson's Rule:

$$\int x^2\,dx \approx \tfrac{1}{9}\left[0 + 4\left(\tfrac{1}{3}\right)^2 + 2\left(\tfrac{2}{3}\right)^2 + 4(1)^2 + 2\left(\tfrac{4}{3}\right)^2 + 4\left(\tfrac{5}{3}\right)^2 + 2^2\right]$$

$$\approx \tfrac{1}{9}(0.44444 + 0.88889 + 4 + 3.55556 + 11.11111 + 4) \approx 2.6667.$$

Exact value: $\int_0^2 x^2\,dx = \frac{1}{3}x^3\Big|_0^2 = \frac{8}{3}$.

3. $\Delta x = \frac{b-a}{n} = \frac{1-0}{4} = \frac{1}{4}$, so $x_0 = 0, x_1 = \frac{1}{4}, x_2 = \frac{1}{2}, x_3 = \frac{3}{4}, x_4 = 1$.

Trapezoidal Rule:

$$\int_0^1 x^3\,dx \approx \frac{1}{8}\left[0 + 2\left(\frac{1}{4}\right)^3 + 2\left(\frac{1}{2}\right)^3 + 2\left(\frac{3}{4}\right)^3 + 1^3\right] \approx \frac{1}{8}(0 + 0.03125 + 0.25 + 0.84375 + 1) \approx 0.265625.$$

Simpson's Rule:

$$\int_0^1 x^3\,dx \approx \frac{1}{12}\left[0 + 4\left(\frac{1}{4}\right)^3 + 2\left(\frac{1}{2}\right)^3 + 4\left(\frac{3}{4}\right)^3 + 1\right] \approx \frac{1}{12}(0 + 0.0625 + 0.25 + 1.6875 + 1) \approx 0.25.$$

Exact value: $\int_0^1 x^3\,dx = \frac{1}{4}x^4\Big|_0^1 = \frac{1}{4} - 0 = \frac{1}{4}$.

5. Here $a = 1, b = 2$, and $n = 4$, so $\Delta x = \frac{2-1}{4} = \frac{1}{4} = 0.25$ and $x_0 = 1, x_1 = 1.25, x_2 = 1.5, x_3 = 1.75, x_4 = 2$.

Trapezoidal Rule: $\int_1^2 \frac{1}{x}\,dx \approx \frac{0.25}{2}\left[1 + 2\left(\frac{1}{1.25}\right) + 2\left(\frac{1}{1.5}\right) + 2\left(\frac{1}{1.75}\right) + \frac{1}{2}\right] \approx 0.697$.

Simpson's Rule: $\int_1^2 \frac{1}{x}\,dx \approx \frac{0.25}{3}\left[1 + 4\left(\frac{1}{1.25}\right) + 2\left(\frac{1}{1.5}\right) + 4\left(\frac{1}{1.75}\right) + \frac{1}{2}\right] \approx 0.6933$.

Exact value: $\int_1^2 \frac{1}{x}\,dx = \ln x|_1^2 = \ln 2 - \ln 1 = \ln 2 \approx 0.6931$.

7. $\Delta x = \frac{1}{4}, x_0 = 1, x_1 = \frac{5}{4}, x_2 = \frac{3}{2}, x_3 = \frac{7}{4}, x_4 = 2$.

Trapezoidal Rule: $\int_1^2 \frac{1}{x^2}\,dx \approx \frac{1}{8}\left[1 + 2\left(\frac{4}{5}\right)^2 + 2\left(\frac{2}{3}\right)^2 + 2\left(\frac{4}{7}\right)^2 + \left(\frac{1}{2}\right)^2\right] \approx 0.5090$.

Simpson's Rule: $\int_1^2 \frac{1}{x^2}\,dx \approx \frac{1}{12}\left[1 + 4\left(\frac{4}{5}\right)^2 + 2\left(\frac{2}{3}\right)^2 + 4\left(\frac{4}{7}\right)^2 + \left(\frac{1}{2}\right)^2\right] \approx 0.5004$.

Exact value: $\int_1^2 \frac{1}{x^2}\,dx = -\frac{1}{x}\Big|_1^2 = -\frac{1}{2} + 1 = \frac{1}{2}$.

9. $\Delta x = \frac{b-a}{n} = \frac{4-0}{8} = \frac{1}{2}$, so $x_0 = 0, x_1 = \frac{1}{2}, x_2 = \frac{2}{2}, x_3 = \frac{3}{2}, \ldots, x_8 = \frac{8}{2}$.

Trapezoidal Rule: $\int_0^4 \sqrt{x}\,dx \approx \frac{1/2}{2}\left(0 + 2\sqrt{0.5} + 2\sqrt{1} + 2\sqrt{1.5} + \cdots + 2\sqrt{3.5} + \sqrt{4}\right) \approx 5.26504$.

Simpson's Rule: $\int_0^4 \sqrt{x}\,dx \approx \frac{1/2}{3}\left(0 + 4\sqrt{0.5} + 2\sqrt{1} + 4\sqrt{1.5} + \cdots + 4\sqrt{3.5} + \sqrt{4}\right) \approx 5.30463$.

Exact value: $\int_0^4 \sqrt{x}\,dx \approx \frac{2}{3}x^{3/2}\Big|_0^4 = \frac{2}{3}(8) = \frac{16}{3}$.

11. $\Delta x = \frac{1-0}{6} = \frac{1}{6}$, so $x_0 = 0, x_1 = \frac{1}{6}, x_2 = \frac{2}{6}, \ldots, x_6 = \frac{6}{6}$.

Trapezoidal Rule: $\int_0^1 e^{-x}\,dx \approx \frac{1/6}{2}\left(1 + 2e^{-1/6} + 2e^{-2/6} + \cdots + 2e^{-5/6} + e^{-1}\right) \approx 0.633583$.

Simpson's Rule: $\int_0^1 e^{-x}\,dx \approx \frac{1/6}{3}\left(1 + 4e^{-1/6} + 2e^{-2/6} + \cdots + 4e^{-5/6} + e^{-1}\right) \approx 0.632123$.

Exact value: $\int_0^1 e^{-x}\,dx = -e^{-x}\big|_0^1 = -e^{-1} + 1 \approx 0.632121$.

13. $\Delta x = \frac{1}{4}$, so $x_0 = 0, x_1 = \frac{5}{4}, x_2 = \frac{3}{2}, x_3 = \frac{7}{4}, x_4 = 2$.

Trapezoidal Rule: $\int_1^2 \ln x\,dx \approx \frac{1}{8}\left(\ln 1 + 2\ln\frac{5}{4} + 2\ln\frac{3}{2} + 2\ln\frac{7}{4} + \ln 2\right) \approx 0.38370$.

Simpson's Rule: $\int_1^2 \ln x\,dx \approx \frac{1}{12}\left(\ln 1 + 4\ln\frac{5}{4} + 2\ln\frac{3}{2} + 4\ln\frac{7}{4} + \ln 2\right) \approx 0.38626$.

Exact value: $\int_1^2 \ln x\,dx \approx x(\ln x - 1)|_1^2 = 2(\ln 2 - 1) + 1 = 2\ln 2 - 1 \approx 0.3863$.

15. $\Delta x = \frac{1-0}{4} = \frac{1}{4}$, so $x_0 = 0, x_1 = \frac{1}{4}, x_2 = \frac{2}{4}, x_3 = \frac{3}{4}, x_4 = \frac{4}{4}$.

Trapezoidal Rule: $\int_0^1 \sqrt{1+x^3}\,dx \approx \frac{1/4}{2}\left[\sqrt{1} + 2\sqrt{1+\left(\frac{1}{4}\right)^3} + \cdots + 2\sqrt{1+\left(\frac{3}{4}\right)^3} + \sqrt{2}\right] \approx 1.1170.$

Simpson's Rule:

$\int_0^1 \sqrt{1+x^3}\,dx \approx \frac{1/4}{3}\left[\sqrt{1} + 4\sqrt{1+\left(\frac{1}{4}\right)^3} + 2\sqrt{1+\left(\frac{2}{4}\right)^3} + \cdots + 4\sqrt{1+\left(\frac{3}{4}\right)^3} + \sqrt{2}\right] \approx 1.1114.$

17. $\Delta x = \frac{2-0}{4} = \frac{1}{2}$, so $x_0 = 0, x_1 = \frac{1}{2}, x_2 = \frac{2}{2}, x_3 = \frac{3}{2}, x_4 = \frac{4}{2}$.

Trapezoidal Rule:

$$\int_0^2 \frac{1}{\sqrt{x^3+1}}\,dx = \frac{1/2}{2}\left[1 + \frac{2}{\sqrt{\left(\frac{1}{2}\right)^3+1}} + \frac{2}{\sqrt{(1)^3+1}} + \frac{2}{\sqrt{\left(\frac{3}{2}\right)^3+1}} + \frac{1}{\sqrt{(2)^3+1}}\right] \approx 1.3973.$$

Simpson's Rule:

$$\int_0^2 \frac{1}{\sqrt{x^3+1}}\,dx = \frac{1/2}{3}\left[1 + \frac{4}{\sqrt{\left(\frac{1}{2}\right)^3+1}} + \frac{2}{\sqrt{(1)^3+1}} + \frac{4}{\sqrt{\left(\frac{3}{2}\right)^3+1}} + \frac{1}{\sqrt{(2)^3+1}}\right] \approx 1.4052.$$

19. $\Delta x = \frac{2}{4} = \frac{1}{2}$, so $x_0 = 0, x_1 = \frac{1}{2}, x_2 = 1, x_3 = \frac{3}{2}, x_4 = 2$.

Trapezoidal Rule: $\int_0^2 e^{-x^2}\,dx = \frac{1}{4}\left[e^{-0} + 2e^{-(1/2)^2} + 2e^{-1} + 2e^{-(3/2)^2} + e^{-4}\right] \approx 0.8806.$

Simpson's Rule: $\int_0^2 e^{-x^2}\,dx = \frac{1}{6}\left[e^{-0} + 4e^{-(1/2)^2} + 2e^{-1} + 4e^{-(3/2)^2} + e^{-4}\right] \approx 0.8818.$

21. $\Delta x = \frac{2-1}{4} = \frac{1}{4}$, so $x_0 = 1, x_1 = \frac{5}{4}, x_2 = \frac{6}{4}, x_3 = \frac{7}{4}, x_4 = \frac{8}{4}$.

Trapezoidal Rule: $\int_1^2 x^{-1/2}e^x\,dx = \frac{1/4}{2}\left[e + \frac{2e^{5/4}}{\sqrt{5/4}} + \cdots + \frac{2e^{7/4}}{\sqrt{7/4}} + \frac{e^2}{\sqrt{2}}\right] \approx 3.7757.$

Simpson's Rule: $\int_1^2 x^{-1/2}e^x\,dx = \frac{1/4}{3}\left[e + \frac{4e^{5/4}}{\sqrt{5/4}} + \cdots + \frac{4e^{7/4}}{\sqrt{7/4}} + \frac{e^2}{\sqrt{2}}\right] \approx 3.7625.$

23. a. Here $a = -1$, $b = 2$, $n = 10$, and $f(x) = x^5$. Thus, $f'(x) = 5x^4$ and $f''(x) = 20x^3$. Because $f'''(x) = 60x^2 > 0$ on $(-1, 0) \cup (0, 2)$, we see that $f''(x)$ is increasing on $(-1, 0)$ and $(0, 2)$. Thus, we can take $M = f''(2) = 20\left(2^3\right) = 160$. Using Formula (7), we see that the maximum error incurred is $\frac{M(b-a)^3}{12n^2} = \frac{160[2-(-1)]^3}{12(100)} = 3.6.$

b. We compute $f'''(x) = 60x^2$ and $f^{(4)}(x) = 120x$. $f^{(4)}$ is clearly increasing on $(-1, 2)$, so we can take $M = f^{(4)}(2) = 240$. Therefore, using Formula (8), we see that an error bound is $\frac{M(b-a)^3}{180n^4} = \frac{240(3)^5}{180\left(10^4\right)} \approx 0.0324.$

25. a. Here $a = 1$, $b = 3$, $n = 10$, and $f(x) = \frac{1}{x}$. We find $f'(x) = -\frac{1}{x^2}$ and $f''(x) = \frac{2}{x^3}$. Because $f'''(x) = -\frac{6}{x^4} < 0$ on $(1, 3)$, we see that f'' is decreasing there. We may take $M = f''(1) = 2$. Using Formula (7), we find the error bound $\frac{M(b-a)^3}{12n^2} = \frac{2(3-1)^3}{12(100)} \approx 0.013.$

b. $f'''(x) = -\frac{6}{x^4}$ and $f^{(4)}(x) = \frac{24}{x^5}$. $f^{(4)}(x)$ is decreasing on $(1, 3)$, so we can take $M = f^{(4)}(1) = 24$. Using Formula (8), we find the error bound $\frac{24(3-1)^5}{180(10^4)} \approx 0.00043$.

27. a. Here $a = 0$, $b = 2$, $n = 8$, and $f(x) = (1+x)^{-1/2}$. We find $f'(x) = -\frac{1}{2}(1+x)^{-3/2}$ and $f''(x) = \frac{3}{4}(1+x)^{-5/2}$. Because f'' is positive and decreasing on $(0, 2)$, we see that $|f''(x)| \le \frac{3}{4}$. So the maximum error is $\frac{\frac{3}{4}(2-0)^3}{12(8)^2} = 0.0078125$.

b. $f'''(x) = -\frac{15}{8}(1+x)^{-7/2}$ and $f^{(4)}(x) = \frac{105}{16}(1+x)^{-9/2}$. Because $f^{(4)}$ is positive and decreasing on $(0, 2)$, we find $|f^{(4)}(x)| \le \frac{105}{16}$. Therefore, the maximum error is $\frac{\frac{105}{16}(2-0)^5}{180(8)^4} = 0.000285$.

29. The distance covered is given by

$$d = \int_0^2 V(t)\,dt = \frac{1/4}{2}\left[V(0) + 2V\left(\tfrac{1}{4}\right) + \cdots + 2V\left(\tfrac{7}{4}\right) + V(2)\right]$$

$$= \tfrac{1}{8}[19.5 + 2(24.3) + 2(34.2) + 2(40.5) + 2(38.4) + 2(26.2) + 2(18) + 2(16) + 8] \approx 52.84, \text{ or } 52.84 \text{ miles.}$$

31. $\Delta x = \frac{30-0}{10} = 3$. Trapezoidal Rule:

$$A = \tfrac{1}{30}\int_0^{30} f(x)\,dx$$

$$\approx \left(\tfrac{3}{2}\right)\left(\tfrac{1}{30}\right)[66 + 2(68 + 72 + 72 + 70 + 64 + 60 + 62 + 62 + 56) + 60] \approx 64.9°\text{F}.$$

Simpson's Rule:

$$A = \tfrac{1}{30}\int_0^{30} f(x)\,dx \approx \left(\tfrac{1}{30}\right)\left(\tfrac{3}{3}\right)[66 + 4(68) + 2(72) + 4(72) + 2(70) + 4(64)$$
$$+ 2(60) + 4(62) + 2(62) + 4(56) + 60] \approx 64.73°\text{F}.$$

33. $\frac{1}{13}\int_0^{13} f(t)\,dt = \left(\frac{1}{13}\right)\left(\frac{1}{2}\right)\{13.2 + 2[14.8 + 16.6 + 17.2 + 18.7 + 19.3 + 22.6 + 24.2 + 25$
$+ 24.6 + 25.6 + 26.4 + 26.6] + 26.6\} \approx 21.65$, or 21.65 mpg.

35. The total projected cost over the years from 2006 through 2014 is approximately $\int_1^9 C(t)\,dt$. Using the trapezoidal rule with $n = 8$, we have

$$\int_1^9 C(t)\,dt \approx \left(\tfrac{9-1}{8}\right)\left(\tfrac{1}{2}\right)[1.00 + 2(1.39) + 2(1.81) + 2(2.31) + 2(2.90) + 2(3.61) + 2(4.48) + 2(5.54) + 6.8]$$
$$= 25.94.$$

Thus, the total projected cost is approximately $25.94 billion.

37. The average petroleum reserves from 2002 through 2012 were $A = \frac{1}{10-0}\int_0^{10} S(t)\,dt = \frac{1}{10}\int_0^{10} \frac{720t^2 + 3480}{t^2 + 6.3}\,dt$. Using the Trapezoidal Rule with $a = 0$, $b = 10$, and $n = 10$, so that $\Delta t = \frac{10-0}{10} = 1$, we have $t_0 = 0$, $t = 1, \ldots$, $t_{10} = 10$. Thus,

$$A = \tfrac{1}{10}\int_0^{10} S(t)\,dt = \left(\tfrac{1}{10}\right)\left(\tfrac{1}{2}\right)[S(0) + 2S(1) + 2S(2) + \cdots + 2S(9) + S(10)]$$

$$\approx \tfrac{1}{20}[552.38 + 2(575.34 + 617.48 + 650.98 + 672.65 + 686.26+$$
$$+ 695.04 + 700.90 + 704.98 + 707.90) + 710.07] \approx 664.27,$$

or approximately 664.27 million barrels.

39. The value of Ivan's account at the end of 5 years is given by the future value of the income stream, that is, by $FV = e^{0.04(5)} \int_0^5 t^{1.5} e^{-0.02t} e^{-0.04t}\,dt = e^{0.2} \int_0^5 t^{1.5} e^{-0.06t}\,dt$. To approximate the integral using the Trapezoidal Rule with $n = 10$, we first find $\Delta t = \frac{5-0}{10} = \frac{1}{2}$. Thus, with $t_0 = 0$, $t_1 = \frac{1}{2}, \ldots, t_{10} = 5$, we have

$$\int_0^5 t^{1.5} e^{-0.06t}\,dt \approx \frac{1/2}{2}\left[0 + 2(0.5)^{1.5} e^{-0.06(0.5)} + 2(1.0)^{1.5} e^{-0.06(1.0)} + \cdots + 2(4.5)^{1.5} e^{-0.06(4.5)} + (5.0)^{1.5} e^{-0.06(5.0)}\right]$$
$$= \frac{1}{4}\left[0 + 2(0.34310 + 0.94176 + 1.67900 + 2.50859 + 3.40225 + 4.34019 + 5.30762 + 6.29302 + 7.28718) + 8.28260\right]$$
$$\approx 18.12201.$$

Thus, $FV \approx e^{0.2}(18.12201) \approx 22.13427$, and at the end of 5 years, Ivan's account will be worth about \$22,134.

41. We solve the equation $8 = \sqrt{0.01x^2 + 0.11x + 38}$, finding $64 = 0.01x^2 + 0.11x + 38$, $0.01x^2 + 0.11x - 26 = 0$, $x^2 + 11x - 2600 = 0$, and so $x = \dfrac{-11 \pm \sqrt{121 + 10{,}400}}{2} \approx 45.786$ (we choose the positive root). Therefore, $PS = (8)(45.786) - \int_0^{45.786} \sqrt{0.01x^2 + 0.11x + 38}\,dx$. We have $\Delta x = \dfrac{45.786}{8} = 5.72$, so $x_0 = 0$, $x_1 = 5.72$, $x_2 = 11.44, \ldots, x_8 = 45.79$.

a. $PS = 366.288 - \frac{5.72}{2}\left[\sqrt{38} + 2\sqrt{0.01(5.72)^2 + 0.11(5.72) + 38} + \cdots + \sqrt{0.01(45.79)^2 + 0.11(45.79) + 38}\right] \approx 51{,}558$, or \$51,558.

b. $PS = 366.288 - \frac{5.72}{3}\left[\sqrt{38} + 4\sqrt{0.01(5.72)^2 + 0.11(5.72) + 38} + \cdots + \sqrt{0.01(45.79)^2 + 0.11(45.79) + 38}\right] \approx 51{,}708$, or \$51,708.

43. The Gini Index is given by $L = 2\int_0^1 [x - f(x)]\,dx = 2\int_0^1 x\,dx - 2\int_0^1 f(x)\,dx = 1 - 2\int_0^1 f(x)\,dx$. But

$$\int_0^1 f(x)\,dx \approx \frac{0.1}{3}\left[0.01 + 4(0.02) + 2(0.04) + 4(0.08) + 2(0.13) + 4(0.20) + 2(0.27) + 4(0.36) + 2(0.47) + 4(0.60) + 1\right] = 0.262,$$

so $I \approx 1 - 2(0.262) \approx 0.48$.

45. Observe that if $p = 69.7$, then $x = 12$, so $PS = 69.7(12) - \int_0^{12} S(x)\,dx$. To estimate the integral using Simpson's Rule with $n = 6$, we find $\Delta x = \frac{12-0}{6} = 2$. Thus, with $x_0 = 0$, $x_1 = 2, \ldots, x_6 = 12$, we have

$$\int_0^{12} S(x)\,dx \approx \frac{2}{3}\left[f(0) + 4f(2) + 2f(4) + 4f(6) + 2f(8) + 4f(10) + f(12)\right]$$
$$= \frac{2}{3}\left[20.0 + 4(21.5) + 2(25.0) + 4(31.2) + 2(40.3) + 4(53.0) + 69.7\right]$$
$$\approx 428.7333.$$

Thus, the producers' surplus is $PS \approx 69.7(12) - 428.7333 \approx 407.6667$, or approximately \$407,667.

47. The required rate of flow is

$$R = (4.2)\,(\text{area of cross-section})$$
$$\approx (4.2)\left(\frac{6}{3}\right)[0.8 + 4(1.2) + 2(3) + 4(4.1) + 2(5.8) + 4(6.6) + 2(6.8) + 4(7) + 2(7.2) + 4(7.4) + 2(7.8) + 4(7.6) + 2(7.4) + 4(7) + 2(6.6) + 4(6) + 2(5.1) + 4(4.3) + 2(3.2) + 4(2.2) + 1.1]$$
$$= 2698.92, \text{ or } 2698.92\ \text{ft}^3/\text{sec}.$$

49. $\Delta x = \frac{21-19}{10} = 0.2$, so $x_0 = 19, x_1 = 19.2, x_2 = 19.4, \ldots, x_{10} = 21$.

$$P = \frac{100}{2.6\sqrt{2\pi}} \int_{19}^{21} e^{-0.5[(x-20)/2.6]^2}\,dx$$
$$\approx \frac{100}{2.6\sqrt{2\pi}} \left(\frac{0.2}{3}\right) \left\{e^{-0.5[(19-20)/2.6]^2} + 4e^{-0.5[(19.2-20)/2.6]^2} + \cdots + e^{-0.5[(21-20)/2.6]^2}\right\} \approx 29.94, \text{ or } 30\%.$$

51. False. The number n must be even for Simpson's Rule.

53. True. Using Formula (8), we see that the error incurred in the approximation is zero since, in this situation $f^{(4)}(x) = 0$ for all x in $[a, b]$.

7.4 Improper Integrals

Problem-Solving Tips

1. The improper integral on the left-hand side of the equation $\int_{-\infty}^{\infty} f(x)\,dx = \int_{-\infty}^{c} f(x)\,dx + \int_{-c}^{\infty} f(x)\,dx$ is convergent only if both integrals on the right-hand side of the equation converge.

2. It is often convenient to choose $c = 0$ in the formula given in Tip 1.

Concept Questions page 539

1. a. $\int_a^{\infty} f(x)\,dx = \lim_{b\to\infty} \int_a^b f(x)\,dx$

b. $\int_{-\infty}^{b} f(x)\,dx = \lim_{a\to-\infty} \int_a^b f(x)\,dx$

c. $\int_{-\infty}^{\infty} f(x)\,dx = \int_{-\infty}^{c} f(x)\,dx + \int_c^{\infty} f(x)\,dx$, where c is any real number.

Exercises page 539

1. $A = \displaystyle\int_2^{\infty} \frac{dx}{(x-1)^{3/2}} = \lim_{b\to\infty} \int_2^b (x-1)^{-3/2}\,dx = \lim_{b\to\infty} \left[-\frac{2}{(x-1)^{1/2}}\right]_2^b = \lim_{b\to\infty} \left[-\frac{2}{(b-1)^{1/2}} + 2\right] = 2.$

3. $A = \int_0^{\infty} e^{-2x}\,dx = \lim_{b\to\infty} \int_0^b e^{-2x}\,dx = \lim_{b\to\infty} \left(-\frac{1}{2}e^{-2x}\right)\Big|_0^b = \lim_{b\to\infty} \left(-\frac{1}{2}e^{-2b} + \frac{1}{2}\right) = \frac{1}{2}.$

5. The required area is given by $\displaystyle\int_3^{\infty} \frac{2}{x^2}\,dx = \lim_{b\to\infty} \int_3^b \frac{2}{x^2}\,dx = \lim_{b\to\infty} \left(-\frac{2}{x}\right)\Big|_3^b = \lim_{b\to\infty} \left(-\frac{2}{b} + \frac{2}{3}\right) = \frac{2}{3}.$

7. $A = \displaystyle\int_3^{\infty} \frac{1}{(x-2)^2}\,dx = \lim_{b\to\infty} \int_3^b (x-2)^{-2}\,dx = \lim_{b\to\infty} \left(-\frac{1}{x-2}\right)\Big|_3^b = \lim_{b\to\infty} \left(-\frac{1}{b-2} + 1\right) = 1.$

9. $A = \displaystyle\int_1^{\infty} \frac{1}{x^{3/2}}\,dx = \lim_{b\to\infty} \int_1^b x^{-3/2}\,dx = \lim_{b\to\infty} \left(-\frac{2}{\sqrt{x}}\right)\Big|_1^b = \lim_{b\to\infty} \left(-\frac{2}{\sqrt{b}} + 2\right) = 2.$

11. $A = \displaystyle\int_0^{\infty} \frac{1}{(x+1)^{5/2}}\,dx = \lim_{b\to\infty} \int_1^b (x+1)^{-5/2}\,dx = \lim_{b\to\infty} \left[-\frac{2}{3}(x+1)^{-3/2}\right]_0^b$

$= \displaystyle\lim_{b\to\infty} \left[-\frac{2}{3(b+1)^{3/2}} + \frac{2}{3}\right] = \frac{2}{3}.$

13. $A = \int_{-\infty}^{2} e^{2x}\,dx = \lim_{a\to-\infty} \int_a^2 e^{2x}\,dx = \lim_{a\to-\infty} \frac{1}{2}e^{2x}\Big|_a^2 = \lim_{a\to-\infty} \left(\frac{1}{2}e^4 - \frac{1}{2}e^{2a}\right) = \frac{1}{2}e^4.$

15. Using symmetry, the required area is given by $2\int_0^\infty \frac{x}{(1+x^2)^2}\,dx = 2\lim_{b\to\infty}\int_0^\infty \frac{x}{(1+x^2)^2}\,dx$.

To evaluate the indefinite integral $\int \frac{x}{(1+x^2)^2}\,dx$, put $u = 1+x^2$, so $du = 2x\,dx$ and $x\,dx = \frac{1}{2}\,du$. Then $\int \frac{x}{(1+x^2)^2}\,dx = \frac{1}{2}\int \frac{du}{u^2} = -\frac{1}{2u} + C = -\frac{1}{2(1+x^2)} + C$. Therefore, $2\lim_{b\to\infty}\int_0^b \frac{x}{(1+x^2)}\,dx = \lim_{b\to\infty}\left[-\frac{1}{1+x^2}\right]_0^b = \lim_{b\to\infty}\left[-\frac{1}{1+b^2}+1\right] = 1$.

17. a. $I(b) = \int_0^b \sqrt{x}\,dx = \frac{2}{3}x^{3/2}\Big|_0^b = \frac{2}{3}b^{3/2}$.

b. $\lim_{b\to\infty} I(b) = \lim_{b\to\infty}\frac{2}{3}b^{3/2} = \infty$.

19. $\int_1^\infty \frac{3}{x^4}\,dx = \lim_{b\to\infty}\int_1^b 3x^{-4}\,dx = \lim_{b\to\infty}\left(-\frac{1}{x^3}\right)\Big|_1^b = \lim_{b\to\infty}\left(-\frac{1}{b^3}+1\right) = 1$.

21. $A = \int_4^\infty \frac{2}{x^{3/2}}\,dx = \lim_{b\to\infty}\int_4^b 2x^{-3/2}\,dx = \lim_{b\to\infty}\left(-4x^{-1/2}\right)\Big|_4^b = \lim_{b\to\infty}\left(-\frac{4}{\sqrt{b}}+2\right) = 2$.

23. $\int_1^\infty \frac{4}{x}\,dx = \lim_{b\to\infty}\int_1^b \frac{4}{x}\,dx = \lim_{b\to\infty} 4\ln x\big|_1^b = \lim_{b\to\infty}(4\ln b) = \infty$.

25. $\int_{-\infty}^0 (x-2)^{-3}\,dx = \lim_{a\to-\infty}\int_a^0 (x-2)^{-3}\,dx = \lim_{a\to-\infty}\left[-\frac{1}{2(x-2)^2}\right]_a^0 = -\frac{1}{8}$.

27. $\int_1^\infty \frac{1}{(2x-1)^{3/2}}\,dx = \lim_{b\to\infty}\int_1^b (2x-1)^{-3/2}\,dx = \lim_{b\to\infty}\left[-\frac{1}{(2x-1)^{1/2}}\right]_1^b = \lim_{b\to\infty}\left(-\frac{1}{\sqrt{2b-1}}+1\right) = 1$.

29. $\int_0^\infty e^{-x}\,dx = \lim_{b\to\infty}\int_0^b e^{-x}\,dx = \lim_{b\to\infty}\left(-e^{-x}\right)\big|_0^b = \lim_{b\to\infty}\left(-e^{-b}+1\right) = 1$.

31. $\int_{-\infty}^0 e^{2x}\,dx = \lim_{a\to-\infty}\frac{1}{2}e^{2x}\Big|_a^0 = \lim_{a\to-\infty}\left(\frac{1}{2}-\frac{1}{2}e^{2a}\right) = \frac{1}{2}$.

33. We use the substitution $u=\sqrt{x}$: $\int_1^\infty \frac{e^{\sqrt{x}}}{\sqrt{x}}dx = \lim_{b\to\infty}\int_1^b \frac{e^{\sqrt{x}}}{\sqrt{x}}\,dx = \lim_{b\to\infty}\left(-2e^{\sqrt{x}}\right)\Big|_1^b = \lim_{b\to\infty}\left(2e^{\sqrt{b}}-2e\right) = \infty$, and so the improper integral diverges.

35. Integrating by parts, we have $\int_{-\infty}^0 xe^x\,dx = \lim_{a\to-\infty}\int_a^0 xe^x\,dx = \lim_{a\to-\infty}(x-1)e^x\big|_a^0 = \lim_{a\to-\infty}\left[-1+(a-1)e^a\right] = -1$.

37. $\int_{-\infty}^\infty x\,dx = \lim_{a\to-\infty}\frac{1}{2}x^2\Big|_a^0 + \lim_{b\to\infty}\frac{1}{2}x^2\Big|_0^b$, both of which diverge, and so the integral diverges.

39. $\int_{-\infty}^{\infty} x^3 \left(1+x^4\right)^{-2} dx = \int_{-\infty}^{0} x^3 \left(1+x^4\right)^{-2} dx + \int_0^{\infty} x^3 \left(1+x^4\right)^{-2} dx$

$$= \lim_{a \to -\infty} \int_a^0 x^3 \left(1+x^4\right)^{-2} dx + \lim_{b \to \infty} \int_0^b x^3 \left(1+x^4\right)^{-2} dx$$

$$= \lim_{a \to -\infty} \left[-\tfrac{1}{4}\left(1+x^4\right)^{-1}\right]_a^0 + \lim_{b \to \infty} \left[-\tfrac{1}{4}\left(1+x^4\right)^{-1}\right]_0^b$$

$$= \lim_{a \to -\infty} \left[-\frac{1}{4} + \frac{1}{4\left(1+a^4\right)}\right] + \lim_{b \to \infty} \left[-\frac{1}{4\left(1+b^4\right)} + \frac{1}{4}\right] = -\frac{1}{4} + \frac{1}{4} = 0.$$

41. $\int_{-\infty}^{\infty} xe^{1-x^2} dx = \lim_{a \to -\infty} \int_a^0 xe^{1-x^2} dx + \lim_{b \to \infty} \int_0^b xe^{1-x^2} dx = \lim_{a \to -\infty} -\tfrac{1}{2}e^{1-x^2}\Big|_a^0 + \lim_{b \to \infty} -\tfrac{1}{2}e^{1-x^2}\Big|_0^b$

$$= \lim_{a \to -\infty} \left(-\tfrac{1}{2}e + \tfrac{1}{2}e^{1-a^2}\right) + \lim_{b \to \infty} \left(-\tfrac{1}{2}e^{1-b^2} + \tfrac{1}{2}e\right) = 0.$$

43. $\displaystyle\int_{-\infty}^{\infty} \frac{e^{-x}}{1+e^{-x}} dx = \lim_{a \to -\infty} \left[-\ln\left(1+e^{-x}\right)\right]_a^0 + \lim_{b \to \infty} \left[-\ln\left(1+e^{-x}\right)\right]_0^b = \infty$, so the integral diverges.

45. First, we find the indefinite integral $I = \displaystyle\int \frac{dx}{x \ln^3 x}$. Let $u = \ln x$, so $du = \dfrac{1}{x}\,dx$. Then $I = \displaystyle\int \frac{du}{u^3} = -\frac{1}{2u^2} + C = -\frac{1}{2\ln^2 x} + C$, so $\displaystyle\int_e^{\infty} \frac{dx}{x \ln^3 x} = \lim_{b \to \infty} \int_e^b \frac{dx}{x \ln^3 x} = \lim_{b \to \infty} \left(-\frac{1}{2\ln^2 x}\right)\Bigg|_e^b = \lim_{b \to \infty} \left[-\frac{1}{2(\ln b)^2} + \frac{1}{2}\right] = \frac{1}{2}$, and so the given integral converges.

47. We want the present value PV of a perpetuity with $m = 1$, $P = 1500$, and $r = 0.05$. We find $PV = \frac{(1)(1500)}{0.05} = 30{,}000$, or \$30,000.

49. Using Formula (16) with $m = 1$, $P = 30{,}000$ and $r = 0.04$, we see that the present value of Heidi's investment is $PV = \frac{30{,}000}{0.04} = 750{,}000$, or \$750,000.

51. Integrating by parts, we find

$$PV = \int_0^{\infty} (20+t)\,e^{-0.04t}\,dt = \lim_{b \to \infty} \int_0^b 20e^{-0.04t}\,dt + \lim_{b \to \infty} \int_0^b te^{-0.04t}\,dt$$

$$= \lim_{b \to \infty} \left(-200e^{-0.4t}\right)\Big|_0^b + \lim_{b \to \infty} \left[100\,(-0.1t-1)\,e^{-0.4t}\right]_0^b = 1125, \text{ or } \$1{,}125{,}000.$$

53. True. $\int_a^{\infty} f(x)\,dx = \int_a^b f(x)\,dx + \int_b^{\infty} f(x)\,dx$, so if $\int_a^{\infty} f(x)\,dx$ exists, then $\int_b^{\infty} f(x)\,dx = \int_a^{\infty} f(x)\,dx - \int_a^b f(x)\,dx$.

55. False. Let $f(x) = \begin{cases} e^{-2x} & \text{if } x \le 0 \\ e^{-x} & \text{if } x > 0 \end{cases}$. Then $\int_{-\infty}^{\infty} f(x)\,dx = \int_{-\infty}^{0} e^{2x}\,dx + \int_0^{\infty} e^{-x}\,dx = \frac{1}{2} + 1 = \frac{3}{2}$. But $2\int_0^{\infty} f(x)\,dx = 2\int_0^{\infty} e^{-x}\,dx = 2$.

57. a. $CV \approx \displaystyle\int_0^{\infty} Re^{-it}\,dt = \lim_{b \to \infty} \int_0^b Re^{-it}\,dt = \lim_{b \to \infty} \left(-\frac{R}{i}e^{-it}\right)\Bigg|_0^b = \lim_{b \to \infty} \left(-\frac{R}{i}e^{-ib} + \frac{R}{i}\right) = \frac{R}{i}$.

b. $CV = \frac{10{,}000}{0.06} \approx 166{,}667$, or \$166,667.

59. $\int_{-\infty}^{b} e^{px}\,dx = \lim_{a\to-\infty} \int_{a}^{b} e^{px}\,dx = \lim_{a\to-\infty} \left(\frac{1}{p}e^{px}\right)\Big|_{a}^{b} = \lim_{a\to-\infty} \left(\frac{1}{p}e^{pb} - \frac{1}{p}e^{pa}\right) = -\frac{1}{p}e^{pa}$ if $p > 0$. If $p \le 0$, the integral diverges.

7.5 Applications of Calculus to Probability

Concept Questions page 551

1. See the definition on page 542 of the test. For example, $f(x) = \frac{3}{125}x^2$ on the interval $[0, 5]$.

2. a. See the definition on page 548 of the text.

b. $E(x) = 1/k$

Exercises page 551

1. $f(x) = \frac{1}{16}x \ge 0$ on $[2, 6]$. Next $\int_2^6 \frac{1}{16}x\,dx = \frac{1}{32}x^2\Big|_2^6 = \frac{1}{32}(36 - 4) = 1$, and so f is a probability density function on $[2, 6]$.

3. $f(x) = \frac{3}{8}x^2$ is nonnegative on $[0, 2]$. Next, we compute $\int_0^2 \frac{3}{8}x^2\,dx = \frac{1}{8}x^3\Big|_0^2 = \frac{1}{8}(8) = 1$, and so f is a probability density function on $[0, 2]$.

5. $\int_0^1 20(x^3 - x^4)\,dx = 20\left(\frac{1}{4}x^4 - \frac{1}{5}x^5\right)\Big|_0^1 = 20\left(\frac{1}{4} - \frac{1}{5}\right) = 20\left(\frac{1}{20}\right) = 1$. Furthermore, $f(x) = 20(x^3 - x^4) = 20x^3(1 - x) \ge 0$ on $[0, 1]$. Therefore, f is a density function on $[0, 1]$, as asserted.

7. Clearly $f(x) \ge 0$ on $[1, 4]$. Next, $\int_1^4 f(x)\,dx = \frac{3}{14}\int_1^4 x^{1/2}\,dx = \left(\frac{3}{14}\right)\left(\frac{2}{3}x^{3/2}\right)\Big|_1^4 = \frac{1}{7}(8 - 1) = 1$, and so f is a probability density function on $[1, 4]$.

9. First, note that $f(x) \ge 0$ on $[0, \infty)$. Next, let $I = \int x(x^2+1)^{-3/2}\,dx$. Integrate I using the substitution $u = x^2 + 1$, so $du = 2x\,dx$. Then $I = \frac{1}{2}\int u^{-3/2}du = \frac{1}{2}\left(-2u^{-1/2}\right) + C = -\frac{1}{\sqrt{u}} + C = -\frac{1}{\sqrt{x^2+1}} + C$.

Therefore, $\displaystyle\int_0^\infty \frac{x\,dx}{(x^2+1)^{3/2}} = \lim_{b\to\infty}\left(-\frac{1}{\sqrt{x^2+1}}\right)\Big|_0^b = \lim_{b\to\infty}\left(-\frac{1}{\sqrt{b^2+1}} + 1\right) = 1$, completing the proof.

11. $\int_1^4 k\,dx = kx|_1^4 = 3k = 1$ implies that $k = \frac{1}{3}$.

13. $\int_0^4 k(4 - x)\,dx = k\int_0^4 (4 - x)\,dx = k\left(4x - \frac{1}{2}x^2\right)\Big|_0^4 = k(16 - 8) = 8k = 1$ implies that $k = \frac{1}{8}$.

15. $\int_0^4 kx^{1/2}\,dx = \frac{2}{3}kx^{3/2}\Big|_0^4 = \frac{16}{3}k = 1$ implies that $k = \frac{3}{16}$.

17. $\displaystyle\int_1^\infty \frac{k}{x^3}\,dx = \lim_{b\to\infty}\int_1^b kx^{-3}\,dx = \lim_{b\to\infty}\left(-\frac{k}{2x^2}\right)\Big|_1^b = \lim_{b\to\infty}\left(-\frac{k}{2b^2} + \frac{k}{2}\right) = \frac{k}{2} = 1$ implies that $k = 2$.

19. a. $P(2 \le X \le 4) = \int_2^4 \frac{1}{12}x\,dx = \frac{1}{24}x^2\Big|_2^4 = \frac{1}{24}(16 - 4) = \frac{1}{2}$.

b. $P(1 \le X \le 4) = \int_1^4 \frac{1}{12}x\,dx = \frac{1}{24}x^2\Big|_1^4 = \frac{1}{24}(16 - 1) = \frac{5}{8}$.

c. $P(X \geq 2) = \int_2^5 \frac{1}{12}x\,dx = \frac{1}{24}x^2\Big|_2^5 = \frac{1}{24}(25-4) = \frac{7}{8}$.

d. $P(X = 2) = \int_2^2 \frac{1}{12}x\,dx = 0$.

21. a. $P(-1 \leq X \leq 1) = \int_{-1}^1 \frac{3}{32}(4-x^2)\,dx = \frac{3}{32}\left(4x - \frac{1}{3}x^3\right)\Big|_{-1}^1 = \frac{3}{32}\left[\left(4-\frac{1}{3}\right) - \left(-4+\frac{1}{3}\right)\right] = \frac{11}{16}$.

b. $P(X \leq 0) = \int_{-2}^0 \frac{3}{32}(4-x^2)\,dx = \frac{3}{32}\left(4x - \frac{1}{3}x^3\right)\Big|_{-2}^0 = \frac{3}{32}\left[0 - \left(-8+\frac{8}{3}\right)\right] = \frac{1}{2}$.

c. $P(X > -1) = \int_{-1}^2 \frac{3}{32}(4-x^2)\,dx = \frac{3}{32}\left(4x - \frac{1}{3}x^3\right)\Big|_{-1}^2 = \frac{3}{32}\left[\left(8-\frac{8}{3}\right) - \left(-4+\frac{1}{3}\right)\right] = \frac{27}{32}$.

d. $P(X = 0) = \int_0^0 \frac{3}{32}(4-x^2)\,dx = 0$.

23. a. $P(X \geq 4) = \int_4^9 \frac{1}{4}x^{-1/2}\,dx = \frac{1}{2}x^{1/2}\Big|_4^9 = \frac{1}{2}(3-2) = \frac{1}{2}$.

b. $P(1 \leq X < 8) = \int_1^8 \frac{1}{4}x^{-1/2}\,dx = \frac{1}{2}x^{1/2}\Big|_1^8 = \frac{1}{2}\left(2\sqrt{2}-1\right) \approx 0.9142$.

c. $P(X = 3) = \int_3^3 \frac{1}{4}x^{-1/2}\,dx = 0$.

d. $P(X \leq 4) = \int_1^4 \frac{1}{4}x^{-1/2}\,dx = \frac{1}{2}x^{1/2}\Big|_1^4 = \frac{1}{2}(2-1) = \frac{1}{2}$.

25. a. $P(0 \leq X \leq 4) = \int_0^4 4xe^{-2x^2}\,dx = -e^{-2x^2}\Big|_0^4 = -e^{-32} + 1 \approx 1$.

b. $P(X \geq 1) = \int_1^\infty 4xe^{-2x^2}\,dx = \lim_{b\to\infty} \int_1^b 4xe^{-2x^2}\,dx = \lim_{b\to\infty} \left(-e^{-2x^2}\right)\Big|_1^b = \lim_{b\to\infty}\left(-e^{-2b^2} + e^{-2}\right) = e^{-2}$
≈ 0.1353.

27. $\mu = \int_3^6 \frac{1}{3}x\,dx = \frac{1}{6}x^2\Big|_3^6 = \frac{1}{6}(36-9) = \frac{9}{2}$.

29. $\mu = \int_0^5 \frac{3}{125}x^3\,dx = \frac{3}{500}x^4\Big|_0^5 = \frac{15}{4}$.

31. $\mu = \int_1^5 \frac{3}{32}x(x-1)(5-x)\,dx = \frac{3}{32}\int_1^5 \left(-x^3+6x^2-5x\right)dx = \frac{3}{32}\left(-\frac{1}{4}x^4 + 2x^3 - \frac{5}{2}x^2\right)\Big|_1^5$
$= \frac{3}{32}\left[\left(-\frac{625}{4} + 250 - \frac{125}{2}\right) - \left(-\frac{1}{4} + 2 - \frac{5}{2}\right)\right] = 3$.

33. $\mu = \displaystyle\int_1^8 \frac{8}{7x}\,dx = \frac{8}{7}\ln x\Big|_1^8 = \frac{8}{7}\ln 8 \approx 2.3765$.

35. $\mu = \int_1^4 \frac{3}{14}x^{3/2}dx = \frac{3}{35}x^{5/2}\Big|_1^4 = \frac{3}{35}(32-1) = \frac{93}{35}$.

37. $\mu = \displaystyle\int_1^\infty \frac{3}{x^3}\,dx = \lim_{b\to\infty}\int_1^b 3x^{-3}\,dx = \lim_{b\to\infty}\left(-\frac{3}{2x^2}\right)\Big|_1^b = \lim_{b\to\infty}\left(-\frac{3}{2b^2} + \frac{3}{2}\right) = \frac{3}{2}$.

39. $\mu = \int_0^\infty \frac{1}{4}xe^{-x/4}\,dx = \lim_{b\to\infty}\int_0^b \frac{1}{4}xe^{-x/4}\,dx = \lim_{b\to\infty} 4\left(-\frac{1}{4} - 1\right)e^{-x/4}\Big|_0^b = \lim_{b\to\infty}\left[4\left(-\frac{1}{4}b - 1\right)e^{-b/4} + 4\right] = 4$.

41. a. Here $k = \frac{1}{15}$, so $f(x) = \frac{1}{15}e^{(-1/15)x}$.

b. The probability is $\int_{10}^{12} \frac{1}{15}e^{(-1/15)x}\,dx = -e^{(-1/15)x}\Big|_{10}^{12} = -e^{-12/15} + e^{-10/15} \approx 0.06$.

c. The probability is
$\int_{15}^{\infty} \frac{1}{15} e^{(-1/15)x}\,dx = \lim_{b\to\infty} \int_{15}^{b} \frac{1}{15} e^{(-1/15)x}\,dx = \lim_{b\to\infty} \left[-e^{(-1/15)x}\right]_{15}^{b} = \lim_{b\to\infty} \left[-e^{(-1/15)b} + e^{-1}\right] \approx 0.37.$

43. a. $P(X \le 30) = \int_0^{30} 0.02e^{-0.02x}\,dx = -e^{-0.02x}\Big|_0^{30} = -e^{-0.6} + 1 \approx 0.4512.$

b. $P(40 \le X \le 60) = \int_{40}^{60} 0.02e^{-0.02x}\,dx = -e^{-0.02x}\Big|_{40}^{60} = -e^{-1.2} + e^{-0.8} \approx 0.1481.$

c. $P(X \ge 70) = \int_{70}^{\infty} 0.02e^{-0.02x}\,dx = \lim_{b\to\infty} \int_{70}^{b} 0.02e^{-0.02x}\,dx = \lim_{b\to\infty} \left(-e^{-0.02x}\right)\Big|_{70}^{b} = \lim_{b\to\infty} \left(-e^{-0.02b} + e^{-1.4}\right)$
$= e^{-1.4} \approx 0.2466.$

45. a. The probability density function is $0.001e^{-0.001x}$. The required probability is
$P(600 \le x \le 800) = 0.001 \int_{600}^{800} e^{-0.001x}\,dx = -e^{-0.001x}\Big|_{600}^{800} = -e^{-0.8} + e^{-0.6} \approx 0.099.$

b. The probability is
$P(x \ge 1200) = 0.001 \int_{1200}^{\infty} e^{-0.001x}\,dx = 0.001 \lim_{b\to\infty} \int_{1200}^{b} e^{-0.001x}\,dx = \lim_{b\to\infty} \left(-e^{-0.001x}\right)\Big|_{1200}^{b}$
$= \lim_{b\to\infty} -e^{-0.001b} + e^{-1.2} \approx 0.30.$

47. Here $f(x) = \frac{1}{30} e^{-x/30}$. Thus, $P(x \ge 120) = \frac{1}{30} \int_{120}^{\infty} e^{-x/30}\,dx = \lim_{b\to\infty} \left(-e^{-x/30}\right)\Big|_{120}^{b} = e^{-120/30} \approx 0.02.$

49. Let X be a random variable that denotes the waiting time. Then X is uniformly distributed over the interval $[0, 15]$. So the associated probability density function is $f(x) = \frac{1}{15}$, $0 \le X \le 15$.

a. The desired probability is $P(X \ge 5) = P(5 \le X \le 15) = \int_5^{15} \frac{1}{15}\,dx = \frac{1}{15}x\Big|_5^{15} = \frac{10}{15} = \frac{2}{3}.$

b. The desired probability is $P(5 \le X \le 8) = \int_5^8 \frac{1}{15}\,dx = \frac{1}{15}x\Big|_5^8 = \frac{1}{5}.$

51. Let X denote the number of minutes past 8:00 P.M. that Joan arrives for the show. Since X is uniformly distributed over the interval $[0, 30]$, the uniform density function associated with this problem is $f(x) = \frac{1}{30}$, $0 \le x \le 30$.

a. In this case, Joan must arrive between 8:10 P.M. and 8:15 P.M. or between 8:25 P.M. and 8:30 P.M. Therefore, the required probability is $P(10 < X < 15) + P(25 < X < 30) = \int_{10}^{15} \frac{1}{30}\,dx + \int_{20}^{30} \frac{1}{30}\,dx = \frac{1}{6} + \frac{1}{6} = \frac{1}{3}.$

b. In this case, Joan must arrive between 8:00 P.M. and 8:05 P.M. or between 8:15 P.M. and 8:20 P.M. Therefore, the required probability is $P(0 < X < 5) + P(15 < X < 20) = \int_0^5 \frac{1}{30}\,dx + \int_{15}^{20} \frac{1}{30}\,dx = \frac{1}{6} + \frac{1}{6} = \frac{1}{3}.$

53. $\mu = \displaystyle\int_1^3 t \cdot \frac{9}{4t^3}\,dt = \frac{9}{4}\int_1^3 t^{-2}\,dt = -\frac{9}{4t}\Big|_1^3 = -\frac{9}{4}\left(\frac{1}{3} - 1\right) = \frac{3}{2}$, so the expected reaction time is 1.5 seconds.

55. $\mu = \int_0^3 x \cdot \frac{2}{9}x(3 - x)\,dx = \frac{2}{9}\int_0^3 (3x^2 - x^3)\,dx = \frac{2}{9}\left(x^3 - \frac{1}{4}x^4\right)\Big|_0^3 = \frac{2}{9}\left(27 - \frac{81}{4}\right) = 1.5$, so the expected amount of snowfall is 1.5 ft.

57. $\mu = \int_0^{\infty} t \cdot 9(9 + t^2)^{-3/2}\,dt = \lim_{b\to\infty} \int_0^b 9t(9 + t^2)^{-3/2}\,dt = \lim_{b\to\infty} \left[(9)\left(\frac{1}{2}\right)(-2)(9 + t^2)^{-1/2}\right]_0^b$
$= \lim_{b\to\infty} \left[-\dfrac{9}{\sqrt{9 + b^2}} + 3\right] = 3$, so the plasma TVs are expected to last 3 years.

59. We need to find $E(X) = \int_0^\infty x\,f(x)\,dx = \frac{1}{10}\int_0^\infty xe^{-x/10}\,dx = \frac{1}{10}\lim_{b\to\infty}\int_0^b xe^{-x/10}\,dx$. Using integration by parts or Formula (23) from Section 7.2, we find that

$$\int xe^{-x/10}\,dx = \frac{1}{\left(-\frac{1}{10}\right)^2}\left(-\frac{x}{10}-1\right)e^{-x/10} + C = -100\left(\frac{x+10}{10}\right)e^{-x/10} + C = -10\,(x+10)\,e^{-x/10}.$$

Using this result, we have

$E(x) = \frac{1}{10}(-10)\lim_{b\to\infty}\left[(x+10)\,e^{-x/10}\right]_0^b = -\lim_{b\to\infty}\left(be^{-b/10} + 10e^{-b/10} - 10\right) = (-1)(-10) = 10$, or 10 minutes.

61. Refer to Exercise 60. The required average waiting time is $E(X) = \frac{1}{2}(a-b) = \frac{1}{2}(0+15) = \frac{15}{2}$, or $7\frac{1}{2}$ minutes.

63. Because f is a probability density function on $[0,1]$, we have

$\int_0^1 \left(ax^2+bx\right)dx = \left(\frac{1}{3}ax^3+\frac{1}{2}bx^2\right)\Big|_0^1 = \frac{1}{3}a+\frac{1}{2}b = 1$ and

$\int_0^1 \left(ax^3+bx^2\right)dx = \left(\frac{1}{4}ax^4+\frac{1}{3}bx^3\right)\Big|_0^1 = \frac{1}{4}a+\frac{1}{3}b = 0.6$. Solving the system of equations, we find $a = -2.4$ and $b = 3.6$.

65. We require that $\int_0^\infty e^{-ax}(bx+c)\,dx = 1$. Consider $I = \int e^{-ax}(bx+c)\,dx$. Let $u = bx+c$ and $dv = e^{-ax}\,dx$, so $du = b\,dx$ and $v = -\frac{1}{a}e^{-ax}$. Thus, $I = -\frac{1}{a}(bx+c)\,e^{-ax} + \frac{b}{a}\int e^{-ax}\,dx = -\frac{1}{a}(bx+c)\,e^{-ax} - \frac{b}{a^2}e^{-ax} + C$.

Thus,

$$\int_0^\infty e^{-ax}(bx+c)\,dx = \lim_{L\to\infty}\int_0^L e^{-ax}(bx+c)\,dx = \lim_{L\to\infty}\left\{\left[-\frac{1}{a}(bL+C)\,e^{-aL} + \frac{b}{a^2}e^{-aL}\right] - \left(-\frac{c}{a}-\frac{b}{a^2}\right)\right\}$$
$$= \frac{b}{a^2}+\frac{c}{a} = \frac{b+ac}{a^2}.$$

Because $f(x) = e^{-ax}(bx+c)$ must be nonnegative, we see that $a > 0$, $b \geq 0$, $c \geq 0$, and $\dfrac{b+ac}{a^2} = 1$; that is, $b+ac = a^2$.

67. True. Observe that $P(x < a) = \int_{-\infty}^a f(x)\,dx$ and $P(x > b) = \int_b^\infty f(x)\,dx$, so

$P(x<a) + P(a<x<b) + P(x>b) = \int_{-\infty}^a f(x)\,dx + \int_a^b f(x)\,dx + \int_b^\infty f(x)\,dx = \int_{-\infty}^\infty f(x)\,dx = 1$.

Therefore, $P(x<a) + P(x>b) = 1 - \int_a^b f(x)\,dx$.

69. False. The expected value of x is $\int_a^b xf(x)\,dx$.

CHAPTER 7 Concept Review Questions page 556

1. product, $uv - \int v\,du$, u, easy to integrate

3. $\dfrac{\Delta x}{2}\left[f(x_0)+2f(x_1)+2f(x_2)+\cdots+2f(x_{n-1})+f(x_n)\right]$, $\dfrac{M(b-a)^3}{12n^2}$

5. $\lim_{a\to-\infty}\int_a^b f(x)\,dx$, $\lim_{b\to\infty}\int_a^b f(x)\,dx$, $\int_{-\infty}^c f(x)\,dx + \int_c^\infty f(x)\,dx$

CHAPTER 7 Review Exercises page 556

1. Let $u = 2x$ and $dv = e^{-x}\,dx$, so $du = 2\,dx$ and $v = -e^{-x}$. Then $\int 2xe^{-x}\,dx = uv - \int v\,du = -2xe^{-x} + 2\int e^{-x}\,dx = -2xe^{-x} - 2e^{-x} + C = -2(1+x)e^{-x} + C$.

3. Let $u = \ln 5x$ and $dv = dx$, so $du = \frac{1}{x}\,dx$ and $v = x$. Then $\int \ln 5x\,dx = x\ln 5x\,dx - \int dx = x\ln 5x - x + C = x(\ln 5x - 1) + C$.

5. Let $u = x$ and $dv = e^{-2x}\,dx$, so $du = dx$ and $v = -\frac{1}{2}e^{-2x}$. Then $\int_0^1 xe^{-2x}\,dx = -\frac{1}{2}xe^{-2x}\Big|_0^1 + \frac{1}{2}\int_0^1 e^{-2x}\,dx = -\frac{1}{2}e^{-2} - \left(\frac{1}{4}e^{-2x}\right)\Big|_0^1 = -\frac{1}{2}e^{-2} - \frac{1}{4}e^{-2} + \frac{1}{4} = \frac{1}{4}\left(1 - 3e^{-2}\right)$.

7. $f(x) = \displaystyle\int f'(x)\,dx = \int \frac{\ln x}{\sqrt{x}}\,dx$. To evaluate the integral, we use parts with $u = \ln x$, $dv = x^{-1/2}\,dx$, $du = \frac{1}{x}\,dx$ and $v = 2x^{1/2}$. Then

$$\int \frac{\ln x}{x^{1/2}}\,dx = 2x^{1/2}\ln x - \int 2x^{-1/2}\,dx = 2x^{1/2}\ln x - 4x^{1/2} + C = 2x^{1/2}(\ln x - 2) + C = 2\sqrt{x}(\ln x - 2) + C.$$

But $f(1) = -2$, giving $2\sqrt{1}(\ln 1 - 2) + C = -2$, so $C = 2$. Therefore, $f(x) = 2\sqrt{x}(\ln x - 2) + 2$.

9. Using Formula (4) with $a = 3$, $b = 2$, and $u = x$, we have

$$\int \frac{x^2}{(3+2x)^2}\,dx = \frac{1}{8}\left(3 + 2x - \frac{9}{3+2x} - 6\ln|3+2x|\right) + C.$$

11. Using Formula (24) with $a = 4$, $n = 2$, and $u = x$, we have $\int x^2e^{4x}\,dx = \frac{1}{4}x^2e^{4x} - \frac{1}{2}\int xe^{4x}\,dx$. Now we use Formula (23) to obtain $\int x^2e^{4x}\,dx = \frac{1}{4}x^2e^{4x} - \frac{1}{2}\left[\frac{1}{16}(4x-1)e^{4x}\right] + C = \frac{1}{32}\left(8x^2 - 4x + 1\right)e^{4x} + C$.

13. Using Formula (17) with $a = 2$ and $u = x$, we have $\displaystyle\int \frac{dx}{x^2\sqrt{x^2-4}} = \frac{\sqrt{x^2-4}}{4x} + C$.

15. $\int_0^\infty e^{-2x}\,dx = \lim\limits_{b\to\infty}\int_0^b e^{-2x}\,dx = \lim\limits_{b\to\infty}\left(-\frac{1}{2}e^{-2x}\right)\Big|_0^b = \lim\limits_{b\to\infty}\left(-\frac{1}{2}e^{-2b} + \frac{1}{2}\right) = \frac{1}{2}$.

17. $\displaystyle\int_3^\infty \frac{2}{x}\,dx = \lim_{b\to\infty}\int_3^b \frac{2}{x}\,dx = \lim_{b\to\infty} 2\ln x\Big|_3^b = \lim_{b\to\infty}(2\ln b - 2\ln 3) = \infty$.

19. $\displaystyle\int_2^\infty \frac{dx}{(1+2x)^2} = \lim_{b\to\infty}\int_2^b (1+2x)^{-2}\,dx = \lim_{b\to\infty}\left(\frac{1}{2}\right)(-1)(1+2x)^{-1}\Big|_2^b = \lim_{b\to\infty}\left[-\frac{1}{2(1+2b)} + \frac{1}{2(5)}\right] = \frac{1}{10}$.

21. $\Delta x = \frac{b-a}{n} = \frac{3-1}{4} = \frac{1}{2}$, so $x_0 = 1$, $x_1 = \frac{3}{2}$, $x_2 = 2$, $x_3 = \frac{5}{2}$, $x_4 = 3$.

Trapezoidal Rule: $\displaystyle\int_1^3 \frac{dx}{1+\sqrt{x}} \approx \frac{\frac{1}{2}}{2}\left[\frac{1}{2} + \frac{2}{1+\sqrt{1.5}} + \frac{2}{1+\sqrt{2}} + \frac{2}{1+\sqrt{2.5}} + \frac{1}{1+\sqrt{3}}\right] \approx 0.8421$.

Simpson's Rule $\displaystyle\int_1^3 \frac{dx}{1+\sqrt{x}} \approx \frac{\frac{1}{2}}{3}\left[\frac{1}{2} + \frac{4}{1+\sqrt{1.5}} + \frac{2}{1+\sqrt{2}} + \frac{4}{1+\sqrt{2.5}} + \frac{1}{1+\sqrt{3}}\right] \approx 0.8404$.

23. $\Delta x = \frac{1-(-1)}{4} = \frac{1}{2}$, so $x_0 = -1, x_1 = -\frac{1}{2}, x_2 = 0, x_3 = \frac{1}{2}, x_4 = 1$.

Trapezoidal Rule: $\int_{-1}^{1} \sqrt{1+x^4}\,dx \approx \frac{0.5}{2}\left[\sqrt{2} + 2\sqrt{1+(-0.5)^4} + 2 + 2\sqrt{1+(0.5)^4} + \sqrt{2}\right] \approx 2.2379$.

Simpson's Rule: $\int_{-1}^{1} \sqrt{1+x^4}\,dx \approx \frac{0.5}{3}\left[\sqrt{2} + 4\sqrt{1+(-0.5)^4} + 2 + 4\sqrt{1+(0.5)^4} + \sqrt{2}\right] \approx 2.1791$.

25. a. Here $a = 0$, $b = 1$, and $f(x) = \dfrac{1}{x+1}$. We have $f'(x) = -\dfrac{1}{(x+1)^2}$ and $f''(x) = \dfrac{2}{(x+1)^3}$.

Because f'' is positive and decreasing on $(0, 1)$, it attains its maximum value of 2 at $x = 0$, so we take $M = 2$. Using Formula (7) from Section 7.3, we see that the maximum error incurred is

$$\frac{M(b-a)^3}{12n^2} = \frac{2(1^3)}{12(8^2)} = \frac{1}{384} \approx 0.002604.$$

b. We compute $f'''(x) = -\dfrac{6}{(x+1)^4}$ and $f^{(4)}(x) = \dfrac{24}{(x+1)^5}$. Because $f^{(4)}(x)$ is positive and decreasing on $(0, 1)$, we take $M = 24$. The maximum error is $\dfrac{24(1^5)}{180(8^4)} = \dfrac{1}{30720} \approx 0.000033$.

27. $f(x) \geq 0$ on $[0, 3]$. Next,

$$\frac{1}{9}\int_0^3 x\sqrt{9-x^2}\,dx = \frac{1}{9}\int_0^3 x(9-x^2)^{1/2}\,dx = \left(\frac{1}{9}\right)\left(-\frac{1}{2}\right)\left(\frac{2}{3}\right)(9-x^2)^{3/2}\Big|_0^3 = -\frac{1}{27}(9-x^2)^{3/2}\Big|_0^3$$
$$= 0 + \frac{1}{27}(9)^{3/2} = 1.$$

29. a. $\displaystyle\int_1^4 \frac{k}{\sqrt{x}}dx = k\int_1^4 x^{-1/2}\,dx = 2k\sqrt{x}\Big|_{x=1}^{x=4} = 2k(2-1) = 2k = 1$, so $k = \frac{1}{2}$.

b. $\displaystyle\int_2^3 \frac{\frac{1}{2}}{\sqrt{x}}\,dx = \frac{1}{2}\int_2^3 x^{-1/2}\,dx = 2\left(\frac{1}{2}\right)\sqrt{x}\Big|_2^3 = \sqrt{3} - \sqrt{2} = 0.3178.$

31. Since f is a probability density function on $[0, 1]$, we have $\int_0^1 f(x)\,dx = 1$, so $\int_0^1 (a + bx^2)\,dx = \left[ax + \frac{1}{3}bx^3\right]_0^1 = a + \frac{1}{3}b = 1$. Next $E(X) = 0.6$ gives $E(x) = \int_0^1 xf(x)\,dx = \int_0^1 x(a+bx^2)\,dx = \int_0^1 (ax + bx^3)\,dx = \left[\frac{1}{2}ax^2 + \frac{1}{4}bx^4\right]_0^1 = \frac{1}{2}a + \frac{1}{4}b = 0.6 = \frac{3}{5}$. So we have a system of equations in a and b: $a + \frac{1}{3}b = 1$ and $\frac{1}{2}a + \frac{1}{4}b = \frac{3}{5}$. The first equation gives $a = 1 - \frac{1}{3}b$. Substituting into the second equation gives $\frac{1}{2}(1 - \frac{1}{3}b) + \frac{1}{4}b = \frac{3}{5}$, so $\frac{1}{2} - \frac{1}{6}b + \frac{1}{4}b = \frac{3}{5}$, $30 - 10b + 15b = 36$, $5b = 6$, and $b = \frac{6}{5}$. Using the first equation in the original system, we find that $a = 1 - \frac{1}{3}b = 1 - \frac{1}{3}\left(\frac{6}{5}\right) = \frac{3}{5}$. Therefore, $a = \frac{3}{5}$ and $b = \frac{6}{5}$.

33. $P(0 \leq X \leq 2) = \int_0^2 \frac{1}{16}(6x - x^2)\,dx = \frac{1}{16}\left(3x^2 - \frac{1}{3}x^3\right)\Big|_0^2 = \frac{1}{16}\left(12 - \frac{8}{3}\right) = \frac{7}{12} \approx 0.5833$.

35. The producer's surplus is given by $PS = \overline{p}\,\overline{x} - \int_0^{\overline{x}} s(x)\,dx$, where $\overline{x}$ is found by solving the equation $2\sqrt{25+x^2} = 26$. Thus, $\sqrt{25+x^2} = 13$, $25 + x^2 = 169$, and $x = \pm 12$. Therefore, $\overline{x} = 12$, and so $PS = (26)(12) - 2\int_0^{12}(25+x^2)^{1/2}\,dx$. Using Formula (7) from Section 7.2 with $a = 5$, we obtain

$$PS = (26)(12) - 2\int_0^{12}(25+x^2)^{1/2}\,dx = 312 - 2\left[\frac{1}{2}x(25+x^2)^{1/2} + \frac{25}{2}\ln\left|x + (25+x^2)^{1/2}\right|\right]_0^{12}$$
$$= 312 - 2\left[6(13) + \frac{25}{2}\ln(12+13) - \frac{25}{2}\ln 5\right] \approx 115.76405, \text{ or } \$1{,}157{,}641.$$

37. If $p = 30$, we have $2\sqrt{325 - x^2} = 30$, $\sqrt{325 - x^2} = 15$, $325 - x^2 = 225$, $x^2 = 100$, and so $x = \pm 10$. The equilibrium point is thus $(10, 30)$, and $CS = \int_0^{10} 2\sqrt{325 - x^2}\,dx - (30)(10)$. To approximate the integral using Simpson's Rule with $n = 10$, we have $\Delta x = \frac{10-0}{10} = 1$, so $x_0 = 0, x_1 = 1, x_2 = 2, \ldots, x_{10} = 10$. Thus,
$2\int_0^{10} \sqrt{325 - x^2}\,dx \approx \frac{2}{3}\left(\sqrt{325} + 4\sqrt{325 - 1} + 2\sqrt{325 - 4} + \cdots + 4\sqrt{325 - 81} + \sqrt{325 - 100}\right)$

≈ 341.1, so $CS \approx 341.1 - 300 \approx 41.1$, or \$41,100.

39. Think of the upper curve as the graph of f and the lower curve as the graph of g. Then the required area is given by $A = \int_0^{150} [f(x) - g(x)]\,dx = \int_0^{150} h(x)\,dx$,where $h = f - g$. Using Simpson's Rule,
$A \approx \frac{15}{3}[h(0) + 4h(1) + 2f(2) + 4f(3) + 2f(4) + \cdots + 4f(9) + f(10)]$
$= 5[0 + 4(25) + 2(40) + 4(70) + 2(80) + 4(90) + 2(65) + 4(50) + 2(60) + 4(35) + 0] = 7850$,
or 7850 ft^2.

CHAPTER 7 Before Moving On... page 558

1. Let $u = \ln x$ and $dv = x^2\,dx$, so $du = \frac{1}{x}\,dx$ and $v = \frac{1}{3}x^3$. Then
$\int x^2 \ln x\,dx = \frac{1}{3}x^3 \ln x - \int \frac{1}{3}x^2\,dx = \frac{1}{3}x^3 \ln x - \frac{1}{9}x^3 + C = \frac{1}{9}x^3(3\ln x - 1) + C$.

2. $I = \int \frac{dx}{x^2\sqrt{8 + 2x^2}}$. Let $u = \sqrt{2}x$, so $du = \sqrt{2}\,dx$ and $dx = \frac{\sqrt{2}}{2}\,du$. Then

$I = \frac{\sqrt{2}}{2}\int \frac{du}{\frac{1}{2}u^2\sqrt{8 + u^2}} = \sqrt{2}\int \frac{du}{u^2\sqrt{\left(2\sqrt{2}\right)^2 + u^2}}$. Using Formula (11) from Section 7.2 with $a = 2\sqrt{2}$ and

$x = u$, $I = \sqrt{2}\int \frac{du}{u^2\sqrt{\left(2\sqrt{2}\right)^2 + u^2}} = \sqrt{2}\left(-\frac{\sqrt{8 + u^2}}{8u}\right) + C = -\frac{\sqrt{8 + 2x^2}}{8x} + C$.

3. $n = 5$, so $\Delta x = \frac{4-2}{5} = 0.4$ and $x_0 = 2, x_1 = 2.4, x_2 = 2.8, x_3 = 3.2, x_4 = 3.6, x_5 = 4$. Thus,
$\int_2^4 \sqrt{x^2 + 1}\,dx \approx \frac{0.4}{2}[f(2) + 2f(2.4) + 2f(2.8) + 2f(3.2) + 2f(3.2) + 2f(3.6) + f(4)]$
$= 0.2[2.23607 + 2(2.6) + 2(2.97321) + 2(3.35261) + 2(3.73631) + 4.12311] \approx 6.3367$.

4. $n = 6$, so $\Delta x = \frac{3-1}{6} = \frac{1}{3}$ and $x_0 = 1, x_1 = \frac{4}{3}, x_2 = \frac{5}{3}, x_3 = 2, x_4 = \frac{7}{3}, x_5 = \frac{8}{3}, x_6 = 3$.
$\int_1^3 e^{0.2x}\,dx \approx \frac{1/3}{3}\left[f(1) + 4f\left(\frac{4}{3}\right) + 2f\left(\frac{5}{3}\right) + 4f(2) + 2f\left(\frac{7}{3}\right) + 4f\left(\frac{8}{3}\right) + f(3)\right]$
$\approx \frac{1}{9}[1.2214 + 4(1.30561) + 2(1.39561) + 4(1.49182) + 2(1.59467) + 4(1.7046) + 1.82212] \approx 3.0036$.

5. $\int_1^\infty e^{-2x}\,dx = \lim_{b\to\infty} \int_1^b e^{-2x}\,dx = \lim_{b\to\infty}\left(-\frac{1}{2}e^{-2x}\Big|_1^b\right) = \lim_{b\to\infty}\left(-\frac{1}{2}e^{-2b} + \frac{1}{2}e^{-2}\right) = \frac{1}{2}e^{-2} = \frac{1}{2e^2}$.

6. a. $f(x) \geq 0$ on $[0, 8]$ and $\int_0^8 \frac{5}{96}x^{2/3}\,dx = \frac{5}{96}\left(\frac{3}{5}x^{5/3}\right)\Big|_0^8 = \frac{5}{96}\left(\frac{3}{5}\right)(8^{5/3}) = 1$. Therefore, f is a probability density function on $[0, 8]$.

b. $P(1 \leq X \leq 8) = \int_1^8 \frac{5}{96}x^{2/3}\,dx = \frac{5}{96}\left(\frac{3}{5}x^{5/3}\right)\Big|_1^8 = \frac{1}{32}(32 - 1) = \frac{31}{32}$.

8 CALCULUS OF SEVERAL VARIABLES

8.1 Functions of Several Variables

Concept Questions page 567

1. A function of two variables is a rule that assigns to each point (x, y) in a subset of the plane a unique number $f(x, y)$. For example, $f(x, y) = x^2 + 2y^2$ has the whole xy-plane as its domain.

3. a. The graph of $f(x, y)$ is the set $S = \{(x, y, z) \mid z = f(x, y), (x, y) \in D\}$, where D is the domain of f.

b. The level curve of f is the projection onto the xy-plane of the trace of $f(x, y)$ in the plane $z = k$, where k is a constant in the range of f.

Exercises page 567

1. $f(x, y) = 2x + 3y - 4$, so $f(0, 0) = 2(0) + 3(0) - 4 = -4$, $f(1, 0) = 2(1) + 3(0) - 4 = -2$, $f(0, 1) = 2(0) + 3(1) - 4 = -1$, $f(1, 2) = 2(1) + 3(2) - 4 = 4$, and $f(2, -1) = 2(2) + 3(-1) - 4 = -3$.

3. $f(x, y) = x^2 + 2xy - x + 3$, so $f(1, 2) = 1^2 + 2(1)(2) - 1 + 3 = 7$, $f(2, 1) = 2^2 + 2(2)(1) - 2 + 3 = 9$, $f(-1, 2) = (-1)^2 + 2(-1)(2) - (-1) + 3 = 1$, and $f(2, -1) = 2^2 + 2(2)(-1) - 2 + 3 = 1$.

5. $g(s, t) = 3s\sqrt{t} + t\sqrt{s} + 2$, so $g(1, 2) = 3(1)\sqrt{2} + 2\sqrt{1} + 2 = 4 + 3\sqrt{2}$, $g(2, 1) = 3(2)\sqrt{1} + \sqrt{2} + 2 = 8 + \sqrt{2}$, $g(0, 4) = 0 + 0 + 2 = 2$, and $g(4, 9) = 3(4)\sqrt{9} + 9\sqrt{4} + 2 = 56$.

7. $h(s, t) = s\ln t - t\ln s$, so $h(1, e) = \ln e - e\ln 1 = \ln e = 1$, so $h(e, 1) = e\ln 1 - \ln e = -1$, and $h(e, e) = e\ln e - e\ln e = 0$.

9. $g(r, s, t) = re^{s/t}$, so $g(1, 1, 1) = e$, $g(1, 0, 1) = 1$, and $g(-1, -1, -1) = -e^{-1/(-1)} = -e$.

11. $f(x, y) = 2x + 3y$. The domain of f is the set of all ordered pairs (x, y), where x and y are real numbers.

13. $h(u, v) = \dfrac{uv}{u - v}$. The domain is all real values of u and v except those satisfying the equation $u = v$.

15. $g(r, s) = \sqrt{rs}$. The domain of g is the set of all ordered pairs (r, s) satisfying $rs \geq 0$, that is the set of all ordered pairs whose members have the same sign (allowing zeros).

17. $h(x, y) = \ln(x + y - 5)$. The domain of h is the set of all ordered pairs (x, y) such that $x + y > 5$.

19. The graph shows level curves of $z = f(x, y) = 2x + 3y$ for $z = -2, -1, 0, 1$, and 2.

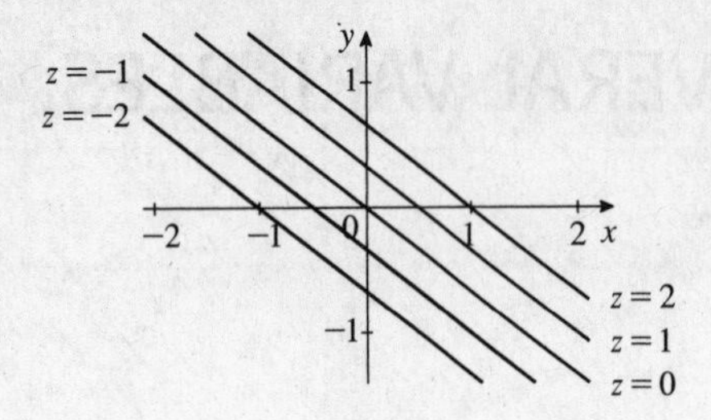

21. The graph shows level curves of $z = f(x, y) = 2x^2 + y$ for $z = -2, -1, 0, 1$, and 2.

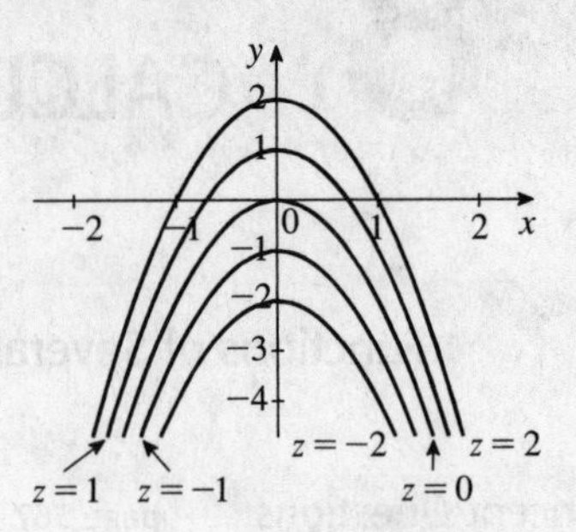

23. The graph shows level curves of $z = f(x, y) = \sqrt{16 - x^2 - y^2}$ for $z = 0, 1, 2, 3$, and 4.

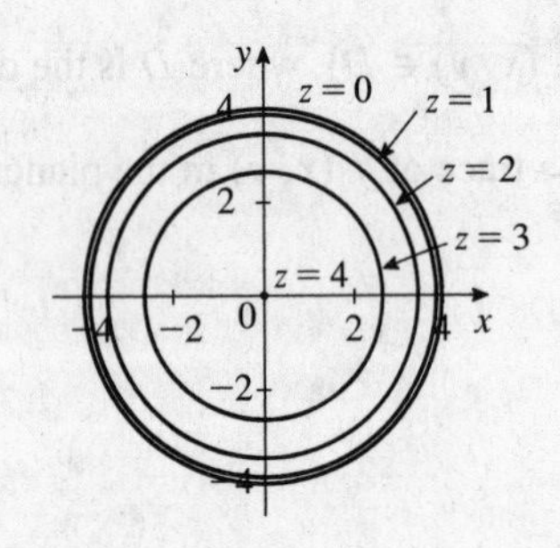

25. The level curves of f have equations $f(x, y) = \sqrt{x^2 + y^2} = C$. An equation of the curve containing the point (3, 4) satisfies $\sqrt{3^2 + 4^2} = C$, so $C = \sqrt{9 + 16} = 5$. Thus, an equation is $\sqrt{x^2 + y^2} = 5$.

27. (b)

29. No. Suppose the level curves $f(x, y) = c_1$ and $f(x, y) = c_2$ intersect at a point (x_0, y_0) and $c_1 \neq c_2$. Then $f(x_0, y_0) = c_1$ and $f(x_0, y_0) = c_2$ where $c_1 \neq c_2$. Thus, f takes on two distinct values at (x_0, y_0), contradicting the definition of a function.

31. $V = f(1.5, 4) = \pi (1.5)^2 (4) = 9\pi$, or 9π ft^3.

33. a. P and E are real numbers with $E \neq 0$.

b. $R = \frac{205.56}{13.09} \approx 15.704$.

35. a. $M = \dfrac{80}{(1.8)^2} = 24.69$.

b. We must have $\dfrac{w}{(1.8)^2} < 25$; that is, $w < 25(1.8)^2 = 81$. Thus, the maximum weight is 81 kg.

37. a. $C(x, y) = 200x + 120y + 20{,}000$.

b. The domain of C is the set of all x and y such that $x \geq 0$ and $y \geq 0$.

c. The total cost is $C(1000, 200) = 200(1000) + 120(200) + 20{,}000$, or \$244,000.

39. a. $R(x, y) = xp + yq = x\left(200 - \frac{1}{5}x - \frac{1}{10}y\right) + y\left(160 - \frac{1}{10}x - \frac{1}{4}y\right) = -\frac{1}{5}x^2 - \frac{1}{4}y^2 - \frac{1}{5}xy + 200x + 160y$.

b. The domain of R is the set of all points (x, y) satisfying $200 - \frac{1}{5}x - \frac{1}{10}y \geq 0$, $160 - \frac{1}{10}x - \frac{1}{4}y \geq 0$, $x \geq 0$, and $y \geq 0$.

41. a. $R(x, y) = xp + yq = 20x - 0.005x^2 - 0.001xy + 15y - 0.001xy - 0.003y^2$

$= -0.005x^2 - 0.003y^2 - 0.002xy + 20x + 15y.$

b. Because p and q must both be nonnegative, the domain of R is the set of all ordered pairs (x, y) for which $20 - 0.005x - 0.001y \geq 0$, $15 - 0.001x - 0.003y \geq 0$, $x \geq 0$, and $y \geq 0$.

43. a. The domain of V is the set of all ordered pairs (P, T) where P and T are positive real numbers.

b. $V = \dfrac{30.9\,(273)}{760} \approx 11.10$ liters.

45. a. The domain of W is the set of all ordered pairs (L, G) for which $L > 0$ and $G > 0$.

b. The approximate weight of Sue's catch is $W = \dfrac{(20)(12)^2}{800} = 3.6$, or 3.6 lb.

47. The output is $f(32, 243) = 100\,(32)^{3/5}\,(243)^{2/5} = 100\,(8)\,(9) = 7200$, or \$7200 billion.

49. The number of suspicious fires is $N(100, 20) = \dfrac{100\left[1000 + 0.03\left(100^2\right)(20)\right]^{1/2}}{[5 + 0.2\,(20)]^2} = 103.29$, or about 103.

51. a. If $r = 4\%$, then $P = f(300000, 0.04, 30) = \dfrac{300{,}000\,(0.04)}{12\left[1 - \left(1 + \frac{0.04}{12}\right)^{-360}\right]} \approx 1432.25$, or \$1432.25. If $r = 6\%$,

then $P = f(300000, 0.06, 30) = \dfrac{300{,}000\,(0.06)}{12\left[1 - \left(1 + \frac{0.06}{12}\right)^{-360}\right]} \approx 1798.65$, or \$1798.65.

b. $P = f(300000, 0.06, 20) = \dfrac{300{,}000\,(0.06)}{12\left[1 - \left(1 + \frac{0.06}{12}\right)^{-240}\right]} \approx 2149.29$, or \$2149.29.

53. $f(20, 40, 5) = \sqrt{\dfrac{2\,(20)\,(40)}{5}} = \sqrt{320} \approx 17.9$, or approximately 18 bicycles.

55. For yacht A, we have $f(20.95, 277.3, 17.56) = \dfrac{20.95 + 1.25\,(277.3)^{1/2} - 9.80\,(17.56)^{1/3}}{0.388} \approx 41.993$. Because this is less than 42, yacht A satisfies the formula.

For yacht B, we have $f(21.87, 311.78, 22.48) = \dfrac{21.87 + 1.25\,(311.78)^{1/2} - 9.80\,(22.48)^{1/3}}{0.388} \approx 41.967$. Because this is less than 42, yacht B satisfies the formula as well.

57. The level curves of V have equation $\dfrac{kT}{P} = C$, where C is a positive constant. The level curves are the family of straight lines $T = \dfrac{C}{k}P$ lying in the first quadrant, because k, T, and P are positive. Every point on the level curve $V = C$ gives the same volume C.

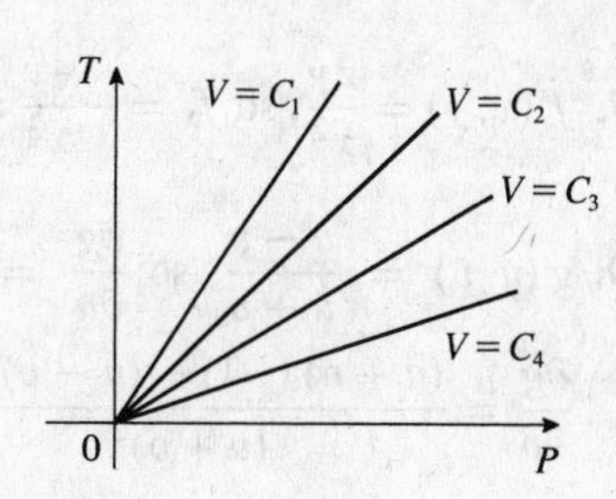

59. False. Let $h(x, y) = xy$. Then there is no pair of functions f and g such that $h(x, y) = f(x) + g(y)$.

61. False. Because $x^2 - y^2 = (x+y)(x-y)$, we see that $x^2 - y^2 = 0$ if $y = \pm x$. Therefore, the domain of f is $\{(x,y) \mid y \neq \pm x\}$.

63. False. Take $f(x,y) = \sqrt{x^2+y^2}$, $P_1(-1,1)$, and $P_2(1,1)$. Then $f(x_1,y_1) = f(-1,1) = \sqrt{(-1)^2+1^2} = \sqrt{2}$ and $f(x_2,y_2) = f(1,1) = \sqrt{1^2+1^2} = \sqrt{2}$. So $f(x_1,y_1) = f(x_2,y_2)$, but $P(x_1,y_1) \neq P(x_2,y_2)$.

8.2 Partial Derivatives

Problem-Solving Tips

1. The expressions f_{xy} and f_{yx} denote the second partial derivatives of the function $f(x,y)$. Note that when this notation is used, the differentiation is carried out in the order in which x and y appear (left to right).

2. The notation $\dfrac{\partial^2 f}{\partial y \partial x}$ and $\dfrac{\partial^2 f}{\partial x \partial y}$ is also used to denote the second partial derivatives of the function $f(x,y)$, but in this case the differentiation is carried out in reverse order (right to left).

Concept Questions page 582

1. a. $\dfrac{\partial f}{\partial x}(a,b) = \dfrac{\partial f}{\partial x}(x,y)\Big|_{(a,b)} = \left[\lim\limits_{h \to 0} \dfrac{f(x+h,y) - f(x,y)}{h}\right]_{(a,b)}$

b. See pages 572–576 of the text.

3. f_{xx}, f_{yy}, f_{xy}, and f_{yx}.

Exercises page 582

1. a. $f(x,y) = x^2 + 2y^2$, so $f_x(2,1) = 4$ and $f_y(2,1) = 4$.

b. $f_x(2,1) = 4$ says that the slope of the tangent line to the curve of intersection of the surface $z = x^2 + 2y^2$ and the plane $y = 1$ at the point $(2,1,6)$ is 4. $f_y(2,1) = 4$ says that the slope of the tangent line to the curve of intersection of the surface $z = x^2 + 2y^2$ and the plane $x = 2$ at the point $(2,1,6)$ is 4.

c. $f_x(2,1) = 4$ says that the rate of change of $f(x,y)$ with respect to x with y held fixed with a value of 1 is 4 units per unit change in x. $f_y(2,1) = 4$ says that the rate of change of $f(x,y)$ with respect to y with x held fixed with a value of 2 is 4 units per unit change in y.

3. $f(x,y) = 2x + 3y + 5$, so $f_x = 2$ and $f_y = 3$.

5. $g(x,y) = 2x^2 + 4y + 1$, so $g_x = 4x$ and $g_y = 4$.

7. $f(x,y) = \dfrac{2y}{x^2}$, so $f_x = -\dfrac{4y}{x^3}$ and $f_y = \dfrac{2}{x^2}$.

9. $g(u,v) = \dfrac{u-v}{u+v}$, so $\dfrac{\partial g}{\partial u} = \dfrac{(u+v)(1) - (u-v)(1)}{(u+v)^2} = \dfrac{2v}{(u+v)^2}$ and $\dfrac{\partial g}{\partial v} = \dfrac{(u+v)(-1) - (u-v)(1)}{(u+v)^2} = -\dfrac{2u}{(u+v)^2}$.

11. $f(s,t) = (s^2 - st + t^2)^3$, so $f_s = 3(s^2 - st + t^2)^2(2s - t)$ and $f_t = 3(s^2 - st + t^2)^2(2t - s)$.

13. $f(x,y) = (x^2+y^2)^{2/3}$, so $f_x = \frac{2}{3}(x^2+y^2)^{-1/3}(2x) = \frac{4}{3}x(x^2+y^2)^{-1/3}$ and $f_y = \frac{4}{3}y(x^2+y^2)^{-1/3}$.

15. $f(x,y) = e^{xy+1}$, so $f_x = ye^{xy+1}$ and $f_y = xe^{xy+1}$.

17. $f(x,y) = x\ln y + y\ln x$, so $f_x = \ln y + \dfrac{y}{x}$ and $f_y = \dfrac{x}{y} + \ln x$.

19. $g(u,v) = e^u \ln v$, so $g_u = e^u \ln v$ and $g_v = \dfrac{e^u}{v}$.

21. $f(x,y,z) = xyz + xy^2 + yz^2 + zx^2$, so $f_x = yz + y^2 + 2xz$ and $f_y = xz + 2xy + z^2$, $f_z = xy + 2yz + x^2$.

23. $h(r,s,t) = e^{rst}$, so $h_r = ste^{rst}$, $h_s = rte^{rst}$, and $h_t = rse^{rst}$.

25. $f(x,y) = x^2y + xy^2$, so $f_x(1,2) = \left.(2xy+y^2)\right|_{(1,2)} = 8$ and $f_y(1,2) = \left.(x^2+2xy)\right|_{(1,2)} = 5$.

27. $f(x,y) = x\sqrt{y} + y^2 = xy^{1/2} + y^2$, so $f_x(2,1) = \left.\sqrt{y}\right|_{(2,1)} = 1$ and $f_y(2,1) = \left.\left(\dfrac{x}{2\sqrt{y}} + 2y\right)\right|_{(2,1)} = 3$.

29. $f(x,y) = \dfrac{x}{y}$, so $f_x(1,2) = \left.\dfrac{1}{y}\right|_{(1,2)} = \dfrac{1}{2}$ and $f_y(1,2) = -\left.\dfrac{x}{y^2}\right|_{(1,2)} = -\dfrac{1}{4}$.

31. $f(x,y) = e^{xy}$, so $f_x(1,1) = \left.ye^{xy}\right|_{(1,1)} = e$ and $f_y(1,1) = \left.xe^{xy}\right|_{(1,1)} = e$.

33. $f(x,y,z) = x^2yz^3$, so $f_x(1,0,2) = \left.2xyz^3\right|_{(1,0,2)} = 0$, $f_y(1,0,2) = \left.x^2z^3\right|_{(1,0,2)} = 8$, and $f_z(1,0,2) = \left.3x^2yz^2\right|_{(1,0,2)} = 0$.

35. $f(x,y) = x^2y + xy^3$, so $f_x = 2xy + y^3$ and $f_y = x^2 + 3xy^2$. Therefore, $f_{xx} = 2y$, $f_{xy} = 2x + 3y^2 = f_{yx}$, and $f_{yy} = 6xy$.

37. $f(x,y) = x^2 - 2xy + 2y^2 + x - 2y$, so $f_x = 2x - 2y + 1$ and $f_y = -2x + 4y - 2$. Therefore, $f_{xx} = 2$, $f_{xy} = -2 = f_{yx}$, and $f_{yy} = 4$.

39. $f(x,y) = (x^2+y^2)^{1/2}$, so $f_x = \frac{1}{2}(x^2+y^2)^{-1/2}(2x) = x(x^2+y^2)^{-1/2}$ and $f_y = y(x^2+y^2)^{-1/2}$. Therefore,

$$f_{xx} = (x^2+y^2)^{-1/2} + x\left(-\tfrac{1}{2}\right)(x^2+y^2)^{-3/2}(2x) = (x^2+y^2)^{-1/2} - x^2(x^2+y^2)^{-3/2}$$
$$= (x^2+y^2)^{-3/2}(x^2+y^2-x^2) = \frac{y^2}{(x^2+y^2)^{3/2}},$$
$$f_{xy} = x\left(-\tfrac{1}{2}\right)(x^2+y^2)^{-3/2}(2y) = -\frac{xy}{(x^2+y^2)^{3/2}} = f_{yx}, \text{ and}$$
$$f_{yy} = (x^2+y^2)^{-1/2} + y\left(-\tfrac{1}{2}\right)(x^2+y^2)^{-3/2}(2y) = (x^2+y^2)^{-1/2} - y^2(x^2+y^2)^{-3/2}$$
$$= (x^2+y^2)^{-3/2}(x^2+y^2-y^2) = \frac{x^2}{(x^2+y^2)^{3/2}}.$$

41. $f(x,y) = e^{-x/y}$, so $f_x = -\dfrac{1}{y}e^{-x/y}$ and $f_y = \dfrac{x}{y^2}e^{-x/y}$. Therefore, $f_{xx} = \dfrac{1}{y^2}e^{-x/y}$,

$$f_{xy} = -\frac{x}{y^3}e^{-x/y} + \frac{1}{y^2}e^{-x/y} = \left(\frac{-x+y}{y^3}\right)e^{-x/y} = f_{yx}, \text{ and } f_{yy} = -\frac{2x}{y^3}e^{-x/y} + \frac{x^2}{y^4}e^{-x/y} = \frac{x}{y^3}\left(\frac{x}{y} - 2\right)e^{-x/y}.$$

43. a. $f(x,y) = 20x^{3/4}y^{1/4}$, so $f_x(256,16) = 15\left(\frac{y}{x}\right)^{1/4}\Big|_{(256,16)} = 15\left(\frac{16}{256}\right)^{1/4} = 15\left(\frac{2}{4}\right) = 7.5$ and

$f_y(256,16) = 5\left(\frac{x}{y}\right)^{3/4}\Big|_{(256,16)} = 5\left(\frac{256}{16}\right)^{3/4} = 5(80) = 40.$

b. Yes.

45. $P(x,y) = x^2 + 5x + 2xy + 3y^2 + 2y$, so $P_x(x,y) = 2x + 5 + 2y$ and $P_y(x,y) = 6y + 2 + 2x$. Thus, $P_x(400,300) = 2(400) + 5 + 2(300) = 1405$ and $P_y(400,300) = 6(300) + 2 + 2(400) = 2602$. Thus, with labor at 400 work-hours per day and capital expenditure at $300/day:

- an increase of 1 work-hour per day with capital expenditure held fixed results in a production increase of approximately 1405 candles per day;
- an increase of $1 per day in capital expenditure with labor held fixed results in a production increase of approximately 2602 candles per day.

47. $p(x,y) = 200 - 10\left(x - \frac{1}{2}\right)^2 - 15(y-1)^2$, so $\frac{\partial p}{\partial x}(0,1) = -20\left(x - \frac{1}{2}\right)\Big|_{(0,1)} = 10$. At the location $(0,1)$ in the figure, the price of land is increasing by $10 per square foot per mile to the east. $\frac{\partial p}{\partial y}(0,1) = -30(y-1)|_{(0,1)} = 0$, so at the point $(0,1)$ in the figure, the price of land is unchanging with respect to north-south change.

49. $P(x,y) = -0.02x^2 - 15y^2 + xy + 39x + 25y - 20{,}000$, so $\frac{\partial P}{\partial x}(4000,150) = (-0.04x + y + 39)|_{(4000,150)} = -0.04(4000) + 150 + 39 = 29$. This says that when inventory is $4,000,000 and floor space is 150,000 ft², monthly profit increases by $29 per thousand dollars increase in inventory. $\frac{\partial P}{\partial y}(4000,150) = (-30y + x + 25)|_{(4000,150)} = -475$, and this says that with the same inventory and floor space as above, monthly profit decreases by $475 per thousand-square-foot increase in floor space. Similarly, $\frac{\partial P}{\partial x}(5000,150) = (-0.04x + y + 39)|_{(5000,150)} = -0.04(5000) + 150 + 39 = -11$ and $\frac{\partial P}{\partial y}(5000,150) = (-30y + x + 25)|_{(5000,150)} = -30(150) + 5000 + 25 = 525.$

51. $N(x,y) = \dfrac{120\sqrt{1000+0.03x^2y}}{(5+0.2y)^2}$, so

$$\frac{\partial N}{\partial x} = \frac{\partial}{\partial x}\frac{120\left(1000+0.03x^2y\right)^{1/2}}{(5+0.2y)^2} = \frac{120\left(\frac{1}{2}\right)\left(1000+0.03x^2y\right)^{-1/2}(0.06xy)}{(5+0.2y)^2}.$$ Thus,

$\dfrac{\partial N}{\partial x}(100,20) = \left.\dfrac{3.6xy}{(5+0.2y)^2\sqrt{1000+0.03x^2y}}\right|_{(100,20)} \approx 1.06$. This means that with the level of reinvestment held constant at 20 cents per dollar deposited, the number of suspicious fires will grow at the rate of approximately 1 fire per increase of 1 person per census tract when the number of people per census tract is 100. Next,

$$\frac{\partial N}{\partial y}(100,20) = 120\frac{\partial}{\partial y}\left[\left(1000+0.03x^2y\right)^{1/2}(5+0.2y)^{-2}\right]\Bigg|_{(100,20)}$$

$$= 120\left[\tfrac{1}{2}\left(1000+0.03x^2y\right)^{-1/2}\left(0.03x^2\right)(5+0.2y)^{-2} + \left(1000+0.03x^2y\right)^{1/2}(-2)(5+0.2y)^{-3}(0.2)\right]\Bigg|_{(100,20)}$$

$$= \left.\frac{9x^2-1.08x^2y-48{,}000}{(5+0.2y)^3\sqrt{1000+0.03x^2y}}\right|_{(100,20)} \approx -2.85$$

which tells us that if the number of people per census tract is constant at 100 per tract, the number of suspicious fires decreases at a rate of approximately 2.9 per increase of 1 cent per dollar deposited for reinvestment when the level of reinvestment is 20 cents per dollar deposited.

53. a. The rate of change of T with respect to t with s held constant is given by $\dfrac{\partial T}{\partial t}(t,s) = \lim\limits_{h\to 0}\dfrac{T(t+h,s)-T(t,s)}{h}$.

So if h is small, then $\dfrac{\partial T}{\partial t}(t,s) \approx \dfrac{T(t+h,s)-T(t,s)}{h}$. In particular, with $t=34$, $s=25$, and $h=2$, we see that the required rate of change is $\dfrac{\partial T}{\partial t}(34,25) \approx \dfrac{T(36,25)-T(34,25)}{2} = \dfrac{24-21.4}{2} = 1.3$ °F/°F.

b. The required rate of change is given by

$$\frac{\partial T}{\partial s}(34,25) \approx \frac{T(34,30)-T(34,25)}{5} = \frac{20.3-21.4}{5} = -0.22 \text{ °F/mi/h}.$$

55. $E = \left(1-\dfrac{v}{V}\right)^{0.4}$.

a. $\dfrac{\partial E}{\partial V} = 0.4\left(1-\dfrac{v}{V}\right)^{-0.6}\dfrac{\partial}{\partial V}\left(-\dfrac{v}{V}\right) = 0.4\left(1-\dfrac{v}{V}\right)^{-0.6}\left(\dfrac{v}{V^2}\right) = \dfrac{0.4v}{V^2\left(1-\dfrac{v}{V}\right)^{0.6}} > 0$. This says that for constant v, an increase in V increases the engine efficiency, which is to be expected.

b. $\dfrac{\partial E}{\partial v} = 0.4\left(1-\dfrac{v}{V}\right)^{-0.6}\dfrac{\partial}{\partial v}\left(-\dfrac{v}{V}\right) = -\dfrac{0.4v}{V\left(1-\dfrac{v}{V}\right)^{0.6}} < 0$. This says that for constant V, an increase in v decreases the engine efficiency.

57. $f(p,q) = 10{,}000 - 10p + 0.2q^2$, so $g(p,q) = 5000 + 0.8p^2 - 20q$. $\dfrac{\partial f}{\partial q} = 0.4q > 0$ and $\dfrac{\partial g}{\partial p} = 1.6p > 0$, and so the two products are substitute commodities.

59. We have $\frac{1}{5}x + \frac{1}{10}y = 200 - p$ and $x + \frac{1}{2}y = 1000 - 5p$. Thus, $\frac{1}{10}x + \frac{1}{4}y = 160 - q$ and $x + \frac{5}{2}y = 1600 - 10q$. Then $2y = 600 + 5p - 10q$, so $y = \frac{1}{2}(600 + 5p - 10q) = g(p, q)$. Also, $x = 1000 - 5p - \left(\frac{1}{2}\right)\left(\frac{1}{2}\right)(600 + 5p - 10q) = 1000 - 5p - 150 - \frac{5}{4}p + \frac{5}{2}q = 850 - \frac{25}{4}p + \frac{5}{2}q = f(p, q)$.

Next, we compute $\dfrac{\partial f}{\partial q} = \dfrac{5}{2} > 0$ and $\dfrac{\partial g}{\partial p} = \dfrac{5}{2} > 0$, and so they are substitute commodities.

61. We have $\dfrac{\partial x}{\partial p} = \dfrac{\partial}{\partial p}\left[(3q)\left(1 + p^2\right)^{-1}\right] = (3q)(-1)\left(1 + p^2\right)^{-2}(2p) = -\dfrac{6pq}{\left(1 + p^2\right)^2}$ and $\dfrac{\partial x}{\partial q} = \dfrac{3}{1 + p^2}$. If $p = 5$ and $q = 4$, we have $x = f(5, 4) = \dfrac{3(4)}{1 + 5^2} = \dfrac{12}{26} = \dfrac{6}{13}$, $\dfrac{\partial x}{\partial p} = -\dfrac{6(5)(4)}{\left(1 + 5^2\right)^2} = -\dfrac{30}{169}$, and $\dfrac{\partial x}{\partial q} = \dfrac{3}{1 + 5^2} = \dfrac{3}{26}$.

Therefore, $E_p = -\dfrac{p\frac{\partial x}{\partial p}}{x} = -\dfrac{5\left(-\frac{30}{169}\right)}{\frac{6}{13}} \approx 1.923$ and $E_q = -\dfrac{q\frac{\partial x}{\partial q}}{x} = -\dfrac{4\left(\frac{3}{26}\right)}{\frac{6}{13}} = -1$. Thus, an increase of 1% in the price of butter will result in a drop of approximately 1.9% in demand for butter (with the price of margarine held fixed at \$4/lb), while an increase of 1% in the price of margarine will result in the same percentage increase in the demand for butter (with the price of butter held fixed at \$5/lb).

63. $V = \dfrac{kT}{P}$, so $\dfrac{\partial V}{\partial T} = \dfrac{k}{P}$; $T = \dfrac{VP}{k}$, so $\dfrac{\partial T}{\partial P} = \dfrac{V}{k} = \dfrac{T}{P}$; and $P = \dfrac{kT}{V}$, so $\dfrac{\partial P}{\partial V} = -\dfrac{kT}{V^2} = -kT\dfrac{P^2}{(kT)^2} = -\dfrac{P^2}{kT}$.

Therefore $\dfrac{\partial V}{\partial T} \cdot \dfrac{\partial T}{\partial P} \cdot \dfrac{\partial P}{\partial V} = \dfrac{k}{P}\left(\dfrac{T}{P}\right)\left(-\dfrac{P^2}{kT}\right) = -1$.

65. $\dfrac{\partial P}{\partial x} = \dfrac{\partial}{\partial x}kx^{\alpha}y^{1-\alpha} = k\alpha x^{\alpha-1}y^{1-\alpha} = k\alpha\left(\dfrac{y}{x}\right)^{1-\alpha}$ and $\dfrac{\partial P}{\partial y} = k(1 - \alpha)x^{\alpha}y^{-\alpha} = k(1 - \alpha)\left(\dfrac{x}{y}\right)^{\alpha}$.

Therefore, $x\dfrac{\partial P}{\partial x} + y\dfrac{\partial P}{\partial y} = \dfrac{k\alpha xy^{1-\alpha}}{x^{1-\alpha}} + \dfrac{k(1 - \alpha)yx^{\alpha}}{y^{\alpha}} = k\alpha x^{\alpha}y^{1-\alpha} + k(1 - \alpha)x^{\alpha}y^{1-\alpha} = kx^{\alpha}y^{1-\alpha} = P$, as was to be shown.

67. False. Let $f(x, y) = xy^{1/2}$. Then $f_x = y^{1/2}$ is defined at $(0, 0)$, but $f_y = \frac{1}{2}xy^{-1/2} = \dfrac{x}{2y^{1/2}}$ is not defined at $(0, 0)$.

69. True. Since $f_x(x, y) = 0$ for all x and y, we see that f is independent of x. So there exists a function $g(y)$ such that $f(x, y) = g(y)$. Then $f_y(x, y) = \frac{d}{dy}g(y) = g'(y) = 0$ for all y, implying that $g(y) = C$, a constant. Thus, $f(x, y) = C$.

71. True. See page 580 of the text.

Using Technology page 587

1. 1.3124, 0.4038. **3.** −1.8889, 0.7778. **5.** −0.3863, −0.8497.

8.3 Maxima and Minima of Functions of Several Variables

Problem-Solving Tips

1. To find the relative extrema of a function of several variables, first find the critical points of $f(x, y)$ by solving the simultaneous equations $f_x = 0$ and $f_y = 0$, then use the second derivative test to classify those points.

2. To use the second derivative test, first evaluate the function $D(x, y) = f_{xx} f_{yy} - f_{xy}^2$ for each critical point found in Tip 1.

- If $D(a, b) > 0$ and $f_{xx}(a, b) < 0$, then $f(x, y)$ has a relative maximum at the point (a, b).
- If $D(a, b) > 0$ and $f_{xx}(a, b) > 0$, then $f(x, y)$ has a relative minimum at the point (a, b).
- If $D(a, b) < 0$ then $f(x, y)$ has neither a relative maximum nor a relative minimum at the point (a, b).
- If $D(a, b) = 0$, then the test is inconclusive.

Concept Questions page 594

1. a. A function $f(x, y)$ has a relative maximum at (a, b) if $f(a, b)$ is the largest value of $f(x, y)$ for all (x, y) near (a, b).

b. $f(a, y)$ has an absolute maximum at (a, b) if $f(a, b)$ is the largest value of $f(x, y)$ for all (x, y) in the domain of f.

3. See the procedure on page 590 of the text.

Exercises page 595

1. $f(x, y) = 1 - 2x^2 - 3y^2$. To find the critical points of f, we solve the system $\begin{cases} f_x = -4x = 0 \\ f_y = -6y = 0 \end{cases}$ obtaining $(0, 0)$ as the only critical point of f. Next, $f_{xx} = -4$, $f_{xy} = 0$, and $f_{yy} = -6$. In particular, $f_{xx}(0, 0) = -4$, $f_{xy}(0, 0) = 0$, and $f_{yy}(0, 0) = -6$, giving $D(0, 0) = (-4)(-6) - 0^2 = 24 > 0$. Because $f_{xx}(0, 0) < 0$, the Second Derivative Test implies that $(0, 0)$ gives rise to a relative maximum of f. Finally, the relative maximum value of f is $f(0, 0) = 1$.

3. $f(x, y) = x^2 - y^2 - 2x + 4y + 1$. To find the critical points of f, we solve the system $\begin{cases} f_x = 2x - 2 = 0 \\ f_y = -2y + 4 = 0 \end{cases}$ obtaining $x = 1$ and $y = 2$, so $(1, 2)$ is the only critical point of f. $f_{xx} = 2$, $f_{xy} = 0$, and $f_{yy} = -2$, so $D(x, y) = f_{xx} f_{yy} - f_{xy}^2 = -4$. In particular, $D(1, 2) = -4 < 0$, so $(1, 2)$ gives a saddle point of f. Because $f(1, 2) = 4$, the saddle point is $(1, 2, 4)$.

5. $f(x, y) = x^2 + 2xy + 2y^2 - 4x + 8y - 1$. To find the critical points of f, we solve the system $\begin{cases} f_x = 2x + 2y - 4 = 0 \\ f_y = 2x + 4y + 8 = 0 \end{cases}$ obtaining $(8, -6)$ as the critical point of f. Next, $f_{xx} = 2$, $f_{xy} = 2$, and $f_{yy} = 4$. In particular, $f_{xx}(8, -6) = 2$, $f_{xy}(8, -6) = 2$, and $f_{yy}(8, -6) = 4$, giving $D = 2(4) - 4 = 4 > 0$. Because $f_{xx}(8, -6) > 0$, $(8, -6)$ gives rise to a relative minimum of f. The relative minimum value of f is $f(8, -6) = -41$.

7. $f(x,y) = 2x^3 + y^2 - 9x^2 - 4y + 12x - 2$. To find the critical points of f, we solve the system $\begin{cases} f_x = 6x^2 - 18x + 12 = 0 \\ f_y = 2y - 4 = 0 \end{cases}$ The first equation is equivalent to $x^2 - 3x + 2 = 0$, or $(x-2)(x-1) = 0$, giving $x = 1$ or 2. The second equation of the system gives $y = 2$. Therefore, there are two critical points, $(1,2)$ and $(2,2)$. Next, we compute $f_{xx} = 12x - 18 = 6(2x-3)$, $f_{xy} = 0$, and $f_{yy} = 2$.

At the point $(1,2)$, $f_{xx}(1,2) = 6(2-3) = -6$, $f_{xy}(1,2) = 0$, and $f_{yy}(1,2) = 2$, so $D(1,2) = (-6)(2) - 0 = -12 < 0$ and we conclude that $(1,2)$ gives a saddle point of f. Because $f(1,2) = -1$, the saddle point is $(1,2,-1)$.

At the point $(2,2)$, $f_{xx}(2,2) = 6(4-3) = 6$, $f_{xy}(2,2) = 0$, and $f_{yy}(2,2) = 2$, so $D(2,2) = (6)(2) - 0 = 12 > 0$. Because $f_{xx}(2,2) > 0$, we see that $(2,2)$ gives a relative minimum with value $f(2,2) = -2$.

9. $f(x,y) = x^3 + y^2 - 2xy + 7x - 8y + 4$. To find the critical points of f, we solve the system $\begin{cases} f_x = 3x^2 - 2y + 7 = 0 \\ f_y = 2y - 2x - 8 = 0 \end{cases}$ Adding the two equations gives $3x^2 - 2x - 1 = (3x+1)(x-1) = 0$. Therefore, $x = -\frac{1}{3}$ or 1. Substituting each of these values of x into the second equation gives $y = \frac{11}{3}$ and $y = 5$, respectively. Therefore, $\left(-\frac{1}{3}, \frac{11}{3}\right)$ and $(1,5)$ are critical points of f. Next, $f_{xx} = 6x$, $f_{xy} = -2$, and $f_{yy} = 2$, so $D(x,y) = 12x - 4 = 4(3x-1)$. Then $D\left(-\frac{1}{3}, \frac{11}{3}\right) = 4(-1-1) = -8 < 0$, and so $\left(-\frac{1}{3}, \frac{11}{3}\right)$ gives a saddle point. Because $f\left(-\frac{1}{3}, \frac{11}{3}\right) = -\frac{319}{27}$, the saddle point is $\left(-\frac{1}{3}, \frac{11}{3}, -\frac{319}{27}\right)$. Next, $D(1,5) = 4(3-1) = 8 > 0$, and since $f_{xx}(1,5) = 6 > 0$, we see that $(1,5)$ gives a relative minimum with value $f(1,5) = -13$.

11. $f(x,y) = x^3 - 3xy + y^3 - 2$. To find the critical points of f, we solve the system $\begin{cases} f_x = 3x^2 - 3y = 0 \\ f_y = -3x + 3y^2 = 0 \end{cases}$

The first equation gives $y = x^2$, and substituting this into the second equation gives $-3x + 3x^4 = 3x(x^3 - 1) = 0$. Therefore, $x = 0$ or 1. Substituting these values of x into the first equation gives $y = 0$ and $y = 1$, respectively. Therefore, $(0,0)$ and $(1,1)$ are critical points of f. Next, we find $f_{xx} = 6x$, $f_{xy} = -3$, and $f_{yy} = 6y$, so $D = f_{xx}f_{yy} - f_{xy}^2 = 36xy - 9$. Because $D(0,0) = -9 < 0$, we see that $(0,0)$ gives a saddle point of f. Because $f(0,0) = -2$, the saddle point is $(0,0,-2)$. Next, $D(1,1) = 36 - 9 = 27 > 0$, and since $f_{xx}(1,1) = 6 > 0$, we see that $f(1,1) = -3$ is a relative minimum value of f.

13. $f(x,y) = xy + \dfrac{4}{x} + \dfrac{2}{y}$. Solving the system of equations $\begin{cases} f_x = y - \dfrac{4}{x^2} = 0 \\ f_y = x - \dfrac{2}{y^2} = 0 \end{cases}$ we obtain $y = \dfrac{4}{x^2}$.

Therefore, $x - 2\left(\dfrac{x^4}{16}\right) = 0$ and $8x - x^4 = x(8 - x^3) = 0$, so $x = 0$ or $x = 2$. Because $x = 0$ is not in the domain of f, $(2,1)$ is the only critical point of f. Next, $f_{xx} = \dfrac{8}{x^3}$, $f_{xy} = 1$, and $f_{yy} = \dfrac{4}{y^3}$. Therefore, $D(2,1) = \left.\left(\dfrac{32}{x^3y^3} - 1\right)\right|_{(2,1)} = 4 - 1 = 3 > 0$ and $f_{xx}(2,1) = 1 > 0$, so the relative minimum value of f is $f(2,1) = 2 + \frac{4}{2} + \frac{2}{1} = 6$.

15. $f(x, y) = x^2 - e^{y^2}$. Solving the system of equations $\begin{cases} f_x = 2x = 0 \\ f_y = -2ye^{y^2} = 0 \end{cases}$ we obtain $x = 0$ and $y = 0$.

Therefore, $(0, 0)$ is the only critical point of f. Next, $f_{xx} = 2$, $f_{xy} = 0$, and $f_{yy} = -2e^{y^2} - 4y^2e^{y^2}$, so $D(0, 0) = \left[-4e^{y^2}(1 + 2y^2)\right]_{(0,0)} = -4(1) < 0$, and we conclude that $(0, 0)$ gives a saddle point. Because $f(0, 0) = -1$, the saddle point is $(0, 0, -1)$.

17. $f(x, y) = e^{x^2+y^2}$. Solving the system $\begin{cases} f_x = 2xe^{x^2+y^2} = 0 \\ f_y = 2ye^{x^2+y^2} = 0 \end{cases}$ we see that $x = 0$ and $y = 0$

(recall that $e^{x^2+y^2} \neq 0$). Therefore, $(0, 0)$ is the only critical point of f. Next, we compute $f_{xx} = 2e^{x^2+y^2} + 2x(2x)e^{x^2+y^2} = 2(1 + 2x^2)e^{x^2+y^2}$, $f_{xy} = 2x(2y)e^{x^2+y^2} = 4xye^{x^2+y^2}$, and $f_{yy} = 2(1 + 2y^2)e^{x^2+y^2}$. In particular, at the point $(0, 0)$, $f_{xx}(0, 0) = 2$, $f_{xy}(0, 0) = 0$, and $f_{yy}(0, 0) = 2$. Therefore, $D(0, 0) = (2)(2) - 0 = 4 > 0$. Because $f_{xx}(0, 0) > 0$, we conclude that $(0, 0)$ gives rise to a relative minimum of f. The relative minimum value is $f(0, 0) = 1$.

19. $f(x, y) = \ln(1 + x^2 + y^2)$. We solve the system of equations $\begin{cases} f_x = \dfrac{2x}{1 + x^2 + y^2} = 0 \\ f_y = \dfrac{2y}{1 + x^2 + y^2} = 0 \end{cases}$ obtaining

$x = 0$ and $y = 0$. Therefore, $(0, 0)$ is the only critical point of f. Next,

$$f_{xx} = \frac{(1 + x^2 + y^2)2 - (2x)(2x)}{(1 + x^2 + y^2)^2} = \frac{2 + 2y^2 - 2x^2}{(1 + x^2 + y^2)^2},\ f_{yy} = \frac{(1 + x^2 + y^2)2 - (2y)(2y)}{(1 + x^2 + y^2)^2} = \frac{2 + 2x^2 - 2y^2}{(1 + x^2 + y^2)^2},$$

and $f_{xy} = -2x(1 + x^2 + y^2)^{-2}(2y) = -\dfrac{4xy}{(1 + x^2 + y^2)^2}$. Therefore,

$$D(x, y) = \frac{(2 + 2y^2 - 2x^2)(2 + 2x^2 - 2y^2)}{(1 + x^2 + y^2)^4} - \frac{16x^2y^2}{(1 + x^2 + y^2)^4}.$$ Because $D(0, 0) = 4 > 0$ and $f_{xx}(0, 0) = 2 > 0$, $f(0, 0) = 0$ is a relative minimum value.

21. $P(x) = -0.2x^2 - 0.25y^2 - 0.2xy + 200x + 160y - 100x - 70y - 4000$

$= -0.2x^2 - 0.25y^2 - 0.2xy + 100x + 90y - 4000.$

Thus, $\begin{cases} P_x = -0.4x - 0.2y + 100 = 0 \\ P_y = -0.5y - 0.2x + 90 = 0 \end{cases}$ implies that $\begin{cases} 4x + 2y = 1000 \\ 2x + 5y = 900 \end{cases}$ Solving, we find $x = 200$ and $y = 100$. Next, $P_{xx} = -0.4$, $P_{yy} = -0.5$, $P_{xy} = -0.2$, and $D(200, 100) = (-0.4)(-0.5) - (-0.2)^2 > 0$. Because $P_{xx}(200, 100) < 0$, we conclude that $(200, 100)$ is a relative maximum of P. Thus, the company should manufacture 200 finished and 100 unfinished units per week. The maximum profit is $P(200, 100) = -0.2(200)^2 - 0.25(100)^2 - 0.2(100)(200) + 100(200) + 90(100) - 4000 = 10{,}500$, or \$10,500.

23. $p(x,y) = 200 - 10\left(x - \frac{1}{2}\right)^2 - 15(y-1)^2$. Solving the system of equations $\begin{cases} p_x = -20\left(x - \frac{1}{2}\right) = 0 \\ p_y = -30(y-1) = 0 \end{cases}$ we obtain $x = \frac{1}{2}$ and $y = 1$. We conclude that the only critical point of f is $\left(\frac{1}{2}, 1\right)$. Next, $p_{xx} = -20$, $p_{xy} = 0$, and $p_{yy} = -30$, so $D\left(\frac{1}{2}, 1\right) = (-20)(-30) = 600 > 0$. Because $p_{xx} = -20 < 0$, we conclude that $f\left(\frac{1}{2}, 1\right)$ gives a relative maximum. We conclude that the price of land is highest at $\left(\frac{1}{2}, 1\right)$.

25. a. $R(p,q) = xp + yq = (6400 - 400p - 200q)p + (5600 - 200p - 400q)q = 400\left(-p^2 - q^2 - pq + 16p + 14q\right)$.

b. To find the critical point of R, we solve the system $\begin{cases} R_p = 400(-2p - q + 16) = 0 \\ R_q = 400(-p - 2q + 14) = 0 \end{cases}$ obtaining $p = 6$ and $q = 4$. So the sole critical point of R is $(6, 4)$. Next, we find $R_{pp} = -800$, $R_{pq} = -400$, and $R_{qq} = -800$. Then $D(6,4) = (-800)(-800) - (-400)^2 = 480{,}000 > 0$. Since $R_{pp}(6,4) < 0$, we see that $(6, 4)$ does yield a relative (and therefore, the absolute) maximum of R. We conclude that the supermarket should charge \$6/lb for the ground sirloin and \$4/lb for the ground beef. The quantity of sirloin sold per week will be $x = 6400 - 400(6) - 200(4) = 3200$, or 3200 lb, and the quantity of ground beef sold per week will be $y = 5600 - 200(6) - 400(4) = 2800$, or 2800 lb. The maximum revenue is $R(6,4) = 400\left[-(6)^2 - (4)^2 - 4(6)(4) + 16(6) + 14(4)\right] = 30{,}400$, or \$30,400.

27. a. $R(x,y) = px + qy = (3000 - 20x - 10y)x + (4000 - 10x - 30y)y = -20x^2 - 30y^2 - 20xy + 3000x + 4000y$.

b. $C(x,y) = 400x + 500y + 20{,}000$.

c. $P(x,y) = R(x,y) - C(x,y) = \left(-20x^2 - 30y^2 - 20xy + 3000x + 4000y\right) - (400x + 500y + 20{,}000) = -20x^2 - 30y^2 - 20xy + 2600x + 3500y - 20{,}000$.

d. To find the critical point of P, we solve $\begin{cases} P_x = -40x - 20y + 2600 = 0 \\ P_y = -20x - 60y + 3500 = 0 \end{cases}$ or $\begin{cases} 4x + 2y = 260 \\ 2x + 6y = 350 \end{cases}$ finding $x = 43$ and $y = 44$. Next, we find $P_{xx} = -40$, $P_{xy} = -20$, and $P_{yy} = -60$. Thus, $D(43,44) = (-40)(-60) - (-20)^2 = 2000 > 0$. Since $D(43,44) > 0$ and $P_{xx}(43,44) < 0$, we see that $(43, 44)$ yields a relative (and therefore absolute) maximum of P. So the company should produce 43,000 tubes of regular and 44,000 tubes of whitening toothpaste. (Recall that x and y are measured in thousands .) The maximum weekly profit will be $P(43,44) = -20(43)^2 - 30(44)^2 - 20(43)(44) + 2600(43) + 3500(44) - 20{,}000 = 112{,}900$, or \$112,900.

29. The sums of the squares of the distances from the proposed site of the radio station to the three communities is

$$D = f(x,y) = [d(P,A)]^2 + [d(P,B)]^2 + [d(P,C)]^2$$
$$= \left[(x-2)^2 + (y-4)^2\right] + \left[(x-20)^2 + (y-8)^2\right] + \left[(x-4)^2 + (y-24)^2\right]$$

$$\left.\begin{array}{l} f_x(x,y) = 2(x-2) + 2(x-20) + 2(x-4) = 0 \\ f_y(x,y) = 2(y-4) + 2(y-8) + 2(y-24) = 0 \end{array}\right\} \Rightarrow x = \tfrac{26}{3} \text{ and } y = 12, \text{ so } \left(\tfrac{26}{3}, 12\right) \text{ is a critical point of } f.$$

Since it is clear that D must attain a minimum, we see that the station should be located at $\left(\frac{26}{3}, 12\right)$.

31. Solving the equation $xyz = 108$ for x, we have $z = 108/(xy)$. Substituting this value of z into the expression for S, we obtain $S = f(x, y) = xy + 2y[108/(xy)] + 2x[108/(xy)] = xy + (216/x) + (216/y)$. To minimize f, we first find the critical points of f. To do this, we solve the system $\begin{cases} f_x = y - (216/x^2) = 0 \\ f_y = x - (216/y^2) = 0 \end{cases}$ Solving the first equation for y, we obtain $y = 216/x^2$. Substituting this value into the second equation then yields $x - 216(x^2/216)^2 = 0$, $x - (x^4/216) = 0$, and $x(216 - x^3) = 0$, from which we deduce that $x = 0$, or $x = 6$. We reject the root $x = 0$, since it is not in the domain of f. Next, substituting $x = 6$ into the expression for y obtained earlier, we find $y = 6$. Thus, the point $(6, 6)$ is the only critical point of f. We then compute $f_{xx} = 432/x^3$, $f_{xy} = 1$, $f_{yy} = 432/y^3$. In particular, $f_{xx}(6, 6) = 2$, $f_{xy}(6, 6) = 1$, and $f_{yy}(6, 6) = 2$. Thus, $D = (2)(2) - 1 = 3 > 0$. Because $f_{xx}(6, 6) > 0$, we conclude that $(6, 6)$ gives rise to a relative minimum of f. Substituting these values of x and y into the expression for z yields $z = \frac{108}{6 \cdot 6} = 3$. Therefore, the required dimensions of the box are $6'' \times 6'' \times 3''$.

33. The volume is given by $V = xyz = xz(130 - 2x - 2z) = 130xz - 2x^2z - 2xz^2$. Solving the system of equations $\begin{cases} V_x = 130z - 4xz - 2z^2 = 0 \\ V_z = 130x - 2x^2 - 4xz = 0 \end{cases}$ we obtain $(130 - 4x - 2z)z = 0$, giving $130 - 4x - 2z = 0$, and $(130 - 4z - 2x)x = 0$, giving $130 - 2x - 4z = 0$. Thus, we have $\begin{cases} 130 - 4x - 2z = 0 \\ 260 - 4x - 8z = 0 \end{cases}$ giving $130 - 6z = 0$ and $z = \frac{65}{3}$. Therefore, $x = \frac{1}{4}\left(130 - 2 \cdot \frac{65}{3}\right) = \frac{65}{3}$ and $y = 130 - 2x - 2z = 130 - 2\left(\frac{65}{3}\right) - 2\left(\frac{65}{3}\right) = \frac{130}{3}$, and so $(x, z) = \left(\frac{65}{3}, \frac{65}{3}\right)$ is the critical point of V. Next, $V_{xx} = -4z$, $V_{zz} = -4x$, $V_{xz} = 130 - 4x - 4z$, and $D\left(\frac{65}{3}, \frac{65}{3}\right) = -4\left(\frac{65}{3}\right)(-4)\left(\frac{65}{3}\right) - \left(130 - 4 \cdot \frac{65}{3} - 4 \cdot \frac{65}{3}\right)^2 > 0$ and $V_{xx}\left(\frac{65}{3}, \frac{65}{3}\right) < 0$. We conclude that the dimensions yielding the maximum volume are $21\frac{2}{3}'' \times 43\frac{1}{3}'' \times 21\frac{2}{3}''$.

35. Because $V = xyz$, $z = \dfrac{48}{xy}$. Then the amount of material used in the box is given by $S = xy + 2xz + 3yz = xy + \dfrac{48}{xy}(2x + 3y) = xy + \dfrac{96}{y} + \dfrac{144}{x}$. Solving the system of equations $\begin{cases} S_x = y - \dfrac{144}{x^2} = 0 \\ S_y = x - \dfrac{96}{y^2} = 0 \end{cases}$ we have $y = \dfrac{144}{x^2}$. Therefore, $x - \dfrac{96x^4}{144^2} = 0$, $144^2x - 96x^4 = 0$, $96x(216 - x^3) = 0$, and so $x = 0$ or $x = 6$. We reject $x = 0$, so $x = 6$ and $y = \frac{144}{36} = 4$. Next, $S_{xx} = \dfrac{288}{x^3}$, $S_{yy} = \dfrac{192}{y^3}$, and $S_{xy} = 1$. At the point $(6, 4)$, $D(x, y) = \left(\dfrac{288 \cdot 192}{x^3y^3} - 1\right)\bigg|_{(6,4)} = \dfrac{288(192)}{216(64)} - 1 = 3 > 0$ and $S_{xx} > 0$. We conclude that the function is minimized when the dimensions of the box are $6'' \times 4'' \times 2''$.

37. False. Let $f(x, y) = -x^2 - y^2 + 4xy$. Then setting $\begin{cases} f_x(x, y) = -2x + 4y = 0 \\ f_y(x, y) = -2y + 4x = 0 \end{cases}$ we find that $(0, 0)$ is the only critical point of f. Next, $f_{xx} = -2$, $f_{xy} = 4$, and $f_{yy} = -2$. Because $D(x, y) = f_{xx}f_{yy} - f_{xy}^2 = (-2)(-2) - 4^2 = -12 < 0$, we see that $(0, 0, 0)$ is a saddle point.

39. False. Take $f(x) = \dfrac{1}{x^2+1}$ and $g(y) = -y^2 - 1$. Then $h(x, y) = \dfrac{1}{x^2+1} - y^2 - 1$, so $h_x = -\dfrac{2x}{(x^2+1)^2}$ and $h_y = -2y$. Therefore, $(0, 0)$ is a critical point of h. $h_{xx} = -2\left[\dfrac{1}{(x^2+1)^2} - \dfrac{4x^3}{(x^2+1)^3}\right]$, $h_{xy} = 0$, and $h_{yy} = -2$, so $D(0, 0) = h_{xx}(0)\, h_{yy}(0) - h_{xy}^2(0, 0) = (-2)(-2) - 0 = 4 > 0$. Because $h_{yy}(0, 0) = -2 < 0$, we see that $(0, 0)$ gives a relative maximum of h.

41. True. $h(x, y) = f(x) + g(y)$, so $h_x(x, y) = f'(x)$ and $h_y(x, y) = g'(y)$. Because a and b are critical numbers of f and g, respectively, $f'(a) = g'(b) = 0$ and (a, b) is a critical point of h. Next, $h_{xx}(x, y) = f''(x)$, $h_{xy}(x, y) = 0$, and $h_{yy}(x, y) = g''(y)$, so $D(a, b) = f''(a)\, g''(b) - 0 > 0$ and $h_{xx}(a, b) = f''(a) \neq 0$. Therefore, h has a relative extremum.

8.4 The Method of Least Squares

Problem-Solving Tips

You will find it helpful to organize the data in a least-squares problem in the form of a table like those in Examples 2 and 3 in the text.

Concept Questions

page 604

1. a. A scatter diagram is a graph showing the data points that describe the relationship between the two variables x and y.

b. The least squares line is the straight line that best fits a set of data points when the points are scattered about a straight line.

3. a. Since an error term in the sum could be nonnegative or nonpositive, the sum of the errors could be very small, and even be zero, even if the points are widely scattered about the least-squares line.

b. Using the absolute values of the errors does eliminate the problems mentioned in part (a), but introduces another disadvantage—the absolute value function is not differentiable.

Exercises

page 604

1. a. We first summarize the data.

	x	y	x^2	xy
	1	4	1	4
	2	6	4	12
	3	8	9	24
	4	11	16	44
Sum	10	29	30	84

b.

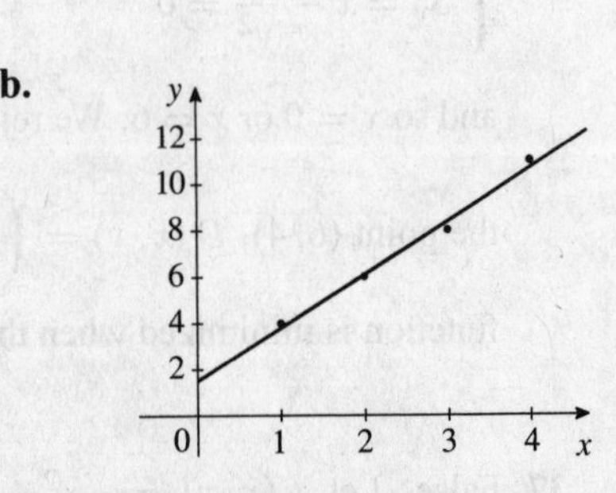

The normal equations are $4b + 10m = 29$ and $10b + 30m = 84$. Solving this system of equations, we obtain $m = 2.3$ and $b = 1.5$, so an equation is $y = 2.3x + 1.5$.

3. a. We first summarize the data.

	x	y	x^2	xy
	1	4.5	1	4.5
	2	5	4	10
	3	3	9	9
	4	2	16	8
	4	3.5	16	14
	6	1	36	6
Sum	20	19	82	51.5

The normal equations are $6b + 20m = 19$ and $20b + 82m = 51.5$. The solutions are $m \approx -0.7717$ and $b \approx 5.7391$, so the required equation is $y = -0.772x + 5.739$.

b.

5. a. We first summarize the data:

	x	y	x^2	xy
	1	3	1	3
	2	5	4	10
	3	5	9	15
	4	7	16	28
	5	8	25	40
Sum	15	28	55	96

b.

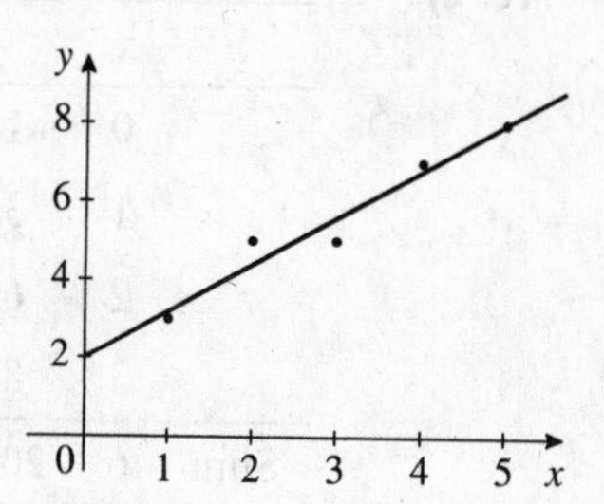

The normal equations are $55m + 15b = 96$ and $15m + 5b = 28$. Solving, we find $m = 1.2$ and $b = 2$, so the required equation is $y = 1.2x + 2$.

7. a. We first summarize the data:

	x	y	x^2	xy
	4	0.5	16	2
	4.5	0.6	20.25	2.7
	5	0.8	25	4
	5.5	0.9	30.25	4.95
	6	1.2	36	7.2
Sum	25	4	127.5	20.85

The normal equations are $5b + 25m = 4$ and $25b + 127.5m = 20.85$. The solutions are $m = 0.34$ and $b = -0.9$, so the required equation is $y = 0.34x - 0.9$.

b.

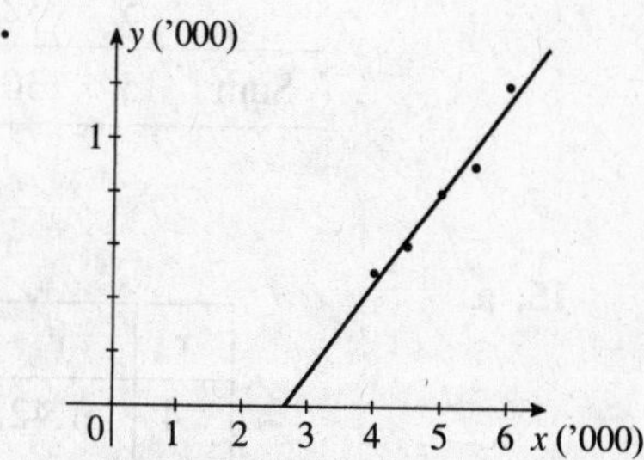

c. If $x = 6.4$, then $y = 0.34(6.4) - 0.9 = 1.276$, and so 1276 completed applications can be expected.

9. a. We first summarize the data:

	x	y	x^2	xy
	1	436	1	436
	2	438	4	876
	3	428	9	1284
	4	430	16	1720
	5	426	25	2138
Sum	15	2158	55	6446

The normal equations are $5b + 15m = 2158$ and $15b + 55m = 6446$. Solving this system, we find $m = -2.8$ and $b = 440$. Thus, the equation of the least-squares line is $y = -2.8x + 440$.

b.

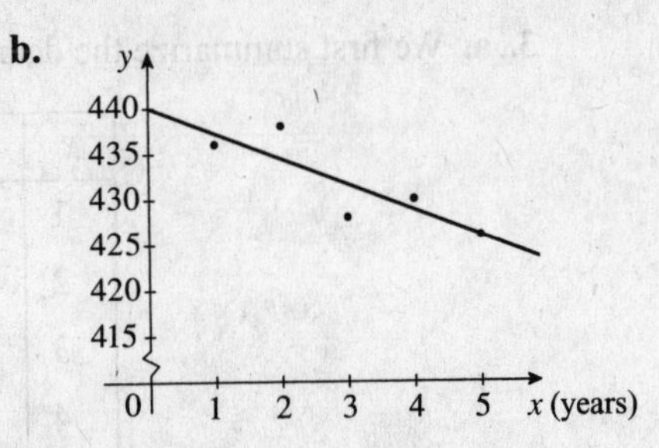

c. Two years from now, the average SAT verbal score in that area will be $y = -2.8\,(7) + 440 = 420.4$.

11. a.

	x	y	x^2	xy
	0	154.5	0	0
	1	381.8	1	381.8
	2	654.5	4	1309
	3	845	9	2535
Sum	6	2035.8	14	4225.8

The normal equations are $4b + 6m = 2035.8$ and $6b + 14m = 4225.8$. The solutions are $m = 234.42$ and $b = 157.32$, so the required equation is $y = 234.4x + 157.3$.

b. The projected number of Facebook users is $f\,(7) = 234.4\,(7) + 157.3 = 1798.1$, or approximately 1798.1 million.

13. a.

	x	y	x^2	xy
	1	20	1	20
	2	24	4	48
	3	26	9	78
	4	28	16	112
	5	32	25	160
Sum	15	130	55	418

The normal equations are $5b + 15m = 130$ and $15b + 55m = 418$. The solutions are $m = 2.8$ and $b = 17.6$, and so an equation of the line is $y = 2.8x + 17.6$.

b. When $x = 8$, $y = 2.8\,(8) + 17.6 = 40$. Hence, the state subsidy is expected to be \$40 million for the eighth year.

15. a.

	x	y	x^2	xy
	4	1.42	16	5.68
	5	1.73	25	8.65
	6	1.98	36	11.88
	7	2.32	49	16.24
	8	2.65	64	21.2
Sum	30	10.1	190	63.65

The normal equations are $5b + 30m = 10.1$ and $30b + 190m = 63.65$. The solutions are $m \approx 0.305$ and $b \approx 0.19$, so the required equation is $y = 0.305x + 0.19$.

b. The rate of change is given by the slope of the least-squares line, that is, approximately \$0.305 billion/yr.

c. $f\,(10) = 0.305\,(10) + 0.19 = 3.24$, or \$3.24 billion

17.

x	y	x^2	xy
0	82.0	0	0
1	84.7	1	84.7
2	86.8	4	173.6
3	89.7	9	269.1
4	91.8	16	367.2
Sum 10	435	30	894.6

The normal equations are $5b + 10m = 435$ and $10b + 30m = 894.6$. The solutions are $m = 2.46$ and $b = 82.08$, so the required equation is $y = 2.46x + 82.1$.

b. The estimated number of credit union members in 2013 is $f(5) = 2.46(5) + 82.1 = 94.4$, or approximately 94.4 million.

19. a.

x	y	x^2	xy
0	29.4	0	0
1	32.2	1	32.2
2	34.8	4	69.6
3	37.7	9	113.1
4	40.4	16	161.6
Sum 10	174.5	30	376.5

The normal equations are $5b + 10m = 174.5$ and $10b + 30m = 376.5$. The solutions are $m = 2.75$ and $b = 29.4$, so $y = 2.75x + 29.4$.

b. The average rate of growth of the number of subscribers from 2006 through 2010 was 2.75 million per year.

21. a.

x	y	x^2	xy
0	6.4	0	0
1	6.8	1	6.8
2	7.1	4	14.2
3	7.4	9	22.2
4	7.6	16	30.4
Sum 10	35.3	30	73.6

The normal equations are $5b + 10m = 35.3$ and $10b + 30m = 73.6$. The solutions are $m = 0.3$ and $b = 6.46$, so the required equation is $y = 0.3x + 6.46$.

b. The rate of change is given by the slope of the least-squares line, that is, approximately \$0.3 billion/yr.

23. a. We summarize the data at right. The normal equations are $6b + 39m = 195.5$ and $39b + 271 = 1309$. The solutions are $b = 18.38$ and $m = 2.19$, so the required least-squares line is given by $y = 2.19x + 18.38$.

b. The average rate of increase is given by the slope of the least-squares line, namely \$2.19 billion/yr.

c. The revenue from overdraft fees in 2011 is $y = 2.19(11) + 18.38 = 42.47$, or approximately \$42.47 billion.

	x	y	x^2	xy
	4	27.5	16	110
	5	29	25	145
	6	31	36	186
	7	34	49	238
	8	36	64	288
	9	38	81	342
Sum	39	195.5	271	1309

25. a.

x	y	x^2	xy
0	60	0	0
2	74	4	148
4	90	16	360
6	106	36	636
8	118	64	944
10	128	100	1280
12	150	144	1800
Sum 42	726	364	5168

The normal equations are $7b + 42m = 726$ and $42b + 364m = 5168$. The solutions are $m \approx 7.25$ and $b \approx 60.21$, so the required equation is $y = 7.25x + 60.21$.

b. $y = 7.25\,(11) + 60.21 = 139.96$, or \$139.96 billion.

c. \$7.25 billion/yr.

27. False. See Example 1 on page 600 of the text.

29. True.

Technology Exercises page 609

1. $y = 2.3596x + 3.8639$

3. $y = -1.1948x + 3.5525$

5. a. $22.3x + 143.5$ **b.** \$22.3 billion/yr **c.** \$366.5 billion

7. a. $y = 1.5857t + 6.6857$ **b.** \$19.4 billion

9. a. $y = 1.751x + 7.9143$ **b.** \$22 billion

8.5 Constrained Maxima and Minima and the Method of Lagrange Multipliers

Problem-Solving Tips

1. The method of Lagrange multipliers allows us to find the critical points of the function $f(x, y)$ subject to the constraint $g(x, y) = 0$ (if extrema exist). However, it does not tell us whether those critical points lead to relative extremum. Instead we rely on the geometric or physical nature of the problem to classify these points.

2. To use the method of Lagrange multipliers, first form the function $F(x, y, \lambda) = f(x, y) + \lambda g(x, y)$ and then solve the system of equations $f(x) = \begin{cases} F_x = 0 \\ F_y = 0 \\ F_\lambda = 0 \end{cases}$ for all values of x, y and λ. The solutions are candidates for extrema of f.

Concept Questions page 618

1. A constrained relative extremum of f is an extremum of f subject to a constraint of the form $g(x, y) = 0$.

Exercises page 618

1. $f(x,y) = x^2 + 3y^2$. We form the Lagrangian function $F(x,y,\lambda) = x^2 + 3y^2 + \lambda(x + y - 1)$ and solve the system $\begin{cases} F_x = 2x + \lambda = 0 \\ F_y = 6y + \lambda = 0 \\ F_\lambda = x + y - 1 = 0 \end{cases}$ Solving the first and the second equations for x and y in terms of λ, we obtain $x = -\frac{\lambda}{2}$ and $y = -\frac{\lambda}{6}$ which, upon substitution into the third equation, yields $-\frac{\lambda}{2} - \frac{\lambda}{6} - 1 = 0$ or $\lambda = -\frac{3}{2}$. Therefore, $x = \frac{3}{4}$ and $y = \frac{1}{4}$, which gives the point $\left(\frac{3}{4}, \frac{1}{4}\right)$ as the sole critical point of f. Thus, $\left(\frac{3}{4}, \frac{1}{4}\right) = \frac{3}{4}$ is a minimum of f.

3. $f(x,y) = 2x + 3y - x^2 - y^2$. We form the Lagrangian function $F(x,y,\lambda) = 2x + 3y - x^2 - y^2 + \lambda(x + 2y - 9)$ and solve the system $\begin{cases} F_x = 2 - 2x + \lambda = 0 \\ F_y = 3 - 2y + 2\lambda = 0 \\ F_\lambda = x + 2y - 9 = 0 \end{cases}$ Solving the first equation for λ, we obtain $\lambda = 2x - 2$. Substituting into the second equation, we have $3 - 2y + 4x - 4 = 0$, or $4x - 2y - 1 = 0$. Adding this equation to the third equation in the system, we have $5x - 10 = 0$, or $x = 2$. Therefore, $y = \frac{7}{2}$ and $f\left(2, \frac{7}{2}\right) = -\frac{7}{4}$ is the maximum value of f.

5. $f(x,y) = x^2 + 4y^2$. We form the Lagrangian function $F(x,y,\lambda) = x^2 + 4y^2 + \lambda(xy - 1)$ and solve the system $\begin{cases} F_x = 2x + \lambda y = 0 \\ F_y = 8y + \lambda x = 0 \\ F_\lambda = xy - 1 = 0 \end{cases}$ Multiplying the first and second equations by x and y, respectively, and subtracting the resulting equations, we obtain $2x^2 - 8y^2 = 0$, or $x = \pm 2y$. Substituting this into the third equation gives $2y^2 - 1 = 0$ or $y = \pm\frac{\sqrt{2}}{2}$. We conclude that $f\left(-\sqrt{2}, -\frac{\sqrt{2}}{2}\right) = f\left(\sqrt{2}, \frac{\sqrt{2}}{2}\right) = 4$ is the minimum value of f.

7. $f(x,y) = x + 5y - 2xy - x^2 - 2y^2$. We form the Lagrangian function $F(x,y,\lambda) = x + 5y - 2xy - x^2 - 2y^2 + \lambda(2x + y - 4)$ and solve the system $\begin{cases} F_x = 1 - 2y - 2x + 2\lambda = 0 \\ F_y = 5 - 2x - 4y + \lambda = 0 \\ F_\lambda = 2x + y - 4 = 0 \end{cases}$ Solving the last two equations for x and y in terms of λ, we obtain $y = \frac{1}{3}(1 + \lambda)$ and $x = \frac{1}{6}(11 - \lambda)$ which, upon substitution into the first equation, yields $1 - \frac{2}{3}(1 + \lambda) - \frac{1}{3}(11 - \lambda) + 2\lambda$, so $1 - \frac{2}{3} - \frac{2}{3}\lambda - \frac{11}{3} + \frac{1}{3}\lambda + 2\lambda = 0$. Hence, $\lambda = 2$, so $x = \frac{3}{2}$ and $y = 1$. The maximum of f is $f\left(\frac{3}{2}, 1\right) = \frac{3}{2} + 5 - 2\left(\frac{3}{2}\right) - \left(\frac{3}{2}\right)^2 - 2 = -\frac{3}{4}$.

9. $f(x,y)=xy^2$. We form the Lagrangian $F(x,y,\lambda)=xy^2+\lambda(9x^2+y^2-9)$ and solve the system

$$\begin{cases} F_x = y^2+18\lambda x=0 \\ F_y=2xy+2\lambda y=0 \\ F_\lambda=9x^2+y^2-9=0 \end{cases}$$

The first equation gives $\lambda=-\dfrac{y^2}{18x}$. Substituting into the second gives

$2xy+2y\left(-\dfrac{y^2}{18x}\right)=0$, or $18x^2y-y^3=y\left(18x^2-y^2\right)=0$, giving $y=0$ or $y=\pm3\sqrt{2}x$. If $y=0$, then the third equation gives $9x^2-9=0$, so $x=\pm1$. Therefore, the points $(-1,0)$, $(1,0)$, $\left(-\frac{\sqrt{3}}{3},-\sqrt{6}\right)$, $\left(-\frac{\sqrt{3}}{3},\sqrt{6}\right)$, $\left(\frac{\sqrt{3}}{3},-\sqrt{6}\right)$ and $\left(\frac{\sqrt{3}}{3},\sqrt{6}\right)$ give extreme values of f subject to the given constraint. Evaluating $f(x,y)$ at each of these points, we see that $f\left(\frac{\sqrt{3}}{3},-\sqrt{6}\right)=f\left(\frac{\sqrt{3}}{3},\sqrt{6}\right)=2\sqrt{3}$ is the maximum value of f.

11. $f(x,y)=xy$. We form the Lagrangian function $F(x,y,\lambda)=xy+\lambda\left(x^2+y^2-16\right)$ and solve the system

$$\begin{cases} F_x=y+2\lambda x=0 \\ F_y=x+2\lambda y=0 \\ F_\lambda=x^2+y^2-16=0 \end{cases}$$

Solving the first equation for λ and substituting this value into the second equation yields $x-2\left(\dfrac{y}{2x}\right)y=0$, or $x^2=y^2$. Substituting the last equation into the third equation in the system, yields $x^2+x^2-16=0$, or $x^2=8$, that is, $x=\pm2\sqrt{2}$. The corresponding values of y are $y=\pm2\sqrt{2}$. Therefore the critical points of F are $\left(\pm2\sqrt{2},\pm2\sqrt{2}\right)$ (all four possibilities). Evaluating f at each of these points, we find that $f\left(-2\sqrt{2},2\sqrt{2}\right)=f\left(2\sqrt{2},-2\sqrt{2}\right)=-8$ are relative minima and $f\left(-2\sqrt{2},-2\sqrt{2}\right)=f\left(2\sqrt{2},2\sqrt{2}\right)=8$ are relative maxima.

13. $f(x,y)=xy^2$. We form the Lagrangian function $F(x,y,\lambda)=xy^2+\lambda\left(x^2+y^2-1\right)$ and solve the system

$$\begin{cases} F_x=y^2+2x\lambda=0 \\ F_y=2xy+2y\lambda=0 \\ F_\lambda=x^2+y^2-1=0 \end{cases}$$

We find that either $x=\pm\frac{\sqrt{3}}{3}$ and $y=\pm\frac{\sqrt{6}}{3}$, or $x=\pm1$ and $y=0$. Evaluating f at each of the critical points $\left(\pm\frac{\sqrt{3}}{3},\pm\frac{\sqrt{6}}{3}\right)$ (all four possibilities) and $(\pm1,0)$, we find that $f\left(-\frac{\sqrt{3}}{3},\pm\frac{\sqrt{6}}{3}\right)=-\frac{2\sqrt{3}}{9}$ are relative minima and $f\left(\frac{\sqrt{3}}{3},\pm\frac{\sqrt{6}}{3}\right)=\frac{2\sqrt{3}}{9}$ are relative maxima.

15. $f(x,y,z)=x^2+y^2+z^2$. We form the Lagrangian function $F(x,y,z,\lambda)=x^2+y^2+z^2+\lambda(3x+2y+z-6)$ and solve the system

$$\begin{cases} F_x=2x+3\lambda=0 \\ F_y=2y+2\lambda=0 \\ F_z=2z+\lambda=0 \\ F_\lambda=3x+2y+z-6=0 \end{cases}$$

The third equation gives $\lambda=-2z$. Substituting into the first two equations, we obtain

$$\begin{cases} 2x-6z=0 \\ 2y-4z=0 \end{cases}$$

Thus, $x=3z$ and $y=2z$. Substituting into the fourth equation yields $9z+4z+z-6=0$, or $z=\frac{3}{7}$. Therefore, $x=\frac{9}{7}$ and $y=\frac{6}{7}$, and so $f\left(\frac{9}{7},\frac{6}{7},\frac{3}{7}\right)=\frac{18}{7}$ is the minimum value of f.

17. $P(x, y) = -0.2x^2 - 0.25y^2 - 0.2xy + 100x + 90y - 4000$. We want to maximize P subject to the constraint $x + y = 200$. The Lagrangian function is $F(x, y, \lambda) = -0.2x^2 - 0.25y^2 - 0.2xy + 100x + 90y - 4000 + \lambda(x + y - 200)$. We solve the system

$$\begin{cases} F_x = -0.4x - 0.2y + 100 + \lambda = 0 \\ F_y = -0.5y - 0.2x + 90 + \lambda = 0 \\ F_\lambda = x + y - 200 = 0 \end{cases}$$

Subtracting the first equation from the second yields $0.2x - 0.3y - 10 = 0$, or $2x - 3y - 100 = 0$. Multiplying the third equation in the system by 2 and subtracting the resulting equation from the last equation, we find $-5y + 300 = 0$, so $y = 60$. Thus, $x = 140$ and the company should make 140 finished and 60 unfinished units.

19. Suppose each of the sides made of pine board is x feet long and those of steel are y feet long. Then $xy = 800$. The cost is $C = 12x + 3y$ and is to be minimized subject to the condition $xy = 800$. We form the Lagrangian function $F(x, y, \lambda) = 12x + 3y + \lambda(xy - 800)$ and solve the system

$$\begin{cases} F_x = 12 + \lambda y = 0 \\ F_y = 3 + \lambda x = 0 \\ F_\lambda = xy - 800 = 0 \end{cases}$$

Multiplying the first equation by x and the second equation by y and subtracting the resulting equations, we obtain $12x - 3y = 0$, or $y = 4x$. Substituting this into the third equation of the system, we obtain $4x^2 - 800 = 0$, so $x = \pm 10\sqrt{2}$. Because x must be positive, we take $x = 10\sqrt{2}$, so $y = 40\sqrt{2}$ and the dimensions are approximately 14.14 ft by 56.56 ft.

21. Let the dimensions of the box (in feet) be $x \times y \times z$. We want to maximize $f(x, y, z) = xyz$ subject to the constraint $g(x, y, z) = xyz + 2xz + 2yz - 12 = 0$. The Lagrangian function is $F(x, y, z, \lambda) = xyz + \lambda(xy + 2xz + 2yz - 48)$. To find the critical points of F, we solve the system

$$\begin{cases} F_x = yz + \lambda y + 2\lambda z = 0 & (1) \\ F_y = xz + \lambda x + 2\lambda z = 0 & (2) \\ F_z = xy + 2\lambda x + 2\lambda y = 0 & (3) \\ F_\lambda = xy + 2xz + 2yz - 12 = 0 & (4) \end{cases}$$

Multiplying (1) by x and (2) by y, we obtain

$$\begin{cases} xyz + \lambda xy + 2\lambda xz = 0 & (5) \\ xyz + \lambda xy + 2\lambda yz = 0 & (6) \end{cases}$$

Subtracting (5) from (6), we have $2\lambda z(y - x) = 0$. Since λ and z cannot be zero, we see that $y = x$. Next, multiplying (3) by z, we obtain $xyz + 2\lambda xz + 2\lambda yz = 0$ (7). Subtracting (6) from (7) gives $\lambda x(2z - y) = 0$, so $z = \frac{1}{2}y$. Substituting $x = y$ and $z = \frac{1}{2}y$ into (4) gives $y^2 + 2y\left(\frac{1}{2}y\right) + 2y\left(\frac{1}{2}y\right) - 12 = 0$, so $3y^2 = 12$ and $y = 2$. Thus $x = 2$ and $z = 1$. The dimensions of the required box are therefore $2' \times 2' \times 1'$.

23. Let the dimensions of the box (in feet) be $x \times y \times z$. We want to maximize $f(x, y, z) = xyz$ subject to the constraint $g(x, y, z) = 2x + y + 2z - 108 = 0$. The Lagrangian function is $F(x, y, \lambda) = xyz + \lambda(2x + y + 2z - 108)$. To find the critical points of F, we solve the system

$$\begin{cases} F_x = yz + 2\lambda = 0 & (1) \\ F_y = xz + \lambda = 0 & (2) \\ F_z = xy + 2\lambda = 0 & (3) \\ F_\lambda = 2x + y + 2z - 108 = 0 & (4) \end{cases}$$

Multiplying (1) by x and (2) by y, we obtain

$$\begin{cases} xyz + 2\lambda x = 0 & (5) \\ xyz + \lambda y = 0 & (6) \end{cases}$$

Subtracting (5) from (6), we have $\lambda(y - 2x) = 0$. Since $\lambda \neq 0$, we see that $y = 2x$. Next, multiplying (3) by z, we obtain $xyz + 2\lambda z = 0$ (7). Subtracting (5) from (7) gives $2\lambda(z - x) = 0$, so $z = x$, and substituting $y = 2x$ and $z = x$ into (4) gives $2x + 2x + 2x = 108$, so $6x = 108$, giving $x = 18$. Thus, $y = 2(18) = 36$ and $z = 18$, and so the required dimensions are $18'' \times 18'' \times 36''$.

25. We want to minimize the function $C(r, h) = 8\pi rh + 6\pi r^2$ subject to the constraint $\pi r^2 h - 64 = 0$. We form the Lagrangian function $F(r, h, \lambda) = 8\pi rh + 6\pi r^2 - \lambda(\pi r^2 h - 64)$ and solve the system

$$\begin{cases} F_r = 8\pi h + 12\pi r - 2\lambda\pi rh = 0 \\ F_h = 8\pi r - \lambda\pi r^2 = 0 \\ F_\lambda = \pi r^2 h - 64 = 0 \end{cases}$$

Solving the second equation for λ yields $\lambda = 8/r$, which when substituted into the first equation yields $8\pi h + 12\pi r - 2\pi rh\left(\frac{8}{r}\right) = 0$, $12\pi r = 8\pi h$, and $h = \frac{3}{2}r$. Substituting this value of h into the third equation of the system, we find $3r^2\left(\frac{3}{2}r\right) = 64$, $r^3 = \frac{128}{3\pi}$, so $r = \frac{4}{3}\sqrt[3]{\frac{18}{\pi}}$ and $h = 2\sqrt[3]{\frac{18}{\pi}}$.

27. Let the box have dimensions x by y by z feet. Then $xyz = 4$. We want to minimize $C = 2xz + 2yz + \frac{3}{2}(2xy) = 2xz + 2yz + 3xy$. We form the Lagrangian function $F(x, y, z, \lambda) = 2xz + 2yz + 3xy + \lambda(xyz - 4)$ and solve the system

$$\begin{cases} F_x = 2z + 3y + \lambda yz = 0 \\ F_y = 2z + 3x + \lambda xz = 0 \\ F_z = 2x + 2y + \lambda xy = 0 \\ F_\lambda = xyz - 4 = 0 \end{cases}$$

Multiplying the first, second, and third equations by x, y, and z respectively, we have

$$\begin{cases} 2xz + 3xy + \lambda xyz = 0 \\ 2yz + 3xy + \lambda xyz = 0 \\ 2xz + 2yz + \lambda xyz = 0 \end{cases}$$

The first two equations imply that $2z(x - y) = 0$. Because $z \neq 0$, we see that $x = y$. The second and third equations imply that $x(3y - 2z) = 0$ or $x = \frac{3}{2}y$. Substituting these values into the fourth equation in the system, we find $y^2\left(\frac{3}{2}y\right) = 4$, so $y^3 = \frac{8}{3}$. Therefore, $y = \frac{2}{3^{1/3}} = \frac{2}{3}\sqrt[3]{9}$, $x = \frac{2}{3}\sqrt[3]{9}$, and $z = \sqrt[3]{9}$, and the dimensions (in feet) are $\frac{2}{3}\sqrt[3]{9} \times \frac{2}{3}\sqrt[3]{9} \times \sqrt[3]{9}$.

29. Let x, y, and z denote the length, width, and height of the box. We can assume without loss of generality that the cost of the material for constructing the sides and top is \$1/ft^2. Then the total cost is $C = f(x,y,z) = 3xy + 2xz + 2yz$. We want to minimize f subject to the constraint $g(x,y,z) = xyz - 16 = 0$. We form the Lagrangian $F(x,y,z,\lambda) = 3xy + 2xz + 2yz - \lambda(xyz - 16)$ and solve the system
$$\begin{cases} F_x = 3y + 2z - \lambda yz = 0 \\ F_y = 3x + 2z - \lambda xz = 0 \\ F_z = 2x + 3y - \lambda xy = 0 \\ F_\lambda = xyz - 16 = 0 \end{cases}$$
From the first and second equations, we find $\lambda = \dfrac{3y+2z}{yz} = \dfrac{3x+2z}{xz} \Rightarrow 3xyz + 2xz^2 = 3xyz + 2yz^2 \Rightarrow x = y$. From the second and third equations, we have $\lambda = \dfrac{3x+2z}{xz} = \dfrac{2x+2y}{xy} \Rightarrow 3x^2y + 2xyz = 2x^2z + 2xyz \Rightarrow$ $z = \frac{3}{2}y$. Substituting into the fourth equation, we have $y(y)\left(\frac{3}{2}y\right) = 16 \Rightarrow y^3 = \frac{32}{3} \Rightarrow y = \frac{2}{3}\sqrt[3]{36}$. Thus, $x = \frac{2}{3}\sqrt[3]{36}$ and $z = \sqrt[3]{36}$. The dimensions of the box are $\frac{2}{3}\sqrt[3]{36}' \times \frac{2}{3}\sqrt[3]{36}' \times \sqrt[3]{36}'$.

31. We want to maximize $P(x,y)$ subject to $g(x,y) = px + qy - C = 0$. First, we form the Lagrangian $F(x,y) = P(x,y) + \lambda(px + qy - C)$. Next, we solve the system $F_x = P_x + \lambda p = 0$, $F_y = Py + \lambda q = 0$, $F_\lambda = px + qy - C = 0$. From the first two equations, we see that $\lambda = -\dfrac{P_x}{p} = -\dfrac{P_y}{q}$, so if (x^*, y^*) gives rise to a relative maximum value of P subject to the constraint $g(x,y) = 0$, then $\dfrac{P_x(x^*,y^*)}{p} = \dfrac{P_y(x^*,y^*)}{q}$ or $\dfrac{P_x(x^*,y^*)}{P_y(x^*,y^*)} = \dfrac{p}{q}$.

33. We want to maximize $f(x,y) = 100x^{3/4}y^{1/4}$ subject to $100x + 200y = 200{,}000$. We form the Lagrangian function $F(x,y,\lambda) = 100x^{3/4}y^{1/4} + \lambda(100x + 200y - 200{,}000)$ and solve the system
$$\begin{cases} F_x = 75x^{-1/4}y^{1/4} + 100\lambda = 0 \\ F_y = 25x^{3/4}y^{-3/4} + 200\lambda = 0 \\ F_\lambda = 100x + 200y - 200{,}000 = 0 \end{cases}$$
The first two equations imply that $150x^{-1/4}y^{1/4} - 25x^{3/4}y^{-3/4} = 0$ or, upon multiplying by $x^{1/4}y^{3/4}$, $150y - 25x = 0$, which implies that $x = 6y$. Substituting this value of x into the third equation of the system, we have $600y + 200y - 200{,}000 = 0$, giving $y = 250$, and therefore $x = 1500$. So to maximize production, he should buy 1500 units of labor and 250 units of capital.

35. We use the result of Exercise 33 with $P(x,y) = f(x,y) = 100x^{3/4}y^{3/4}$, $P = 100$, $q = 200$, and $C = 200{,}000$. Here $P_x(x,y) = f_x(x,y) = 100\left(\frac{3}{4}x^{-1/4}y^{1/4}\right) = 75\left(\dfrac{y}{x}\right)^{1/4}$ and $P_y(x,y) = f_y(x,y) = 100\left(\frac{1}{4}x^{3/4}y^{-3/4}\right) = 25\left(\dfrac{x}{y}\right)^{3/4}$. Thus, $\dfrac{P_x(x,y)}{P_y(x,y)} = \dfrac{p}{q}$ gives $\dfrac{75(y/x)^{1/4}}{25(x/y)^{3/4}} = \dfrac{100}{200}$, $\dfrac{3y^{1/4}y^{3/4}}{x^{1/4}x^{3/4}} = \dfrac{1}{2}$, $\dfrac{y}{x} = \dfrac{1}{6}$, and $y = \frac{1}{6}x$. Substituting this into the constraint equation $100x + 200y = 200{,}000$ yields $100x + \frac{200}{6}x = 200{,}000$, $600x + 200x = 12{,}000{,}000$, and $x = 1500$, and so $y = \frac{1500}{6} = 250$. Therefore, 1500 units should be expended on labor and 250 units on capital, as obtained earlier.

37. We use the result of Exercise 36 with $P(x,y) = f(x,y) = ax^b y^{1-b}$. Here $P_x(x,y) = abx^{b-1}y^{1-b}$ and $P_y(x,y) = a(1-b)x^b y^{-b}$, so $\dfrac{P_x(x,y)}{P_y(x,y)} = \dfrac{abx^{b-1}y^{1-b}}{a(1-b)x^b y^{-b}} = \dfrac{bx^{-1}y}{1-b} = \dfrac{by}{(1-b)x}$. At the production level with minimum cost, we have $\dfrac{by}{(1-b)x} = \dfrac{p}{q}$, so $y = \dfrac{(1-b)px}{bq}$. Substituting this into the equation $ax^b y^{1-b} = k$, we obtain $ax^b\left[\dfrac{(1-b)px}{bq}\right]^{1-b} = k$, whence $ax^b\left[\dfrac{(1-b)p}{bq}\right]^{1-b} x^{1-b} = k$, $ax\left[\dfrac{(1-b)p}{bq}\right]^{1-b} = k$, and

$x = \dfrac{k}{a}\left[\dfrac{bq}{(1-b)p}\right]^{1-b}$, so $y = \dfrac{(1-b)p}{bq}\cdot\dfrac{k}{a}\left[\dfrac{bq}{(1-b)p}\right]^{1-b} = \dfrac{k}{a}\left[\dfrac{bq}{(1-b)p}\right]^{-b} = \dfrac{k}{a}\left[\dfrac{(1-b)p}{bq}\right]^{b}$. Thus,

$x^* p = \dfrac{kp}{a}\left[\dfrac{bq}{(1-b)p}\right]^{1-b}$ should be spent on labor and $y^* q = \dfrac{kq}{a}\left[\dfrac{1-b}{bq}p\right]^b$ should be spent on capital.

39. False. See Example 1.

41. True. We form the Lagrangian function $F(x,y,\lambda) = f(x,y) + \lambda g(x,y)$. Then $F_x = 0$, $F_y = 0$, and $F_\lambda = 0$ at (a,b) and $f_x(a,b) + \lambda g_x(a,b) = 0$, so $f_x(a,b) = -\lambda g_x(a,b)$, and $f_y(a,b) + \lambda(a,b) = 0$, so $f_y(a,b) = -\lambda g_y(a,b)$ and $g(a,b) = 0$.

8.6 Double Integrals

Concept Questions page 630

1. An iterated integral is a single integral such as $\int_a^b f(x,y)\,dx$, where we think of y as a constant. It is evaluated as follows: $\int_R \int f(x,y)\,dA = \int_c^d \left[\int_a^b f(x,y)\,dx\right] dy$.

3. $\int_R \int f(x,y)\,dA = \int_c^d \left[\int_{h_1(y)}^{h_2(y)} f(x,y)\,dx\right] dy$.

5. The average value is $\dfrac{\int_R \int f(x,y)\,dA}{\int_R \int dA}$.

Exercises page 631

1. $\int_1^2 \int_0^1 (y+2x)\,dy\,dx = \int_1^2 \left(\frac{1}{2}y^2 + 2xy\right)\Big|_{y=0}^{y=1} dx = \int_1^2 \left(\frac{1}{2} + 2x\right) dx = \left(\frac{1}{2}x + x^2\right)\Big|_1^2 = 5 - \frac{3}{2} = \frac{7}{2}$.

3. $\int_{-1}^1 \int_0^1 xy^2\,dy\,dx = \int_{-1}^1 \frac{1}{3}xy^3\Big|_{y=0}^{y=1} dx = \int_{-1}^1 \frac{1}{3}x\,dx = \frac{1}{6}x^2\Big|_{-1}^1 = \frac{1}{6} - \left(\frac{1}{6}\right) = 0$.

5. $\displaystyle\int_{-1}^2 \int_1^{e^3} \frac{x}{y}\,dy\,dx = \int_{-1}^2 x\ln y\Big|_{y=1}^{y=e^3} dx = \int_{-1}^2 x\ln e^3\,dx = \int_{-1}^2 3x\,dx = \frac{3}{2}x^2\Big|_{-1}^2 = \frac{3}{2}(4) - \frac{3}{2}(1) = \frac{9}{2}$.

7. $\int_{-2}^0 \int_0^1 4xe^{2x^2+y}\,dx\,dy = \int_{-2}^0 e^{2x^2+y}\Big|_{x=0}^{x=1} dy = \int_{-2}^0 (e^{2+y} - e^y)\,dy = (e^{2+y} - e^y)\Big|_{-2}^0 = (e^2 - 1) - (e^0 - e^{-2})$
$= e^2 - 2 + e^{-2} = (e^2 - 1)(1 - e^{-2})$.

9. $\int_0^1 \int_1^e \ln y\,dy\,dx = \int_0^1 (y\ln y - y)\big|_{y=1}^{y=e} dx = \int_0^1 dx = 1$.

11. $\int_0^1 \int_0^x (x+2y)\,dy\,dx = \int_0^1 (xy+y^2)\big|_{y=0}^{y=x}\,dx = \int_0^1 2x^2\,dx = \frac{2}{3}x^3\Big|_0^1 = \frac{2}{3}$.

13. $\int_1^3 \int_0^{x+1} (2x+4y)\,dy\,dx = \int_1^3 (2xy+2y^2)\big|_{y=0}^{y=x+1}\,dx = \int_1^3 \left[2x(x+1)+2(x+1)^2\right]dx = \int_1^3 (4x^2+6x+2)\,dx$

$= \left(\frac{4}{3}x^3+3x^2+2x\right)\Big|_1^3 = (36+27+6)-\left(\frac{4}{3}+3+2\right) = \frac{188}{3}$.

15. $\int_0^4 \int_0^{\sqrt{y}} (x+y)\,dx\,dy = \int_0^4 \left(\frac{1}{2}x^2+xy\right)\Big|_{x=0}^{x=\sqrt{y}}\,dy = \int_0^4 \left(\frac{1}{2}y+y^{3/2}\right)dy = \left(\frac{1}{4}y^2+\frac{2}{5}y^{5/2}\right)\Big|_0^4 = 4+\frac{64}{5} = \frac{84}{5}$.

17. $\int_0^2 \int_0^{\sqrt{4-y^2}} y\,dx\,dy = \int_0^2 xy\big|_{x=0}^{x=\sqrt{4-y^2}}\,dy = \int_0^2 y\sqrt{4-y^2}\,dy = -\frac{1}{2}\left(\frac{2}{3}\right)(4-y^2)^{3/2}\Big|_0^2 = \frac{1}{3}\left(4^{3/2}\right) = \frac{8}{3}$.

19. $\int_0^1 \int_0^x 2xe^y\,dy\,dx = \int_0^1 2xe^y\big|_{y=0}^{y=x}\,dx = \int_0^1 (2xe^x-2x)\,dx = 2(x-1)e^x - x^2\big|_0^1 = (-1)+2 = 1$.

21. $\int_0^1 \int_x^{\sqrt{x}} ye^x\,dy\,dx = -\int_0^1 \int_{\sqrt{x}}^x ye^x\,dy\,dx = \int_0^1 \left(-\frac{1}{2}y^2e^x\right)\Big|_{y=\sqrt{x}}^{y=x}\,dx = -\frac{1}{2}\int_0^1 (x^2e^x - xe^x)\,dx$

$= -\frac{1}{2}\left(x^2e^x\big|_0^1 - 2\int_0^1 xe^x\,dx - \int_0^1 xe^x\,dx\right) = -\frac{1}{2}\left(x^2e^x\big|_0^1 - 3\int_0^1 xe^x\,dx\right)$

$= -\frac{1}{2}\left(x^2e^x - 3xe^x + 3e^x\right)\Big|_0^1 = -\frac{1}{2}(e-3e+3e-3) = \frac{1}{2}(3-e)$.

23. $\int_0^1 \int_{2x}^2 e^{y^2}\,dy\,dx = \int_0^2 \int_0^{y/2} e^{y^2}\,dx\,dy = \int_0^2 xe^{y^2}\Big|_{x=0}^{x=y/2}\,dy = \int_0^2 \frac{1}{2}ye^{y^2}\,dy = \frac{1}{4}e^{y^2}\Big|_0^2 = \frac{1}{4}\left(e^4-1\right)$.

25. $\int_0^2 \int_{y/2}^1 ye^{x^3}\,dx\,dy = \int_0^1 \int_0^{2x} ye^{x^3}\,dy\,dx = \int_0^1 \frac{1}{2}y^2e^{x^3}\Big|_{y=0}^{y=2x}\,dx = \int_0^1 2x^2e^{x^3}\,dx = \frac{2}{3}e^{x^3}\Big|_0^1 = \frac{2}{3}(e-1)$.

27. $V = \int_0^4 \int_0^3 \left(4-x+\frac{1}{2}y\right)dx\,dy = \int_0^4 \left(4x-\frac{1}{2}x^2+\frac{1}{2}xy\right)\Big|_{x=0}^{x=3}\,dy = \int_0^4 \left(\frac{15}{2}+\frac{3}{2}y\right)dy$

$= \left(\frac{15}{2}y+\frac{3}{4}y^2\right)\Big|_0^4 = 42$.

29. $V = \int_0^2 \int_0^{3-(3/2)z} (6-2y-3z)\,dy\,dz = \int_0^2 (6y-y^2-3yz)\big|_{y=0}^{y=3-(3/2)z}\,dz$

$= \int_0^2 \left[6\left(3-\frac{3}{2}z\right)-\left(3-\frac{3}{2}z\right)^2-3\left(3-\frac{3}{2}z\right)z\right]dz = \left[-2\left(3-\frac{3}{2}z\right)^2-\frac{2}{9}\left(3-\frac{3}{2}z\right)^3-\frac{9}{2}z^2+\frac{3}{2}z^3\right]_0^2$

$= (-18+12)-(-18+6) = 6$.

31. $V = \iint_R f(x,y)\,dA = \int_0^1 \int_0^{2-2y} (4-x^2-y^2)\,dx\,dy$

$= \int_0^1 \left[(4-y^2)x - \frac{1}{3}x^3\right]_{x=0}^{x=2-2y}\,dy$

$= \int_0^1 \left[(4-y^2)(2-2y) - \frac{1}{3}(2-2y)^3\right]dy$

$= \int_0^1 \left(\frac{14}{3}y^3 - 10y^2 + \frac{16}{3}\right)dy = \left(\frac{7}{6}y^4 - \frac{10}{3}y^3 + \frac{16}{3}y\right)\Big|_0^1 = \frac{19}{6}$

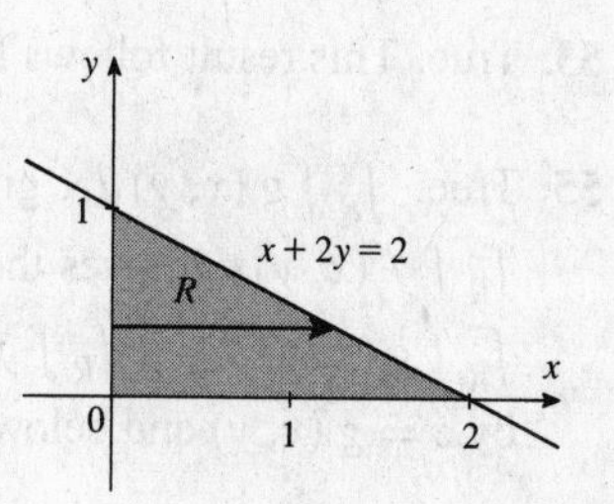

33. $V = \iint_R f(x,y)\,dA = \int_0^2 \int_0^2 2e^{-x}e^{-y}\,dx\,dy = \int_0^2 \left[-2e^{-x}e^{-y}\right]_{x=0}^{x=2}\,dy = \int_0^2 \left(-2e^{-2}e^{-y}+2e^{-y}\right)dy$

$= \left(2e^{-2}e^{-y}-2e^{-y}\right)\big|_0^2 = \frac{2\left(e^2-1\right)^2}{e^4}$

35. $V = \int_0^2 \int_0^{2x} (2x + y)\,dy\,dx = \int_0^2 \left(2xy + \frac{1}{2}y^2\right)\Big|_0^{2x} dx = \int_0^2 (4x^2 + 2x^2)\,dx = \int_0^2 6x^2\,dx = 2x^3\Big|_0^2 = 16.$

37. $V = \int_0^1 \int_0^{-x+1} e^{x+2y}\,dy\,dx = \int_0^1 \frac{1}{2}e^{x+2y}\Big|_{y=0}^{y=-x+1} dx = \frac{1}{2}\int_0^1 (e^{-x+2} - e^x)\,dx = \frac{1}{2}\left(-e^{-x+2} - e^x\right)\Big|_0^1$
$= \frac{1}{2}\left(-e - e + e^2 + 1\right) = \frac{1}{2}\left(e^2 - 2e + 1\right) = \frac{1}{2}(e-1)^2.$

39. $V = \displaystyle\int_0^4 \int_0^{\sqrt{x}} \frac{2y}{1+x^2}\,dy\,dx = \int_0^4 \frac{y^2}{1+x^2}\Big|_0^{\sqrt{x}} dx = \int_0^4 \frac{x}{1+x^2}\,dx = \frac{1}{2}\ln\left(1+x^2\right)\Big|_0^4 = \frac{1}{2}(\ln 17 - \ln 1) = \frac{1}{2}\ln 17.$

41. $V = \int_0^4 \int_0^{\sqrt{16-x^2}} x\,dy\,dx = \int_0^4 xy\Big|_{y=0}^{y=\sqrt{16-x^2}} dx = \int_0^4 x\left(16-x^2\right)^{1/2} dx = \left(-\frac{1}{2}\right)\left(\frac{2}{3}\right)\left(16-x^2\right)^{3/2}\Big|_0^4$
$= \frac{1}{3}(16)^{3/2} = \frac{64}{3}.$

43. $A = \frac{1}{1/2}\int_0^1 \int_0^x (x+2y)\,dy\,dx = 2\int_0^1 \left(xy + y^2\right)\Big|_0^x dx = 2\int_0^1 \left(x^2 + x^2\right) dx = 4\int_0^1 x^2\,dx = \frac{4}{3}x^3\Big|_0^1 = \frac{4}{3}.$

45. The area of R is $\frac{1}{2}$, so the average value of f is
$\frac{1}{1/2}\int_0^1 \int_0^x e^{-x^2}\,dy\,dx = 2\int_0^1 e^{-x^2}y\Big|_{y=0}^{y=x} dx = 2\int_0^1 xe^{-x^2}\,dx = -e^{-x^2}\Big|_0^1 = -e^{-1} + 1 = 1 - \frac{1}{e}.$

47. By elementary geometry, the area of the region is $4 + \frac{1}{2}(2)(4) = 8$. Therefore, the required average value is
$A = \frac{1}{8}\int_1^3 \int_0^{2x} \ln x\,dy\,dx = \frac{1}{8}\int_1^3 (\ln x)\,y\Big|_{y=0}^{y=2x} dx = \frac{1}{4}\int_1^3 x\ln x\,dx = \frac{1}{4}\left(\frac{1}{4}x^2\right)(2\ln x - 1)\Big|_1^3$ (by parts)
$= \frac{9}{16}(2\ln 3 - 1) - \frac{1}{16}(-1) = \frac{1}{8}(9\ln 3 - 4).$

49. The average population density inside R is $\frac{43{,}329}{20} \approx 2166$ people per square mile.

51. The average weekly profit is
$P = \frac{1}{(20)(20)}\int_{100}^{120}\int_{180}^{200} \left(-0.2x^2 - 0.25y^2 - 0.2xy + 100x + 90y - 4000\right) dx\,dy$
$= \frac{1}{400}\int_{100}^{120} \left(-\frac{1}{15}x^3 - 0.25y^2x - 0.1x^2y + 50x^2 + 90xy - 4000x\right)\Big|_{x=180}^{x=200} dy$
$= \frac{1}{400}\int_{100}^{120} \left(-144{,}533.33 - 5y^2 - 760y + 380{,}000 + 1800y - 80{,}000\right) dy$
$= \frac{1}{400}\int_{100}^{120} \left(155{,}466.67 - 5y^2 + 1040y\right) dy = \frac{1}{400}\left(155{,}466.67y - \frac{5}{3}y^3 + 520y^2\right)\Big|_{100}^{120}$
$= \frac{1}{400}(3{,}109{,}333.40 - 1{,}213{,}333.30 + 2{,}288{,}000) \approx 10{,}460$, or \$10,460.

53. True. This result follows from the definition.

55. True. $\int_R\int g(x,y)\,dA$ gives the volume of the solid bounded above by the surface $z = g(x,y)$ and $\int_R\int f(x,y)\,dA$ gives the volume of the solid bounded above by the surface $z = f(x,y)$. Therefore, $\int_R\int g(x,y)\,dA - \int_R\int f(x,y)\,dA = \int_R\int \left[g(x,y) - f(x,y)\right] dA$ gives the volume of the solid bounded above by $z = g(x,y)$ and below by $z = f(x,y)$.

CHAPTER 8 Concept Review Questions page 635

1. xy, ordered pair, real number, $f(x,y)$

3. $z = f(x,y)$, f, surface

5. constant, x

7. $\leq$, (a, b), $\leq$, domain

9. scatter, minimizing, least-squares, normal

11. volume, solid

CHAPTER 8 Review Exercises page 635

1. $f(x, y) = \dfrac{xy}{x^2 + y^2}$, so $f(0, 1) = 0$, $f(1, 0) = 0$, $f(1, 1) = \dfrac{1}{1+1} = \dfrac{1}{2}$, and $f(0, 0)$ does not exist because the point $(0, 0)$ does not lie in the domain of f.

3. $h(x, y, z) = xye^z + \dfrac{x}{y}$, so $h(1, 1, 0) = 1 + 1 = 2$, $h(-1, 1, 1) = -e - 1 = -(e + 1)$, and $h(1, -1, 1) = -e - 1 = -(e + 1)$.

5. $f(x, y) = \dfrac{x - y}{x + y}$, so $D = \{(x, y) \mid y \neq -x\}$.

7. $f(x, y, z) = \dfrac{xy\sqrt{z}}{(1 - x)(1 - y)(1 - z)}$. The domain of f is the set of all ordered triples (x, y, z) of real numbers such that $z \geq 0$, $x \neq 1$, $y \neq 1$, and $z \neq 1$.

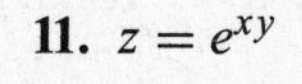

9. $z = y - x^2$

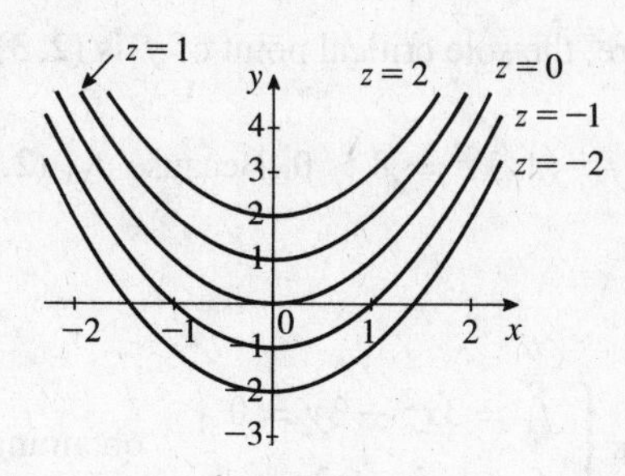

11. $z = e^{xy}$

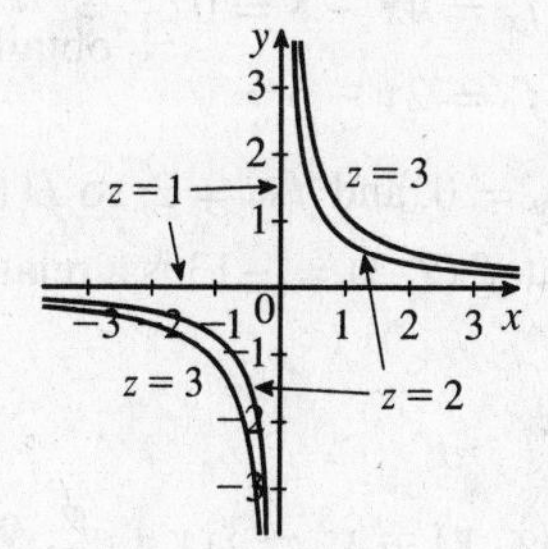

13. $f(x, y) = x\sqrt{y} + y\sqrt{x}$, so $f_x = \sqrt{y} + \dfrac{y}{2\sqrt{x}}$ and $f_y = \dfrac{x}{2\sqrt{y}} + \sqrt{x}$.

15. $f(x, y) = \dfrac{x - y}{y + 2x}$, so $f_x = \dfrac{(y + 2x) - (x - y)(2)}{(y + 2x)^2} = \dfrac{3y}{(y + 2x)^2}$ and $f_y = \dfrac{(y + 2x)(-1) - (x - y)}{(y + 2x)^2} = \dfrac{-3x}{(y + 2x)^2}$.

17. $h(x, y) = (2xy + 3y^2)^5$, so $h_x = 10y(2xy + 3y^2)^4$ and $h_y = 10(x + 3y)(2xy + 3y^2)^4$.

19. $f(x, y) = (x^2 + y^2)e^{x^2+y^2}$, so $f_x = 2xe^{x^2+y^2} + (x^2 + y^2)(2x)e^{x^2+y^2} = 2x(x^2 + y^2 + 1)e^{x^2+y^2}$ and $f_y = 2ye^{x^2+y^2} + (x^2 + y^2)(2y)e^{x^2+y^2} = 2y(x^2 + y^2 + 1)e^{x^2+y^2}$.

21. $f(x, y) = \ln\left(1 + \dfrac{x^2}{y^2}\right)$, so $f_x = \dfrac{2x/y^2}{1 + (x^2/y^2)} = \dfrac{2x}{x^2 + y^2}$ and $f_y = \dfrac{-2x^2/y^3}{1 + (x^2/y^2)} = -\dfrac{2x^2}{y(x^2 + y^2)}$.

23. $f(x,y) = x^4 + 2x^2y^2 - y^4$, so $f_x = 4x^3 + 4xy^2$ and $f_y = 4x^2y - 4y^3$. Therefore, $f_{xx} = 12x^2 + 4y^2$, $f_{xy} = 8xy = f_{yx}$, and $f_{yy} = 4x^2 - 12y^2$.

25. $g(x,y) = \dfrac{x}{x+y^2}$, so $g_x = \dfrac{(x+y^2)-x}{(x+y^2)^2} = \dfrac{y^2}{(x+y^2)^2}$ and

$g_y = \dfrac{-2xy}{(x+y^2)^2}$. Therefore, $g_{xx} = -2y^2(x+y^2)^{-3} = -\dfrac{2y^2}{(x+y^2)^3}$,

$$g_{xy} = \frac{(x+y^2)\,2y - y^2(2)(x+y^2)\,2y}{(x+y^2)^4} = \frac{2(x+y^2)(xy+y^3-2y^3)}{(x+y^2)^4} = \frac{2y(x-y^2)}{(x+y^2)^3} = g_{yx}, \text{ and}$$

$$g_{yy} = \frac{(x+y^2)^2(-2x) + 2xy(2)(x+y^2)\,2y}{(x+y^2)^4} = \frac{2x(x^2+y^2)(-x-y^2+4y^2)}{(x+y^2)^4} = \frac{2x(3y^2-x)}{(x+y^2)^3}$$

27. $h(s,t) = \ln\left(\dfrac{s}{t}\right)$. Write $h(s,t) = \ln s - \ln t$. Then $h_s = \dfrac{1}{s}$ and $h_t = -\dfrac{1}{t}$, so $h_{ss} = -\dfrac{1}{s^2}$, $h_{st} = h_{ts} = 0$, and $h_{tt} = \dfrac{1}{t^2}$.

29. $f(x,y) = 2x^2 + y^2 - 8x - 6y + 4$. To find the critical points of f, we solve the system $\begin{cases} f_x = 4x - 8 = 0 \\ f_y = 2y - 6 = 0 \end{cases}$ obtaining $x = 2$ and $y = 3$. Therefore, the sole critical point of f is $(2,3)$. Next, $f_{xx} = 4$, $f_{xy} = 0$, and $f_{yy} = 2$, so $D(2,3) = f_{xx}(2,3)\,f_{yy}(2,3) - f_{xy}(2,3)^2 = 8 > 0$. Because $f_{xx}(2,3) > 0$, we see that $f(2,3) = -13$ is a relative minimum value.

31. $f(x,y) = x^3 - 3xy + y^2$. We solve the system of equations $\begin{cases} f_x = 3x^2 - 3y = 0 \\ f_y = -3x + 2y = 0 \end{cases}$ obtaining $x^2 - y = 0$, and so $y = x^2$. Then $-3x + 2x^2 = 0$, $x(2x-3) = 0$, and so $x = 0$ or $x = \frac{3}{2}$. The corresponding values of y are $y = 0$ and $y = \frac{9}{4}$, so the critical points are $(0,0)$ and $\left(\frac{3}{2}, \frac{9}{4}\right)$. Next, $f_{xx} = 6x$, $f_{xy} = -3$, and $f_{yy} = 2$, and so $D(x,y) = 12x - 9 = 3(4x-3)$. Therefore, $D(0,0) = -9$, and so $(0,0)$ is a saddle point and $f(0,0) = 0$. $D\left(\frac{3}{2}, \frac{9}{4}\right) = 3(6-3) = 9 > 0$ and $f_{xx}\left(\frac{3}{2}, \frac{9}{4}\right) > 0$, and so $f\left(\frac{3}{2}, \frac{9}{4}\right) = \frac{27}{8} - \frac{81}{8} + \frac{81}{16} = -\frac{27}{16}$ is a relative minimum value.

33. $f(x,y) = f(x,y) = e^{2x^2+y^2}$. To find the critical points of f, we solve the system $\begin{cases} f_x = 4xe^{2x^2+y^2} = 0 \\ f_y = 2ye^{2x^2+y^2} = 0 \end{cases}$ giving $(0,0)$ as the only critical point of f. Next, $f_{xx} = 4\left(e^{2x^2+y^2} + 4x^2e^{2x^2+y^2}\right) = 4(1+4x^2)\,e^{2x^2+y^2}$, $f_{xy} = 8xye^{2x^2+y^2} = f_{yx}$, and $f_{yy} = 2(1+2y^2)\,e^{2x^2+y^2}$, so $D = f_{xx}(0,0)\,f_{yy}(0,0) - f_{xy}^2(0,0) = (4)(2) - 0 = 8 > 0$. Because $f_{xx}(0,0) > 0$, we see that $(0,0)$ gives a relative minimum of f. The minimum value of f is $f(0,0) = e^0 = 1$.

35. We form the Lagrangian function $F(x, y, \lambda) = -3x^2 - y^2 + 2xy + \lambda(2x + y - 4)$. Next, we solve the system

$$\begin{cases} F_x = 6x + 2y + 2\lambda = 0 \\ F_y = -2y + 2x + \lambda = 0 \\ F_\lambda = 2x + y - 4 = 0 \end{cases}$$

Multiplying the second equation by 2 and subtracting the resulting equation from the first equation yields $6y - 10x = 0$ so $y = \frac{5}{3}x$. Substituting this value of y into the third equation of the system gives $2x + \frac{5}{3}x - 4 = 0$, so $x = \frac{12}{11}$ and consequently $y = \frac{20}{11}$. Therefore, $\left(\frac{12}{11}, \frac{20}{11}\right)$ gives the maximum value $f\left(\frac{12}{11}, \frac{20}{11}\right) = -\frac{32}{11}$ for f subject to the given constraint.

37. The Lagrangian function is $F(x, y, \lambda) = 2x - 3y + 1 + \lambda(2x^2 + 3y^2 - 125)$. Next, we solve the system of equations

$$\begin{cases} F_x = 2 + 4\lambda x = 0 \\ F_y = -3 + 6\lambda y = 0 \\ F_\lambda = 2x^2 + 3y^2 - 125 = 0 \end{cases}$$

Solving the first equation for x gives $x = -\frac{1}{2}\lambda$, and the second equation gives $y = \frac{1}{2}\lambda$. Substituting these values of x and y into the third equation gives $2\left(-\frac{1}{2\lambda}\right)^2 + 3\left(\frac{1}{2\lambda}\right)^2 - 125 = 0$, so $\frac{1}{2\lambda^2} + \frac{3}{4\lambda^2} - 125 = 0$, $2 + 3 - 500\lambda^2 = 0$, and so $\lambda = \pm\frac{1}{10}$. Therefore, $x = \pm 5$ and $y = \pm 5$, and so the critical points of f are $(-5, 5)$ and $(5, -5)$. Next, we compute $f(-5, 5) = 2(-5) - 3(5) + 1 = -24$ and $f(5, -5) = 2(5) - 3(-5) + 1 = 26$. We conclude that f has a maximum value of 26 at $(5, -5)$ and a minimum value of -24 at $(-5, 5)$.

39. $\int_{-1}^{2}\int_{2}^{4}(3x - 2y)\,dx\,dy = \int_{-1}^{2}\left(\frac{3}{2}x^2 - 2xy\right)\Big|_{x=2}^{x=4}dy = \int_{-1}^{2}[(24 - 8y) - (6 - 4y)]\,dy = \int_{-1}^{2}(18 - 4y)\,dy$
$= (18y - 2y^2)\big|_{-1}^{2} = (36 - 8) - (-18 - 2) = 48.$

41. $\int_0^1\int_{x^3}^{x^2} 2x^2y\,dy\,dx = \int_0^1\left[x^2y^2\right]_{y=x^3}^{y=x^2}dx = \int_0^1 x^2(x^4 - x^6)\,dx = \int_0^1(x^6 - x^8)\,dx = \left[\frac{1}{7}x^7 - \frac{1}{9}x^9\right]_0^1 = \frac{1}{7} - \frac{1}{9} = \frac{2}{63}.$

43. $\int_0^2\int_0^1(4x^2 + y^2)\,dy\,dx = \int_0^2\left(4x^2y + \frac{1}{3}y^3\right)\Big|_{y=0}^{y=1}dx = \int_0^2\left(4x^2 + \frac{1}{3}\right)dx = \left(\frac{4}{3}x^3 + \frac{1}{3}x\right)\Big|_0^2 = \frac{32}{3} + \frac{2}{3} = \frac{34}{3}.$

45. The area of R is $\int_0^2\int_{x^2}^{2x}dy\,dx = \int_0^2 y\big|_{y=x^2}^{y=2x}dx = \int_0^2(2x - x^2)\,dx = \left(x^2 - \frac{1}{3}x^3\right)\Big|_0^2 = \frac{4}{3}$. Thus,

$AV = \frac{1}{4/3}\int_0^2\int_{x^2}^{2x}(xy + 1)\,dy\,dx = \frac{3}{4}\int_0^2\left(\frac{1}{2}xy^2 + y\right)\Big|_{x^2}^{2x}dx = \frac{3}{4}\int_0^2\left(-\frac{1}{2}x^5 + 2x^3 - x^2 + 2x\right)dx$
$= \frac{3}{4}\left(-\frac{1}{12}x^6 + \frac{1}{2}x^4 - \frac{1}{3}x^3 + x^2\right)\Big|_0^2 = \frac{3}{4}\left(-\frac{16}{3} + 8 - \frac{8}{3} + 4\right) = 3.$

47. a. $R(x, y) = px + qy = -0.02x^2 - 0.2xy - 0.05y^2 + 80x + 60y$.

b. The domain of R is the set of all points satisfying $0.02x + 0.1y \le 80$, $0.1x + 0.05y \le 60$, $x \ge 0$, and $y \ge 0$.

c. $R(100, 300) = -0.02(100)^2 - 0.2(100)(300) - 0.05(300)^2 + 80(100)^2 + 60(300) = 15{,}300$, giving revenue of \$15,300 realized from the sale of 100 sixteen-speed and 300 ten-speed electric blenders.

49. a. We summarize the data at right. The normal equations are $5b + 25m = 2011$ and $25b + 165m = 10{,}383$, and the solutions are $b = 361.2$ and $m = 8.2$. Therefore, the least-squares line has equation $y = 8.2x + 361.2$.

	x	y	x^2	xy
	1	369	1	369
	3	390	9	1170
	5	396	25	1980
	7	420	49	2940
	9	436	81	3924
Sum	25	2011	165	10,383

b. The average daily viewing time in 2014 (when $x = 11$) is $y = 8.2(11) + 361.2 = 451.4 = 7.52$, or 7 hr 31 min.

51. a. We summarize the calculations as follows:

	x	y	x^2	xy
	0	547.2	0	0
	1	638.9	1	638.9
	2	750.1	4	1500.2
	3	861.2	9	2583.6
	4	929.8	16	3719.2
Sum	10	3727.2	30	8441.9

The normal equations are $5b + 10m = 3727.2$ and $10b + 30m = 8441.9$. The solutions are $m = 547.94$ and $b = 98.75$, so the required equation is $y = 98.75x + 547.94$.

b. The estimated number of users in 2014 is $f(7) = 98.75(7) + 547.95 = 1239.19$, or approximately 1,239.2 million.

53. We want to minimize $C(x, y) = 3(2x) + 2(x) + 3y = 8x + 3y$ subject to $xy = 303{,}750$. The Lagrangian function is $F(x, y, \lambda) = 8x + 3y + \lambda(xy - 303{,}750)$, so we solve the system

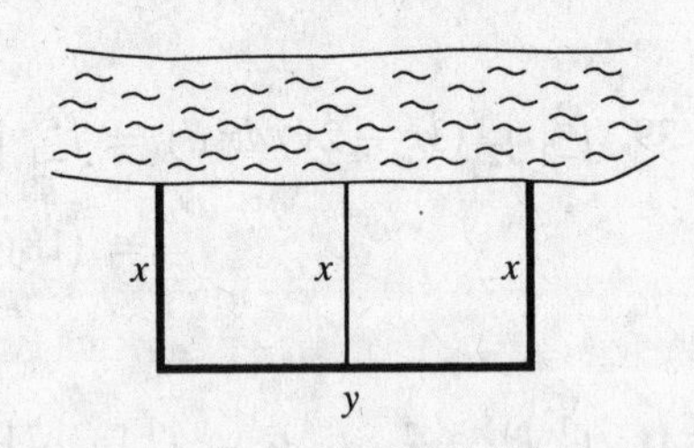

$$\begin{cases} F_x = 8 + \lambda y = 0 \\ F_y = 3 + \lambda x = 0 \\ F_\lambda = xy - 303{,}750 = 0 \end{cases}$$

Solving the first equation for y gives $y = -\frac{8}{\lambda}$. The second equation gives $x = -\frac{3}{\lambda}$. Substituting these values into the third equation gives $\left(-\frac{3}{\lambda}\right)\left(-\frac{8}{\lambda}\right) = 303{,}750$, so or $\lambda = \pm\frac{2}{225}$. Therefore, $x = 337.5$ and $y = 900$, and so the required dimensions of the pasture are 337.5 yd by 900 yd.

55. We want to maximize the function Q subject to the constraint $x + y = 100$. We form the Lagrangian function $f(x, y, \lambda) = x^{3/4}y^{1/4} + \lambda(x + y - 100)$. To find the critical points of F, we solve

$$\begin{cases} F_x = \frac{3}{4}\left(\frac{y}{x}\right)^{1/4} + \lambda = 0 \\ F_y = \frac{1}{4}\left(\frac{x}{y}\right)^{3/4} + \lambda = 0 \\ F_\lambda = x + y - 100 = 0 \end{cases}$$

Solving the first equation for λ and substituting this value into the second equation yields $\frac{1}{4}\left(\frac{x}{y}\right)^{3/4} - \frac{3}{4}\left(\frac{y}{x}\right)^{1/4} = 0$, $\left(\frac{x}{y}\right)^{3/4} = 3\left(\frac{y}{x}\right)^{1/4}$, and so $x = 3y$. Substituting this value of x into the third equation, we have $4y = 100$, so $y = 25$ and $x = 75$. Therefore, 75 units should be spent on labor and 25 units on capital.

CHAPTER 8 Before Moving On... page 638

1. In order for $f(x,y) = \dfrac{\sqrt{x}+\sqrt{y}}{(1-x)(2-y)}$ to be defined, we must have $x \geq 0$, $y \geq 0$, $x \neq 1$ and $y \neq 2$. Therefore, the domain of f is $D = \{(x,y) \mid x \geq 0, y \geq 0, x \neq 1, y \neq 2\}$.

2. $f(x,y) = x^2y + e^{xy}$, so $f_x = 2xy + ye^{xy}$, $f_y = x^2 + xe^{xy}$, $f_{xx} = 2y + y^2e^{xy}$, $f_{xy} = 2x + (1+xy)e^{xy} = f_{yx}$, and $f_{yy} = x \cdot xe^{xy} = x^2e^{xy}$.

3. $f(x,y) = 2x^3 + 2y^3 - 6xy - 5$. Solving $f_x = 6x^2 + 6y = 6(x^2 - y^2) = 0$ and $f_y = 6y^2 - 6x = 6(y^2 - x) = 0$ simultaneously gives $y = x^2$ and $x = y^2$. Therefore, $x = x^4$, $x^4 - x = x(x^3 - 1) = 0$, and so $x = 0$ or 1. The critical points of f are $(0,0)$ and $(1,1)$. $f_{xx} = 12x$, $f_{xy} = -6$, and $f_{yy} = 12y$, so $D(x,y) = 144x^2 + 144y^2 - 36$. In particular, $D(0,0) = -36 < 0$, and so $(0,0)$ does not give a relative extremum; and $D(1,1) = 252 > 0$ and $f_{xx}(1,1) = 12 > 0$, and so $f(1,1)$ gives a relative minimum value of $f(1,1) = 2(1)^3 + 2(1)^3 - 6(1)(1) - 5 = -7$.

4. We summarize the data at right. The normal equations are $5b + 11m = 36.8$ and $11b + 39m = 111.1$. Solving, we find $m \approx 2.04$ and $b \approx 2.88$. Thus, the least-squares line has equation $y = 2.04x + 2.88$.

	x	y	x^2	xy
	0	2.9	0	0
	1	5.1	1	5.1
	2	6.8	4	13.6
	3	8.8	9	26.4
	5	13.2	25	66
Sum	11	36.8	39	113.1

5. $F(x,y,\lambda) = 3x^2 + 3y^2 + 1 + \lambda(x + y - 1)$, so we solve the system $\begin{cases} F_x = 6x + \lambda = 0 \\ F_y = 6y + \lambda = 0 \\ F_\lambda = x + y - 1 = 0 \end{cases}$ We find $\lambda = -6x - 6y$, so $y = x$. Substituting into the third equation gives $2x = 1$, so $x = \frac{1}{2}$ and $y = \frac{1}{2}$. Therefore, $\left(\frac{1}{2}, \frac{1}{2}, \frac{5}{2}\right)$ is the required minimum.

6. $\iint_R (1 - xy)\,dA = \int_0^1 \int_{x^2}^x (1 - xy)\,dy\,dx = \int_0^1 \left(y - \frac{1}{2}xy^2\right)\Big|_{y=x^2}^{y=x} dx = \int_0^1 \left(x - \frac{1}{2}x^3 - x^2 + \frac{1}{2}x^5\right) dx$

$$= \left(\frac{1}{2}x^2 - \frac{1}{8}x^4 - \frac{1}{3}x^3 + \frac{1}{12}x^6\right)\Big|_0^1 = \frac{1}{2} - \frac{1}{8} - \frac{1}{3} + \frac{1}{12} = \frac{1}{8}.$$

CHAPTER 8 Before Moving On... page 658

1. In order for $f(x, y) = \dfrac{\sqrt{x} + \sqrt{y}}{(x-1)(2-y)}$ to be defined, we must have $x \geq 0$, $y \geq 0$, $x \neq 1$ and $y \neq 2$. Therefore, the domain of f is $D = \{(x, y) \mid x \geq 0, y \geq 0, x \neq 1, y \neq 2\}$.

2. $f(x, y) = x^2y + e^{xy}$, so $f_x = 2xy + ye^{xy}$, $f_y = x^2 + xe^{xy}$, $f_{xx} = 2y + y^2e^{xy}$, $f_{xy} = 2x + (1 + xy)e^{xy} = f_{yx}$, and $f_{yy} = x^2e^{xy}$.

3. $f(x, y) = 2x^3 + 2y^3 - 6xy - 5$. Solving $f_x = 6x^2 - 6y = 6(x^2 - y) = 0$ and $f_y = 6y^2 - 6x = 6(y^2 - x) = 0$ simultaneously gives $y = x^2$ and $x = y^2$. Therefore, $x = x^4 \Rightarrow x^4 - x = x(x^3 - 1) = 0$ and so $x = 0$ or 1. The critical points of f are $(0, 0)$ and $(1, 1)$. $f_{xx} = 12x$, $f_{xy} = -6$, and $f_{yy} = 12y$, so $D(x, y) = 144xy - 36$. In particular, $D(0, 0) = -36 < 0$ and so $(0, 0)$ does not give a relative extremum, and $D(1, 1) = 108 > 0$ and $f_{xx}(1, 1) = 12 > 0$, and so $f(1, 1)$ gives a relative minimum value of $f(1, 1) = 2(1)^3 + 2(1)^3 - 6(1)(1) - 5 = -7$.

4. We summarize the data at right. The normal equations are $5b + 11m = 36.8$ and $11b + 39m = 111.1$. Solving, we find $m \approx 2.04$ and $b \approx 2.88$. Thus, the least-squares line has equation $y = 2.04x + 2.88$.

xy	x^2	y	x	
0	0	2.9	0	
5.1	1	5.1	1	
13.6	4	6.8	2	
26.4	9	8.8	3	
66	25	13.2	5	
111.1	39	36.8	11	Sum

5. $F(x, y, \lambda) = 3x^2 + 3y^2 + 1 + \lambda(x + y - 1)$, so we solve the system $\begin{cases} F_x = 6x + \lambda = 0 \\ F_y = 6y + \lambda = 0 \\ F_\lambda = x + y - 1 = 0 \end{cases}$. We find $\lambda = -6x = -6y$, so $y = x$. Substituting into the third equation gives $2x = 1$, so $x = \frac{1}{2}$ and $y = \frac{1}{2}$. Therefore, $\left(\frac{1}{2}, \frac{1}{2}\right)$ is the required minimum.

6. $\iint_R (1 - xy)\, dA = \int_0^1 \int [illegible] (1 - xy)\, dy\, dx = [illegible]$

$= [illegible]$